Modern Climatology

Modern Climatology

Editor:

Quan Cui, Ph.D.

ArclerPress

www.arclerpress.com

Modern Climatology

Editor: Quan Cui, Ph.D.

708 3rd Avenue, 6th Floor
New York
NY 10017
United States of America
www.arclerpress.com

ISBN: 978-1-68094-476-1

Library of Congress Control Number: 2016951163

About the Editor

Quan Cui, Ph.D.

Quan Cui obtained her PhD from Beijing Normal University in 2015. She worked as environmental impact assessment engineer for years. Aiming at linking science to policy, her study interest spans from ecosystem modelling to water resources management. She published peer-reviewed articles about using machine learning models in inflow forecasting, CO_2 management and watershed ecosystem health assessment.

Contents

List of Contributors

Madan Mohan Dey
Aquaculture/Fisheries Center of Excellence, University of Arkansas at Pine Bluff, United States

Mark W. Rosegrant
Environment and Production Technology Division, International Food Policy Research Institute, United States

Kamal Gosh
School of Fisheries, Aquaculture and Aquatic Sciences, Auburn University, United States

Oai Li Chen
Aquaculture/Fisheries Center of Excellence, University of Arkansas at Pine Bluff, United States

Rowena Valmonte-Santos
Environment and Production Technology Division, International Food Policy Research Institute, United States

Ly Mouhamed
AGRHYMET Regional Center, Niamey, Niger

Seydou B. Traore
AGRHYMET Regional Center, Niamey, Niger

Agali Alhassane
AGRHYMET Regional Center, Niamey, Niger

Benoît Sarr
AGRHYMET Regional Center, Niamey, Niger

David R. Easterling
Center for Weather and Climate, NOAA's National Centers for Environmental Information, Asheville, NC, United States

Kenneth E. Kunkel
Cooperative Institute for Climate and Satellites, NC State University, NOAA's National Centers for Environmental Information, Asheville, NC, United States

Michael F. Wehner
Lawrence Berkeley National Laboratory, Berkeley, CA, United States

Liqiang Sun
Cooperative Institute for Climate and Satellites, NC State University, NOAA's National Centers for Environmental Information, Asheville, NC, United States

Jordan S. Read
Center for Integrated Data Analytics, U.S. Geological Survey, Middleton, WI, USA

Luke A. Winslow
Center for Limnology, University of Wisconsin - Madison, Madison, WI, USA

Gretchen J.A. Hansen
Center for Limnology, University of Wisconsin - Madison, Madison, WI, USA
Wisconsin Department of Natural Resources, Bureau of Science Services, Madison, WI, USA

Jamon Van Den Hoek
NASA Goddard Space Flight Center, Biospheric Sciences Branch, Code 618.0, Greenbelt, MD, USA

Paul C. Hanson
Center for Limnology, University of Wisconsin - Madison, Madison, WI, USA

Louise C. Bruce
School of Earth & Environment, The University of Western Australia, Perth, Australia

Corey D. Markfort
IIHR-Hydroscience and Engineering, Civil and Environmental Engineering, The University of Iowa, Iowa City, IA, USA

Nobuhito Mori
Disaster Prevention Research Institute, Kyoto University, Uji, Kyoto, Japan

Tetsuya Takemi
Disaster Prevention Research Institute, Kyoto University, Uji, Kyoto, Japan

Sika Gbegbelegbe
International Maize and Wheat Improvement Center (CIMMYT), Kenya Office, PO Box 1041, Village Market, Nairobi, Kenya

Uran Chung
International Maize and Wheat Improvement Center (CIMMYT), Apdo. Postal 6-641, CP 06600 Mexico DF, Mexico

Bekele Shiferaw
International Maize and Wheat Improvement Center (CIMMYT), Kenya Office, PO Box 1041, Village Market, Nairobi, Kenya

Siwa Msangi
International Food Policy Research Institute (IFPRI), Washington, DC, USA

Kindie Tesfaye
International Maize and Wheat Improvement Center (CIMMYT), Ethiopia Office, PO Box 5689, Addis Ababa, Ethiopia

Milton S. Speer
Climate Change Research Centre, Faculty of Science, The University of New South Wales, Sydney, NSW 2052, Australia

Arun Kumar Taxak
Department of Hydrology, IIT Roorkee, 247667 Uttarakhand, India

A.R. Murumkar
Department of Hydrology, IIT Roorkee, 247667 Uttarakhand, India

D.S. Arya
Department of Hydrology, IIT Roorkee, 247667 Uttarakhand, India

Nabil Laanaia
CNRM, UMR3589 (Météo-France, CNRS), 42 Avenue Gaspard Coriolis, Toulouse, France

Dominique Carrer
CNRM, UMR3589 (Météo-France, CNRS), 42 Avenue Gaspard Coriolis, Toulouse, France

Jean-Christophe Calvet
CNRM, UMR3589 (Météo-France, CNRS), 42 Avenue Gaspard Coriolis, Toulouse, France

Christian Pagé
CECI, CERFACS – CNRS, 42 Avenue Gaspard Coriolis, Toulouse, France

Uran Chung
International Maize and Wheat Improvement Center (CIMMYT), Mexico DF., Mexico

Sika Gbegbelegbe
International Maize and Wheat Improvement Center (CIMMYT), Mexico DF., Mexico

Bekele Shiferaw
Partnership for Economic Policy (PEP), Nairobi, Kenya

Richard Robertson
International Food Policy Research Institute (IFPRI), Washington, DC, USA

Jin I. Yun
Kyung Hee University, Yongin 446-701, South Korea

Kindie Tesfaye
International Maize and Wheat Improvement Center (CIMMYT), Mexico DF., Mexico

Gerrit Hoogenboom
Washington State University, Prosser, Washington, USA

Kai Sonder
International Maize and Wheat Improvement Center (CIMMYT), Mexico DF., Mexico

Peter E. Robins
Centre for Applied Marine Sciences, Bangor University, LL59 5AB, United Kingdom

Martin W. Skov
School of Ocean Sciences, Bangor University, LL59 5AB, United Kingdom

Matt J. Lewis
School of Ocean Sciences, Bangor University, LL59 5AB, United Kingdom

Luis Giménez
School of Ocean Sciences, Bangor University, LL59 5AB, United Kingdom

Alan G. Davies
Centre for Applied Marine Sciences, Bangor University, LL59 5AB, United Kingdom

Shelagh K. Malham
Centre for Applied Marine Sciences, Bangor University, LL59 5AB, United Kingdom

Simon P. Neill
School of Ocean Sciences, Bangor University, LL59 5AB, United Kingdom

James E. McDonald
School of Biological Sciences, Bangor University, LL57 2UW, United Kingdom

Timothy A. Whitton
Centre for Applied Marine Sciences, Bangor University, LL59 5AB, United Kingdom

Suzanna E. Jackson
School of Ocean Sciences, Bangor University, LL59 5AB, United Kingdom

Colin F. Jago
Centre for Applied Marine Sciences, Bangor University, LL59 5AB, United Kingdom
School of Ocean Sciences, Bangor University, LL59 5AB, United Kingdom
School of Biological Sciences, Bangor University, LL57 2UW, United Kingdom

Akio Kitoh
University of Tsukuba, Tsukuba, Ibaraki 305-8572, Japan

Hirokazu Endo
Meteorological Research Institute, Tsukuba, Ibaraki 305-0052, Japan

Shada Elalem
Department of Civil Engineering, College of Engineering and Applied Science, University of Colorado Denver, Denver, 1200 Larimer Street, Denver, CO 80204, USA

Indrani Pal
Department of Civil Engineering, College of Engineering and Applied Science, University of Colorado Denver, Denver, 1200 Larimer Street, Denver, CO 80204, USA

Vinit K. Jain
Research Scholar, Department of Hydrology, Indian Institute of Technology, Roorkee, India

Rajendra P. Pandey
National Institute of Hydrology, Roorkee, India

Manoj K. Jain
Department of Hydrology, Indian Institute of Technology, Roorkee, India

Hi-Ryong Byun
Department of Environmental Atmospheric Sciences, Pukyong National University, Korea

Jurgen D. Garbrecht
Research Hydraulic Engineer, USDA, Agricultural Research Service, Grazinglands Research Laboratory, 7207 West Cheyenne Street, El Reno, OK, United States

Mark A. Nearing
Research Agricultural Engineer, USDA, Agricultural Research Service, Southwest Watershed Research, Tucson, AZ, United States

Jean L. Steiner
Supervisory Soil Scientist/Laboratory Director, USDA, Agricultural Research Service, Grazinglands Research Laboratory, El Reno, OK, United States

Xunchang J. Zhang
Research Hydrologist, USDA, Agricultural Research Service, Grazinglands Research Laboratory, El Reno, OK, United States

Mary H. Nichols
Research Hydraulic Engineer, USDA, Agricultural Research Service, Southwest Watershed Research, Tucson, AZ, United States

Rashid Mahmood
Department of Irrigation and Drainage, Faculty of Agricultural Engineering and Technology, University of Agriculture, Faisalabad, Pakistan
Institute of Geographic Science and Natural Resources Research / Key Laboratory of Water Cycle and Related Land
Surface Processes, Chinese Academy of Sciences, Beijing, 100101, China

Mukand S Babel
Water Engineering and Management, Asian Institute of Technology, Pathumthani, Thailand

Shaofeng JIA
Department of Irrigation and Drainage, Faculty of Agricultural Engineering and Technology, University of Agriculture, Faisalabad, Pakistan
Institute of Geographic Science and Natural Resources Research / Key Laboratory of Water Cycle and Related Land Surface Processes, Chinese Academy of Sciences, Beijing, 100101, China

Junguo Liu
School of Nature Conservation, Beijing Forestry University, Beijing, 100083, China

Giri Kattel
School of Nature Conservation, Beijing Forestry University, Beijing, 100083, China
Self-Sustaining Regions Research and Innovation Initiative, Collaborative Research Network (CRN), Federation University Australia, Mt. Helen Campus, Victoria 3350, Australia

Hans Peter H. Arp
Norwegian Geotechnical Institute, Sognsveien 72, NO-0855, Oslo, Norway

Hong Yang
CEES, Department of Biosciences, University of Oslo, Blindern, Oslo 0316, Norway

Mark D. Svoboda
National Drought Mitigation Center, University of Nebraska-Lincoln, School of Natural Resources, 3310 Holdrege Street, Lincoln, NE 68583-0988, USA

Brian A. Fuchs
National Drought Mitigation Center, University of Nebraska-Lincoln, School of Natural Resources, 3310 Holdrege Street, Lincoln, NE 68583-0988, USA

Chris C. Poulsen
National Drought Mitigation Center, University of Nebraska-Lincoln, School of Natural Resources, 3310 Holdrege Street, Lincoln, NE 68583-0988, USA

Jeff R. Nothwehr
National Drought Mitigation Center, University of Nebraska-Lincoln, School of Natural Resources, 3310 Holdrege Street, Lincoln, NE 68583-0988, USA

Sorin Ionuț Dascălu
Meteo Romania (National Meteorological Administration), Bucharest, Romania

Mădălina Gothard
Meteo Romania (National Meteorological Administration), Bucharest, Romania

Roxana Bojariu
Meteo Romania (National Meteorological Administration), Bucharest, Romania

Marius-Victor Birsan
Meteo Romania (National Meteorological Administration), Bucharest, Romania

Roxana Cică
Meteo Romania (National Meteorological Administration), Bucharest, Romania

Ruxandra Vintilă
National Research and Development Institute for Soil Science, Agrochemistry and Environmental Protection, Bucharest, Romania

Mary-Jeanne Adler
National Institute for Hydrology and Water Management, Bucharest, Romania

Viorel Chende
National Institute for Hydrology and Water Management, Bucharest, Romania

Rodica-Paula Mic
National Institute for Hydrology and Water Management, Bucharest, Romania

Joeseph P. Smith
Cooperative Institute for Limnology and Ecosystems Research, University of Michigan, Ann Arbor, MI 48109, USA

Timothy S. Hunter
Great Lakes Environmental Research Laboratory, National Oceanic and Atmospheric Administration, Ann Arbor, MI 48108, USA

Anne H. Clites
Great Lakes Environmental Research Laboratory, National Oceanic and Atmospheric Administration, Ann Arbor, MI 48108, USA

Craig A. Stow
Great Lakes Environmental Research Laboratory, National Oceanic and Atmospheric Administration, Ann Arbor, MI 48108, USA

Tad Slawecki
LimnoTech, Ann Arbor, MI 48108, USA

Glenn C. Muhr
Great Lakes Environmental Research Laboratory, National Oceanic and Atmospheric Administration, Ann Arbor, MI 48108, USA

Andrew D. Gronewold
Great Lakes Environmental Research Laboratory, National Oceanic and Atmospheric Administration, Ann Arbor, MI 48108, USA
Department of Civil and Environmental Engineering, University of Michigan, Ann Arbor, MI 48109, USA

Preface

Climatology is more than just the study of weather. Climatology is the scientific study of climates, which is defined as the mean weather conditions over a period of time. A branch of study within atmospheric sciences, it also takes into account the variables and averages of short-term and long-term weather conditions. Climatology is different than meteorology and can be divided into different areas of study. Various approaches to this field can be taken, including paleoclimatology, which focuses on studying the climate over the course of the Earth's existence by examining records of tree rings, rocks and sediment, and ice cores. Historical climatology focuses primarily on climate changes throughout history and the effects of the climate on people and events over time. Though both climatology and meteorology are areas of study that are considered branches of similar areas of study, climatology differs from meteorology because its focus is on averages of weather and climatic conditions over a long period of time.

The study of climatology has been changed in the past decades because people had realized climate is not a stable science that only fluctuated within the boundaries of historical records. As an interdisciplinary science of physical geography, atmospheric science and biochemistry, modern climatology paid more attention to how the climate responds to drivers like human activities and how they interact with each other. Unlike the classical climatology focused on certain regional boundaries, scales of modern climatology models in this book are both regional and global, with purposes of understanding the dynamics of the weather system and forecasting future climate. This book also includes topics such as climate data analysis, long-tern forecasting models, dust storm, draught, coastal hazards, the impact of climate change and ecosystem responds, and climate extremes quantifying studies. The modern era of climatology is echoed in this book. On the one hand it offers a broad synoptic perspective but also considers the regional standpoint, as it is this that affects what people need from climatology. Aspects on the topic of climate change - what is often considered a contradiction in terms - is also addressed. It is all too evident these days that what recent work in climatology has revealed carries profound implications for economic and social policy.

Editor

Quan Cui, Ph.D.

INTRODUCTION

Climatology or climate science is the study of climate, scientifically defined as weather conditions averaged over a period of time. This modern field of study is regarded as a branch of the atmospheric sciences and a subfield of physical geography, which is one of the Earth sciences. Climatology now includes aspects of oceanography and biogeochemistry. Basic knowledge of climate can be used within shorter term weather forecasting using analog techniques such as the El Niño–Southern Oscillation (ENSO), the Madden–Julian oscillation (MJO), the North Atlantic oscillation (NAO), the Northern Annular Mode (NAM) which is also known as the Arctic oscillation (AO), the Northern Pacific (NP) Index, the Pacific decadal oscillation (PDO), and the Interdecadal Pacific Oscillation (IPO). Climate models are used for a variety of purposes from study of the dynamics of the weather and climate system to projections of future climate.

Approaches

Climatology is approached in a variety of ways. Paleoclimatology seeks to reconstruct past climates by examining records such as ice cores and tree rings (dendroclimatology). Paleotempestology uses these same records to help determine hurricane frequency over millennia. The study of contemporary climates incorporates meteorological data accumulated over many years, such as records of rainfall, temperature and atmospheric composition. Knowledge of the atmosphere and its dynamics is also embodied in models, either statistical or mathematical, which help by integrating different observations and testing how they fit together. Modeling is used for understanding past, present and potential future climates. Historical climatology is the study of climate as related to human history and thus focuses only on the last few thousand years.

Climate research is made difficult by the large scale, long time periods, and complex processes which govern climate. Climate is governed by physical laws which can be expressed as differential equations. These equations are coupled and nonlinear, so that approximate solutions are obtained by using numerical methods to create global climate models. Climate is sometimes modeled as a stochastic process but this is generally accepted as an approximation to processes that are otherwise too complicated to analyze.

Models

Climate models use quantitative methods to simulate the interactions of the atmosphere, oceans, land surface, and ice. They are used for a variety of purposes from study of the dynamics of the weather and climate system to projections of future climate. All climate models balance, or very nearly balance, incoming energy as short wave (including visible) electromagnetic radiation to the earth with outgoing energy as long wave (infrared) electromagnetic radiation from the earth. Any unbalance results in a change in the average temperature of the earth.

The most talked-about models of recent years have been those relating temperature to emissions of carbon dioxide. These models predict an upward trend in the surface temperature record, as well as a more rapid increase in temperature at higher latitudes.

Models can range from relatively simple to quite complex:

- A simple radiant heat transfer model that treats the earth as a single point and averages outgoing energy
- this can be expanded vertically (radiative-convective models), or horizontally
- finally, (coupled) atmosphere–ocean–sea ice global climate models discretize and solve the full equations for mass and energy transfer and radiant exchange.

What does a Climatologist do?

A climatologist is a scientist who studies the climate. This field in the sciences is related to meteorology, the study of weather, except that it looks at long-term trends and the history of the climate, rather than examining weather systems in the short term like meteorologists do. These scientists can work in a variety of settings, including government agencies, nonprofit organizations interested in the climate, and even archeology departments at major colleges and universities. Someone who specializes in ancient climates, incidentally, is known as a paleoclimatologist.

There are a number of tools at the disposal of climatologists that allow them to study weather patterns and the climate. Like meteorologists, they use satellites to look at things like cloud cover, and also to compare historic cloud cover, snow pack, and glacier sizes with the modern day to look for climate trends. They also examine samples of ice cores taken from the poles to look for traces of compounds trapped in ancient snow to learn about what the climate was like in the past. Others do the same with samples of mud from long-established lakes.

Mud and ice cores can contain a wealth of information. For example, paleoclimatogists can look at the pollens found in ice cores to learn about which plants thrived when, looking for signs of plants that rely on specific climactic conditions for survival. Mud and ice cores can also contain traces of volcanic ash from eruptions, along with dissolved gases associated with changes in the Earth's atmosphere.

Climatologists also study oceans, using a variety of techniques to monitor ocean temperature, ocean currents, and changes in salinity. Volcanic eruptions are of interest as well, because they can alter the climate and provide information about how volcanoes have behaved in the past. Someone who works in this field can also use some more unexpected tools. For example, some climatology researchers have studied landscape paintings to look at the historical levels of cloud cover and light in addition to searching for signs of indicator plant species that could provide information about the climate at the time that the painting was created.

A climatologist can use the information he or she collects to create climate models using computer software. These models can used to demonstrate historic events involving the Earth's climate, and also to forecast future events on the basis of information about the past. Experts can also manipulate their models to show how various changes can impact the future of the climate, and they can demonstrate how small changes like seemingly minor rises in temperature can have a ripple effect on the planet.

How Do I Become a Climatologist?

A climatologist is a scientist who focuses on the study of long-term weather patterns throughout the world. In order to become a climatologist, you will need to complete a high school education and develop a keen interest in science, particularly meteorology, geology, and related fields. Try to earn good grades in high school so you can be accepted to a strong college program; you will need at least a bachelor's degree in order to become a climatologist, and for some jobs in the field, you will need a master's degree or PhD.

To determine what level of education you will need to become a climatologist, it helps to read a few different job descriptions within the field. Entry-level positions may only require a bachelor's degree, while more advanced positions may require higher degrees. Think about your long-term career goals so you can make a strong decision about your schooling. You can always earn a bachelor's degree, work in the field, and then go back to school later on if you decide you want to further your education and, subsequently, your career.

It is important, of course, to choose a strong college program if you want to become a climatologist. Apply only to accredited institutions that offer coursework in the field in which you are interested, and find out if the school offers job placement services or internship opportunities. Be sure, too, that you can afford the tuition and fees to attend a particular institution.

Taking part in an internship while you are in college is a great way to develop professional relationships within the field, get real work experience that will help you build your resume, and learn relevant skills that will come in handy when you find full-time work. Some schools will help you find internship opportunities, but in most cases, you will need to research various companies or entities that offer internships and apply yourself.

Once you graduate from college with a bachelor's degree, you can take steps to become a climatologist by researching various climatology opportunities in your area. If you are willing to relocate, more job opportunities will open up for you as well. You may find a job with a private company, a non-profit organization, or even within a government agency. Be prepared to start at the bottom and work your way up through the ranks, learning as much as possible from more experienced climatologists.

Climate Conditions

Climate conditions typically refer to various aspects and patterns of weather in a given area, and the potential consequences and affects that such weather can create. The area in which such conditions may be considered can be relatively small, though accurate understanding of climate in any area typically considers worldwide conditions as well. These conditions are often used as the basis upon which weather predictions and disaster warnings are formed, and provide possible causal or corollary data for events that occur. Climate conditions can refer to the actual weather itself, as well as possible results of the weather such as droughts.

One of the simplest explanations of the term is that they are the weather conditions found in a given area. This can include basic aspects of weather such as wind, rain, and snow, or somewhat more complicated elements of weather systems such as tornadoes, hurricanes, droughts, and rising sea levels. Understanding, analyzing, and predicting climate changes and conditions often involves a variety of aspects of weather and how air moves and acts across the surface of the Earth. Atmospheric pressure, precipitation, and jet streams are all considered and utilized to better understand climate conditions and how they create other effects.

The nature of weather conditions throughout the world is such that any one area is typically affected by the weather in other areas. This means that even though someone may look at a weather forecast for a particular town or city at any position in the world, the process used in predicting that weather typically involves understanding climate conditions in other areas as well. Weather forecasting often involves analyzing the conditions in other areas, and then using past behaviors to predict how the weather is likely to develop in the future.

Climate conditions for a given area also often involve secondary aspects caused by weather changes and patterns. Droughts, for example, are often considered part of the climate of an area, since they are frequently caused by a lack of rainfall or diminished snow and rain in tributary areas that feed local water resources. This is similar to regions that may have high risk levels of wildfires, which can be reflected by long periods of dry weather, followed by electrical storms that may start a fire. Climate conditions are especially important for understanding how massive weather systems such as tropical storms are likely to behave over a given time, allowing scientists to more effectively warn people about intense weather that may be heading toward them.

Climate Change

Climate change occurs when the Earth's average temperature changes dramatically over time. As little as one or two degrees can be considered dramatic change because the Earth's ecosystem depends on a very delicate balance, and even small shifts can have a far-reaching impact. A drop in average temperature can also be considered climate change, but in modern times people using the term are usually talking about global warming.

One reason that climate change has become a popular and, at times, controversial topic is that many people believe it is mostly the result of human activity. Burning fossil fuels, such as oil and coal, produces byproducts such as carbon dioxide gas. Since there are not enough plants on Earth to quickly transform all this emitted carbon dioxide into oxygen, the gas remains in the atmosphere. Through a process known as the greenhouse effect, the carbon dioxide traps solar heat, which leads to the warming of the planet.

Other causes of climate change that can be traced back to humans include deforestation, or the widespread cutting of trees, and methane gas production. Methane is typically produced in large amounts by mining, large scale livestock farms, rice paddies and landfills. The commercial use of fertilizers that give off nitrous oxide also contributes to pollutant levels.

Many people believe the effects of climate change can already be seen, in the melting of permafrost near the North Pole and the rise of sea levels. Rising ocean levels cause concern about shrinking coastlines and island land masses. A warmer climate can also cause more severe weather to occur, because weather phenomena, such as hurricanes, gain strength from hot, moist air.

Some say evidence of global warming can also be found in reduced wildlife populations. Some animal species, such as the polar bear, are slowly losing their icy habitats and have shown smaller populations over recent years. For this reason, many wildlife groups want the polar bear to be added to endangered species lists.

The concept of climate change is not a new one. The ice age of long ago is well documented and was another form of climate shift, one not brought on by humans. Modern climate shifts that are the result of human behavior may be positively affected by altering personal lifestyles. While some people consider global warming to be only a theory, it is becoming more widely accepted that the planet's climate is shifting and that people are not blameless.

Many simple lifestyle changes that people can make to help combat climate change center on conserving energy resources. Actions such as turning off unnecessary lights, buying used items rather than new ones and using public transportation or a bicycle instead of driving cars can make a difference. Recycling as many goods and materials as possible is another helpful way to conserve. If humans ignore climate change, it may continue to accelerate and drastically change the planet in both predicted and unforeseen ways.

What are the Different Causes of Climate Change?

Climate change is a natural process that causes global temperatures to fluctuate over a certain period. Some people believe that unnatural factors are also causes of climate change. Theories that climate fluctuations are occurring at concerning paces have sparked an increased desire to understand the factors.

A phenomenon known as the greenhouse effect is believed to be one of the major causes of climate change. The greenhouse effect occurs when certain gases, such as carbon dioxide, water vapor, and methane, are heavily concentrated in the atmosphere. It is believed that these gases absorb radiation that is emitted from the Earth's surface. This causes that radiation to be trapped in the atmosphere. As a result, the Earth's temperature rises.

Carbon dioxide is considered one of the biggest perpetrators of this global warming process. Many scientists believe that the oceans are natural keepers of carbon dioxide. The amount that can be stored is affected by temperature. As the Earth becomes warmer, theories suggest that the oceans can store less carbon dioxide. When these increasing amounts of gas are released, they cannot escape the atmosphere. This process of release and trapping becomes a cycle.

Humans are believed to be aggravating causes of climate change due to a number of activities. Burning fossil fuels such as oil, natural gas, and coal is one of them. Burning these substances produces more carbon dioxide. It is assumed that limiting the use of fossil fuels has a positive effect on global warming.

Aerosols are another human aggravation to climate change. Aerosols are small particles in the air that can impact the temperature up or down, depending on the type. Some aerosols, such as the type produced by volcanoes, cause cooling and are not related to human activity. Other aerosols, such as smog and smoke from forest fires, result from people. These aerosols are said to have warming effects because they absorb the sun's energy.

It was once believed that the sun's output varied only slightly. Newer theories based on satellites containing radiometers suggest that the radiation from the sun may have a significantly greater variance. If varying output from the sun is proven to be a constant reality instead of a trend, it would only be logical to expect climate change, according to some scientists.

The causes of climate change is a subject that is surrounded by much debate. The issue has been analyzed by a variety of professionals. Some of them wholly reject the idea that any significant climate changes are taking place. Others warn that the rate of change must be minimized for the sake of human livelihood.

Climate Map

A climate map depicts variations in climate over a geographical area. It may cover the entire planet, a single continent or a smaller region. Climate maps may show overall climate according to a classification system based on averages or a single factor, such as temperature or precipitation. They can also show seasonal variations, projected future climates or reconstructions of past climates.

Various climate classification systems can be employed to produce a world climate map defining distinct climatic regions, but the most widely used method was developed in the early 1900s by the German-Russian climatologist Wladimir Koppen. It is based on annual and monthly average temperature and precipitation, and the resulting regions correlate closely with vegetation zones. This system was frequently modified by Koppen himself and, later, by others.

The Koppen system divides climates into five main categories, denoted by upper case letters A to E. "A" represents a humid tropical climate, with high temperatures and high rainfall throughout the year; "B" is a dry climate, with low rainfall throughout the year, subdivided into type "S" which is semi-arid and type "W" which is arid; "C" is a humid mid-latitude climate; "D" is a continental climate with relatively low rainfall and large seasonal temperature variations and "E" is a polar climate, with low temperatures throughout the year. Sometimes an "H" category is added to denote a highland climate, which applies to high mountain regions such as the Himalayas.

A second, lowercase, letter was added to create subcategories based on the seasonal pattern of precipitation, for example "s" denotes a dry summer season. Not all of these sub-categories apply to every main climate type. A third lowercase letter was added to further subdivide some climate categories according to seasonal temperature patterns, for example "a" indicates a hot summer, with average temperatures above 72°F (22°C). Again, not all of these apply to every climate type. In all, this system creates 30 climate types that can be illustrated in different colors in a climate regions map.

Specific seasons can be illustrated in a climate map, showing, for example, variability in temperature or precipitation through the year. Lines can be drawn connecting points which have the same value for a given factor to produce a "contour" map showing high and low areas for this factor. For example, lines connecting points of equal temperature are called isotherms, so a map with isotherms can show at a glance the variations in temperature within a region, or the planet as a whole. A series of maps for different times of year can illustrate seasonal variations.

Concern about climate change has led to the creation of climate maps showing possible future climates. These use projections based on the estimated overall temperature increases that would result from various levels of carbon dioxide production. One such map shows the effects of an overall rise in global temperature of 39.2°F (4°C). It is thought that temperatures would increase more over land than over the oceans and that increases would be greater at high latitudes than at low latitudes. These maps can also illustrate changing geography due to sea level rise.

It is possible to reconstruct past climates using fossil and other evidence. For example, analysis of pollen grains in ancient soil samples can indicate the types of vegetation present when the soil was formed, giving a good indication of the type of climate up to a few tens of thousands of years ago. Going further back into the past, fossil evidence of plant remains, examples of glaciation — indicating a cold climate — or wind erosion — indicating a dry climate — can provide information about climates going back hundreds of millions of years, enabling past climate maps to be created.

Climate maps have even been produced for the planet Mars. They are much simpler than maps illustrating the Earth's climate as there are no oceans and no precipitation. Nevertheless, information on topography, albedo, the presence of ice and evidence of wind speed and direction, combined with latitude, has enabled the production of a reasonably detailed climate map of the Martian surface.

What Is Climate Variability?

Climate is the measure of weather patterns over a long period of time, and inherent to climate are changes, both long-term and short-term. Short-term climate changes represent periodic or intermittent changes that occur, and this is termed "climate variability." These short-term changes might include floods, drought, temperature changes or oscillating weather patterns such as the effects of El Niño or La Niña. In the most general sense, climate variability is thought of as the deviations in climate statistics over a long period of time. Accurately identifying and understanding climate variations is important to recognizing and understanding their effect on humans.

The term "climate" is a term that is used to describe the average mix of meteorological conditions in a geographical location over the long term. Scientists determine the climate of a geographic location by compiling statistics over an extended time period, usually several decades or longer. Such statistics include the value, variance and probabilities associated with meteorological conditions such as temperature, humidity, precipitation and wind.

When scientists study climate over the long term, they often find inconsistencies in weather when compared to the average. For example, climates that normally are wet might experience dry spells or even droughts because of a change in weather patterns. Although these short-term changes are inherent, they don't mean that the climate has changed. Instead, they simply are a deviation from the region's accustomed climate. Scientists assign the term "climate variability" to identify such anomalies, which usually last for less than a decade.

There are many reasons why climates deviate from the norm. One of the most distinguishable natural variations in climate occurs with the El Niño-southern oscillation (ENSO) circumstance. ENSO explains the interaction of the Pacific Ocean with the atmosphere, creating global climate deviations. Every few years, there are changes in the ocean surface temperature and other meteorological elements around the Pacific equator. Colder temperatures characterize La Niña, and warmer temperatures are the signature of the El Niño cycle. The different temperatures produce variations in tropical rainfall, agitating climates all over the world.

The study of climate variations is important to scientists because of its effect on people. If scientists can identify patterns or discern the effect on weather attributed to climate variability, people usually can acclimate. To illustrate, scientist have placed an observation system in the tropical Pacific that allows forecasting of the El Niño affects several years before they manifest in the global climate. This information can help calculate resulting conditions such as intensified storms or drought conditions ahead of time.

By contrast, when scientists fail to understand climate variability, people usually suffer. For example, the Dustbowl of the 1930s is believed to have been caused by climate variability associated with warmer ocean temperatures over the Atlantic Ocean and La Niña effects in the Pacific Ocean. If these effects had been known ahead of time, rather than decades later, scientists could have warned about the resulting consequences.

CHAPTER 1

ECONOMIC IMPACT OF CLIMATE CHANGE AND CLIMATE CHANGE ADAPTATION STRATEGIES FOR FISHERIES

Madan Mohan Dey, Mark W. Rosegrant, Kamal Gosh, Oai Li Chen, Rowena Valmonte-Santos

Keywords: Supply-demand model, Fisheries, Climate change, Fiji, Solomon Islands, Vanuatu, Timor-Leste

INTRODUCTION

There is concern around the globe about the likely biophysical and economic impacts that the future changes to climate will have on aquaculture and fisheries (see, for example, [1], [2], [3], [4], [5], [6], [7], [8], [9], [10], [11], [12], [13] and [14]). The biophysical impact of climate change on aquaculture and fisheries has been closely studied by a number of groups (see, for example, [4], [5] and [6], [15], [16], [17], [18], [19] and [20]). This and other papers of this special section of Marine Policy contribute to the growing, but still limited, body of knowledge on economic impact of climate change on aquaculture and fisheries. The geographic focus of this study is four Pacific coral triangle (CT) countries (Fiji, Solomon Islands, Vanuatu and Timor-Leste), which are highly exposed and vulnerable to the impacts of climate change.

The two broad approaches used to analyze the economic impact of climate change in fisheries and aquaculture are bioeconomic modeling and market supply and demand[21]. One benefit of bioeconomic modeling is that it can incorporate climate change scenarios by altering an appropriate set of ecosystem or population parameters. However, much of the required information on resource and ecosystem interactions is not available for Pacific Island countries. The market supply–demand approach represents climate change in terms of supply shocks, and works out its economic consequences using the microeconomic tools of supply and demand. Briones [22], Delgado et al. [23], World Bank [24] have effectively used this approach to analyze the "fisheries collapse"/"climate change" scenarios. One main advantage of the market supply–demand approach is that it incorporates price responses of market agents (such as producers and consumers) and interactions across sectors (for example, different fish species groups). Some studies [22] and [25] incorporated both bioeconomic and supply–demand approaches in analyzing the effect of climate change on fish production and consumption. Given the scarcity of data regarding the biological and physical responses to climate change in the Pacific Islands, this study has developed and used a market fish supply–demand model for Fiji, Solomon Islands, Vanuatu and Timor-Leste.

The paper is organized as follows. Section 2 discusses the modeling methodology and describes the data. Section 3 reports the country-specific empirical results for medium term (2035) and long term (2050) under baseline scenarios. Finally, Section 4 contains concluding remarks and policy implications for the economies of the four Pacific CT countries.

OVERVIEW OF THE MODEL

Conceptual Framework

The study has followed a three-step procedure to measure the potential impacts of various climate change adaptation strategies: (1) development of a baseline model of the fisheries sector for each country; (2) collection and construction of data sets for model parameters, exogenous variables, and ex ante impact indicators of various climate change adaptation

Citation: Madan Mohan Dey , Mark W. Rosegrant, Kamal Gosh, Oai Li Chen, Rowena Valmonte-Santos , Economic impact of climate change and climate change adaptation strategies for fisheries, http://dx.doi.org/10.1016/j.marpol.2015.12.011.

strategies; and (3) analysis of the overall impact of climate change adaptation strategies by incorporating the ex ante impact indicators into the fisheries sector model.

Fig. 1 represents the economic framework used in each of the four countries in the study (Fiji, Solomon Islands, Vanuatu and Timor-Leste), for assessing the impact of climate change adaptation strategies. As shown in Fig. 1, the framework considers three supply scenarios: scenario I (S_1)—before climate change; scenario II (S_2)—with climate change effects, but no adaptation strategy; and scenario III (S_3)—after adopting climate change adaptation strategies.

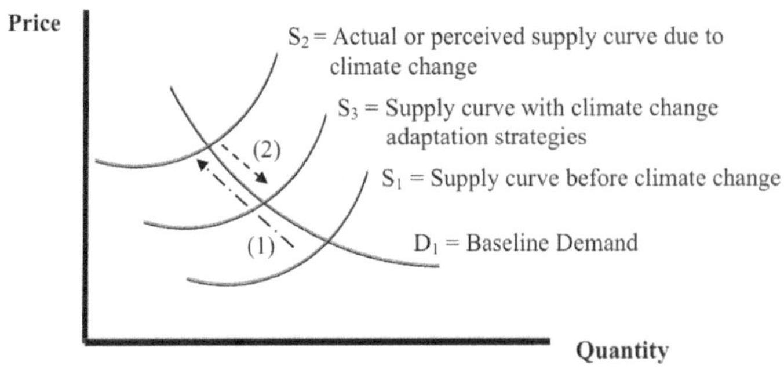

Figure 1: Shifts of supply curve due to climate change and adaptation strategies.

The supply curve for capture fisheries and aquaculture under scenario 1 (no climate change effect) is depicted by S_1. The actual (or perceived) supply curve of farmers/fishers may shift to the left from the S_1 to the S_2 position as a result of the realized (or potential) negative impacts of climate change (scenario II). These impacts may be a reduction of catch per unit of effort, lower productivity, and higher cost per unit of fish produced or caught, among others. It is, however, important noting that the Fifth Assessment Report of the Intergovernmental Panel on Climate Change indicates that the impact of climate change on fisheries is uneven [26]; and as such, one can expect that natural productivity and fish stocks to increase for some species and decline for others. Impact of climate change on seafood supply is, therefore, indeterminate and climate change may also have positive effects on some sub-sectors of aquaculture/fisheries; in that case, the supply curve may shift to the right (i.e., the position of S_2 will be at the right side of S_1).

Nature of supply curve differs between aquaculture and capture fisheries sub-sectors. Output, or total catch, of capture fisheries can increase or decrease due to changes in stock abundance and effort. Higher level of stock abundance (biomass) yields higher catch per effort, given all other factors held constant. Supply curve of a particular capture fishery under a specific scenario (S_1 or S_2) assumes a particular stock size (biomass), and underlying the changes in the supply curve of capture fisheries (from S_1 to S_2) there will be changes in biomass.

The adoption of various climate change adaptation strategies is expected to reduce the negative effects of climate change. As a result, the supply curve from capture fisheries and aquaculture (scenario III) may shift from the S_2 position to S_3. Comparison of S_1 and S_2 shows the impact of climate change, while comparison of S_2 and S_3 shows the effect of climate change adaptation strategies.

This study has considered three time periods: current (represented by the average data of the three most recent available years 2006–2009), medium (2035), and long term (2050). Each of these periods is represented by a specific demand curve. As shown inFig. 1, D_1 denotes the demand situation in the immediate past (2006–2009). The demand is likely to shift upward with the increase in population and income from D_1 to D_2in 2035 and D_3 in 2050. D_2 and D_3 have not shown in the graph for ease of presentation.

Algebraic representation of the model

The fisheries sector model developed to evaluate the impact of climate change adaptation strategies is based on the modified balance-of-trade function approach suggested by Martin and Alston [27] and used by Dey et al. [28] and [29]. The basic form of the model formulated for this study for a national economy can be expressed as:

equation(1)

$$B^h = e(\mathbf{P}, u^h) - g(\mathbf{P}, \mathbf{V}, \lambda) - f \quad (1)$$

where

e=expenditure function of a national economy,

P=vector of domestic prices of different fish groups,

u^h=level of utility exogenously specified to define Hicksian money-metric measures of welfare change,

g=function that defines maximum profit generated from capture fisheries and aquaculture in the national economy,

V=vector of fixed factors,

λ=vector of technology and fisheries management variables representing the state of climate change and adaptation strategy, and

f=financial inflow (outflow) to (from) the non-fisheries/aquaculture sector.

The modified balance of trade function (B) provides a money metric measure of overall welfare. The impact of the climate change on welfare is obtained from Eq. (1) by comparing the net expenditures required to achieve a given level of utility (u^h) under the "no climate change" (scenario I), and the "climate change" (scenario II) situations. Similarly, the impact of the climate change adaptation strategy (new management practices/technologies) on welfare is obtained from Eq. (1) by comparing the net expenditures required to achieve a given level of utility (u^h) under the "climate change, but no climate change adaptation strategy" (λ_0) (scenario

II), and under the "climate change adaptation strategy" (λ_1) (scenario III).

It is assumed that the climate change adaptation strategy leads to an improvement from a situation yielding utility level u^0 to one yielding utility level u^1. An equivalent variation measure of the welfare change due to climate change adaptation is defined with the utility level in the expenditure function held constant at the level achieved after implementing the climate change adaptation strategy (u^1):

equation(2)

$$B_1^1 - B_0^1 = B(\mathbf{P}_1, \mathbf{V}_1, \lambda_1, u^1) - B(\mathbf{P}_0, \mathbf{V}_0, \lambda_0, u^1) \quad (2)$$

A compensating variation version of the welfare measure is same as defined in Eq. (2), except that the utility level is held at u^0 rather than u^1. The modified balance-of-trade function defined in Eqs. (1) and (2) shares the fundamental parameters of the behavioral system; it includes the parameters of the consumer expenditure system and the profit function.

The supply and supply systems of each of the country-specific models consider six groups: tuna, other oceanic finfish, coastal finfish, coastal invertebrates, freshwater finfish, and freshwater invertebrates. These fish groups are found in oceanic, coastal, and freshwater areas, as well as those produced in aquaculture.

It is assumed that fish production in a country can be represented by a normalized quadratic profit function. The profit function subsumes the effects of fixed factors and intermediate inputs into the constant terms, and the quadratic profit function can be expressed in this multiple product case as [27]:

equation(3)

$$\pi = a_n p_n' + \sum_{i=1}^{n-1} a_i p_i' + 0.5 \sum_{i=1}^{n-1} \sum_{j=1}^{n-1} b_{ij} \left[\left(p_i' p_j' \right) / \left(p_n' \right)^2 \right] \quad (3)$$

where

π = profit (not normalized),

p_i' = effective nominal price of the ith fish group, and

p_n' = effective nominal price of the numéraire fish group.

The effective price of fish group i is defined as:

equation(4)

$$p_i' = p_i \lambda_i \quad (4)$$

where

p_i = nominal price of the ith group of fish, and

λi = an index of climate change and/or climate change adaptation strategies (new management practices/technologies) in farming/fishing of fish group i, set to unity in the base period.

For a specific scenario (I, II or III discussed earlier), the profit function represented in Eq.(3) assumes a given level of exogenous factors (Z) such as technology, stock size (for capture fisheries) and other biological and environmental factors. These exogenous factors are expected to change over time. Biological conditions (like stock abundance) and technology have been modeled as exogenous components that shift the level of production. It is considered that profits are an increasing function of the exogenous factors (Z); for capture fisheries, higher stock abundance will give higher profits.

The new management practices/technologies are assumed to be a disembodied, output-augmenting technical change. The technical change increases the "effective" quantity of fish associated with a given physical quantity, and there is a corresponding change in the effective price of the fish. This specification of technical change in the profit function leads to proportional shifts in the resulting supply curves and a shift in the intercept on the price axis.

This climate change adaptation index (λ) represents a proportional shift down of the supply function involving two proportional shifts: a λ percentage proportional shift in the quantity direction, and a λ percentage proportional shift in the price direction. [1] The proportional shift in the quantity direction results from an increase in the output of product or species i achieved by reducing the output of other products or species, known as the competitiveness effect [27] and [30]. The proportional shift in the price direction results from increases in the output of one product or species (say, i) without any reduction in the output of other products or species, referred to in the literature as a pure productivity effect [27] and [30]. A λ percent output augmented by technical change, as assumed in this paper, will yield an increase in output that is larger than λ percent.

Following the envelope theorem, the supply functions for different fish groups are obtained by differentiating the profit function (Eq. (3)) with respect to effective prices. For each non-numéraire fish group, the equations can be written as:

equation(5)

$$q_i' = a_i' + \sum_{j=1}^{n-1} b_{ij} \left(\frac{p_j'}{p_n'} \right) \quad (5)$$

While for the numéraire group, the supply function takes the form:

equation(6)

$$q_n' = a_n' - 0.5 \sum_{i=1}^{n-1} \sum_{j=1}^{n-1} b_{ij} \left[\left(p_i' p_j' \right) / \left(p_n' \right)^2 \right] \quad (6)$$

On the consumption side, an Almost Ideal Demand System (AIDS) representation (Deaton and Muellbauer [34]) was used. The AIDS share equation can be expressed as:

equation(7)

$$w_i = \alpha_i + \sum_{j=1}^{n} \varphi_{ij} \ln p_j + \beta_i \ln \left(\frac{e}{P} \right)$$

where

e=total expenditure on fish,

p_j=price of the jth group of fish,

w_i=share of total fish expenditure allocated to the ith fish group.

The price index (P) is:

equation(8)

$$\ln P = \alpha_0 + \sum_{j=1}^{n} \varphi_j \ln p_j + \frac{1}{2} \sum_{j=1}^{n} \sum_{j=1}^{n} \varphi_j \ln p_i \ln p_j \quad (8)$$

The AIDS expenditure function can be written as:

equation(9)

$$In\ e = InP + u\beta_0 \sum_{j=1}^{n} Inp_k^{\beta k} \qquad (9)$$

The final equation needed to complete the model is the income–expenditure equation for equating production income with expenditure and any inflow (outflow) to (from) non-fishery/aquaculture activities:

equation(10)

$$e - \pi - f = 0 \qquad (10)$$

MODEL PARAMETERS AND DATA

The country models require (i) baseline data on demand, supply, and prices for each fish type; (ii) parameters for the fish supply and demand equations (Eqs. (5), (6), (7),(8) and (9)); and (iii) data on exogenous variables, including climate change trend and climate change adaptation strategies. Baseline data on demand, supply, and prices for various fish types were collected from various secondary sources and surveys of local markets in each country. As no published estimates of demand and supply elasticities for the fisheries sector are available, and the resources and time needed to estimate these elasticities were lacking, a subjective elicitation process and secondary data were used to make benchmark estimates. This process, using expert opinion surveys (EOSs) and focus group discussions (FGDs), is described in more detail in the subsequent country specific papers of this special section. The preliminary benchmark estimates were presented to various experts for validation in intensive validation meetings.

The slopes of the supply functions (Eqs. (5) and (6)) were inferred from the local price elasticities of supply, the relevant base period price, and quantity variables. The intercepts were then obtained by subtraction, and were used to calibrate the production system to the base data set. Following Martin and Alston [27], the demand system (Eqs.(8) and (9)) was parameterized based on market price and income elasticities of demand for various groups of fish; base period data on price, quantity, and expenditure shares of different groups of fish; and using the symmetry and homogeneity, and adding up restrictions of the demand system. The intercepts of the demand equations were also obtained by subtraction to calibrate the demand system to the base data set. The value of "u" is set at 0.5, consistent with other analyses.

As indicated earlier, the models developed and reported in this paper are country-specific fish sector models designed to analyze the effects of climate change and related adaptation strategies in individual countries. Global prices of different fish groups were used as exogenous variables in the modeling exercise. The most recent global projections of fish prices generated by the International Food Policy Institute (IFPRI) and research partners (see [24] and [35]) are used in the models. Given global fish prices and other exogenous variables, the models endogenously determine domestic fish prices, supply, demand, export, and import of various fish categories for each of the study countries.

For each time period, the model was run for a baseline scenario and for various adaptation strategies. The baseline scenario denotes the most likely case that was identified with respect to trends in the exogenous variables (including population, income growth, technological change in aquaculture/fisheries, climate change, etc.). For climate change trajectories under baseline scenarios (λ_0), this paper has followed low (B1) and high (A2) emission scenarios originally reported in the Intergovernmental Panel on Climate Change Special Report on Emissions Scenarios [36] and subsequently used by Bell et al. [5]. However, countries differ in assumptions regarding the baseline scenario and adaptation strategies. The country-specific values of the shift in supply curve (%) from current (2006–2009) to 2035 and 2050 due to climate Change (λ_0) in Fiji, Solomon Islands, Vanuatu and Timor-Leste are reported in Table 1.

Table 1: Shift in supply curve[a] (%) from Current (2006–2009) to 2035 and 2050 due to climate change in Fiji, Solomon Islands, Vanuatu and Timor-Leste

Species group	2035				2050			
	Fiji	Solomon Islands	Vanuatu	Timor-Leste	Fiji	Solomon Islands	Vanuatu	Timor-Leste
Tuna	15	3	10	0	15	3	10	0
Other oceanic finfish	15	3	10	0	15	3	10	0
Coastal finfish	−5	−5	−10	−5	−15	−10	−15	−10
Coastal invertebrates	−5	−5	−10	−5	−15	−10	−15	−10
Freshwater finfish	25	12.5	25	50	75	25	50	100
Freshwater invertebrates	25	25	25	50	75	25	50	100

This shift in supply curve for the baseline scenario has been denoted in Eq. (2) as (λ_0).

Sources: Authors. Calculated based on secondary literature [5], [37], [38], [39] and [40], and primary data (EOS, FGD, post-survey validation meeting, expert consultations).

The projection of the Spatial Ecosystem and Population Dynamics Model (SEAPODYM)[2] was used as a reference for the likely effects of climate change on tuna catch under a relatively low emission (B1 scenario) and a relatively high emission scenarios (A2 scenario). The preliminary simulations of the potential impact of global warming on tuna populations using SEAPODYM are presently available only for skipjack and bigeye tuna, and are reported in [5] and [37]. Among the four main species of tuna (albacore, bigeye, skipjack and yellowfin), skipjack tuna is the highest contributor (more than 70%) of total tuna catch in the Pacific region.

The results of SEAPODYM model show that the concentrations of skipjack and bigeye tuna are likely to be located further east than in the past due to climate change. As reported in Lehodey et al. [37], catch of skipjack tuna in Fiji is likely to increase by 25.8%, relative to 1980–2000 (20 years) average, in 2035. The corresponding increase in Vanuatu is 18.4%. Tuna catch is expected to increase by 3.2% in Solomon Islands in 2035 [5]. This paper has used more conservative estimates of the likely effect of climate change on tuna catch (please see Table 1).

This paper has used Pratchett et al. [38], Gehrke et al. [39], and Pickering et al. [40] as the main references for projected direct effects of climate change in coastal fisheries, freshwater fisheries and aquaculture production, respectively. The reported of values of (λ_0) were adjusted based on experts consultation. It is important to note that many studies on climate change impact have used expert consultation as a way to collect relevant information (see, for example, [38] and [41]).

Ex ante indicators of the direct impact of climate change adaptation strategies were constructed using data collected through FGDs and EOSs. The values of these indicators (λi) were computed by comparing the cost per unit of fish produced (cost per unit of catch for capture fisheries) and/or volume of fish produced/captured between scenarios II and III, as discussed earlier.

Tuna supply: catch by National and Foreign Flights

Broadly speaking, there are three types of tuna and oceanic catches: catch by domestic/national fleets in national waters, catch by domestic fleets in international waters, and catch by foreign fleets in national waters. Because of the focus of this and the subsequent related papers on national food security, the supply volumes used in the analysis include catch by national fleets in both national and international waters, but do not include catch by foreign fleets in national waters. The models developed and reported in this paper project the behavior of fish production, consumption and trade sectors of four pacific countries under alternative scenarios. Tuna catch by foreign vessels is not determined by the economic factors of supply and demand that are analyzed in these papers. Rather, foreign catch is determined by international and regional agreements, including regional agreements like the Pacific Islands Forum Fisheries Agency (FFA) and the Parties to the Nauru Agreement (PNA). Relative contributions of tuna catch by foreign fleets and that of national fleets in most of the case-study countries vary substantially over period, partly due to changes in bilateral/regional/international agreements. Therefore, catch by foreign fleets was treated as a residual sector, influenced by domestic demand and supply (which includes catches by national fleets outside national waters). The available data on tuna catch by foreign vessels supports this interpretation.

Among the four case study countries, foreign vessel catch in national waters is not very substantial in Fiji and Vanuatu. In fact, the total tuna catch by national fleets of Fiji and Vanuatu are higher than the total tuna catch in their respective national waters. For example, in 2008 total tuna catch by Fijian national fleets was 11,656 t whereas total tuna catch in Fijian national waters was only 8536 t. Similarly, in 2008 total catch by Vanuatu national fleets was 47,234 t, but the total tuna catch in Vanuatu national waters was only 6021 t (FFA database). Among the four case-study countries, tuna catch by foreign fleets is a significant source of government revenue only in Solomon Islands.

For Fiji and Vanuatu, where tuna catch by their national fleets are higher than the tuna catch in their respective national waters, the projections of tuna catch are within the sustainable limits of tuna population, particularly given the significant increase in tuna population projected under climate change. Solomon Islands and Timor-Leste are keen to develop their domestic tuna industry, and their aim is to access some of the tuna resources currently utilized by foreign vessels. The models for Solomon Islands and Timor-Leste reflect that reality, as also shown in the individual country papers of this special section. The projected tuna catches for Solomon Islands and Timor-Leste reported in this and the subsequent papers represent only a small share of total tuna resources available in their respective national waters.

Overall Impact Assessment

The ex ante indicators of the direct impact of climate change adaptation strategies were used in the fisheries sector model to estimate the effects of various adaptation strategies on fish supply, demand, price, and trade. The specified climate change adaptation strategies were used in the fish sector model to estimate the economic (welfare) gain to society. The fish sector model (Eqs. 3–10, which include behavioral models of supply relations, Marshallian demand equations, market-clearing conditions, and income–expenditure conditions) was solved using linear program solver in Excel 4.0.

An equivalent variation version of welfare measure was computed using Eq. (2), which can be written as follows:

$$B_1^1 - B_0^1 = [e(\mathbf{P}_1, u^1) - e(\mathbf{P}_0, u^1)] - [g(\mathbf{P}_1, \mathbf{V}_1, \lambda_1) - g(\mathbf{P}_0, \mathbf{V}_0, \lambda_0)] - [f_1 - f_0] \tag{11}$$

In Eq. (11), the total welfare change is equal to (i) the equivalent variation for the change in consumer welfare due to price change induced by the adoption of adaptation strategies (the first term in brackets) minus (ii) the increase in producer profit due to the adoption of adaptation strategies (the second term in brackets), minus (iii) changes in inflow (outflow) to (from) non-fishery/aquaculture activities (the last term in brackets).

The model (Eqs. (3)–(10)) was solved for three different time periods: current (2006–2009), medium term (2035) and long term (2050). The supply equations Eqs. (3)–(6)were shifted with various time period-specific values of λ_i, and the demand equations (Eqs. (7)–(9)) were shifted with time period-specific income and population. The values of fish supply, demand, net trade, and domestic fish prices for various fish species groups were estimated in two time periods (2035 and 2050).

This paper followed comparative static analysis. Its focus was not to generate projections for key variables in the fisheries sector, but to examine how projections respond to various developments or events within and outside of the fisheries sector. The study accomplished this by implementing various country-specific climate change adaptation strategies. As in any other modeling effort, the outcomes are likely to be influenced by the assumptions that are adopted. For this reason, this study implemented a baseline scenario, and each of the adaptation scenarios, for two growth rates of per capita real income.

BASELINE RESULTS

Fiji

For each time period (2035 and 2050), two baseline (most plausible) scenarios were implemented representing two annual growth rates of real per capita income: medium (1% per year) and high (2% per year) growth of real per capita income.[3] The model considered a population of 977,586 in 2035 and 1,060,706 in 2050 [43].

Fig. 2 shows the projected quantities of production and consumption of different fish groups in Fiji for 2035 and 2050 under two baseline scenarios, with respective 1% and 2% annual growth rates of real per capita income. It indicates that oceanic production is expected to grow to some extent, and that supply is likely to be higher than demand during the projected years. However, projections are quite different for coastal fisheries. The model projects that coastal production will decline over time while demand for coastal fish will likely increase further. Though the supply from freshwater areas is projected to expand substantially, its share is expected to remain small. The projected supplies of different fish types are plausible, given the past trends. The main reason for the decline in supply from coastal areas is the anticipated negative effects of climate change and other adverse environmental factors.

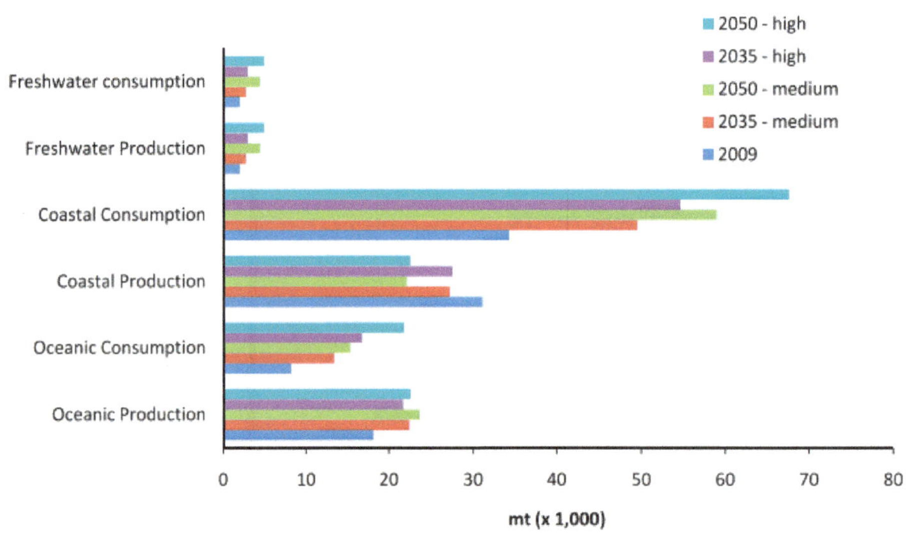

Figure 2: Projected production and consumption of different types of fish under two baseline scenarios (1% and 2% annual growth rates of per capita real income) in Fiji, 2035 and 2050.

Results show that the aggregated fish consumption is expected to rise substantially in 2035 and 2050, and the demand for fish is likely to be higher than the production. These results imply that the country may need to import fish to fulfill this deficit in production under the baseline scenario (i.e., without the technological and policy interventions analyzed here).

There are four important points regarding the projections presented in this section: (1) domestic production is projected to grow at a negligible rate; (2) domestic demand is projected to rise over the period; (3) as expected, higher income growth will be accompanied by higher rise in demand for fish; and (4) fish exports are expected to fall, fish imports are expected to increase, and, therefore, net trade (export minus import) is projected to decline over time. These results have serious food security implications, given that poor households mostly rely on coastal finfish for their fish consumption needs.

Solomon Islands

The model predicts that the supply of fish from oceanic and freshwater systems aggregate will increase marginally in the medium term (2035) and long term (2050) under the baseline scenarios. However, the supply from coastal fisheries is likely to decrease over time. On the other hand, the demand for different types of fish will increase over time as a result of an increase in population and per capita real income (Fig. 3). As expected, demand will be higher, with higher growth (3% per annum) in per capita real income. Results show that total demand is likely to surpass total domestic fish production in the long term (2050). Among various species groups, a major part of this increased demand will be for oceanic species, such as tuna. The modeling results reported in this paper imply that increases in domestic supply of tuna and other oceanic species will be helpful to meet the growing demand for this category of fish. If Solomon Islands cannot catch more oceanic fish than otherwise harvested by foreign vessels, the country may have to import fish in large volumes to meet the projected demand.

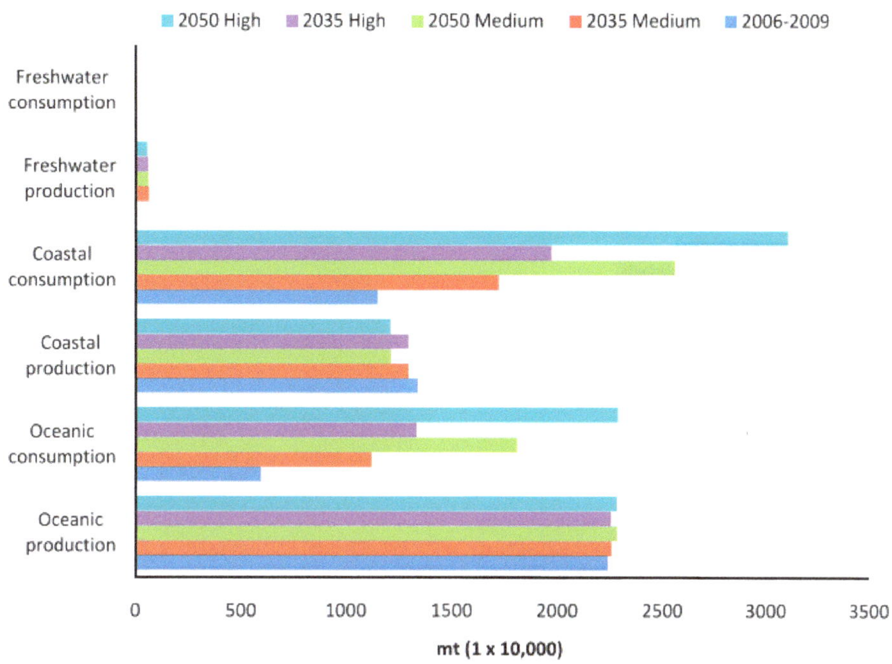

Figure 3: Projected production and consumption of different types of fish under two baseline scenarios (2% and 3% annual growth rate of per capita real income) in Solomon Islands, 2035 and 2050.

It is important to note that foreign vessels currently catch most of the tuna caught in Solomon Islands, and almost all of these fish are exported by foreign companies (not by Solomon Islands). The domestic production and net trade (difference between production and consumption) reported in this section cover only the domestic industry. Income from foreign vessels is considered as an outside income.

Vanuatu

Fig. 4 shows the projected production, consumption, and net trade (difference between production and consumption) of different fish groups in Vanuatu for 2035 and 2050 under two baseline scenarios, with 1.5% and 2.5% annual growth rates of real per capita income. As shown in Fig. 4, the performance of various fisheries subsectors will not be uniform. During 2010–2050, production of oceanic fish is expected to increase, but production of coastal fish is projected to decline. Though the consumption of oceanic fish is expected to rise at a faster rate than any other sector, the model predicts that the oceanic fisheries sector will continue to be a net exporter. However, Vanuatu will have to import coastal fish to meet the increasing demand from population and income growth. Given that many of the poorer households rely on coastal fisheries for their consumption needs, this likely scenario has serious food security implications. The model projects that demand for freshwater fish will exceed domestic production in 2035 and 2050 under the baseline scenarios, and Vanuatu will need to rely on imports to meet this demand.

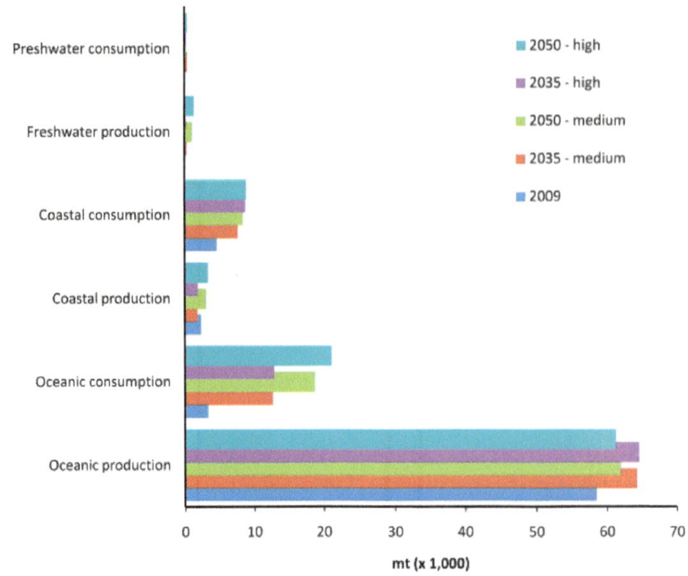

Figure 4: Projected production and consumption of different types of fish under two baseline scenarios (1.5%, 2.5% annual growth rate of per capita real income) in Vanuatu, 2035 and 2050.

Currently, Vanuatu is a net exporter of fish and seafood, with domestic production far exceeding domestic consumption. The model analysis projects that the aggregated fish consumption in Vanuatu will rise substantially in 2035 and 2050, but the country will remain a net exporter by 2050.

Timor-Leste

Fig. 5 shows the projected quantities of production and consumption of different fish groups in Timor-Leste for 2035 and 2050 under two baseline scenarios, with 2% and 3% annual growth rates of real per capita income. The results indicate that fish supplies from oceanic and coastal ecosystems are likely to decrease during 2010–2050. The model predicts that only freshwater ecosystems will be able to supply more fish in the future. Given that oceanic and coastal fisheries supply about 94% of current fish consumption in Timor-Leste, this projected fish supply scenario has serious food security implications for the country.

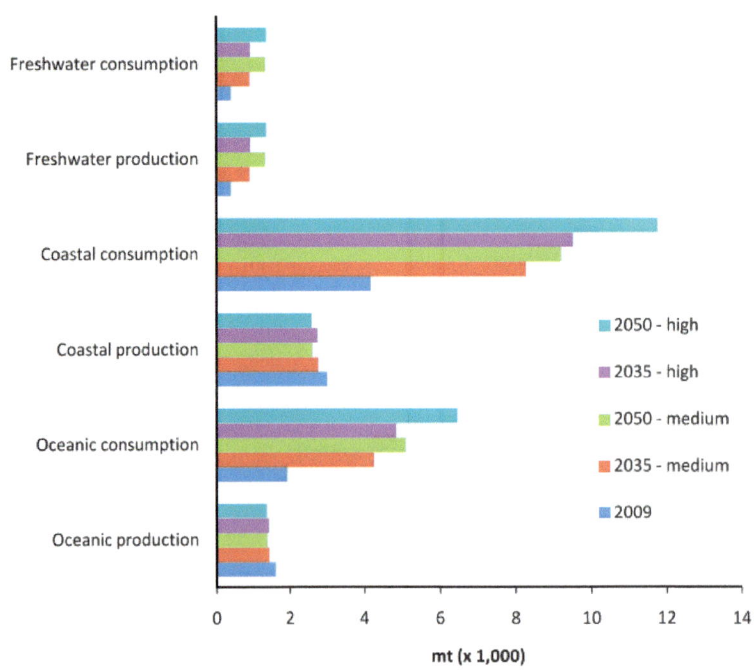

Source: Model projections.

Note: medium = 2% annual growth rate of per capita real income; high = 3% annual growth rate of per capita real income.

Figure 5: Projected production and consumption of different types of fish under two baseline scenarios (2% and 3% annual growth rate of per capita real income) in Timor-Leste, 2035 and 2050.

Currently, Timor-Leste is a net importer of fish and seafood. The model predicts that total fish production will increase only marginally in the medium term (2035) and long term (2050) under the baseline scenarios. Aggregate fish demand is expected to rise substantially over time due to growth in population and real per capita income. This implies that the country will have to import more fish to fill this increasing deficit in domestic fish supply.

CONCLUSIONS AND POLICY IMPLICATIONS

This paper has developed a rigorous analytical tool that is capable of comparative analysis of alternative development scenarios, even under data-scarce situations. The model is utilized to analyze the potential effects of climate change and related adaption strategies in the four PICT countries studied. This paper reports the model and discusses the baseline results. The subsequent three papers [44], [45] and [46] apply the model to analyze the effect of various climate change adaption strategies in the four countries under study.

The baseline results illuminate key differences across the four countries, mainly due to the fish trade regimes prevailing in different countries, the relative importance of various sectors in fish production and consumption, and producers' and consumers' behaviors in various countries as reflected through supply and demand elasticities.

Currently, Fiji, Solomon Islands, and Vanuatu are net exporters of fish and seafood, while Timor-Leste is a net importer of fish.[4] The baseline projections indicate that Vanuatu will continue to be a net fish exporter and Timor-Leste will remain a net fish importer in the long term. For Solomon Islands, domestic demand for fish is likely to surpass the supply from domestic sources (i.e., the domestic fishing industry and aquaculture farms, but not including the catch of foreign fishing fleets) in the long term (2050). It is likely that, with high population and per capita income growth, Fiji may become a net importer of fish in the long run.

Similar to overall fish trade, all countries under consideration, except Timor-Leste, are currently net exporters of tuna and other oceanic fish. The model predicts that Fiji and Vanuatu will continue to be net exporters of oceanic fish in the long run. But Solomon Islands is unlikely to be able to meet domestic demand for oceanic fish from its domestic sources.

The baseline scenarios project that supply of tuna and other oceanic fish from the domestic fishing industry (not considering catch by foreign vessels) will increase over time in Fiji and Vanuatu, but will remain more or less the same in Solomon Islands. Tuna and oceanic fish supply is likely to decrease over time in Timor-Leste. It is important to note that various ecosystem-based tuna models (such as SEAPODYM) show that skipjack tuna is projected to move to the east (Eastern Pacific Ocean), and the biomass of adult tuna will decrease in the West and Central Pacific Ocean [37], consistent with the findings of this study.

With rising per capita income and population, demand for fish and seafood in various Pacific Island countries is expected to increase substantially for the period up to 2050.[5] As indicated earlier, currently many Pacific Island countries (including Fiji, Solomon Islands, and Vanuatu) are net exporters of fish, excluding foreign tuna catch. Without any appropriate policy and/or technological intervention (as reflected in the baseline scenarios), fish exports from these countries are expected to decrease, fish imports are expected to increase, and net exports (export minus import) are likely to decline over time. This will lead to a decline in per capita consumption of domestically produced fish, which has serious negative food security implications for these countries.

In contrast to rising fish demand over time, supply of fish from coastal fisheries is expected to decline due to climate change and other adverse environmental effects. Therefore, there is an urgent need to increase the supply of fish in Pacific Island countries from their domestic sources by reversing the negative trends of coastal fisheries and/or increasing supply from oceanic and freshwater systems.

ACKNOWLEDGMENT

This work was funded by the Asian Development Bank (ADB) Technical Assistance 7753 on "Strengthening Coastal and Marine Resources Management in the Coral Triangle of the Pacific (Phase 2)" and by the United States Agency for International Development (USAID). Inputs and guidance from with the governments, non-government organizations, private sector, coastal communities and other stakeholders in Fiji, Solomon Islands, Vanuatu and Timor-Leste are respectfully appreciated. Comments and suggestions from Dr. Johann Bell, Dr. Carole Engle and Dr Robert Pomeroy are gratefully acknowledged.

REFERENCES

1. E.H. Allison, A.L. Perry, M.C. Badjeck, W.N. Adger, K. Brown, D. Conway, et al., Vulnerability of national economies to the impacts of climate change on fisheries, Fish Fish. 10 (2) (2009) 173–196, Available: ⟨http://onlinelibrary.wi ley.com/doi/10.1111/j.1467-2979.2008.00310.x/abstract⟩.
2. M. Barange, R.I. Perry, Physical and ecological impacts of climate change relevant to marine and inland capture fisheries and aquaculture, in: K. Cochrane, C.D. Young, D. Soto, T. Bahri (Eds.), Climate Change Implications for Fisheries and Aquaculture. Overview of Current Scientific Knowledge, in: FAO Fisheries and Aquaculture Technical Paper, vol. 530, 2009, pp. 7–106. Available: ⟨http:// www.fao.org/fileadmin/user_upload/newsroom/docs/FTP530.pdf⟩.
3. M. Barange, G. Merino, J.L. Blanchard, J. Scholtens, J. Harle, E.H. Allison, et al., Impact of climate change on marine ecosystem production in socities dependent on fisheries, Nat. Clim. Chang. 4 (2014) 211–216.
4. J.D. Bell, M. Kronen, A. Vunisea, W.J. Nash, G. Keeble, A. Demmkea, et al., Planning the use of fish for food security in the Pacific, Mar. Policy 33 (2009) 64–76.
5. J.D. Bell, J.E. Johnson, A.J. Hobday (Eds.), Vulnerability of Tropical Pacific Fisheries and Aquaculture to Climate Change, Secretariat of the Pacific Community, Noumea (New Caledonia), 2011.
6. J.D. Bell, A. Ganachaud, P.C. Gehrke, S.P. Griffiths, A.J. Hobday, O. HoeghGuldberg, et al., Mixed responses of tropical Pacific fisheries and aquaculture to climate

change, Nat. Clim. Chang. 3 (2013) 591–599, http://dx.doi.org/ 10.1038/NCLIMATE1838.

7. K.M. Brander, Global fish production and climate change, Proc. Natl. Acad. Sci. USA 104 (50) (2007) 19709–19714, Available: ⟨http://www.pnas.org/content/104/50/19709.full.pdfþhtml⟩.

8. T. Daw, W.N. Adger, K. Brown, M.C. Badjeck, Climate change and capture fisheries: potential impacts, adaptation and mitigation, in: K. Cochrane, C.D. Young, D. Soto, T. Bahri (Eds.), Climate change implications for fisheries and aquaculture. Overview of current scientific knowledge. in: FAO Fisheries and Aquaculture Technical Paper, vol. 530, 2009, pp. 107–150. Available: ⟨http:// www.fao.org/fileadmin/user_upload/newsroom/docs/FTP530.pdf⟩.

9. N.K. Dulvy, J.D. Reynolds, G.M. Pilling, J.K. Pinnegar, J.S. Phillips, E.H. Allison, M. C. Badjeck, Fisheries management and governance challenges in a climate change, in: OECD. The Economics of Adapting Fisheries to Climate Change, OECD Publishing, Paris, 2010, pp. 31–90, http://dx.doi.org/10.1787/9789264090415-en.

10. S. Hanna, Economic and policy issues related to the impact of climate change on fisheries, in: OECD. The Economics of Adapting Fisheries to Climate Change, OECD Publishing, Paris, 2010, pp. 91–116, http://dx.doi.org/10.1787/9789264090415-en.

11. I. Perry, Potential impacts of climate change on marine wild capture fisheries: an update, J. Agric. Sci. 49 (Suppl. S1) (2011) S63–S75 ⟨⟨http://journals.cambridge.org/action/displayAbstract?fromPage¼online&aid¼8030729⟩⟩.

12. R.I. Perry, R.E. Ommer, M. Barange, F. Werner, The challenge of adapting marine social–ecological systems to the additional stress of climate change, Curr. Opin. Environ. Sustain. 2 (5–6) (2010) 356–363.

13. J.C. Rice, Garcia S.M. Fisheries, food security, climate change, and biodiversity: Characteristics of the sector and perspectives on emerging issues, ICES J. Mar. Sci. 68 (6) (2011) 1343–1353, Available: ⟨http://icesjms.oxfordjournals.org/ content/early/2011/04/21/icesjms.fsr041⟩.

14. U.R. Sumaila, W.W.L. Cheung, V.W.Y. Lam, D. Pauly, S. Herrick, Climate change impacts on the biophysics and economics of world fisheries, Nat. Clim. Chang. 1 (2011) 449–456.

15. M. Barange, W.W.L. Cheung, G. Merino, R.I. Perry, Modelling the potential impacts of climate change and human activities on the sustainability of marine resources, Curr. Opin. Environ. Sustain. 2 (2010) 326–333.

16. W.W.L. Cheung, C. Close, V.W.Y. Lam, R. Watson, D. Pauly, Application of macroecological theory to predict effects of climate change on global fisheries potential, Mar. Ecol. Prog. Ser. 365 (2008) 187–197.

17. W.W.L. Cheung, V.W.Y. Lam, J.L. Sarmiento, K. Kearney, R. Watson, D. Zeller, D. Pauly, Large-scale redistribution of maximum fisheries catch potential in the global ocean under climate change, Glob. Chang. Biol. 16 (1) (2010) 24–35.

18. W.W.L. Cheung, J.L. Sarmiento, J. Dunne, T.L. Frolicher, V.W.Y. Lam, M.L. Deng Palomares, R. Watson, Shrinking of fishes exacerbates impacts of global ocean changes on marine ecosystems, Nat. Clim. Chang. 3 (2013) 254–258.

19. H.O. Pörtner, R. Knust, Climate change affects marine fishes through the oxygen limitation of thermal tolerance, Science 315 (2007) 95–97.

20. H.O. Pörtner, Oxygen- and capacity-limitation of thermal tolerance: a matrix for integrating climate-related stressor effects in marine ecosystems, J. Exp. Biol. 213 (2010) 881–893.

21. R.M. Briones, L. Garces, M. Ahmed, Climate change and small pelagic fisheries in developing Asia: the economic impact on fish producers and consumers, in: R. Hannesson, M. Barange, S. Herrick Jr (Eds.), Climate Change and the Economics of Worlds Fisheries: Examples of Small Pelagic Stocks, Edward Elgar Publishing, Cheltenham, 2006.

22. R.M. Briones, Projecting future fish supplies using stock dynamics and demand, Fish Fish. 7 (2006) 303–315.

23. C. Delgado, N. Wada, M. Rosegrant, S. Meijer, M. Ahmed, Fish to 2020: Supply and Demand in Changing Global Market, International Food Policy Research Institute and WorldFish Center, Washington DC (USA) and Penang (Malaysia), 2003.

24. World Bank, Fish to 2030: Prospects for Fisheries and Aquaculture, World Bank, Washington, DC, 2013.

25. V.W.Y. Lam, W.W.L. Cheung, W. Swartz, U.R. Sumaila, Climate change impacts on fisheries in West Africa: implications for economic, food and nutritional security, Afr. J. Mar. Sci. 34 (1) (2012) 103–117, http://dx.doi.org/10.2989/ 1814232X.2012.673294.

26. H.O. Pörtner, D.M. Karl, P.W. Boyd, W.W.L. Cheung, S.E. Lluch-Cota, Y. Nojiri, et al., Ocean systems, in: C.B. Field, V.R. Barros, D.J. Dokken, K.J. Mach, M. D. Mastrandrea, T.E. Bilir, et al., (Eds.), Climate Change 2014: Impacts, Adaptation, and Vulnerability. Part A: Global and Sectoral Aspects. Contribution of Working Group II to the Fifth Assessment Report of the Intergovernmental Panel on Climate Change, Cambridge University Press, Cambridge (United Kingdom) and New York (USA), 2014, pp. 411–484.

27. W.J. Martin, J.M. Alston, A dual approach to evaluating research benefits in the presence of trade distortions, Am. J. Agric. Econ. 76 (1) (1994) 26–35.

28. M.M. Dey, R.M. Briones, M. Ahmed, Disaggegated analysis of fish supply, demand, and trade in Asia: baseline model and estimation strategy, Aquac. Econ. Manag. 9 (1&2) (2005) 113–140.

29. M.M. Dey, P. Kumar, O.L. Chen, M.A. Khan, N.K. Barik, L. Li, A. Nissapa, N. S. Pham, Potential impact of genetically improved carp strains in Asia, Food Policy 43 (2013) 306–320.

30. W.J. Martin, J.M. Alston, Producer surplus without apology? Evaluating investments in R&D, Econ. Rec. 73 (221) (1997) 146–158.

31. R. Linder, F. Jarreth, Supply shifts and the size of research benefits, Am. J. Agric. Econ. 60 (1978) 48–58.

32. G.W. Norton, J.S. Davis, Evaluating returns to agricultural research: a review, Am. J. Agric. Econ. 63 (1981) 685–699.

33. J.M. Alston, G.N. Norton, P.G. Pardey, Science Under Scarcity: Principles and Practice for Agricultural Research Evaluation and Priority Setting, Cornell University Press, Ithaca (New York), 1995.

34. A. Deaton, J. Muellbauer, An almost ideal demand system, Am. Econ. Rev. 70 (1980) 312–326.

35. M. Kobayashi, S. Msangi, M. Batka, S. Vannuccini, M.M. Dey, J.L. Anderson, Fish to 2030: the role and opportunity for aquaculture, Aquac. Econ. Manag. 19 (3) (2015) 282–300.

36. N. Nakicenovic´´, J. Alcamo, G. Davis, B. De Vries, others. Special Report on Emissions Scenarios, A Special Report of Working Group III of the Intergovernmental Panel on Climate Change. PNNL–SA–39650, Cambridge University Press, New York (United States of America), 2000.

37. P. Lehodey, J. Hampton, R.W. Brill, S. Nicol, I. Senina, B. Calmettes, et al., Vulnerability of oceanic fisheries in the tropical Pacific to climate change, in: J. D. Bell, J.E. Johnson, A.J. Hobday (Eds.), Vulnerability of Tropical Pacific Fisheries and Aquaculture to Climate Change, Secretariat of the Pacific Community, Noumea (New Caledonia), 2011, p. 925 (Chapter 8).

38. M.S. Pratchett, P.L. Munday, N.A.J. Graham, M. Kronen, S. Pinca, K. Friedman, et al., Vulnerability of coastal fisheries in the tropical Pacific to climate change, in: J.D. Bell, J.E. Johnson, A.J. Hobday (Eds.), Vulnerability of Tropical Pacific Fisheries and Aquaculture to Climate Change, Secretariat of the Pacific Community, Noumea (New Caledonia), 2011, p. 925.

39. P.C. Gehrke, M.J. Sheaves, D. Boseto, B.S. Figa, J. Wani, Vulnerability of freshwater and estuarine fisheries in the tropical Pacific to climate change, in: J. D. Bell, J.E. Johnson, A.J. Hobday (Eds.), Vulnerability of Tropical Pacific Fisheries and Aquaculture to Climate Change, Secretariat of the Pacific Community, Noumea (New Caledonia), 2011 (Chapter 10).

40. T.D. Pickering, B. Ponia, C.A. Hair, P.C. Southgate, E.S. Poloczanska, L.D. Patrona, et al., Vulnerability of aquaculture in the tropical Pacific to climate change, in: J.D. Bell, J.E. Johnson, A.J. Hobday (Eds.), Vulnerability of Tropical Pacific Fisheries and Aquaculture to Climate Change, Secretariat of the Pacific Community, Noumea (New Caledonia), 2011, p. 925.

41. J.R.A. Butler, T. Skewes, D. Mitchell, M. Pontio, T. Hills, Stakeholder perceptions of ecosystem service declines in Milne Bay, Papua New Gionea: is human population is more critical driver than climate change? Mar. Policy 46 (2014) 1–13.

42. World Bank, World Development Indicators, World Bank, Washington D.C., 2013, Available: ⟨http://data.worldbank.org/data-catalog/world-developmentindicators⟩.

43. ADB, Food Security and Climate Change: Rethinking the Options, Pacific Studies Series, Manila, 2011 ⟨http://www.adb.org/publications/food-securityand-climate-change-pacific-rethinking-options⟩.

44. M.M. Dey, K. Gosh, R.A. Valmonte-Santos, R. Rosegrant, O.L. Chen, Economic impact of climate change and climate change adaptation strategies for fisheries sector in Fiji, Mar. Policy (2015) (in press, this special section), http://dx. doi.org/10.1016/j.marpol.2015.12.023.

45. M.M. Dey, K. Gosh, R.A. Valmonte-Santos, M. Rosegrant, O.L. Chen, Economic impact of climate change and climate change adaptation strategies for fisheries sector in Solomon Islands: implication for food security, Mar. Policy (2015) (in press, this special issue), http://dx.doi.org/10.1016/j.marpol.2016.01. 004.).

46. M. Rosegant, M.M. Dey, R.A. Valmonte-Santos, O.L. Chen, Economic impact of climate change and climate change adaptation strategies for fisheries sector in Vanuatu and Timor-Leste, Mar. Policy (2015) (in press, this special section), http://dx.doi.org/10.1016/j. marpol.2015.12.010.

47. F. Asche, M.F. Bellemare, C. Roheim, M.D. Smith, S.T. Tveteras, Fair enough? Food security and the international trade of seafood, World Dev. 67 (2015) 151–160.

48. A. Lem, T. Bjorndal, A. Lappo, Economic Analysis of Supply and Demand for Food up to 2030 – Special Focus on Fish and Fishery Products. FAO Fisheries and Aquaculture Circular No. 1089, FAO, Rome (2014), p. 106.

CHAPTER 2

EVOLUTION OF SOME OBSERVED CLIMATE EXTREMES

Ly Mouhamed ,Seydou B. Traore, Agali Alhassane , Benoît Sarr

Keywords: Warming trend, Rainfall decline, Extreme rainfall events, West Africa, Sahel

BACKGROUND

Climates in different parts of the world are characterized by their variability that affects several sectors. This variability has increased since the 1950s, and particularly during the latest decade, mainly because of the increased concentration of anthropogenic greenhouse gases in the atmosphere as argued by some authors (Hegerl et al., 2007,Stott et al., 2011, Trenberth et al., 2007a, Trenberth and Trenberth, 20112012, Min et al., 2011, Solomon et al., 2007, Solomon et al., 2009 and Solomon et al., 2010). In this context, it is argued that surface temperature has increased by 0.7 °C over the last century with a significant warming in many regions; and with the land areas warming faster than the oceans (Trenberth et al., 2007a). In the West African Sahel, an increasing trend was observed in both maximum and minimum temperatures for all the three main ecological zones (Sudanian, Sahelian and Sahelo-Saharan) with minimum temperatures increasing at a faster rate (CEDEAO-ClubSahel/OCDE/CILSS, 2008).

It is also argued that increased temperatures will lead to greater evaporation according to the Clausius–Clapeyron relationship, whereby saturation vapor pressure increases exponentially with temperature. In its turn, the increase of water vapor in the atmosphere affects precipitation events and the risk of flooding. The changing climate is thus perceived through extreme events which tend to modify the magnitude of the predicted climate. This can be supported by the number of flood events having increased on average from less than 2 per year before 1990 to more than 8 or 12 on average per year during the 2000s (Sarr, 2011). Also, the evolution of total annual rainfall has been characterized by a succession of wet years from 1950 to 1969, followed by a period with the persistence of dry years from 1970 to 1993 (L'Hote et al., 2002 and et al.,). This has resulted in a southward movement of isohyets by about 200 km (Diouf et al., 2000). After 1993, another mode of variability, characterized by an alternation between very wet years and very dry years, seems to have started in the region (Ali, 2011). Moreover, the same author indicates that while there is a tendency of persisting drought in the Western Sahel, the East is experiencing gradual return to wetter conditions.

The occurrence of extremes is usually the result of different factors at different time scales. A large amount of the available scientific literature on climate extremes is based on the use of the so-called 'extreme indices', which can either be based on the probability of occurrence of given quantities or on threshold exceedance. In the recent years, the number of publications related to those extremes emerged in some specific countries, in particular in the developing world where, most of the time, data are not available and not always distributed by meteorological services. Especially in Africa, very few studies document such changing climate because of several difficulties in accessing to daily meteorological data (Lamptey, 2009). However the World Meteorological Organization, through the Climate Variability and Predictability (WMO/CLIVAR) Expert Team on Climate Change Detection, Monitoring Indices (ETCCDMI), coordinated a series of workshops in different regions of the world to help scientists and national experts discuss and quality control climate data. Some regional analyses were thus undertaken for the understanding of climate extreme and trends (Easterling et al., 2003, Aguilar et al., 2005,Vincent et al., 2005, Haylock et al., 2006 and

Citation: Ly Mouhamed , Seydou B. Traore, Agali Alhassane , Benoît Sarr , Evolution of some observed climate extremes, http://dx.doi.org/10.1016/j.wace.2013.07.005.

Peterson and Manton, 2008). Recent studies in Africa, when analyzing daily climate in terms of trends and extreme indices revealed some significant increases and decreases in annual precipitation, increases in longest wet spells, increases in high daily precipitation amounts and average rainfall intensity (New et al., 2006, Alexander et al., 2006 and Collins, 2011). Minimum temperature was found to have increased faster than maximum temperature, thus contributing to narrow the diurnal temperature range (Easterling, 1997 and Caesar et al., 2006). Numerous indicators in the extreme indices can be analyzed to further investigate the perception of changed climate in a widespread area in the West African Sahel, where very few studies have documented the changes that occurred in those extremes. The objective of this study is to analyze the evolution of some extreme temperature and precipitation indices over a large area in this region.

METHODS

Study Region

The study was conducted in West Africa over 6 countries, namely Senegal, Mauritania, Mali, Burkina Faso, Niger and Chad, covering three major agro ecological zones: arid, semi arid and sub humid (FAO, 1978). Average annual rainfall varies from 30 to 1000 mm per year (Nicholson, 1993). Mean annual minimum and maximum temperatures are comprised between 16 and 20 °C for the minimum and 27 and 35 °C for the maximum depending on the proximity of the ocean. The rainfall season is unimodal, with the onset of the rains occurring in agricultural areas from May to July and ending in September–October (Sivakumar, 1988 and Camberlin and Diop, 2003). The temporal distribution of temperature is typically bimodal with one maximum in April-May and another one in October, except for coastal zones, particularly in Mauritania and the North of Senegal, where it is unimodal with a maximum occurring in September. The list of the meteorological stations used in this study is given in Table 1.

Table 1: List of meteorological stations in West Africa used in the study

Country	Station name	Longitude	Latitude
Burkina Faso	Bobo-Dioulasso	−4.30	11.17
Burkina Faso	Boromo	−2.92	11.73
Burkina Faso	Dedougou	−3.48	12.47
Burkina Faso	Dori	−0.03	14.03
Burkina Faso	Fada N'Gourma	0.35	12.07
Burkina Faso	Gaoua	−3.18	10.33
Burkina Faso	Ouagadougou	−1.52	12.35
Burkina Faso	Ouahigouya	− 2.43	13.58
Mali	Bamako	−7.95	12.53
Mali	Bougouni	−7.50	11.42
Mali	Kayes	−11.43	14.43
Mali	Kenieba	−11.35	12.80
Mali	Kita	−9.45	13.07
Mali	Segou	− 6.15	13.40
Mali	Sikasso	−5.68	11.35
Mauritania	Aioun	−9.60	16.70
Mauritania	Akjoujt	−14.37	19.75
Mauritania	Atar	−13.07	20.52
Mauritania	Bir Moghrein	−11.62	25.23
Mauritania	BoutilimitT	−14.68	17.53
Mauritania	Kaedi	−13.52	16.13
Mauritania	Kiffa	−11.40	16.63
Mauritania	Nema	−7.27	16.60
Mauritania	Nouadhibou	−17.03	20.93
Mauritania	Nouakchott	−15.95	18.10
Mauritania	Rosso	−15.82	16.50
Mauritania	Tidjikja	−11.43	18.57
Mauritania	Zouerate	−12.70	22.68
Niger	Birni N'Konni	5.28	13.80
Niger	Maine Soroa	11.98	13.23
Niger	Maradi	7.08	13.47
Niger	Niamey	2.17	13.48
Niger	Tahoua	5.30	14.90
Niger	Tillabery	1.45	14.20
Senegal	Bambey	−16.47	14.70
Senegal	Kaolack	−16.07	14.13
Senegal	Matam	−13.25	15.63
Chad	N'Djamena	15.03	12.13

Rainfall and Temperature Data

The data used in this study consist of daily observations of rainfall and temperature available at the AGRHYMET Regional Center for the 1960–2010 period. All the station data were quality controlled using the Rclimdex package (Zhang and Yang, 2004) which is freely downloadable from the ETCCDMI website (http://cccma.seos.uvic.ca/ETCCDMI/index.shtml). The quality control consisted in detecting negative precipitations and minimum temperatures greater than maximum temperatures. To verify the quality of the datasets, simple box plot techniques were also used in order to detect some probable erroneous information in the time series. The result showed some few outliers in the datasets. Stations with high potential of erroneous data and missing data more than 20% were removed from the study

Climate Indices

One of the ways to characterize the frequency, intensity and duration of climate extremes is to calculate climates indices based on daily time series of temperature and rainfall (Karl et al., 1999 and Peterson et al., 2002). International research groups such as the Expert Team on Climate Change Detection Monitoring Indices (ETCCDMI) have undertaken a set a regional analyses for understanding climate extremes and trends (Peterson et al., 2002, Easterling et al., 2003, Vincent and Mekis, 2006 and New et al., 2006). Among the 27 core indices computed on the daily basis by Rclimdex, we selected 9 that we consider most relevant for our study region (Table 2).

Table 2: Definitions of the temperature and rainfall indices used in the study. Tn and Tx are daily minimum and maximum temperature respectively.

Element	Index	Descriptive name	Definition	Units
Tn	Tn10p	Cool night frequency	Percentage of days with TN<10th percentile of 1961–1990	%
Tn	Tn90p	Warm night frequency	Percentage of days with TN>90th percentile of 1961–1990	%
Tx	Tx10p	Cool day frequency	Percentage of days with TX<10th percentile of 1961–1990	%
Tx	Tx90p	Warm day frequency	Percentage of days with TX>90th percentile of 1961–1990	%
Tx	WSDI	Warm spells	Annual count of days with at least 6 consecutive days with TX>90th percentile of 1961–1990	days
Rainfall	RX5day	Maximum 5-day precipitation	Maximum 5-day precipitation	mm
Rainfall	CWD	Consecutive wet days	Maximum number of consecutive wet days	days
Rainfall	R99p	Extremely Wet days	Annual total PRCP when RR>99th percentile	mm
Rainfall	PRCPTOT	Annual Total Rainfall	Annual total PRCP in wet days (RR≥1mm)	mm

Trend Analysis

To estimate the trend in the set of extreme indices time series, an estimator of the slope based on Kendall rank correlation was used (Vincent et al., 2005). In the study, statistical analyses were done over the last 50 years since 1960 including the climatological normal period of 1961–1990. Maps and tables of trends for the period 1961–2010 for some selected indices were constructed by developing scripts with the R open source software (Core Team, 2012).

RESULTS

Temperature Indices

Temperature indices show a general warming trend throughout the West African Sahel during the period from 1960 to 2010 (Fig. 1). Indeed, if one examines the frequency of cools nights (Fig. 1a), it appears that all the stations show a negative slope varying from −0.2 to −0.4, meaning that the nights have become warmer. Warm days and warm spells have also become more frequent (Fig. 1b and c). Except for some few locations, the diurnal temperature range shows a negative trend (Fig. 1d). Given that both maximum and minimum temperatures have increased during the period, this means that the warming has been faster with minimum temperature. Fig. 2 and Fig. 3 illustrate, through some selected locations, that a steady warming took place during the study period. Indeed, while the frequency of cool nights and of cool days decreased, that of warm nights and warm days, and also of warm spells increased. This was the case not only for locations inside the continent, but also for those in coastal areas where temperature regime is different because of the influence of the ocean. In Table 3, one can notice that for the last two decades, the frequency of cool nights decreased about twofold, going from 10.1 to 2.9, 10.3 to 7.4, 10.1 to 4.2, and 10.2 to 04.8, respectively for Nouakchott, Bamako, Ouagadougou and Niamey.

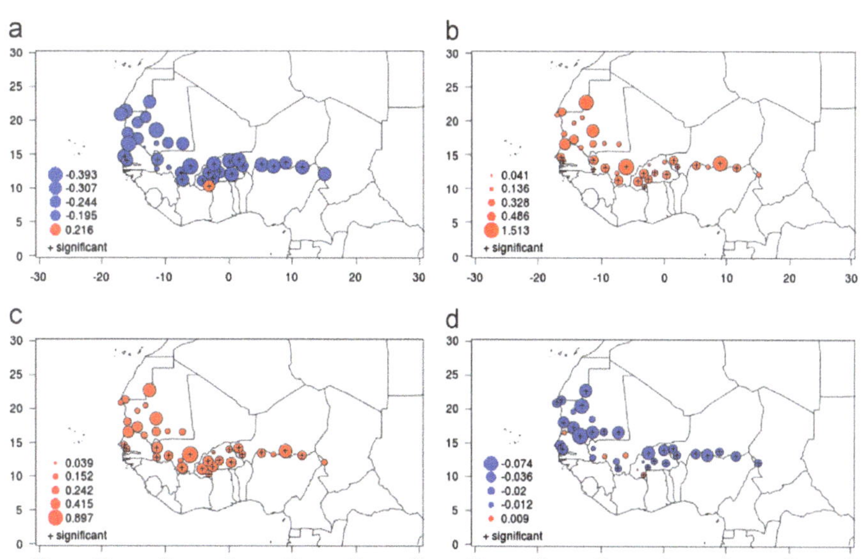

Figure 1: Observed trends in some temperature indices in West Africa from 1960 to 2010. (a) Cool nights (Tn10p), (b), Warm days (Tx90p), (c) Warm Spells (WSDI), and (d) Diurnal temperature range (DTR).

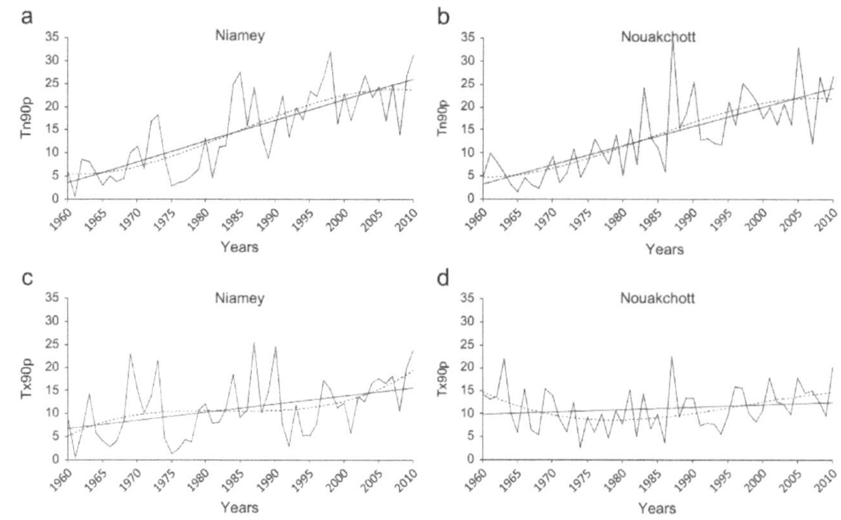

Figure 2: Evolution of the percentage of warm nights (Tn90p) and that of warm days (Tx90p) in Niamey, Niger (a and c) and Nouakchott, Mauritania (b and d) from 1960 to 2010.

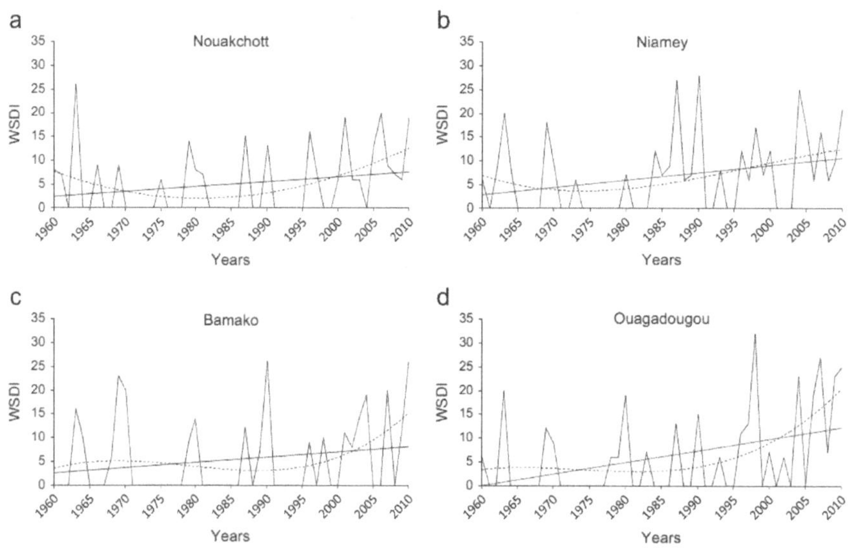

Figure 3: Evolution of Warm spells (WSDI) in some selected locations in West Africa from 1960 to 2010. (a) Nouakchott (Mauritania), (b) Niamey, Niger (c) Bamako, Mali, and (d) Ouagadougou, Burkina Faso.

Table 3: Evolution of the annual temperature indices between the 1961–1990 and 1991–2010 periods in some selected stations in West Africa

Indices	Nouakchott		Bamako		Ouagadougou		Niamey		Units
Periods	1961–1990	1991–2010	1961–1990	1991–2010	1961–1990	1991–2010	1961–1990	1991–2010	s
Cool nights	10.1	2.9	10.3	7.4	10.1	.2	10.2	4.8	%
Warm nights	10.3	19.4	9.9	10.8	10.3	24.3	10.2	22.1	%
Cool days	10.3	7.2	10.2	9.7	10.4	8.0	10.3	8.3	%
Warm days	10.4	12.1	10.1	12.2	10.2	17.4	10.3	12.6	%
Warm spells	3.8	6.7	04.8	6.4	3.6	10.0	05.0	8.5	Days

Rainfall Indices

Trends in rainfall related indices are not as uniform as the ones in temperatures. While a general tendency of decreased annual total rainfall and maximum number of consecutive wet days is observed in the region (Fig. 4b and d), the cumulated rainfall of extremely wet days shows a positive trend in most locations (Fig. 4c, Table 4). However, those observed trends are significant only for a few locations. Moreover, regarding the maximum number of consecutive wet days, if an overall decreasing trend is observed since 1960, one can notice a reversal of this from the late 1980 s in some locations, where an increasing trend can now be observed, except for Nouakchott in Mauritania (Fig. 5). One can also notice from Table 4 that, in some selected locations across the subcontinent, maximum 5-day rainfall amount, total annual rainfall, and the amount of rainfall during extremely wet days have all increased. The maximum number of consecutive wet days, on the other hand, has decreased.

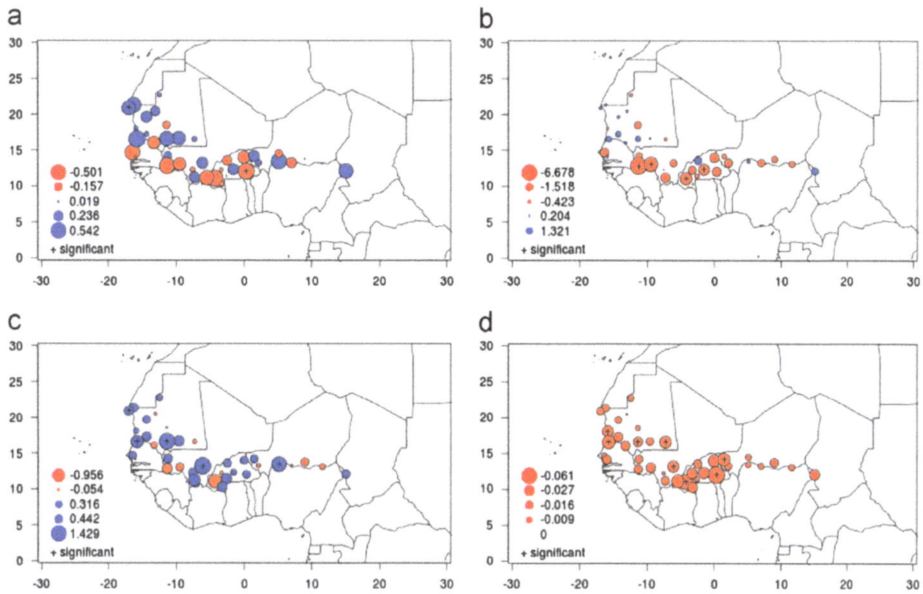

Figure: 4: Observed trends in some rainfall indices in West Africa from 1960 to 2010. (a) 5-day cumulative rainfall, (b) Total annual precipitation, (c) Cumulated rainfall of extremely wet days, and (d) Maximum Consecutive Wet days.

Table 4: Evolution of some rainfall indices between the 1961–1990 and 1991–2010 periods at some selected locations in West Africa

Indices	Nouakchott		Bamako		Ouagadougou		Niamey		N'Djamena		Units
Periods	1961–1990	1991–2010	1961–1990	1991–2010	1961–1990	1991–2010	1961–1990	1991–2010	1961–1990	1991–2010	
Max 5-day rainfall	33.7	44.7	123	117	104	109	86	97	88	103	mm
Extremely wet day total rainfall	5.7	18.6	48.1	62.7	44.4	45.6	30.9	37.8	29.3	45.7	mm
Annual total rainfall	84.6	95.5	913	943	782	724	535	541	505	583	mm
Max consecutive wet days	2.1	1.5	5.0	5.4	5.1	4.6	3.8	3.8	4.2	3.7	days

Figure 5: Evolution of the maximum number of consecutive wet days (CWD) in some selected locations in West Africa from 1960 to 2010. (a) Nouakchott (Mauritania), (b) Niamey (Niger), (c) Bamako (Mali), and(d) Ouagadougou (Burkina Faso).

DISCUSSION

Previous studies in other regions of the world showed the same tendencies in temperature indices we found in the West African Sahel. For example, New et al. (2006)found statistically significant warming trends for most of southern Africa and some few locations in Nigeria and The Gambia during the period from 1961 to 2000. Kruger and Sekele (2013) also found that in South Africa, warm extremes increased and cold extremes decreased for all of the weather stations during the 1962-2009 period. For South America, although Vincent et al. (2005) did not find consistent changes in the indices based on daily maximum temperature, they noticed some significant increasing trends in the percentage of warm nights and decreasing trends in the percentage of cool nights at many stations. Our results in the West African Sahel are consistent with this later finding. The same authors also noticed that stations located in coastal areas have the most significant warming trends. We did not observe this difference among coastal and inland locations in the West African Sahel, probably because of the reduced ocean-land temperature gradient that is characteristic of the monsoon regions as pointed out byTrenberth et al. (2007b) and Giannini et al. (2008).

Regarding rainfall indices, the decrease in annual total rainfall in the sub-region has been documented extensively (L'Hote et al., 2002, et al., and Diouf et al., 2000). As for the increasing trend of the cumulated rainfall of extremely wet days and that of the maximum number of consecutive wet days from the late 1980s, this is in accordance with the findings of Sarr (2011), who also noticed that extreme rainfall events became more frequent in the West African Sahel during the last decade, compared to the 1961–1990 period. Studies in other parts of Africa showed in some cases similar trends, but in other cases the trends were different. For example, Kruger (2006) found that some significant increases and decreases occurred in annual precipitation, some increases in the longest annual wet spells, and some increases in high daily precipitation amounts in South Africa over the 1910–2004 period.

CONCLUSION

This study showed that throughout the West African Sahel, all temperature indices point to a general warming trend since 1960. As for rainfall related indices, although there was a general tendency of decreased annual total rainfall, the observed trends are not as uniform as the ones in temperatures, and some indices clearly indicate that extreme rainfall events have become more frequent during the last decade. These observed trends have serious implications on the economies of countries of this region, most of which are not strong enough to cope with adverse climate effects. Indeed, the observed warming trend means a higher demand on domestic energy consumption for cooling, a higher evaporation rate from water bodies and irrigated crops, and a lower performance of agricultural crops and livestock. On the other hand, an increased frequency of extreme rainfall events such as heavy downpours, long dry or wet spells means more fragile infrastructure and production systems. It is therefore important for policy makers of the region, in order to cope with these changes and avoid additional hardships to their populations, to invest in and promote more energy efficient infrastructure and domestic appliances, including low cost and environmentally friendly energy production systems. For agriculture, research for more heat tolerant crop/animal species and cultivars/breeds should be promoted. The production systems need also be redesigned to account for higher risks of losses due to floods and/or droughts.

ACKNOWLEDGMENTS

This work was realized thanks to the support of the European Commission to the permanent Interstate Committee for drought control in the Sahel (CILSS) through the Global Climate Change Alliance (GCCA) initiative.

REFERENCES

1. Aguilar, E., Peterson, T.C., Obando, P.R., Frutos, R., Retana, J.A., Solera, M., Soley, J., García, I.G., Araujo, R.M., Santos, A.R., Valle, V.E., Brunet, M., Aguilar, L., Álvarez, L., Bautista, M., Castañón, C., Herrera, L., Ruano, E., Sinay, J.J., Sánchez, E., Oviedo, G.I.H., Obed, F., Salgado, J.E., Vázquez, J.L., Baca, M., Gutiérrez, M., Centella, C., Espinosa, J., Martínez, D., Olmedo, B., Espinoza, C.E.O., Núñez, R., Haylock, M., Benavides, H., Mayorga, R., 2005. Changes in precipitation and temperature extremes in Central America and Northern South America, 1961–2003. Journal of Geophysical Research 110, D23107, 10.1029/2005JD006119⟨www.agu.org/ pubs/crossref/2005/2005JD006119.shtml⟩.
2. Alexander, L.V., et al., 2006. Global observed changes in daily climate extremes of temperature and precipitation. Journal of Geophysical Research 111, 10.1029/ 2005JD006290⟨www.agu.org/pubs/crossref/2006/2005JD006290.shtml⟩.
3. Ali, A., 2011. Climate variability and change in the Sahel. Understanding the current situation by observing. Climate change in the Sahel. A challenge for sustainable development. AGRHYMET Monthly Bulletin (Special Issue), pp. 17–20. ⟨http:// www.agrhymet.ne/PDF/Bulletin%20mensuel/specialChCang.pdf⟩.
4. CEDEAO-ClubSahel/OCDE/CILSS, 2008. Climate and Climate Change. The Atlas on Regional Integration in West Africa. Environment Series, p.13, ⟨http://www.oecd.org/fr/csao/publications/40121057.pdf⟩.
5. Caesar, J., Alexander, L., Vose, R., 2006. Large-scale changes in observed daily maximum and minimum temperatures: creation and analysis of a new gridded data set. Journal of Geophysical Research 111, DO5101, 10.1029/2005JD006280 (accessed 11.12.12) ⟨www.agu.org/pubs/crossref/2006/2005JD006280.shtml⟩.
6. Solomon, S., Qin, D., Manning, M., Chen, Z., Marquis, M., Averyt, K.B., Tignor, M., Miller, H.L. (Eds.), 2007. Climate Change: The Physical Science Basis. Contribution of Working Group I to the Fourth Assessment Report of the Intergovernmental Panel on Climate Change. Cambridge University Press, Cambridge, United Kingdom and New York.
7. Camberlin, P., Diop, M., 2003. Application of daily rainfall principal component analysis to the assessment of the rainy season characteristics in Senegal. Climate Research 23, 159–169.

8. Collins, J.M., 2011. Temperature Variability over Africa. Journal of Climate 24, 3649–3666, http://dx.doi.org/10.1175/2011JCLI3753.1 ⟨http://journals.ametsoc.org/doi/abs/10.1175/2011JCLI3753.1⟩.

9. R Core Team, 2012. R: A language and environment for statistical computing. R Foundation for Statistical Computing, Vienna, Austria. ISBN: 3-900051-07-0, ⟨http://www.R-project.org/⟩.

10. Diouf, M., Nonguierma, A., Amani, A., Royer, A., et al., 2000. Lutte contre la sécheresse au Sahel: résultats, acquis et perspectives au Centre régional AGRHYMET. Sécheresse 11 (4), 57–266.

11. Easterling, D.R., 1997. Maximum and minimum temperature trends for the Globe. Science 277, 364–367, http://dx.doi.org/10.1126/science.277.5324.364 ⟨www.sciencemag.org/cgi/doi/10.1126/science.277.5324.364⟩.

12. Easterling, D.R., Alexander, L.V., Mokssit, A., Detemmerman, V., 2003. CCI/CLIVAR Workshop to develop priority climate indices. Bulletin of the American Meteorological Society 8, 1403–1407.

13. FAO, 1978. Report on the agro-ecological zones project. Vol. 1: Results for Africa. World Soil Resources Report 48/1. FAO, Rome, 158 pp.

14. Giannini, A., Biasutti, M., Held, I.M., Sobel., A.H., 2008. A global perspective on African climate. Climatic Change 90, 359–383, http://dx.doi.org/10.1007/ s10584-008-9396-y.

15. Haylock, M.R., Peterson, T.C., Alves, L.M., Ambrizzi, T., Anunciação, Y.M.T., Baez, J., Barros, V.R., Berlato, M.A, Bidegain, M., Coronel, G., Corradi, V., Garcia, V.J., Grimm, A.M., Karoly, D., Marengo, J.A., Marino, M.B., Moncunill, D.F., Nechet, D., Quintana, J., Rebello, E., Rusticucci, M., Santos, J.L., Trebejo, I., Vincent, L.A., 2006. Trends in total and extreme South American rainfall in 1960–2000 and links with sea surface temperature. Journal of Climate 19, 1490–1512, http://dx. doi.org/10.1175/JCLI3695.1. (http://journals.ametsoc.org/doi/abs/10.1175/ JCLI3695).

16. Hegerl, G.C., Zwiers, F.W., Braconnot, P., Gillett, N.P., Luo, Y., Marengo Orsini, J.A., Nicholls, N., Penner, J.E., Stott, P.A., 2007. Understanding and attributing climate change. In: Solomon, S., Qin, D., Manning, M., Chen, Z., Marquis, M., Averyt, K.B., Tignor, M., Miller, H.L. (Eds.), Climate Change: The Physical Science Basis. Contribution of Working Group I to the Fourth Assessment Report of the Intergovernmental Panel on Climate Change. Cambridge University Press, Cambridge, United Kingdom and New York.

17. Karl, T.R., Nicholls, N., Ghazi, A., 1999. CLIVAR/GCOS/WMO workshop on indices and indicators for climate extremes: workshop summary. Climatic Change 42, 3–7.

18. Kruger, A.C., 2006. Observed trends in daily precipitation indices in South Africa: 1910–2004. International Journal of Climatology 26, 2275–2285, http://dx.doi.org/10.1002/joc.1368. (http://doi.wiley.com/10.1002/joc.1368).

19. Kruger, A.C., Sekele, S.S., 2013. Trends in extreme temperature indices in South Africa: 1962–2009. International Journal of Climatology 33, 661–676, http://dx. doi.org/10.1002/joc.3455.

20. Lamptey, B., 2009. An analytical framework for estimating the urban effect on climate. International Journal of Climatology 30, 72–88.

21. L'Hote, Y., Mahé, G., Some, B., Triboulet, J.P., 2002. Analysis of a Sahelian annual rainfall index from 1896 to 2000; the drought continues. Hydrological Sciences —Journal 47 (4), 563–572.

22. Min, S.-K., Zhang, X., Zwiers, F.W., Hegerl, G.C., 2011. Human contribution to moreintense precipitation extremes. Nature 470, 378–381, http://dx.doi.org/10.1038/ nature09763 http://dx.doi.org/10.1038/nature09763.

23. New, M., Hewitson, B., Stephenson, D.B., Tsiga, A., Kruger, A., Manhique, A., Gomez, B., Coelho, C.A.S., Masisi, D.N., Kululanga, E., Mbambalala, E., Adesina, F., Saleh, H., Kanyanga, J., Adosi, J., Bulane, L., Fortunata, L., Mdoka, M.L., Lajoie, R., 2006. Evidence of trends in daily climate extremes over southern and west Africa. Journal of Geophysical Research 111, D14102, 10.1029/2005JD006289 ⟨http://www. agu.org/pubs/crossref/2006/2005JD006289.shtml⟩.

24. Nicholson, S., 1993. An overview of African rainfall fluctuations of the last decade. Journal of Climate 6, 1463–1466. Peterson, T.C., Manton, M.J., 2008. Monitoring changes in climate extremes: a tale of international collaboration. Bulletin of the American Meteorological Society 89, 1266–1271.

25. Peterson, T.C., Taylor, M.A., Demeritte, R., Duncombe, D.L., Burton, S., Thompson, F., Porter, A., Mercedes, M., Villegas, E., Semexant-Fils, R., Klein-Tank, A., Martis, A., Warner, R., Joyette, A., Mills, W., Alexander, L., Gleason, B., 2002. Recent changes in climate extremes in the Caribbean region. Journal of Geophysical Research 107 (D21), 4601, 10.1029/2002JD002251, (16 p.).

26. Sarr, B., 2011. Return of heavy downpours and floods in a context of changing climate. Climate change in the Sahel. A challenge for sustainable development. AGRHYMET Monthly Bulletin (Special Issue), pp. 9–11. ⟨http://www.agrhymet. ne/PDF/Bulletin%20mensuel/specialChCang.pdf⟩.

27. Sivakumar, M.V.K., 1988. Predicting rainy season potential from the onset of rains in southern sahelian and sudanian climatic zones of West Africa. Agricultural and Forest Meteorology 42, 295–305.

28. Solomon, S., Plattner, G.K., Knutti, R., Friedlingstein, P., 2009. Irreversible climate change due to carbon dioxide emissions. Proceedings of the National Academy of Sciences 106, 1704–1709.

29. Solomon, S., Daniel, J.S., Sanford, T.J., Murphy, D.M., Plattner, G.K., Knutti, R., Friedlingstein, P., 2010. Persistence of climate changes due to a range of greenhouse gases. Proceedings of the National Academy of Sciences 107, 18354–18359.

30. Stott, P.A., Jones, G.S., Christidis, N., Zwiers, F.W., Hegerl, G., Shiogama, H., 2011. Single-step attribution of increasing frequencies of very warm regional

temperatures to human influence. Atmospheric Science Letters 12, 220–227.

31. Trenberth, K., 2011. Changes in precipitation with climate change. Climate Research 47, 123–138, http://dx.doi.org/10.3354/cr00953. (accessed 11.12.12) ⟨http://www.int-res.com/abstracts/cr/v47/n1-2/p123-138/⟩.

32. Trenberth, K.E., 2012. Framing the way to relate climate extremes to climate change. Climatic Change 115, 283–290, http://dx.doi.org/10.1007/s10584-012-0441-5 ⟨http:// www.springerlink.com/index/10.1007/s10584-012-0441-5⟩.

33. Trenberth, K.E., Jones, P.D., Ambenje, P., Bojariu, R., Easterling, D., Tank, A.K., Parker, D., Rahimzadeh, F., Renwick, J.A., Rusticucci, M., 2007a. In: Solomon, S., Qin, D., Manning, M., Marquis, M., Averyt, K.B., Tignor, M., Miller, H.L., Chen, Z. (Eds.), Observations: Surface and Atmospheric Climate Change, Chapter 3 of Climate Change 2007: The Physical Science Basis. Contribution of Working Group I to the Fourth Assessment Report of the Intergovernmental Panel on Climate Change. Cambridge University Press, Cambridge, UK and New York, pp. 235–336.

34. Trenberth, K.E., Fasullo, J.T., Kiehl, J., 2007b. Earth's global energy budget. Bulletin of the American Meteorological Society , http://dx.doi.org/10.1175/2008BAMS2634.1.

35. Vincent, L.A., Mekis, E., 2006. Changes in daily and extreme temperature and precipitation indices for Canada over the twentieth century. Atmosphere– Ocean 44, 177–193.

36. Vincent, L.A., Peterson, T.C., Barros, V.R., Marino, M.B., Rusticucci, M., Carrasco, G., Ramirez, E., Alves, L.M., Ambrizzi, T., Berlato, M.A., 2005. Observed trends in indices of daily temperature extremes in South America 1960–2000. Journal of Climate 18, 5011–5023, http://dx.doi.org/10.1175/JCLI3589.1 ⟨http://journals.ametsoc.org/doi/abs/10.1175/JCLI3589.1⟩.

37. Zhang, X., Yang, F., 2004. RClimDex (1.0) User Guide. Climate Research Branch Environment Canada. Downsview, Ontario, Canada.

CHAPTER 3

DETECTION AND ATTRIBUTION OF CLIMATE EXTREMES

David R. Easterlinga, Kenneth E. Kunkelb, Michael F. Wehnerc, Liqiang Sunb

Keywords: Observed climate change, Extremes, Detection, Attribution

INTRODUCTION

Contemporary climate change presents one of the most pressing challenges for human society. As the climate continues to change, the risks associated with climate extremes takes on ever greater importance. Changes in the mean climate, particularly since the middle of the 20th century, have been linked to anthropogenic-induced increases in greenhouse gases (Hegerl et al., 2010). Indeed, a number of recent climate assessments have concluded that observed changes in the climate system over the past century are largely a result of human activities (Seneviratne et al., 2012, Bindoff et al., 2013, Walsh et al., 2014a and Walsh et al., 2014b).

Climate extremes, by definition, are rare events, however climate change has resulted in changes in the occurrence of extreme events (Easterling et al., 2000 and Seneviratne et al., 2012). Climate extremes can result from external forcing of the climate system, such as from increasing greenhouse gases, or natural variability, or more likely some combination of the two. For example, some of the more robust climate change signals related to extremes in both the observed record and in model simulations for the future are decreases in the number of unusually cold days and nights, and increases in the number of unusually warm days and nights (Seneviratne et al., 2012; Min et al., 2013;Kim et al., 2015). Other changes include an increase in the number of heavy precipitation events (Kim et al., 2015) and a likely increase in the incidence of hurricanes in the north Atlantic since about 1970 (Kunkel et al., 2013 and Seneviratne et al., 2012) while longer-term trends in hurricanes remain a subject of inquiry (Landsea, 2015 and Kossin et al., 2015). Once a signal of change in an extreme is found, the question most often becomes how the change is related to human-induced climate change (Hulme, 2014).

Detection of climate change in the observed record refers to the identification of a statistically significant change in some part of the climate system. The change could be in some highly averaged mean quantity or in some measure of extreme weather or climate. Observed climate change over various time scales for many parts of the climate system is well summarized in the IPCC 5th Assessment Report (Bindoff et al., 2013) and continues to be extensively monitored (Blunden and Arndt, 2015). It has been clear for some time that changes in the occurrence of weather and climate extremes are major players in producing changes in the natural environment and society, and these kinds of changes have increasingly been the subject of research papers and scientific assessments (e.g. CCSP, 2008;Seneviratne et al., 2012).

However, it is not enough to show that a change in the climate has occurred; indeed once a change has been detected it is important to attribute that change to some cause. Attribution, especially to human greenhouse gas emissions, lends confidence to model projections of the future driven by anthropogenic forcing as well as predictions of extremes at shorter time scales (Seneviratne and Zwiers, 2015). Attribution also provides information for more robust decisions in adaptation activities related to weather and climate extremes (Sippel et al., 2015). Traditionally, detection and attribution studies focused on mean changes (e.g. Hegerl et al., 2007; Bindoff et al., 2013); however in the past decade or so climate extremes have become a focus of detection and attribution studies. A number of recent papers have included overviews of detection and attribution science related to extremes.

Citation: David R. Easterling, Kenneth E. Kunkel, Michael F. Wehner, Liqiang Sun , Detection and attribution of climate extremes, http://dx.doi.org/10.1016/j.wace.2016.01.001.

Furthermore, when an extreme event occurs climate scientists are increasingly queried by the news media, policy makers, private enterprise and the public as to the likely cause of the event. The question of attribution of these events to human-induced climate change is of particular interest (Stott, 2013, Zwiers, 2013, Hulme, 2014 and Hegerl, 2015). Through the process of answering this question valuable information regarding risk due to climate extremes is provided, which is useful to a wide range of stakeholders for disaster risk reduction activities.

Two schools of thought have emerged in this rapidly developing field and are described inSection 4. The first, referred in this paper as "Oxford", where the technique was first envisioned, quantifies the change in probability of an extreme event of a particular observed magnitude caused by the human alteration of the climate system. The second, referred to here as "Boulder", introduced first in a series of paper by researchers from NOAA›s Earth System Research Laboratory, examines the human induced change in magnitude of an extreme event. While providing different types of information to stakeholders, both these probabilistic and mechanistic schools of thought have been shown to be equivalent.

In this paper we present an overview of practices and challenges related to the detection and attribution of observed changes in climate extremes. In particular we mainly examine temperature and precipitation extremes, while acknowledging that trends in other kinds of extremes, such as tropical cyclones, droughts, and even extreme snow storms may exist and deserve attention. However, we do discuss a newer field of attribution, event attribution, where individual extreme events, such as storms or heatwaves, are analyzed for the express purpose of assigning some measure of the extent to which that event was directly influenced by anthropogenic forcing of the climate system.

DETECTION OF TRENDS IN EXTREME TEMPERATURE AND PRECIPITATION

Extreme weather and climate events are a natural part of the climate system. For example, an examination of the paleoclimate record shows that megadroughts and pluvials have happened in the Western and Central United States throughout the last 2000 years (Woodhouse and Overpeck, 1998). Yet, true climate extremes are rare events. Because of this rarity researchers often relax the definition of extremes in such a way as to increase the number of observations that can be used in a statistical analysis. For example, in the case of studies of changes in the occurrence of hot daily maximum temperature extremes, rather than defining the extreme threshold such that it is observed only once every few years, the definition is often set to a threshold value (e.g. 90th percentile value) that is not truly extreme but produces a larger number of observations that exceed the threshold allowing more robust statistical results.

But what about the data sets used in these analyses? To detect an observed change in the climate system, particularly a change suitable for an attribution study, a data set of sufficient temporal and spatial coverage is necessary. Depending on the climate extreme, there is often a lack of observed climate data to document these events for many parts of the world. If the observations exist they often are not in digital form. Also, although the situation is changing, many countries continue to be reluctant to share them with the research community (Easterling, 2013 and Kunkel and Frankson, 2015).

As noted above, since the analysis of climate extremes often involves examination of the tails of a statistical distribution, a threshold value may be used to determine the number of observations that exceed that value over time creating a time series of exceedance counts. Data quality can impact the counts if there are a number of erroneous values that are not screened out by quality assurance methods, or if the quality assurance methods, which are often more concerned with mean values, are too rigorous and exclude true values. Additional issues include missing data, especially if those missing data would exceed an established threshold or would affect the calculation of the threshold itself. In terms of global analyses, data may be missing for large regions of the globe resulting in a less than true global analysis (Donat et al., 2013). Finally, if longer term data are available they are often observed at weather observing stations, such as at airports, and may be impacted by issues such as urbanization or less than ideal station siting which may result in lower quality data.

The homogeneity of climate data may also impact analyses of climate extremes (Trewin, 2010). Climate data are considered homogenous when all trends and variations are the result of the climate system itself. Inhomogeneities in climate data occur for a variety of reasons. Observing stations often are moved multiple times over longer periods (e.g. 50–100 years) resulting in changes in the local characteristics of the site (e.g. more trees, slight difference in elevation, etc.) Reasons for moves vary but examples include relocation from a city center to an airport, a change in a volunteer observer who also hosts the equipment, or the need to use the site for other purposes. A common inhomogeneity source is urbanization around a station, which will generally cause localized warming, primarily in T_{min} (Karl et al., 1988), the magnitude of which can be several degrees in the largest urban areas. This warming is real and relevant to impacts on urban residents, but will not be representative of real trends at a larger regional scale; thus, for attribution applications, this urban warming should be removed. Changes in instrumentation such as a new type of thermometer, the installation of a wind shield on a raingauge, or changes in observing practices such as the time observations are taken all can result in an inhomogeneous time series. The impact on the observed time series is typically either a discontinuity (jump up or down), or a gradual change that can appear as a trend (Menne and Williams, 2009), either of which can impact the analysis of changes in extremes. Methods for identifying and correcting for inhomogeneities have typically been applied to time series based on longer averaging periods, such as monthly, seasonal, or annual time series (e.g. Easterling and Peterson, 1994, Menne and Williams, 2009). In the past decade or so approaches to assess and correct for inhomogeneities in daily and even sub-daily data have been developed (e.g. Della-Marta and Wanner, 2006, Trewin, 2013), but still have not been widely implemented. However, even without corrections applied to higher temporal resolution data, results of analyses of extremes are consistent with what would be expected based on analyses of mean values (e.g. Alexander et al., 2006, Zwiers et al., 2011, Min et al., 2011).

Incomplete spatial coverage of observing stations for a region or the globe is another potential source of uncertainty. Since there are a number of regions in the world that are not covered in global-scale data sets used for climate analyses, particularly for extremes (Cowtan and Way, 2014), it is unknown how the addition of these regions would impact detection and attribution studies. Even in regions that have observing stations, the question of lower spatial density could prove problematic. Kunkel et al. (2007) used Monte Carlo techniques to examine the impact of lower spatial density of observing stations in the western United States and missing data in detecting changes in heavy precipitation over the contiguous United States. They found that limited spatial density was more important than missing data in detection studies, but that neither issue was severe enough to reduce statistical significance values below standard confidence levels.

Satellite and reanalysis products have the advantage of global coverage. However, both classes of products have significant uncertainties in their representation of extreme daily temperature and precipitation (AghaKourchak et al., 2011). Satellites do not measure these quantities directly, but instead measure radiances in different wavelength bands. Inversion algorithms for satellite data are generally not as well calibrated for large values as they are for average values. For example, AghaKourchak et al. (2011) analyzed the ability of four satellite-based precipitation products to capture precipitation extremes and suggested that extensive efforts are necessary to reliably capture these kinds of events. Reanalysis products are hybrid model-observation based datasets that are produced by assimilating observations into highly constrained climate model simulations. Biases in the reanalysis models, particularly those from coarse horizontal resolution or time-dependent uncertainties due to issues such as observing network changes (Wehner et al., 2014), can be significant for extremes. Station based daily gridded products provide more complete regional coverage to estimate extreme values and can reduce noise that is often inherent in station-based time series (Fischer et al., 2013). The process of gridding a data set can introduce uncertainties in subsequent analyses of extremes, particularly in calculating return periods, but have minimal effect on analyses of long-term trends and inter-annual variability (for more discussion on this and other data issues seeAlexander (2015)). However, as with satellite inversion algorithms, gridding algorithms are also not generally designed for the tails of the distribution. As a result of these factors, extremes from different products can differ substantially over the same locations even when the original observations are closely related (Wehner et al., 2014).

As noted earlier there are still many regions in the world that lack higher temporal resolution climate data that are suitable for examining changes in extremes. Because of the difficulty and/or reluctance for many countries to provide these data, the World Meteorological Organization's Joint WMO CCl/WCRP/JCOMM Expert Team on Climate Change Detection and Indices (ETCCDI) developed a set of climate extremes indices and organized a set of workshops in data poor regions to encourage these countries to calculate the indices at the workshop. The resulting data sets of the indices were then used to construct the HadEX2 data set, which is a quasi-global gridded data set of these indices that can be used in detection and attribution studies (Alexander et al., 2006). A similar data set, GHCNDEX (Donat et al., 2013) has also been constructed by calculating these indices using the GHCN-Daily data set. Although, in general, it lacks the spatial coverage of the HadEX2 data set, it is updated on a regular basis. However, the difference in spatial coverage between the two data sets is not consistent and depends much on the variable of choice. Use of either of these two data sets in detection/attribution analyses presents a tradeoff of either of a lack of coverage, or not being as up to date as possible, both of which increase uncertainty in results.

Fig. 1 shows the pointwise trend in one of these temperature indices, the annual maximum daily high temperature (TXx). With some exceptions, this measure of extreme temperature is largely experiencing increased values. Aggregation into larger regions, a common practice in detection and attribution (D&A) studies, yields greater areas of significance in these trends. Fig. 2, shows the pointwise trend in an extreme precipitation metric, the annual maximum 5-day (pentadal) precipitation (Rx5day). Again, statistical significance is increased with spatial aggregation. As discussed in more detail below, trends in both of these fields have been attributed to anthropogenic changes to the composition of the atmosphere. D&A studies of the ETCCDI indices are still rather limited although further opportunities exist. Fig. 3 shows trends in the consecutive dry days (CDD), a crude measure of drought and/or the dry season. Other more sophisticated measures of meteorological and agricultural drought also offer opportunities (Burke et al., 2006, Wehner et al., 2011, Sheffield et al., 2012 and Dai, 2013). D&A studies of seasonal rather than annual indices can be more insightful as the large scale meteorological patterns behind extreme temperature and precipitation often varies across the annual cycle. It follows that the statistical description as the physical mechanism behind any changes varies seasonally as well.

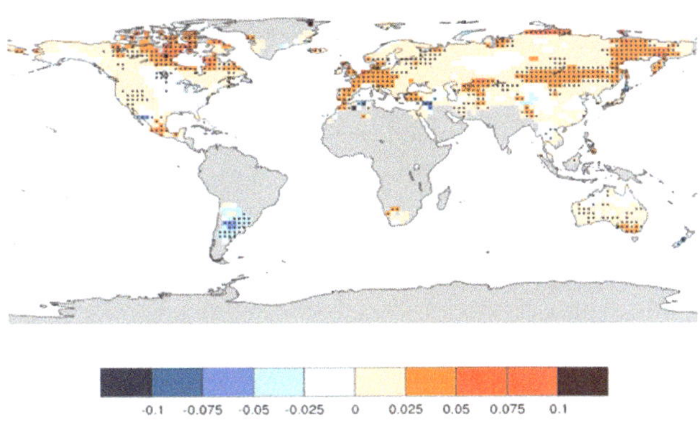

Figure 1: Pointwise linear trend over the 1951–2014 period in annual maximum daily high temperature (TXx) using GHCNDEX on a 2.5°×2.5° latitude/longitude grid (Donat et al., 2013). Units: °C/year, stippling indicates statistically significant trends ($p<0.05$).

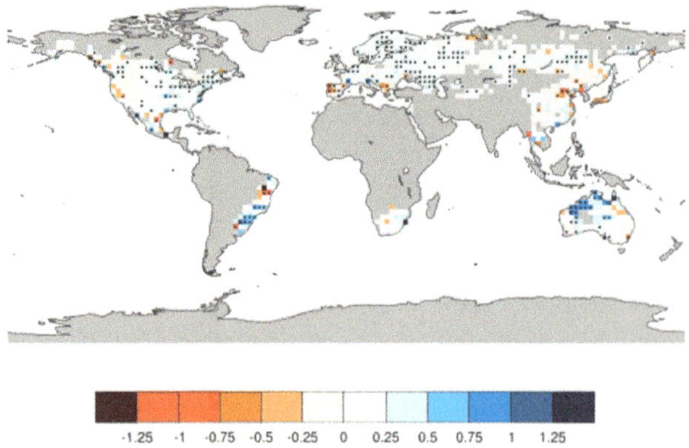

Figure 2: Pointwise linear trend over the 1951–2014 period in annual maximum 5-day total precipitation (Rx5day) using GHCNDEX on a 2.5°×2.5× latitude/longitude grid (Donat et al., 2013). Units: mm/year, stippling indicates statistically significant trends ($p<0.05$).

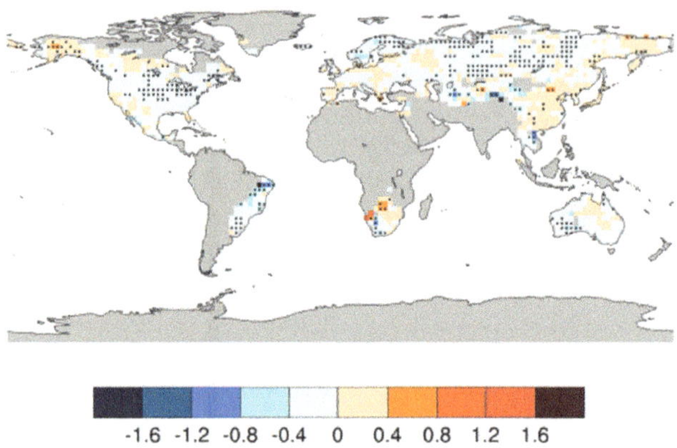

Figure 3: Pointwise linear trend over the 1951–2014 period in the annual consecutive dry days (CDD) using GHCNDEX on a 2.5°×2.5° latitude/longitude grid (Donat et al., 2013). Units: days/year, stippling indicates statistically significant trends ($p<0.05$).

The above examples (Fig. 1, Fig. 2 and Fig. 3) utilized metrics that have an annual time resolution. Calculation of such metrics is convenient but does not necessarily focus the analysis on high-impact extreme events, which in general are rarer than annual occurrence events. However, examination of more rare events usually requires access to the original observational data sets such that the analyses can be customized to focus on metrics more related to impacts. Fig. 4 shows grid box trends for the number of occurrences of 5-day duration cold spells that are colder than the threshold for a 1-in-5yr recurrence. A few features stand out. The direction of trends (downward) is very coherent spatially across the Northern Hemisphere with only a few grid boxes with upward trends. In the Southern Hemisphere, nearly all available grid boxes exhibit downward trends. Despite the relatively small sample size associated with this extremes definition, the downward trends are statistically significant for much of southern Eurasia, western Canada, and Australia. These results clearly illustrate that global warming has been accompanied by a decrease in the number of extreme cold episodes. As with Fig. 1, Fig. 2 and Fig. 3, most of the land areas in the tropics and Southern Hemisphere are missing, due to lack of available data or reluctance of nations to provide their data for general use.

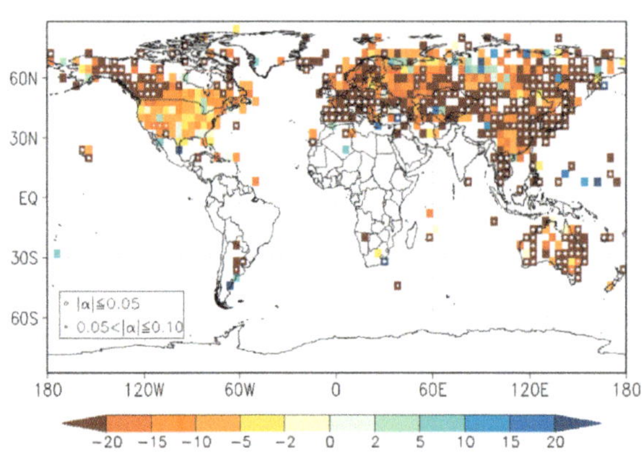

Figure 4: Trend (%/decade) for the period of 1951–2014 in the number of 5-day duration cold spells with mean temperature less than the threshold for a 1-in-5yr recurrence. Grid box (4°×4°) averages calculated from 5084 stations with less than 10% missing daily temperature data for 1951–2014. Grid boxes with statistically significant trends (computed with non-parametric Mann-Kendall test) identified with white dots. Data used are from the Global Historical Climatology Network-Daily data set (Menne et al., 2011).

There are several methods of data analysis to detect trends. Standard linear regression is usually not the preferred method to examine trends because in most cases metrics of extremes are not normally distributed. A common nonparametric approach (e.g.Alexander et al., 2006) is Kendall's tau-based slope estimator (Sen, 1968). Statistical significance is assessed by looking at the sum of the signs of the differences of all possible pairs of data points. The estimate of the magnitude of the trend is the median of all non-zero pairwise trends. Both serial and spatial correlation needs to be taken into account as substantial levels of such correlation are often present in extremes data, a consequence of the large-scale and long-lasting meteorological patterns that are often the physical cause of extreme events.

The Generalized Extreme Value (GEV) distribution has been found to be suitable as a fit to the tails of the distribution for atmospheric variables. In this approach, the original data set is sub-setted by extracting the maxima over each "block" of time, where a block is often a season or year. The probability density distribution of the block maxima, G(y), is given by (Coles, (2001):

$$G(y)=\frac{1}{\sigma}\left(1+\varepsilon\frac{y-\mu}{\sigma}\right)^{-1-\frac{1}{\varepsilon}}\exp\left[-\left(1+\varepsilon\frac{y-\mu}{\sigma}\right)^{-\frac{1}{\varepsilon}}\right],\ \varepsilon>0, y>\mu-\sigma/\varepsilon$$

$$G(y)=\frac{1}{\sigma}\exp\left[-\frac{y-\mu}{\sigma}-\exp\left(-\frac{y-\mu}{\sigma}\right)\right],\ \varepsilon=0$$

$$G(y)=\frac{1}{\sigma}\left(1+\varepsilon\frac{y-\mu}{\sigma}\right)^{-1-\frac{1}{\varepsilon}}\exp\left[-\left(1+\varepsilon\frac{y-\mu}{\sigma}\right)^{-\frac{1}{\varepsilon}}\right],\ \varepsilon<0, y<\mu-\sigma/\varepsilon$$

where ε, μ, and σ are called the shape, location, and scale parameters and has been shown to provide a good fit to seasonal and annual maximum temperatures and precipitation (Zwiers et al., 2010; Wehner, 2013). A trend can be introduced as a trend in the location parameter or log of the shape parameter (Katz, 2010). In Zwiers et al. (2011), trends were studied by assuming that the changes can be expressed by a linear (in time) change in the location parameter with the other parameters remaining constant. The maximum likelihood method was used to find the estimates of the parameters that best fit the observed data.

While the above statistical approaches provide a well-established numerical framework for estimating uncertainties, uncertainties in estimated probabilities for the most extreme events (e.g. 2003 European heat wave, the 2010 Russian heat wave) must be considered very high. Exploring the true probability of the most extreme conditions requires another approach. Climate models offer a possible solution. They are based on the fundamental physical laws governing the climate system. They can produce the dynamic chaotic behavior that is characteristic of the system. In principle then, they can produce events like the singular ones described above and provide insights into their probability. Very long simulations can provide the large sample sizes needed to establish statistical confidence.

But what about detection of changes in other phenomena of the climate system, such as tropical or convective storms? Storms in particular present difficulties in detecting long-term trends. For example, quantifying tropical cyclone numbers, as well as intensities has presented numerous challenges. Observations of tropical cyclones in the North Atlantic in the early part of the observational record (e.g. prior to World War II), occurred only when a storm struck land or was encountered by a ship at sea. Between WWII and the 1960s, aircraft observations were then included, and with the advent of satellites these observations were also included. This change of observing methods through time has resulted in much uncertainty in the annual counts of tropical cyclones in the earlier part of the record (CCSP, 2008 and Vecchi and Knutson, 2011). Fig. 5 shows counts of hurricanes in the Atlantic by year based on the HURDAT data set (Jarvinen et al., 1984 and Landsea, 2004). Vecchi and Knutson (2011) have provided adjustments for numbers of hurricanes for the period prior to 1966 based on ship track densities and sea-surface temperatures. Examining the unadjusted data in Fig. 5 there appears to be a long-term increase in numbers of hurricanes. However, using the adjusted data that account for storms that where likely missed in the pre-1966 period, there is no longer an increasing trend over the entire period (Vecchi and Knutson, 2011). Lastly, if the analysis is restricted to the modern era (satellite era, 1966 to present) confidence is high in the counts of hurricanes and Hartmann et al., (2013) state that there is very high confidence that there has been an increase in numbers of hurricanes over that period (Kossin et al., 2007). Why there has been an increase in hurricanes in this period is still the subject of much debate (Landsea, 2007 and Holland and Webster, 2007).

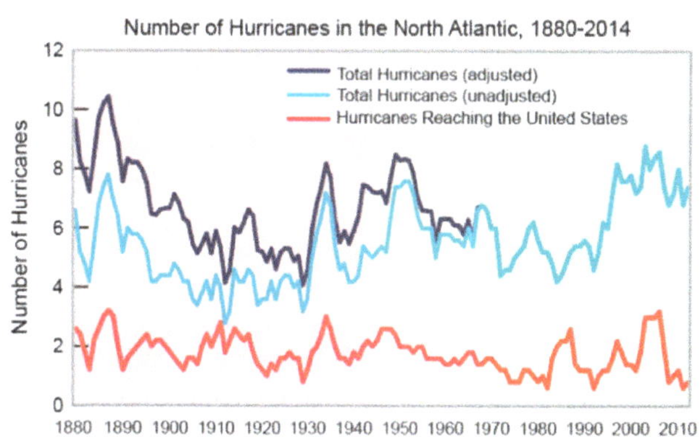

Figure 5: Atlantic hurricane counts 1880–2014. The time series are smoothed using a centered 5 year running mean. Data are from the HURDAT data set (Jarvinen et al., 1984 and Landsea, 2004) and the adjusted data are based on the method described in Vecchi and Knutson (2011). The 1880–1943 (red), 1944–1965 (yellow), and 1966-2014 (green) periods are shaded to indicate the different observing methods used during those periods as discussed in the text. Data are available at http://www3.epa.gov/climatechange/science/indicators/weather-climate/cyclones.html. (For interpretation of the references to color in this figure legend, the reader is referred to the web version of this article.)

Detecting changes in observations of other extremes present their own issues and drought is a good example. With differences in ways drought is defined using indices such as the Palmer Drought Severity Index, or the Standardized Precipitation Index (see Box 3.3 in Seneviratne et al., 2012) and with their reliance on observations of temperature and precipitation to calculate these indices, or the use of modeled data sets (Sheffield et al., 2012) it can be difficult to get coherent conclusions regarding long-term changes, particularly for large regions or the globe (Hartmann et al., 2013).

ATTRIBUTION OF TRENDS IN EXTREME TEMPERATURE AND PRECIPITATION

Attribution is generally defined as the evaluation of the contribution of multiple potential causes to a change in the climate system or to a climate event, with accompanying statistical confidence (Hegerl et al., 2010). "Detection and Attribution" or D&A is often used in the same phrase to describe a specific type of analysis of trends in climatic variables. In such analyses, a detected signal in the observations is sought in climate model simulations. Typically, the first step searches via statistical techniques for a spatially and/or temporally equivalent signal in an ensemble of climate model integrations constructed to realistically simulate the climate system with best estimates of observed external forcing factors. The forcing factors usually include a combination of anthropogenic factors (e.g. greenhouse gases, sulfate and carbonaceous aerosols, ozone, land use etc.) and natural factors (e.g. solar variations, volcanoes). If the signal is not found in these simulations, then the either the observed signal is a natural one and washed out in the chaos of the ensemble of model simulations or the models are not fit for purpose to replicate the observations. If the signal is found in the ensemble of realistic simulations, the next step is to analyze a counterfactual ensemble of simulations. To attribute the signal to human activities, one or more of the anthropogenic forcing factors is held stationary at some estimated preindustrial value in this second set of simulations. If the intended signal is then not found in these simulations, the trend is attributed to human activities.

This general approach is illustrated in Fig. 6, which compares model and observed trends of the annual maximum value of daily maximum temperature (T_{xx}) and the annual minimum value of the daily minimum temperature (T_{nn}) for the coterminous United States for the 50-yr period 1956–2005. The T_{xx} and T_{nn} trends were calculated for simulations from CMIP5 global climate models. Two sets of simulations were analyzed: (1) those driven with both natural and anthropogenic historical forcings ("Hist"; the factual set), and (2) those driven with only natural historical forcings ("Nat"; the counterfactual set). There were 78 Hist simulations from 29 CMIP5 models and 35 Nat simulations from 15 CMIP5 models. Trends were calculated using linear least squares regression. All of the historical simulation trends (red) are positive (warming) for both T_{xx} and T_{nn}. The two distributions are distinct, but there is considerable overlap. About 37% and 54% of the Hist trends are within the envelope of the Nat trends for T_{xx} and T_{nn}, respectively. Observed trends were calculated from long-term stations in the Global Historical Climatology Network-Daily (GHCND) data set (Menne et al., 2012) Stations were selected based on minimal missing data, specifically less than 5% missing daily temperature data for 1956–2005; a total of 2324 stations met this criterion. Time series of T_{xx} and T_{nn} were calculated for each individual station. Then, average time series for 1° by 1° grid boxes were calculated. Finally, national time series were calculated by averaging the grid box time series. The observed T_{xx} trend of +0.01 °C/decade is not statistically significant and is within the range of the both Hist and Nat trends. By contrast, the observed T_{nn} trend of +0.56 °C/decade is highly statistically significant ($p<0.01$) and on the high end of the range of the Hist trends, and well outside the distribution of the Nat simulations. The conclusion from this analysis is that the observed T_{nn} trend is externally forced. However, the lack of an observed T_{xx} trend would appear to be an ambiguous message. It is not inconsistent with external forcing (since a number of Hist simulations show little trend) but little more can be said at this spatial scale. However, trends in any variable at continental scales should be interpreted in the larger context of changes global scale. Statistically significant global increases in both T_{xx} and T_{nn} are seen in the HadEx observations (Brown et al., 2008). The lack of a T_{xx} trend is another facet of the local "warming hole", the lack of summer warming in the much of the central and eastern U.S. which is most apparent in daytime maximum temperatures. The summer (June–August) mean daily maximum temperature trend for the coterminous U.S. for 1956–2005 is a statistically insignificant 0.04 °C/decade. This lack of a detectible change in both average and daily maximum temperatures reveals that the human signal has not yet emerged out of the internal variability at this spatial scale. The very large trend in T_{nn} must also be interpreted in this context. Natural variations may be enhancing, rather than canceling the human induced increase in T_{nn}. In both cases, internal variability at the scale of the coterminous United States is significantly larger than at the global scale, making robust detection and attribution statements more difficult.

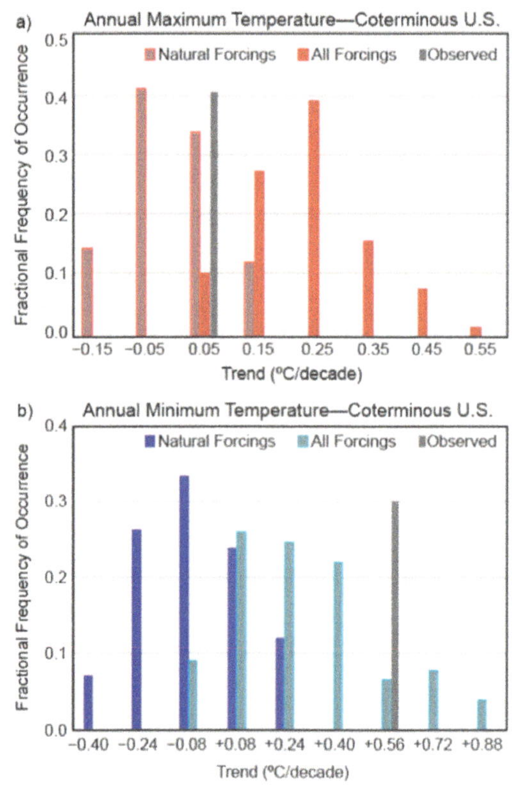

Figure 6: Probability distribution functions of the trends in two extreme temperature metrics for the coterminous United States for 1956–2005. Metrics include (a) annual maximum value of daily maximum temperature; and (b) annual minimum value of daily minimum temperature. The CMIP5 model trend distributions are shown for 78 historical forcing (natural and anthropogenic) simulations from 29 CMIP5 models (red) and for 35 natural forcing only simulations from 15 CMIP5 models (blue). The observed U.S. trend is shown in gray. (For interpretation of the references to color in this figure legend, the reader is referred to the web version of this article.)

Although the human influence on the annual or seasonal average of many different climatic fields have been detected and attributed, the few D&A studies of trends in extremes is limited to the tails in the distribution of short term averages of precipitation and temperature. One of the only attributions of observed trends in extreme precipitation to human changes in atmospheric GHG concentrations was shown by Min et al. (2011). This classical D&A study draws on linear principal component analysis (PCA) based optimal fingerprinting techniques that have been extensively used in attribution studies of the changes in more averaged quantities (Allen and Tett, 1999 and Allen and Stott, 2003). Observed changes in annual maximum daily and pentadal precipitation based on the HadEX (Hadley Centre global land-based gridded climate extremes) data set (Alexander et al., 2006) are analyzed from 1951–2003. Although observations are limited to the Northern Hemisphere and are sparsely covered in many regions, they find that heavy precipitation events have intensified over 2/3 of those observations. Comparing the set of CMIP3 models that provided daily precipitation in the 20th century anthropogenic and natural forcing (20c3m), the 20th century anthropogenic forcing only (ant) and the 20th century natural forcing (nat) experiments, they find that the observed increases in extreme precipitation can be attributed to human causes. Furthermore they find that the magnitude of the simulated increases are significantly smaller than observed suggesting that actual future extreme precipitation increases may be larger than models suggest.

Using a similar optimal fingerprinting technique, Min et al. (2013) examined indices of annual maximum daily high (T_{xx}) and low (T_{nx}) temperatures as well as annual minimum daily high (T_{xn}) and low (T_{nn}) temperatures using the multi-model CMIP3 ensemble. In mid-latitude and polar regions, these indices can be interpreted as measures of summer and winter high temperature extremes and summer and winter low temperature extremes respectively. Observed increases from 1951–2000 are larger in the winter than in the summer in both hemispheres. In this study, detection and attribution statements were attempted for both 5 and 10 year means of these extreme temperature indices both globally and for 5 continental scale regions. Globally, the natural and anthropogenic effects are not only attributable together but can also be separately identified for all 4 indices reflecting the robustness of the forced signal relative to internal chaotic noise. On the continental scale, North America exhibited the most robust D&A results. Consistent with the relative magnitude of the observed changes, winter extreme temperature increases were somewhat more robustly attributable than their summer equivalents. More recent analysis using the CMIP5 models (Kim et al., 2015) shows that the newer generation of models replicate the magnitude of observed trends in these 4 metrics of extreme temperatures significantly better than the previous generation (Shiogama et al., 2006, Christidis et al., 2011 and Min et al., 2013). The removal of two large modes of natural variability, the Arctic and Pacific Decadal Oscillations, further strengthened the robustness of the attribution result.

Christidis et al. (2013) further examined the trend in these extreme temperature indices attributing part of the observed changes in warm days and nights to land use changes. In the tropics, deforestation further increases extreme high temperatures over that due to GHG increases by reducing soil moisture and hence cooling from evapotranspiration. In the mid-latitudes, the reduction in evaporative cooling has less effect than the increase in surface albedo, resulting in a slight reduction in the increase in extreme high temperatures. However, while such a top of canopy analysis is consistent with how the atmosphere interacts with the land surface, the impact of deforestation on the extremely hot temperatures actually experienced by people and animals would not likely be diminished by the replacement of forests by grassland due to the decrease in available shade.

PROBABILISTIC EXTREME EVENT ATTRIBUTION

As Section 3 above describes, detectible trends in certain measures of extreme weather can be attributed to human activities. As individual extreme weather events can have devastating impacts, it is natural to ask to what extent anthropogenic climate change is responsible for the damages. Inherent to this type of question is the causality of specific weather events. However, the climate is a chaotic system and the causal factors behind specific weather events can rarely be described in deterministic terms. Hence, it is often stated in the popular press after a notable extreme weather event that nothing can be said about the role of climate change in that particular event with some caveats that such events can be expected to become more common in the future. This statement is most often patently false. For much can be said about the effect of climate change on many recent extreme weather events in a probabilistic formalism. The rapidly emerging field of Probabilistic Extreme Event Attribution has quantified the effect of climate change on a wide variety of extreme weather (for instance see Peterson et al., 2012, 2013; Herring et al. 2014).

While there are many different approaches to event attribution, we review two alternative, yet complementary approaches here. The earliest efforts to assess the human influence on extreme weather events consider the change in probability, based on climate model simulations of the "world that was" compared to simulations of the "world that might have been" had humans not interfered with the climate system. This "Oxford" school of thought was first proposed by Allen (2003) and implemented by Stott, Stone and Allen (2004) to describe the summer 2003 European heat wave. The chance of this particularly deadly heatwave, associated with up to 70,000 excess deaths (Robine et al., 2008), was found to be at least doubled by anthropogenic changes to the climate system.

In such analyses, the change in probability due to climate change is often expressed as a ratio, originally referred to a "risk ratio" but more precisely termed a "probability ratio" (Fischer and Knutti, 2015)

$$PR = P_{real}/P_{nat} \tag{1}$$

where Preal is the probability of a simulated event of the observed magnitude in the "world that was" simulations and Pnat is its probability in the "world that might have been" simulations. The alternative "fractional attributable risk" (FAR)

$$FAR = 1 - P_{nat}/P_{real} \tag{2}$$

is often used in epidemiology and environmental law (Stone and Allen, 2005) to define how much of a risk is due to a particular forcing agent. Both types of attribution statements are conditioned by the assumptions of the climate model simulations. In the original Stott et al. (2004) study, CMIP3 simulations from the fully coupled ocean-atmosphere HadCM3 model were used. Using this class of models, the attributions statements based on Eqs. (1) and (2) are conditional only on the changes in the external anthropogenic forcing agents, usually atmospheric greenhouse gas, sulfate aerosol and ozone concentrations and sometimes land use changes.

The technique has subsequently been refined by using stand alone atmospheric models with prescribed sea surface temperatures (SST) and prescribed sea ice extent (Pall et al., 2011 and Folland et al., 2014). In such analyses, these surface boundary conditions are derived from actual observations to represent the "world that was". For the counterfactual "world that might have been", the SST and sea ice extent are obtained by subtracting out an estimate of anthropogenic change in those fields. This can be straightforwardly obtained by the linear PCA techniques used in typical D&A analyses (Allen and Tett, 1999 and Allen and Stott, 2003). As there are multiple CMIP3/5 models that are suitable for this task, there can be multiple, equally valid, formulations of the counterfactual simulations. This provides one source of the estimation of the structural uncertainty in PR and FAR based attribution statements. Another source of structural uncertainty is currently being explored in the coordinated C20C+ experiment (Folland et al., 2014) that involves multiple international modeling groups following this atmospheric model based event attribution protocol. In this and related experiments (for instance see http://www.climateprediction.net/weatherathome/), additional conditions are associated with any resulting attribution statements. Principal among these is that the influence of the particular state of the ocean on the simulated probability of the event in question can be quantified. For instance, if the event occurred in a year with a large ENSO event, the SST pattern in the "world that might have been" also has a large ENSO event in that year, albeit cooled by the removal of the human contribution to global warming. Hence, conditions related to the state of the ocean at the time of the event can be incorporated into the attribution statement.

This separation of natural and anthropogenic contributions to extreme events provides some of the motivation behind the alternative "Boulder" school of thought approach to event attribution. This approach begins with a careful deconstruction of the local and large-scale meteorology responsible for the event in the context of observed and simulated trends in the region of interest. The first of these studies examined another deadly heatwave, this time during summer 2010 in Russia (Dole et al., 2011) and was followed by an analysis of the 2011 Texas heatwave and drought (Hoerling et al., 2013). Quantifying the natural and anthropogenic contributions to the magnitude of extreme events provides another perspective on event attribution that may appear to be at odds with the frequency based perspective described above. Because of the asymptotic shape of the tails of probability distributions in questions, small changes in magnitude may result in large changes in frequency (Otto et al., 2012). Hence, it may appear that natural factors play a larger role than anthropogenic factors for some extreme events. However, these factors are not additive but multiplicative. More recently, Trenberth et al. (2015) has argued that anthropogenic changes in the thermodynamics of the climate system are much larger than are changes in atmospheric dynamics and concluded that statements about the human influence on the magnitude of extreme events are more practical and societally relevant than risk attribution statements. Further methodological details of probabilistic extreme event attribution are discussed in Pall et al. (2015).

Event attribution studies are principally based on climate model experiments. Unlike the formal D&A studies performed at near global scales discussed in Section 4, explicit detection exercises are not always performed on the observations alone. However some recent studies (van Oldenborgh et al., 2012; King et al., 2015) have added a detection step at the spatial scales of interest to individual events. This technique uses a peaks over threshold extreme value formalism with a time dependent covariate. King et al. (2015) examined annual mean Central England temperatures using both a large ensemble of atmospheric models and this observationally based method and found that the increased risk for the record 2014 temperatures were entirely consistent between the two approaches.

Uncertainties in the attributed changes in risk (Oxford School) or magnitude (Boulder School) of extreme events arise from several sources. First, observational uncertainties in the actual magnitude of an event can be significant. The median estimate of the risk ratio can be sensitive to this uncertainty, although the lower bound of a confidence interval can be less sensitive (Jeon et al., 2015). Second, while model simulations of the "world that was" are constrained by observations, the counterfactual simulations, especially those using atmospheric only models, are highly determined by estimates of the climate sensitivity to external human forcing agents. Third, given a certain prescribed set of boundary condition changes, the extreme weather response of different also varies. A complete attribution statement must address each of these sources of uncertainty.

Probabilistic extreme event attribution is a rapidly developing field of inquiry with a number of different, if not opposing, views. In addition to the scientific issues raised by such analyses, there are important social, legal and ethical issues that arise. While outside the scope of this review paper, such matters are discussed at length by Hulme (2014). Furthermore, while the number of events studied has been greatly enlarged by the short essays in the Bulletin of the American Meteorological Society (BAMS) supplements (see Herring et al., 2014 and Herring et al., 2015) and range from cold snaps to heat waves and droughts to floods, the events analyzed largely occurred in the industrialized countries where the authors lived. Developing nations are, of course, not immune to the effects of climate change on extreme weather and the impacts there often are more severe and are beginning to draw the attention of the scientific community, as shown by the most recent version of the BAMS supplements (Herring et al., 2015). Systematic or comprehensive global analyses of individual extreme events have not yet been performed. Forecasts of changes in the seasonal risk of extreme events performed in advance using existing seasonal forecasts of sea surface temperatures and sea ice extent are currently possible. Assessment of the skill of such forecasted risk would increase confidence in extreme event attribution statements.

FUTURE DIRECTIONS AND OPPORTUNITIES

High performance computing has recently enabled a new class of climate models that can simulate extreme weather significantly better in multi-decadal integrations than the CMIP5 models. Although still a rapidly developing field, global atmospheric models at resolutions of 25 km can produce tropical cyclones and other intense storms (Walsh et al. 2015). Hence, simulated long period return values of extreme precipitation more closely represent observations (Wehner et al., 2014). The station based raw data that make up observational databases will not significantly improve in quality or coverage in the near future. Uncertainties in gridded observational products are large, even over the well observed North America and Europe (Wehner et al., 2014). Satellite observations are beginning to have long enough periods of record to be useful in detection/attribution research. However, satellites have their own issues that require much post-processing to correct for biases owing to problems such as orbital drift and changes in instrumentation.

There are efforts underway to improve coverage in regions of the globe that are underrepresented in global data sets. The World Meteorological Organization (WMO) has an expert team on the data rescue (ET-DARE) and is cooperating in a number of data rescue initiatives to help digitize and make available observations that continue to exist in country meteorological services archives. In time these kinds of efforts will help fill in data coverage gaps in many parts of the world and should be encouraged.

Advances in extreme value statistical techniques are enabling more robust estimation of long period uncertainties through the use of physical covariates (Sillman et al., 2011). Such reductions in the uncertainty of the statistical fits, will likely improve both detection and attribution of changes in extreme temperature and precipitation.

There is a demand from the journalism community for more rapid assessment of the human contribution to specific weather events while their memory is fresh in the public's mind. The Worldwide Weather Attribution project (http://www.climatecentral.org/wwa) is one such effort. This partnership between Climate Central, the University of Oxford Environmental Change Institute (Oxford ECI), the Royal Netherlands Meteorological Institute (KNMI), the University of Melbourne, and the Red Cross Red Crescent Climate Centre (the Climate Centre) is operationalizing attribution statements for certain classes of extreme weather events. Their first operational statement used two methods to conclude that they are "virtually certain that the heat wave that stretched across much of Europe in early July (2015) was more likely to happen now than in the past due to climate change" (http://www.climatecentral.org/europe-2015-heatwave-climate-change). The statement was made on July 10 during the actual heat wave. The project team came to their conclusion using two methods. The first used the very large ensemble technique described in Section 4. The second used an empirical trend detection technique based on extreme value statistical techniques (van Oldenborgh et al., 2012; King et al., 2015). Detailed statements about specific European cities were also made. In this case, the scientific rationale behind the attribution statements is well established. For other classes of events, particularly strong storms such as hurricanes or mesoscale convective systems, the effect of a warmer climate on event statistics is less well understood. Furthermore, CMIP5 class modeling systems may not be "fit for purpose" to analyze such events requiring that specialized simulations be performed after the event occurrence. Performing and interpreting such custom simulations may likely take longer than the news cycle (typically a few weeks at most).

CONCLUSIONS

This paper has provided an overview of detection and attribution as it relates to extreme events. There is increasing interest by many sectors of society for information about causes of extreme events, particularly if the cause can be linked to human-induced climate change. Furthermore, attribution studies of extremes need to begin considering the impacts of extremes by evaluating exposure and vulnerability in a risk-based framework (Cardona et al., 2012).

The science of attribution, particularly event attribution, is still emerging and for this information to be useful to a wide range of stakeholders, uncertainties in attribution results need to be assessed and articulated in a way that stakeholders can understand. Since attribution science is dependent on climate modeling results, it is important to continue to assess and improve climate models. Similarly, detection studies must have high quality observational data sets, and improvements in spatial and temporal coverage of longer-term climate data sets are also needed. Lastly, since satellite-based data sets of the climate system are now becoming long enough for use in classical style detection/attribution studies, and have proven useful in event attribution, innovative uses of these data must be pursued.

ACKNOWLEDGMENTS

We would like to acknowledge four anonymous reviewers for their comments which helped us produce a much improved manuscript. We also acknowledge the World Meteorological Organization's Expert Team on Climate Change Detection and Indices for developing the indices used in Fig. 1, Fig. 2 and Fig. 3, and www.climdex.org and the University of New South Wales in the construction of Fig. 1, Fig. 2 and Fig. 3. Partial support for DRE, KEK, and LS was provided by the Strategic Environmental Research and Development Program of the , Grant Number W912HQ-15-C-0010.

REFERENCES

1. AghaKourchak, A., Behrangi, A., Sorooshian, S., Hsu, K., Amitai, E., 2011. Evaluation of satellite-retrieved extreme precipitation rates across the central United States. J. Geophys. Res. 116, D2. http://dx.doi.org/10.1029/2010JD014741.
2. Alexander, L.V., et al., 2006. Global observed changes in daily climatic extremes of temperature and precipitation. J. Geophys. Res. 111 (D05109), 2006. http://dx.doi.org/10.1029/2005JD006290.
3. Alexander, L.V., 2015. Global observed long-term changes in temperature and precipitation extremes: a review and update since IPCC AR5. Weather Clim Extremes 11, 4–16.
4. Allen, M.R., 2003. Liability for climate change. Nature 421, 891–892.

5. Allen, M.R., Stott, P.A., 2003. Estimating signal amplitudes in optimal fingerprinting. Part I: theory. Clim. Dyn. 21, 477–491.

6. Allen, M.R., Tett, S.F.B., 1999. Checking for model consistency in optimal fingerprinting.Clim. Dyn. 15, 419–434 1999.

7. Bindoff, N., Stott, P., AchutaRao, K.M., Allen, M., Gillett, N., Gutzler, D., Hansingo, K., Hegerl, G., Hu, Y., Jain, S., Mokhov, I., Overland, J., Perlwitz, J., Sebbari, R., Zhang, X., 2013. Detection and attribution of climate change: from global to regional in: climate change 2013. In: Stocker, T.F., Qin, D., Plattner, G.-K., Tignor, M.,

8. Allen, S.K., Boschung, J., Nauels, A., Xia, Y., Bex, V., Midgley, P.M. (Eds.), The Physical Science Basis. Contribution of Working Group I to the Fifth Assessment. Cambridge University Press, Cambridge, United Kingdom, New York, NY, USA.

9. Blunden, J., Arndt, D.S., 2015. State of the Climate in 2014. Bull. Am. Meteor. Soc. 96,ES1–ES32. http://dx.doi.org/10.1175/2015BAMSStateoftheClimate.1.

10. Burke, E.J., Brown, S.J., Christidis, N., 2006. Modeling the recent evolution of global drought and projections for the twenty-first century with the Hadley Center climate model. J. Hydrometeor. 7, 1113–1125. http://dx.doi.org/10.1175/

11. JHM544.1.

12. Cardona, O.D., van Aalst, M.K., Birkmann, J., Fordham, M., McGregor, G., Perez, R.,Pulwarty, R.S., Schipper, E.L.F., Sinh, B.T., 2012. Determinants of risk: exposure and vulnerability. In: Field, C.B., Barros, V., Stocker, T.F., Qin, D., Dokken, D.J., Ebi,K.L., Mastrandrea, M.D., Mach, K.J., Plattner, G.-K., Allen, S.K., Tignor, M.,Midgley, P.M. (Eds.), Managing the Risks of Extreme Events and Disasters to Advance Climate Change Adaptation. Cambridge University Press, Cambridge,UK, New York, NY, USA, pp. 65–108.

13. CCSP, 2008. Weather and Climate Extremes in a Changing Climate. Regions of Focus: North America, Hawaii, Caribbean, and U.S. Pacific Islands. A Report by theU.S. Climate Change Science Program and the Subcommittee on Global ChangeResearch. Thomas R. Karl, Gerald A. Meehl, Christopher D. Miller, Susan J.Hassol, Anne M. Waple, and William L. Murray, (eds.), Department of Commerce,NOAA's National Climatic Data Center, Washington, D.C., USA, 164 pp. Christidis, N., Stott, P.A., Brown, S.J., 2011. The role of human activity in the recentwarming of extremely warm daytime temperatures. J. Clim. 24, 1922–1930.

14. Christidis, N., Stott, P.A., Hegerl, G.C., Betts, R.A., 2013. The role of land use change inthe recent warming of daily extreme temperatures. Geophys. Res. Lett. 40,589–594. http://dx.doi.org/10.1002/grl.50159.

15. Coles, S., 2001. An introduction to the statistical modeling of extreme values.Springer, London, ISBN: 1-85233-459-2, p. 208.

16. Cowtan, K., Way, R., 2014. Coverage bias in the HadCRUT4 temperature series andits impact on recent temperature trends, Quart. J. R. Meteor. Soc. 140,1935–1944. http://dx.doi.org/10.1002/qj.2297.

17. Dai, A., 2013. Increasing drought under global warming in observations and models. Nature Clim. Chang. 3, 52–58.

18. Della-Marta, P.M., Wanner, H., 2006. A method of homogenizing the extremes andmean of daily temperature measurements. J. Clim. 19, 4179–4197.

19. Dole, R., et al., 2011. Was there a basis for anticipating the 2010 Russian heat wave? Geophys. Res. Lett. 38, L06702.

20. Donat, M.G., Alexander, L.V., Yang, H., Durre, I., Vose, R., Caesar, J., 2013. Global landbased datasets for monitoring climatic extremes. Bull. Am. Meteor. Soc. 94,997–1006. http://dx.doi.org/10.1175/BAMS-D-12-00109.1.

21. Easterling, D.R., Peterson, T., 1994. A new method for detecting undocumented discontinuities in climatological time series. Int. J. Climatol. 15, 369–377.

22. Easterling, D.R., Meehl, G., Changnon, S., Parmesan, C., Karl, T.R., Mearns, L.O., 2000. Climate extremes: observations, modeling, and impacts. Science 289, 2068–2074.

23. Easterling, D.R., 2013. Global datasets for analysis of climate extremes. In: AghaKouchak,A., et al. (Eds.), Extremes in a Changing Climate, Water Science and Technology Library 65. Springer, Dordrecht, pp. 347–361. http://dx.doi.org/10.1007/978-94-007-4479-0_1.

24. Fischer, E.M., Beyerle, U., Knutti, R., 2013. Robust spatially aggregated projections of climate extremes. Nature Clim. Chang. http://dx.doi.org/10.1038/nclimate2051.

25. Fischer, E., Knutti, R., 2015. Anthropogenic contribution to the global occurrence ofheavy precipitation and hot extremes. Nature Clim. Chang . http://dx.doi.org/10.1038/nclimate2617.

26. Folland, C., Stone, D., Frederiksen, C., Karoly, D., Kinter, J., 2014. The International CLIVAR Climate of the 20th Century Plus (C20Cþ) Project: Report of the Sixth Workshop. CLIVAR Exch. 19, 57–59.

27. Hartmann, D.L., Klein Tank, A.M.G., Rusticucci, M., Alexander, L.V., Bronnimann, S., Charabi, Y., Dentener, F.J., Dlugokencky, E.J., Easterling, D.R., Kaplan, A., Soden, B.J., Thorne, P.W., Wild, M., Zhai, P.M., 2013. Observations: Atmosphere and Surface. In: Stocker, T.F., Qin, D., Plattner, G.-K., Tignor, M., Allen, S.K., Boschung, J., Nauels, A., Xia, Y., Bex, V., Midgley, P.M. (Eds.), Climate Change 2013: ThePhysical Science Basis. Contribution of Working Group I to the Fifth AssessmentReport of the Intergovernmental Panel on Climate Change. Cambridge UniversityPress, Cambridge UK, New York, NY, USA.

28. Hegerl, G.C., Zwiers, F.W., Braconnot, P., Gillett, N.P., Luo, Y., Marengo Orsini, J.A.,Nicholls, N., Penner, J.E.,Stott, P.A., 2007. Understanding and Attributing Climate Change. In: Solomon, S., Qin, D., Manning, M., Chen, Z., Marquis, M.,Averyt, K.B., Tignor, M., Miller, H.L. (Eds.), Climate Change 2007: The Physical Science Basis. Contribution of Working Group I to the Fourth Assessment Reportof the Intergovernmental

Panel on Climate Change. Cambridge University Press, Cambridge, United Kingdom, New York, NY, USA.

29. Hegerl, G.C., et al., 2010. Good practice guidance paper on detection and attributionrelated to anthropogenic climate change. In: Stocker, T.G., et al. (Eds.), Meeting report of the Intergovernmental Panel on Climate Change Expert Meeting on Detection and Attribution of Anthropogenic Climate Change. IPCC Working Group I Technical Support Unit, University of Bern, Bern, Switzerland, p. 8 pps.
30. Hegerl, G.C., 2015. Use of models and observations in event attribution. Environ. Res. Lett. 10, 071001. http://dx.doi.org/10.1088/1748-9326/10/7/071001.
31. Herring, S.C., M.P. Hoerling, T.C. Peterson, and P.A. Stott, Eds., 2014. Explaining Extreme Events of 2013 from a Climate Perspective. Bull. Am. Meteor. Soc. Vol. 95, no. (9), pp. S1–S96.
32. Herring, S.C., M.P. Hoerling, J.P. Kossin, T.C. Peterson, and P.A. Stott, Eds., 2015. Explaining Extreme Events of 2014 from a Climate Perspective. Bull. Amer. Meteor. Soc. Vol. 96, no. (12), pp. S1–S172.
33. Hoerling, M., Kumar, A., Dole, R., Nielsen-Gammon, J.W., Eischeid, J., Perlwitz, J., Quan, X.-W., Zhang, T., Pegion, P., Chen, M., 2013. Anatomy of an Extreme Event. J. Clim. 26, 2811–2832. http://dx.doi.org/10.1175/JCLI-D-12-00270.1.
34. Holland, G., Webster, P., 2007. Heightened tropical cyclone activity in the North Atlantic: Natural variability or climate trend? Philos. Trans. R. Soc. London Ser.A 365, 2695–2716.
35. Hulme, M., 2014. Attributing weather extremes to 'climate change': a review. Prog. Phys. Geography. 38, 499–511.
36. Jarvinen, B.R., Neumann, C.J., Davis, M.A.S., 1984. A tropical cyclone data tape for the North Atlantic Basin, 1886-1983: Contents, limitations, and uses. NOAA Technical Memorandum NWS NHC. Vol. 22, p. 21. Available online at: ⟨http://www.nhc.noaa.gov/pdf/NWS-NHC-1988-22.pdf⟩.
37. Karl, T.R., Diaz, H.G., Kukla, G., 1988. Urbanization: its detection and effect in theUnited States climate record. J. Clim. 1, 1099–1123.
38. Katz, R.W., 2010. Statistics of extremes in climate change. Clim. Chang. 100, 71–76.
39. Kim, Y.H., Min, S.K., Zhang, X., Zwiers, F., Alexander, L.V., Donat, M.G., Tung, Y.S.,2015. Attribution of extreme temperature changes during 1951–2010. Clim. Dyn., 1–14. http://dx.doi.org/10.1007/s00382-015-2674-2.
40. Kossin, J.P., et al., 2007. A globally consistent reanalysis of hurricane variability and trends. Geophys. Res. Lett. 34, L04815.
41. Kossin, J.P., Karl, T.R., Knutson, T.R., Emanuel, K.A., Kunkel, K.E., O'Brien, J.J., 2015.Reply to "comments on 'monitoring and understanding trends in extremestorms: state of knowledge'". Bull. Am. Meteor. Soc. 96, 1177–1179.
42. Kunkel, K.E., Karl, T.R., Easterling, D.R., 2007. A monte carlo assessment of uncertainties in heavy precipitation frequency variations. J. Hydrometeor. 8,1152–1160.
43. Kunkel, K.E., Karl, T.R., Brooks, H., Kossin, J., Lawrimore, J., Arndt, D., Bosart, L., Changnon, D., Cutter, S.L., Doesken, N., Emanuel, K., Groisman, P.Ya., Katz, R.W., Knutson, T., O'Brien, J., Paciorek, C.J., Peterson, T., Redmond, K., Robinson, D.,Trapp, J., Vose, R., Weaver, S., Wehner, M., Wolter, K., Wuebbles, D., 2013.
44. Monitoring and understanding changes in extreme storms: state of knowledge.Bull. Am. Meteor. Soc. 94, 499–514. http://dx.doi.org/10.1175/BAMS-D-12-00066.1.
45. Kunkel, K.E., Frankson, R.M., 2015. Global land surface extremes of precipitation: Data limitations and trends. J. Extreme Events . http://dx.doi.org/10.1142/S2345737615500049.
46. Landsea, C.W., 2007. Counting Atlantic tropical cyclones back to 1900. EOS Trans. 88, 197–202.
47. Landsea, C.W., 2015. Comments on "monitoring and understanding trends in extremestorms: state of knowledge". Bull. Am. Meteor. Soc. 96, 1175–1176. http://dx.doi.org/10.1175/BAMS-D-13-00211.1.
48. Landsea, C.W., et al., 2004. The Atlantic hurricane database reanalysis project:Documentation for the 1851-1910 alterations and additions to the HURDAT database. In: Murname, R.J., Liu, K.-B. (Eds.), Hurricanes and Typhoons: Past,Present and Future. Columbia University Press, New York, NY, pp. 177–221.
49. Menne, M., Williams, C., 2009. Homogenization of temperature series via pairwise comparisons. J. Clim. 22, 1700–1717.
50. Menne, M.J., Durre, I., Vose, R.S., Gleason, B.E., Houston, T.G., 2012. An overview of the Global Historical Climatology Network-Daily database. J. Atmos. Ocean. Technol. 29, 897–910. http://dx.doi.org/10.1175/JTECH-D-11-00103.1.
51. Min, S.-K., Zhang, X., Zwiers, F.W., Hegerl, G.C., 2011. Human contribution to moreintense precipitation extremes. Nature 470, 378–381.
52. Min, S.-K., Zhang, X., Zwiers, F.W., Shiogama, H., Tung, Y.-S., Wehner, M., 2013. Multi-Model Detection and Attribution of Extreme Temperature Changes. J. Clim. 26, 7430–7451. http://dx.doi.org/10.1175/JCLI-D-12-00551.1.
53. Otto, F.E.L., Massey, N., van Oldenborgh, G.J., Jones, R.G., Allen, M.R., 2012. Reconciling two approaches to attribution of the 2010 Russian heat wave. Geophys. Res. Lett. 39, L04702.
54. Pall, P., Aina, T., Stone, D.A., Stott, P.A., Nozawa, T., Hilberts, A.G.J., Lohmann, D.,Allen, M.R., 2011. Anthropogenic greenhouse gas contribution to UK autumn flood risk. Nature 470, 382–385.
55. Pall, P., Wehner, Michael, Stone, D.áithí, 2015. Probabilistic Extreme Event Attribution in Dynamics and Predictability of Large-Scale, High-Impact Weather andClimate Events. In: Grotjahn, Richard, Li, Jianping, Swinbank, Richard, Volkert, Hans (Eds.), Cambridge University Press,, Cambridge (in press).

56. Peterson, T.C., Stott, P.A. and Herring, S., eds., 2012. Explaining extreme events of 2011 from a climate perspective. Bull. Amer. Meteor. Soc. Vol. 93, pp. 1041–1067.

57. Peterson, T.C., Hoerling, M.P., Stott, P.A. and Herring, S., eds., 2013. Explaining extreme events of 2012 from a climate perspective. Bull. Amer. Meteor. Soc. Vol. 94,no.(9), pp. S1–S74.

58. Robine, J.-M., Cheung, S.L.K., Le Roy, S., et al., 2008. Death toll exceeded 70,000 in Europe during the summer of 2003. Comptes Rendus Biol. 331, 171–178.

59. Sen, P.K., 1968. Estimates of the regression coefficient based on Kendall's Tau. J. Am. Stat. Assoc. 63, 1379–1389.

60. Seneviratne, S., Nichols, N., Easterling, D., Goodess, C., Kanae, S., Kossin, J., Luo, Y., Marengo, J., McInnes, K., Rahimi, M., Reichstein, M., Sorteberg, A., Vera, C., Zhang, X., 2012. Changes in climate extremes and their impacts on the natural physical environment, chapter 3 in IPCC Special Report on Managing the Risks of Extreme Events and Disasters to Advance Climate Change Adaptation.Cambridge University Press, Cambridge, UK, New York, NY, USA.

61. Seneviratne, S., Zwiers, F., 2015. Attribution and prediction of extreme events: Editorial on the special issue. Weather Clim. Extremes 9, 2–5. http://dx.doi.org/10.1016/j.wace.2015.08.003.

62. Sheffield, J., Wood, E.F., Roderick, M.L., 2012. Little change in global drought overthe past 60 years. Nature 491 (7424), 435–438.

63. Shiogama, H., Christidis, N., Caesar, J., Yokohata, T., Nozawa, T., Emori, S., 2006. Detection of greenhouse gas and aerosol influences on changes in temperatureextremes. SOLA 2, 152–155.

64. Sillman, J., Croci -Maspoli, M., Kallache, M., Katz, R., 2011. Extreme cold winter temperatures in Europe under the influence of north atlantic atmosphericblocking. J. Clim. 24, 5899–5913. http://dx.doi.org/10.1175/2011JCLI4075.1.

65. Sippel, S., Walton, P., Otto, F.E.L., 2015. Stakeholder perspectives on the attribution of extreme weather events: an explorative enquiry. Weather Climate Soc. 7, 224–237. http://dx.doi.org/10.1175/WCAS-D-14-00045.1.

66. Stone, D.A., Allen, M.R., 2005. The end-to-end attribution problem: from emissions to impacts. Clim. Chang. 71, 303–318.

67. Stott, P.A., Stone, D.A., Allen, M.R., 2004. Human contribution to the European heatwave of 2003. Nature 432, 610–614.

68. Stott, et al., 2013. Attribution of weather and climate-related events. in climate science for serving society. In: Asrar, G.R., Hurrell, J.W. (Eds.), Research, Modelling and Prediction Priorities. Springer, Dordrecht, ISBN: 978-94-007-6691-4, pp. 307–337.

69. Trenberth, K.E., Fasullo, J.T., Shepherd, T.G., 2015. Attribution of climate extreme events. Nat. Clim. Chang. . http://dx.doi.org/10.1038/NCLIMATE2657

70. Trewin, B., 2010. Exposure, instrumentation, and observing practice effects on land temperature measurements. Wiley Interdiscip. Rev.: Clim. Chang. 1 (4),490–506.

71. Trewin, B., 2013. A daily homogenized temperature data set for Australia. Int. J. Climatol. . http://dx.doi.org/10.1002/joc.3530

72. Vecchi, G.A., Knutson, T.R., 2011. Estimating annual numbers of Atlantic hurricanes missing from the HURDAT database (1878-1965) using ship track density. J.Clim. 24, 1736–1746. http://dx.doi.org/10.1175/2010JCLI3810.1.

73. Walsh, J., Wuebbles, D., Hayhoe, K., Kossin, J., Kunkel, K., Stephens, G., Thorne, P.,Vose, R., Wehner, M., Willis, J., Anderson, D., Doney, S., Feely, R., Hennon, P.,Kharin, V., Knutson, T., Landerer, F., Lenton, T., Kennedy, J., Somerville, R., 2014a. Ch. 2: Our Changing Climate. Climate Change Impacts in the United States: The Third National Climate Assessment. In: Melillo, J.M., Richmond, Terese (T.C.),Yohe, G.W. (Eds.), U.S. Global Change Research Program, United States 2-1-2-nn.Walsh, K.J.E., Camargo, S., Vecchi, G., Daloz, A.S., Elsner, J., Emanuel, K., Horn, M.,Lim, Y.-K., Roberts, M., Patricola, C., Scoccimarro, E., Sobel, A., Strazzo, S., Villarini,G., Wehner, M., Zhao, M., Kossin, J., LaRow, T., Oouchi, K., Schubert, S., Wang, H., Bacmeister, J., Chang, P., Chauvin, F., Jablonowski, C., Kumar, A., Murakami, H., Ose, T., Reed, K., Saravanan, R., Yamada, Y., Zarzycki, C., Vidale, P.-L., Jonas, J., Henderson, N., 2014b. Hurricanes and climate: the U.S. CLIVAR working group on hurricanes. Bull. Am. Meteor. Soc. . http://dx.doi.org/10.1175/ BAMS-D-13-00242.1

74. Wehner, M.F., Easterling, D.R., Lawrimore, J.H., Heim, R.R., Vose, R.S., Santer, B.D., 2011. Projections of future drought in the continental United States and Mexico.J. Hydrometeor. 12, 1359–1377. http://dx.doi.org/10.1175/2011JHM1351.1.

75. Wehner, M.F., 2013. Very extreme seasonal precipitation in the NARCCAP ensemble:Model performance and projections. Clim. Dyn. 40, 59–80. http://dx.doi.org/10.1007/s00382-012-1393-1.

76. Wehner, M.F., Reed, K.A., Prabhat, F. Li, Bacmeister, J., Chen, C.-T., Paciorek, C., Gleckler, P., Sperber, K., Collins, W.D., Gettelman, A., Jablonowski, C., 2014. The effect of horizontal resolution on simulation quality in the Community Atmospheric Model, CAM5.1. J. Model. Earth Syst. 06, 980–997. http://dx.doi.org/

77. 10.1002/2013MS000276.

78. Woodhouse, C., Overpeck, J., 1998. 2000 years of drought variability in the CentralUnited States. Bull. Am. Meteor. Soc. 79, 2693–2714.

79. Zwiers, F.W., Zhang, X., Feng, Y., 2011. Anthropogenic influence on long returnperiod daily temperature extremes at regional scales. J. Clim. 24, 881–892. http://dx.doi.org/10.1175/2010JCLI3908.1.

80. Zwiers, F.W., et al., 2013. Challenges in estimating and understanding recent changes in the frequency and intensity of extreme climate and weather events. in climate science for serving society. In: Asrar, G.R.,

Hurrell, J.W. (Eds.),Research, Modelling and Prediction Priorities. Springer, Dordrecht, ISBN: 978- 94-007-6691-4, pp. 339–389.

81. van Oldenborgh, G.J., van Urk, A., Allen, M.R., 2012. The absence of a role of climatechange in the 2011 Thailand floods. Bull. Am. Met. Soc 93, 1047–1049. http://dx. doi.org/10.1175/BAMS-D-12-00021.1.

CHAPTER 4

SIMULATING 2368 TEMPERATE LAKES REVEALS WEAK COHERENCE IN STRATIFICATION PHENOLOGY

Jordan S. Read,Luke A. Winslow, Gretchen J.A. Hansen, Jamon Van Den Hoek, Paul C. Hanson, Louise C. Bruce, Corey D. Markfort

Keywords: Limnology, Climate change, Water temperature, Lake temperature modeling, Coherence, Stratification

INTRODUCTION

Lakes are diverse ecosystems that respond in complex ways to the stressors of climate change (Blenckner et al., 2007, Carpenter et al., 1992 and Roach et al., 2013). While lakes have shown regional-scale coherence in temperature (see Livingstone, 2008), biological features and dynamics often are unrelated in neighboring lakes (e.g., Soranno et al., 1999), confounding our understanding of how lakes respond to climate drivers. Interpreting dissimilar responses of lakes to current and future climate is one of the grand challenges of modern limnology.

Lake-specific properties (such as morphometry and surrounding land cover) are recognized controls on the structure and function of aquatic ecosystems, but there is disagreement as to the role they play as mediators of climate variability. Despite large diversity in lake properties, analyses of lake surface temperatures have established strong evidence for among-lake coherence to large-scale climate (Benson et al., 2000,Livingstone and Dokulil, 2001 and Palmer et al., 2014). Conversely, coherence weakens or disappears below the lake surface (Kratz et al., 1998 and Palmer et al., 2014), suggesting there is value in quantifying and explaining among-lake differences in responses to climate. At present, the diversity in lake responses to climate is not quantified; we lack the capacity to observe millions of lakes, and modeling approaches have not filled these information gaps.

Because temperature is a master factor for many aquatic ecosystem processes (Cardoso et al., 2014, Hanson et al., 2011, Magnuson et al., 1979 and Paerl and Huisman, 2008), understanding thermal responses to climate is critical for predicting future biotic change (Sharma et al., 2007). Lake temperatures, stratification, and ice-cover durations are changing for many of the world›s lakes (Livingstone, 2003,Magnuson et al., 2000, Schindler et al., 1990 and Schneider and Hook, 2010), and these changes impact the timing and function of lake ecosystem processes (O'Reilly et al., 2003, Smol et al., 2005 and Weyhenmeyer et al., 1999).

Recent advances in observational capabilities have allowed limnologists to address research questions at larger spatial scales (Heffernan et al., 2014 and Soranno et al., 2010). Federal and state sensor networks, citizen science monitoring programs, and satellite/airborne observations have expanded considerably, contributing "Big Data" to aquatic ecology (Hampton et al., 2013 and Keller et al., 2008). Satellite observations have documented lake warming (Schneider and Hook, 2010) and estimated water quality (Torbick et al., 2013). Gridded time-series data capture retrospective meteorological drivers (Mesinger et al., 2006 and Mitchell et al., 2004) or predictions for future climate (Hostetler et al., 2011 and Wilby et al., 2000) for large numbers of lakes. There are tremendous opportunities to leverage these environmental Big Data for a better understanding of the local context of lakes (e.g., Soranno et al., 2010), and broad-scale hydroclimatic interactions (e.g., Kucharik et al., 2000 and Leonard and Duffy, 2013). Coupling mechanistic modeling with big data may provide means to assess among-lake diversity of lake responses to climate.

At present, two dominant process-based approaches exist for evaluating lake temperature responses at temporal scales appropriate for climate: simulations of generic lakes at broad spatial extents (Fang and Stefan, 2009; e.g., Hostetler and Small, 1999 and Stefan et al., 2001), and detailed modeling

Citation: Jordan S. Read, Luke A. Winslow, Gretchen J.A. Hansen, Jamon Van Den Hoek, Paul C. Hanson, Louise C. Bruce, Corey D. Markfort, Simulating 2368 temperate lakes reveals weak coherence in stratification phenology, http://dx.doi.org/10.1016/j.ecolmodel.2014.07.029.

of small numbers of real lakes (Cahill et al., 2005, Fang et al., 2012 and Hadley et al., 2013). In order to quantify climatically driven regional dynamics and among-lake variability for large numbers of real lakes, a new approach that leverages the advantages of both simulation types is necessary.

Here, we examine the diversity of lake responses to climate by introducing a new methodology that uses freely available, large-scale datasets to run unique hydrodynamic simulations for thousands of lakes. Methods and tools were created explicitly to support future analyses of lakes and their design was informed by the research needs of the limnological community. In order to maximize community use and acceptance, all data, methods, and tools are freely accessible and openly shared. This study leveraged these elements to simulate an unprecedented 2368 lakes for 33 years and contrast individual and population-level lake responses to climate variability.

MATERIALS AND METHODS

Hydrodynamic Model

Because our primary objective was to estimate temporal dynamics in water temperature profiles, a one-dimensional (1D) dynamical model was used to simulate each of the study lakes. The General Lake Model version 1.2.0 (GLM; Hipsey et al., 2013) was chosen for this study. GLM combines fluxes of mass and energy with a Lagrangian layer structure that adapts to changes in vertical gradients, and is freely available to all users. Many energy budget algorithms and mixing schemes in GLM are modern implementations of methods applied in other widely used 1D models (Hamilton and Schladow, 1997 and Mooij et al., 2010).

Heat flux is the primary driver of lake temperature (Wetzel and Likens, 2000), and components of this flux were formatted as time-series input to GLM for the variables of wind speed, air temperature, relative humidity, precipitation, and downwelling long- and shortwave radiation (Fig. 1). At each time step, GLM accounts for energy fluxes into (e.g., downwelling radiation) and out of (e.g., evaporation) the lake, and propagates the resultant temperature changes to various layers according to heat transfer and vertical mixing algorithms. Detailed formulations of these numerical routines can be found inHamilton and Schladow (1997) or Hipsey et al. (2013), which also includes a full parameter list for GLM with recommended default values that are based on field and laboratory studies. Simulations did not include the influence of surface or groundwater flows, as coupling with hydrologic model output was beyond the scope of this study, but the impact of this omission on water temperature simulations was evaluated.

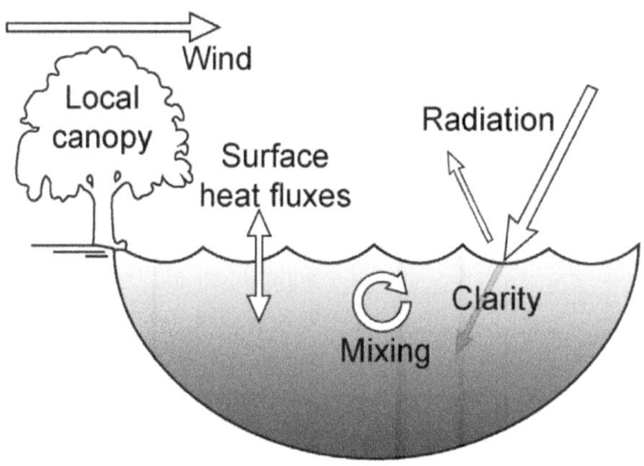

Figure 1: A schematic of drivers and controls on lake temperature and stratification for a one-dimensional hydrodynamic model.

Model Parameterization

Lake-specific data were used for static model inputs of hypsography, water clarity, and local wind sheltering (see below for details on how these were calculated or measured). The effects of these parameters on the hydrodynamic model are as follows: (1) hypsography influences size and depth dependent processes in the model, including the number of vertical layers used by the model, (2) water clarity controls the rate at which the visible wavelengths of solar radiation are attenuated in the water column, and (3) wind sheltering reduces the amount of energy from wind that is available for vertical mixing. Since a generic model parameterization that could be extended to other regions and lake types was desired, the remaining physical coefficients for GLM that parameterize equations of energy and momentum fluxes were set to default values.

The effect of water clarity on water temperature is parameterized in GLM according to the extinction coefficient for shortwave radiation (named 'K_w' in GLM, but we use the more common notation of 'K_d'). The K_d parameter characterizes the exponential decay of light in the water column on a per meter (m^{-1}) basis. The parameter is used to define available light as $E(z) = E_0 \exp(-K_d z)$, where E_0 is the light energy at the surface and $E(z)$ is the remaining energy at depth z. As light is absorbed in the water column, it is converted into thermal energy and warms the waters. Thus, the general thermal effects of a smaller value of K_d (indicative of clearer waters) is to increase the amount of warming of deeper waters and decrease the warming of near-surface waters (Read and Rose, 2013).

Special attention was paid to the parameterization of mixing energy derived from wind speed because canopy and topographic features near the lake edge have a strong influence on physical dynamics in small and medium sized lakes (Markfort et al., 2010). The parameter in GLM that controls the flux of wind-driven mixing energy is the bulk aerodynamic momentum transfer coefficient, or C_M. The C_M parameter for each lake was scaled according to a cubic relationship with the wind sheltering coefficient (W_s; Hondzo and Stefan, 1993 and Markfort et al., 2010), calculated as $C_M = 0.0013\ W_s^{1/3}$. Local canopy, topography, and lake size were included in the formulation of wind-sheltering coefficients for each lake, which were then converted to C_M values (see Van Den Hoek et al., *in review*). Contrary to many other multi-lake modeling efforts that specifically tuned W_s to reduce model error (e.g., Fang et al., 2012), this coefficient was calculated as

$$W_s = \frac{2}{\pi}\cos^{-1}\left(25 \times h_S\sqrt{\left(\frac{\pi}{A}\right)}\right) - \frac{50 \times h_S}{A\sqrt{\pi}}\sqrt{A - 625 \times h_S^2 \times \pi}$$

where hS is the average elevation difference between the top of the surrounding canopy and lake surface (m), and A is the surface area of the lake (m^2). This relationship was formulated according to equation 8 of Markfort et al. (2010). Lakes with a smaller wind sheltering coefficient have a reduced amount of vertical mixing energy available in the water column, and are often more strongly stratified.

Ice-cover Simulations

Preliminary simulations for these lakes were continuous during the time period 1979–2011, but substantial bias in ice-cover duration using GLM's dynamical ice algorithms was discovered that also influenced the timing of springtime lake warming. To remove this bias, start and stop dates for every year were estimated for each lake using the methods of Shuter et al. (2013) for "Freeze-up model" and "Break-up model 1", which were calibrated to ice-cover data from Wisconsin lakes (Benson and Magnuson, 2000). Initial temperature profiles were 4 °C for each year.

Data Sources

Publicly available datasets were used exclusively to parameterize and drive the individual lake models in order to support the reusability of these methods. Datasets with continental or global coverage were used whenever possible, including the North American Land Data Assimilation System (NLDAS-2; ldas.gsfc.nasa.gov/nldas), the National Elevation Dataset (NED; ned.usgs.gov), the ASTER GDEM2 30 m topographic elevation model (asterweb.jpl.nasa.gov), and Landsat 30 m imagery (landsat.usgs.gov) (Fig. 2). These products were used to generate lake-specific parameters or time-series for meteorological forcing, wind sheltering, and lake water clarity (Sup. Table 1).

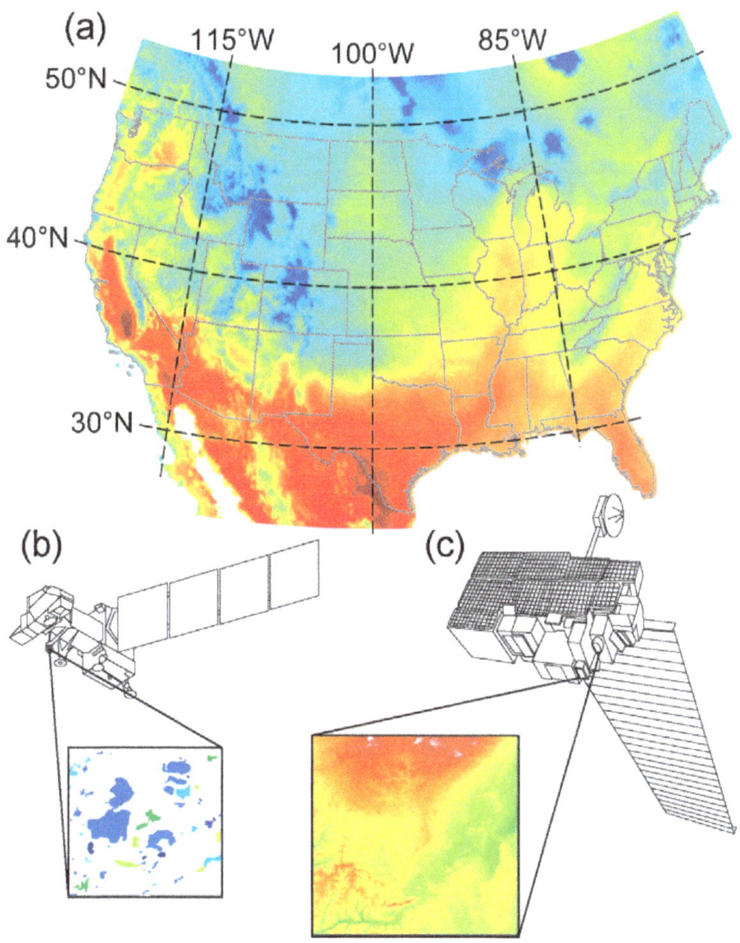

Figure 2: Continental and global-scale data sources are critical inputs for parameterizing and driving thousands of one-dimensional lake models. (a) The NLDAS-2 hourly gridded data for air temperature on June 1st 1979. (b) Landsat imagery was used to estimate water clarity (clarity indices provided by lakesat.org). (c) Topographic elevation data from the ASTER GDEM2 digital elevation model were used to estimate wind sheltering heights within lake buffers.

The lakes chosen for this analysis had at least one fisheries survey or stocking recorded event by the Wisconsin Department of Natural Resources (WI-DNR). Spatial boundaries for the 2368 lakes used here were extracted from the Wisconsin hydrolayer, a statewide, 1:24,000 surface-water dataset (Sup. Table 1).

Depth and clarity data for the 2368 modeled lakes were available from various sources (Sup. Table 1; Table 1). Each study lake had data for maximum depth, and historical bathymetric maps exist for most lakes (dnr.wi.gov/lakes/maps). Hypsography (the relationship between depth and lake area) for 149 randomly chosen lakes was manually digitized from scans of these historical maps. For the remaining lakes, hypsography was estimated as a cone parameterized by the lake surface area and maximum depth. The shoreline development factor (SDF; Kent and Wong, 1982) was calculated for each lake as the ratio of the lake perimeter to the perimeter of a circle with equal area. The SDF is a metric of shoreline complexity, and a high SDF may be indicative of complex lake morphometry that would not be captured in a 1D lake model. Lake water clarity from WI-DNR and lakesat.org provided satellite-derived clarity estimates of Secchi depth (ZSD; see Torbick et al., 2013 for methodology). In situ measured Secchi depths from by WI-DNR monitoring programs and satellite Z_{SD} were converted to a light attenuation coefficient (K_d) using $K_d = 1.7/Z_{SD}$ (Poole and Atkins, 1929). K_d for the model simulations was used as the average of all observations for a given lake. When K_d was unavailable (145 of the 2368 lakes), the median was used (0.7 m^{-1}; Table 1).

Table 1: Lake properties for all simulated lakes, lakes that were used for model validation, and a single lake (Big Muskellunge) used to compare with the population of lakes. K_d is the light attenuation coefficient and SDF is the shoreline development factor, a measure of shoreline complexity. Format shown for multiple numbers is *median* (*25th quantile*, *75th quantile*) or [*minimum*, *maximum*]

Lake dataset	n (#)	Area (km^2)	z_{max}(m)	K_d (m^{-1})	h_s (m)	SDF (-)	Elevation (m)	Latitude (°N)	Longitude (°E)
Simulated lakes	2368	0.308 (0.119, 0.805)	6.71 (3.66, 10.7)	0.699 (0.498, 1.07)	7.24 (5.62, 9.13)	1.65 (1.33, 2.20)	[177.27, 558.01]	[42.497, 46.849]	[−92.770, −86.941]
Validation lakes	434	0.952 (0.435, 2.18)	9.14 (6.10, 14.0)	0.780 (0.503, 1.18)	7.64 (6.11, 9.39)	1.89 (1.52, 2.58)	[180.56, 536.15]	[42.507, 46.387]	[−92.744, −86.986]
Big Muskellunge	1	3.96	21.34	0.36	9.4	2.39	500.88	46.017	−89.615

In this study, 434 lakes had observational data that could be used to validate model performance. Data from citizen monitoring, WI-DNR, and the North Temperate Lakes Long-Term Ecological Research program (NTL-LTER; lter.limnology.wisc.edu) were combined into a water temperature validation dataset containing 140,148 unique measurements. These validation lakes were representative of the various lake types and geographical regions in the full population of simulated lakes (Table 1). A single lake, Big Muskellunge Lake (BM), was also examined in detail. BM is a core study lake in the NTL-LTER program, with data collection beginning in 1981. BM was not chosen to be representative of the remaining lakes, but instead to examine the applicability of extending patterns in climate responses from frequently monitored "sentinel lakes" (such as BM) to the more diverse population of lakes.

Web Geo-processing

Gridded datasets for meteorological forcing and static parameters were spatially weighted and converted into lake-specific GLM inputs. Re-projecting, sub-setting, and transforming gridded data according to overlap with irregular geographical features (i.e., lakes) involves challenging algorithm development and data management techniques. This study leveraged web processing algorithms to reduce these burdens. Two representations of lakes and their surroundings were used to support two different regions of interest for data subsampling: the "lake" was the original polygon from the WI hydrolayer, and the "lake-buffer" was a 100 m wide buffer around the lake. Shapefiles representing all lakes and lake-buffers were hosted on sciencebase.gov, which has an automated web feature service (WFS) extension for shapefiles. WFS is a standard for web-enabled exchange of spatial data that supports filtering of spatial data based on feature attributes.

Following the creation of WFSs for each of the two feature datasets, a geospatial web processing service was used to calculate lake-specific data from the gridded meteorology, canopy cover, and topographic data for each lake. The Geo Data Portal (GDP; Blodgett, 2011) is designed to automate the processing of gridded data according to feature geometries that may differ in spatial resolution and/or native spatial reference systems, and all processing and data exchange are completed via the Web. The GDP extracts data from input source data and returns a post-processed, ASCII file of the results filtered according to the lakes in each WFS. GDP-outputted results were then used either to parameterize the GLM model runs or as model driver data.

Model Post-Processing

Lake simulations were not calibrated to individual lakes, and simulation results were validated against observational data where available. Because observational data were typically available at a lower frequency than daily modeled temperatures, no phenological metrics were used for validation (e.g., the onset of thermal stratification). Four main metrics were used to evaluate simulation quality in terms of goodness of fit (root mean squared error or R^2): Schmidt Stability, epilimnion, hypolimnion, and the full water temperature profiles. Model results were processed and distilled into time-series files of water temperatures and physical metrics. The full water column temperature profiles included a direct comparison between all measured water temperatures (at all depths) and their depth-subsampled equivalents from model outputs. Epilimnion and hypolimnion temperatures were the volumetrically averaged temperatures for each of the two thermal layers, and were calculated only when the lake was stratified. Schmidt Stability was also calculated according to Read et al. (2011). Additionally, the model's representation of stratification was evaluated based on the rate of false absence and false presence compared to observations. Lakes were considered stratified when the difference between surface and bottom water temperatures exceeded 1 °C (Stefan et al., 1996 and Woolway et al., 2014).

Sensitivity Analysis for Selected Model Parameters

Although a detailed uncertainty analysis for all lake modeling inputs was beyond the scope of this paper, the sensitivity of hypolimnetic temperatures to key model parameters was evaluated. Simulated hypolimnion temperatures often have the highest errors compared to shallower depths (Weinberger and Vetter, 2012), and we examined the sensitivity of modeled temperatures to water clarity, lake residence time, and wind sheltering height. All three of these parameters are known to have an influence on hypolimnetic water temperatures (Killworth and Carmack, 1979, Markfort et al., 2010 and Read and Rose, 2013), and were either parameterized via remote sensing data or (in the case of residence time) neglected as model inputs. Using the summer (July–September) mean bottom water temperatures as a response variable, lake-specific parameters were varied within a range of ±50% of their original values.

In the case of residence time, which was not used in the original simulations, an estimate of residence time for each lake (WI-DNR; M. Diebel, *unpublished*) was used to estimate inflow and outflow volumes while holding lake volume at steady state. Daily inflow volume, inflow temperature, and outflow volume (outflow temperature is calculated by the model) were added to each lake model, and modified by ±50% for the sensitivity analysis. Because inflow temperatures of surface waters and groundwater were unknown for the majority of these lakes, two scenarios were used to provide reasonable limits for these thermal inputs: (1) daily air temperatures were used as inflow temperatures, and (2) the seasonally-averaged air temperature was used as a constant inflow temperature.

Lake Modeling Framework

A framework was created for this study that automated the processes of data assimilation, data translation, model parameterization, model execution, and the calculation of derivative model results for each lake (Sup. Table 2; Fig. 3). R version 3.0.1 (R Development Core Team, 2013) was chosen as the scripting language to support these efforts because it is platform independent and open-source, making it particularly conducive to sharing. Documented R packages were developed to facilitate each of the following: automated geo-processing (geoknife), parameterization and analysis with the lake model (glmtools), and post-processing of model results and observational data (rLakeAnalyzer). Each individual R package includes detailed documentation and help files, but the purpose of each package in the modeling framework is described briefly below.

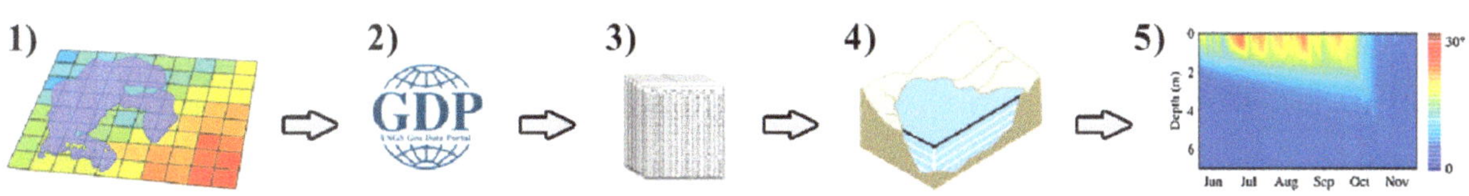

Figure 3: Lake modeling framework includes (1) selecting a lake feature (the surface shape of Big Muskellunge Lake; WI, USA is shown here in blue for example purposes) and gridded data, (2) extracting lake-specific data using the Geo Data Portal, (3) translating time-series and static results from the GDP into model input, (4) driving the lake simulation model, and (5) processing model results for each simulation. The Big Muskellunge Lake outline is used as an example. (For interpretation of the color information in this figure legend, the reader is referred to the web version of the article.)

The geoknife package was created to simplify and automate the execution of GDP processing requests. The glmtools package simplifies interactions between users and the GLM executable model by providing methods for parameterizing the model and analyzing results data (including depth-based subsampling of output files). rLakeAnalyzer calculates derivative physical metrics from water column temperature profiles and standardizes comparisons between model and in situ observational data. rLakeAnalyzer is an R implementation of Read et al. (2011).

To speed up the iteration time of the modeling framework, modeling and processing were parallelized onto a HTCondor computer cluster (Thain et al., 2005). Using a pool of 26 compute nodes, this process reduced the modeling of all 2368 lakes for 33 years from multiple days (using a single computer) to a few hours.

RESULTS

Lake Water Temperature Validation

The root-mean-square error (RMSE) between the model outputs and all temperature observations was 2.78 °C (Fig. 4a; n = 140,148), 1.74 °C for epilimnion temperatures (Fig. 4b; n = 7240), and 3.33 °C for hypolimnion temperatures (Fig. 4c; n = 7240). Model output and field observations agreed on the presence/absence of stratification 81.1% of the time (n = 8690). The rate of false positives (type I errors) was 12.9% (n = 214) and 20% for false negatives (type II errors; n = 1811). Twenty nine of the 434 validation lakes were responsible for over 50% of these mischaracterizations (type I and II errors). Comparisons for modeled versus observed Schmidt Stability had an R^2 of 0.93 (n = 10,715; data not shown).

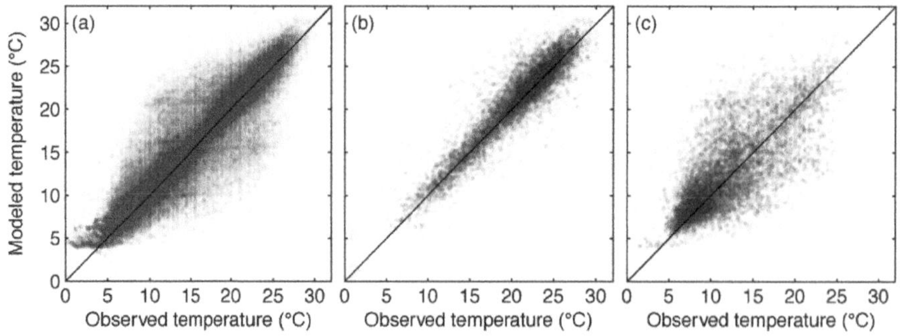

Figure 4: Temperature validation for 434 lakes. (a) A comparison between model output and field temperature observations for all depths had a RMSE of 2.78 °C (n = 140,148). (b) Epilimnion comparisons had a RMSE of 1.74 °C (n = 7240), and (c) RMSE of 3.33 °C for hypolimnion temperatures (n = 7240).

The median RMSE on the full temperature data for individual lakes was 2.44 °C. Lake-specific RMSE was significantly correlated ($p < 0.01$) to several morphological and geographical parameters, including lake depth, elevation, latitude, and longitude, and water clarity (Table 2). Lake area, surrounding canopy height, and the shoreline development factor were not significantly correlated with lake errors (Table 2).

Table 2: Pearson product–moment correlations between lake-specific root mean squared error and parameters.

Parameter	Correlation
z_{max}	0.192*
Area	0.027
h_s	0.029
K_d	−0.130*
Elevation	−0.212*
SDF	−0.039
Latitude	−0.247*
Longitude	0.148*

*$p < 0.01$.

Deep Water Temperature Sensitivity to Model Parameters

Hypolimnetic temperatures had the highest errors among the metrics that were evaluated (see Fig. 4c), and the sensitivity of these deep water temperatures to model inputs for water clarity, lake residence time, and wind sheltering height were evaluated. Deeper water temperatures were more sensitive to changes in water clarity when compared to residence time and wind sheltering height (Fig. 5). Larger lakes were more sensitive to changes in all three parameters (Fig. 5; black line), but lake surface area was significantly positively correlated with maximum depth ($p < 0.01$), which influences the model's layer structure and mixing algorithms.

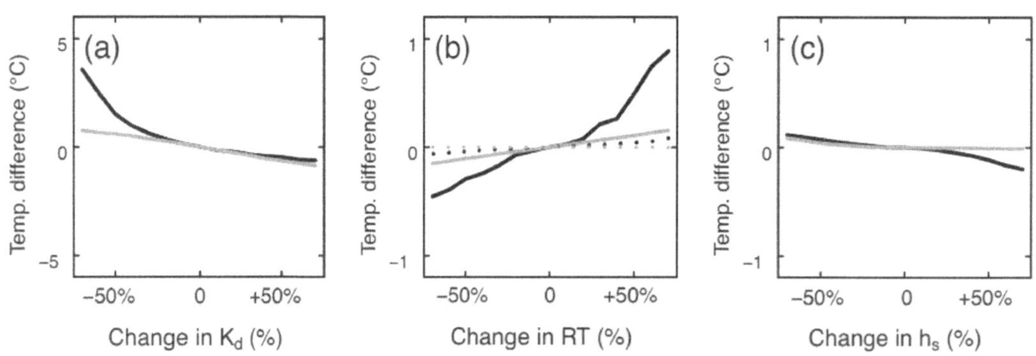

Figure 5: Sensitivity of bottom water temperatures to changes in three model parameters for 1996 simulations. Median change in bottom water temperatures for lakes >100 ha (black line) and <100 ha (gray line), resulting from shifts in (a) the water clarity parameter (K_d), (b) residence time (dashed line is daily inflow temperatures equal to daily air temperatures, and solid line is constant inflow temperatures equal to yearly average air temperature), and (c) lake sheltering height.

DISCUSSION

Simulation quality

The model performed well across a diverse population of lakes during the 33 year simulation period (Fig. 4). Default values were used for most model parameters; with lake-specific values for the light attenuation coefficient, bulk momentum drag coefficient, hypsography, and meteorological drivers. Despite the use of an automated model setup that assimilated broad-scale data and various remotely sensed parameters, the 2.44 °C median RMSE for individual lakes is encouraging. This median lake RSME was only ~1 °C higher than a recent multi-lake modeling study where the investigators used local metrological observations and calibrated simulations individually for each of 28 lakes (median RMSE 1.47 °C; Fang et al., 2012).

Greater RMSEs for lakes with clearer waters were evaluated (see Section 3.2), highlighting the need for high quality data to quantify the K_d parameter. Errors in the value of this parameter from satellite-derived clarity measurements and the omission of important inter- and intra-annual variability in water clarity both likely contribute to substantial temperature errors in deeper waters. The relationship between lake errors with latitude and longitude (elevation was significantly positively correlated with latitude;$p < 0.01$) could be the result of a spatial bias in the input data, or simply an expression of the spatially heterogeneous distribution of Wisconsin lakes.

Model application: lake responses to climate variability

Water temperatures and the timing of stratification were highly variable among years over the 33-year simulation period. A comparison between model outputs from the warmest and coldest climate years based on seasonally averaged statewide air temperatures (Fig. 6a and b; 1998 and 1996 respectively; Wisconsin State climate office) revealed complexities in lake–climate interactions. For example, the date when at least half of the simulation lakes were stably stratified (a minimum of 7 days of sustained stratification) was 27 days earlier in 1998 than in 1996; but the maximum number of lakes stratified on any 1 day was greater for 1996 compared to 1998 (82.5% versus 71.4%; Fig. 6c and d).

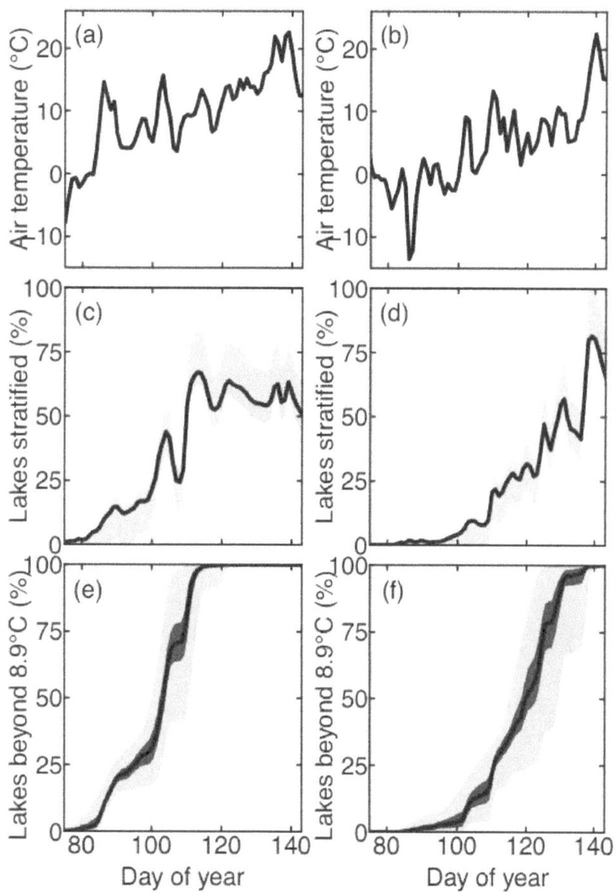

Figure 6: Lake population dynamics for physical metrics during 2 contrasting years. Statewide averaged air temperatures for the warmest (a; 1998) and coldest (b; 1996) years during the simulation period (1979–2011). (c) and (d): Percentage of the 2368 simulated lakes that were stratified during a given day of year in 1998 and 1996, respectively. Extents for the gray bands for (c) and (d) represent the two extreme cases for stratification detection errors: only type II errors occurred, and only type I errors occurred. (e) and (f): percentage of lakes that exceeded the 8.9 °C walleye spawning threshold for a given day of year in 1998 and 1996. Gray and dark gray bands in (e) and (f) represent ranges in the threshold exceedance after randomly applying errors from Fig. 4b for the 95th and 50th percentiles, respectively.

Multi-lake hydrodynamic model output can be used to predict ecosystem-level resilience to climate change, but applications with these data can also focus on responses at the species-level. An example is the examination of the different dates at which lakes exceeded 8.9 °C during these 2 contrasting years. 8.9 °C is an important temperature threshold for Wisconsin lakes, as it is cited as the temperature at which walleye, *Sander vitreus*, begin spawning (Wismer and Christie, 1987). The day of year when all 2368 simulated lakes were projected to pass this threshold was 19 days later in 1996 than in 1998 (Fig. 6e and f). Lake modeling outputs – similar to what we have highlighted here – can help fill observational gaps for physical conditions that may impact the health and function of lake ecosystems.

Individual lakes exhibited different degrees of variability in physical responses to regional climate, and the magnitude of these among-lake differences was dependent on the temperature metric used. Lake-specific variability was quantified by the inter-annual coefficient of variation (CV) for two climatically-relevant physical metrics during 1979–2011: summertime surface temperatures, which are an indicator of lake warming (Schneider and Hook, 2010 and Sharma et al., 2007), and the timing of the onset of stratification, which is important for numerous ecosystem processes and hypothesized to be a sensitive indicator of climate change (Adrian et al., 2009 and Winder and Schindler, 2004). The CVs for average July surface temperatures were similar for most lakes with little spread (Fig. 7a), suggesting that lake-specific properties have relatively little influence on how lakes respond to climate for this metric. Comparatively, the CVs of the timing of stratification differed greatly among lakes (Fig. 7b). These results suggest that climate variability may be differentially mediated for certain lake types, and that determining whether lakes in a region respond coherently or individually depends on the choice of physical metric.

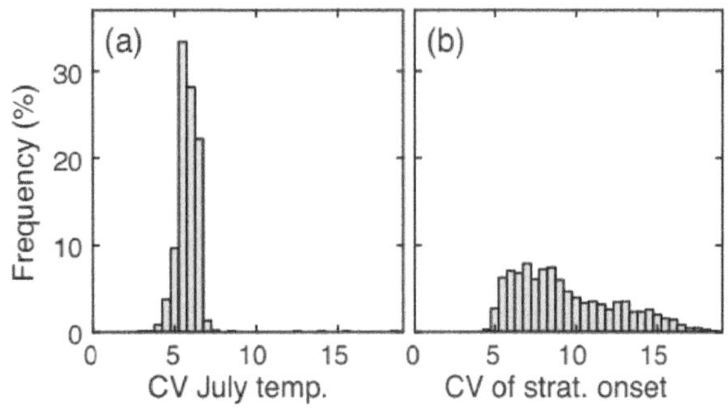

Figure 7: Frequency distributions for lake-specific values of coefficient of variation (CV) for two physical metrics. (a) CV for 33 years of mean July surface water temperatures for each simulated lake (n = 2368). (b) Same as (a) but for the predicted day of year for the onset of stratification (only for lakes that stratified; n = 1213).

MODEL APPLICATION: AMONG-LAKE COHERENCE

These results may help investigators refine expectations for lakes in the future, but the modeling approach shown here also provides an ability to contrast the behavior of individual lakes with regional lake responses. An analysis of coherence was used to determine whether the variability shown in Fig. 7 is structured similarly among lakes or is an individualized expression of climate variability. We followed the methodology ofBenson et al. (2000) for the calculation of among-lake coherence, which evaluated coherence and significance between lakes according to a pairwise Pearson product–moment correlation for the 33 years of simulation data. Big Muskellunge (BM) was chosen to provide a comparison with the other lake responses to climate because of the accuracy of BM model outputs (RMSE = 1.23 °C) and the many long-term lake studies that include this frequently monitored lake (Benson et al., 2000 and Kratz et al., 1998).

Big Muskellunge Lake surface temperatures were highly coherent with the rest of the simulated lakes, and its position among the population of lakes was relatively constant (Fig. 8a and b). In contrast, patterns in stratification were far less synchronous (Fig. 8c and d). These results support earlier work demonstrating greater agreement between temporal patterns for lake surface temperatures than for stratification metrics and the temperatures of deeper waters (Benson et al., 2000, Kratz et al., 1998 and Palmer et al., 2014). Stratification patterns in Big Muskellunge and other frequently monitored lakes may not be representative of stratification dynamics for unobserved lakes that are exposed to the same regional climate.

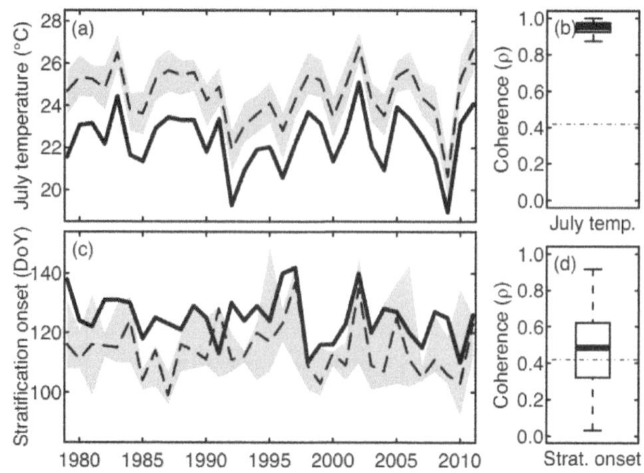

Figure 8: Comparisons between temporal patterns for a single lake (Big Muskellunge Lake, WI USA; BM) and the lake population for two physical metrics derived from model output. (a) Average July surface water temperatures for BM (black line), the median of all lakes (dashed dark gray line; n = 2368), and the interquartile range (gray band). (b) Coherence between BM and all other lakes for July surface water temperatures. Dashed-dot line represents the critical value of for significance in a one-tailed test (p = 0.01). (c) Day of year for the onset of stratification for BM (black line), the median onset date for all lakes that stratified (dashed dark gray line; n = 1213), and the interquartile range of stratified lakes. (d) Same as (b) but for the onset of stratification.

Synchrony in lake surface temperatures and empirical studies that leverage these patterns may have provided limnologists with false confidence in our understanding of lake ecosystems in a changing climate. More ecologically relevant metrics (e.g., stratification and whole-lake temperatures) showed comparatively poor coherence (Fig. 8b versus d), and can express muddled or even divergent responses to climate variability (Benson et al., 2000 and Tanentzap et al., 2008). Among-lake heterogeneity is particularly relevant for understanding how lakes respond to climate change and the types of lakes that will be most affected by climate change. Variability among lakes in response to climate change has important implications for species persistence and dispersal through the provision of microrefugia (Diffenbaugh et al., 2005, Keppel and Wardell-Johnson, 2012 and Suggitt et al., 2011), and is deserving of additional study.

CONCLUSIONS

The General Lake Model (Hipsey et al., 2013) accurately predicted observed surface layer temperatures (RMSE: 1.74 °C) and the presence/absence of stratification (81.1% agreement) when driven and parameterized by large-scale open data. Deeper water temperature simulations were less accurate (RMSE: 3.33 °C for hypolimnion temperatures), but there may be potential for future improvement. Our preliminary sensitivity analysis suggests that improving lake water clarity estimates may yield substantial gains in the quality of deep water simulated temperatures. Additionally, future improvements to the ice algorithms used in GLM are recommended to support continuous simulations for ice-covered lakes. Efforts were made to use the best possible data to parameterize and drive these lake models, but the quality of data varied among lakes and likely contributed to discrepancies in model accuracy. Despite these limitations, both the quality of models and publically available input data are expected to improve in the future.

Recent warming trends in surface temperatures are cause for alarm for many biota or ecosystem processes (Paerl and Huisman, 2008, Schneider and Hook, 2010 and Sharma et al., 2007), but investigators should take caution in extending near-surface temperature patterns into predictions for climate responses of lake hypolimnion temperatures, stratification strength and timing, and whole lake temperatures. Our findings highlighted a contrast between strongly coherent surface temperatures and weakly coherent stratification patterns, suggesting that unified responses of lakes to regional climate may be unlikely for some ecologically relevant metrics. A mechanistic modeling approach, similar to what we have shown here (additionally, see Fang et al., 2012, Hadley et al., 2013 and Weinberger and Vetter, 2012), can be used to refine our expectations for lake ecosystems in a changing climate.

Modeled lake temperature outputs generated at broad spatial scales for a large number of lakes could be useful for a wide range of practical applications. For example, hind-casted lake temperatures can be used to understand mechanisms behind observed changes in fish communities in these Wisconsin lakes, in addition to being used to explain variability in empirical data across the national landscape. Modeling individual lakes within the regional context allows investigators to capture spatial and temporal heterogeneity in individual lake responses. This variability is likely critical to understanding ecosystem responses to climate change that would be masked if "average" values that ignore spatial heterogeneity were used (Ackerly et al., 2010 and Luoto and Heikkinen, 2008).

Similar to alternative 1D models, the utility of the hydrodynamic lake model used here can be extended by coupling additional modeling elements or input data sources. While beyond the scope of this study, combining these simulations with water quality models (Hamilton and Schladow, 1997) or terrestrial models that can dynamically model watershed processes (Kucharik et al., 2000) may improve our understanding of lake ecosystem function. Likewise, including measurements of stream flow (waterdata.usgs.gov) or hydrological modeling input/output could improve simulation quality for lakes with shorter residence times and potentially lead to improvements in stream temperature modeling efforts.

This modeling effort was developed with a foundation of internationally recognized standards for web processing and data transfer, providing a basic methodology that can easily be transferred to a range of different ecological research questions. Whether investigators could benefit from the exploration of thermal habitat change in the millions of lakes in the UGSG›s national hydrography dataset (nhd.usgs.gov) or translate these methods into site-based terrestrial studies, the framework described here could be applied across a wide spectrum of modeling efforts. Additionally, statistically or dynamically downscaled climate projections could be easily exchanged for the NLDAS-2 forcing data and used to predict future scenarios for these ecosystems. These multi-system simulations represent a significant technical advance in limnology and ecological modeling. This framework will provide a useful tool in the interpretation of individual and population-level responses of ecosystems to climate.

ACKNOWLEDGEMENTS

This study was funded by U.S. Geological Survey NCCWSC grant 10909172, NSF grant DEB-0941510 (Global Lake Ecological Observatory Network), the Wisconsin Department of Natural Resources Federal Aid in Sport Fish Restoration (Project F-95-P), the U.S. Geological Survey Center for Integrated Data Analytics, and NSF grants DEB-0822700 (North Temperate Lakes Long-Term Ecological Research), and IIS-1344272 (INSPIRE). Mathew Hipsey and Casper Boon (the Aquatic Ecosystem Dynamics group at The University of Western Australia) shared modeling and development expertise for GLM. We thank Jereme Gaeta, Steven Greb, Matthew Diebel, Alex Latzka, Emily Stanley, Kevin Rose, Dale Robertson, Emily Read, David Blodgett, Ivan Suftin, Jordan Walker, Meredith Pavlick Warren, and Jake Vander Zanden for their contributions to this manuscript. Tim Kratz and Stephen Carpenter helped improve this article with thoughtful reviews. Two anonymous reviewers contributed substantially to the improvement of our original manuscript. We also thank the Ecological Modeling editorial staff.

REFERENCES

1. Ackerly et al., 2010 D.D. Ackerly, S.R. Loarie, W.K. Cornwell, S.B. Weiss, H. Hamilton, R. Branciforte, N.J.B. Kraft The geography of climate change: implications for conservation biogeography Diversity Distribut., 16 (2010), pp. 476–487
2. Adrian et al., 2009 R. Adrian, C.M. O'Reilly, H. Zagarese, S.B. Baines, D.O. Hessen, W. Keller, D.M. Livingstone, R. Sommaruga, D. Straile, E. Van Donk Lakes as sentinels of climate change Limnol. Oceanogr., 54 (2009), p. 2283
3. Benson and Magnuson, 2000 Benson, B., Magnuson, J., 2000. Global lake and river ice phenology database. Boulder, CO: National Snow and Ice Data Center/World Data Center for Glaciology, Boulder, Digital media. Benson et al., 2000
4. B.J. Benson, J.D. Lenters, J.J. Magnuson, M. Stubbs, T.K. Kratz, P.J. Dillon, R.E. Hecky, R.C. Lathrop Regional coherence of climatic and lake thermal variables of four lake districts in the Upper Great Lakes Region of North America Freshwater Biol., 43 (2000), pp. 517–527
5. Blenckner et al., 2007 T. Blenckner, R. Adrian, D.M. Livingstone, E. Jennings, G.A. Weyhenmeyer, D. George, T. Jankowski, M. Järvinen, C.N. Aonghusa, T. Nõges Large-scale climatic signatures in lakes across Europe: a meta-analysis Global Change Biol., 13 (2007), pp. 1314–1326
6. Blodgett, 2011 D.L. Blodgett Description and Testing of the Geo Data Portal: A Data Integration Framework

and Web Processing Services for Environmental Science Collaboration US Department of the Interior, US Geological Survey (2011)

7. Cahill et al., 2005 K.L. Cahill, J.M. Gunn, M.N. Futter Modelling ice cover, timing of spring stratification, and end-of-season mixing depth in small Precambrian Shield lakes Can. J. Fish. Aquat. Sci., 62 (2005), pp. 2134–2142
8. Cardoso et al., 2014 S.J. Cardoso, A. Enrich-Prast, M.L. Pace, F. Roland Do models of organic carbon mineralization extrapolate to warmer tropical sediments? Limnol. Oceanogr., 59 (2014), pp. 48–54
9. Carpenter et al., 1992 S.R. Carpenter, S.G. Fisher, N.B. Grimm, J.F. Kitchell Global change and freshwater ecosystems Annu. Rev. Ecol. Syst. (1992), pp. 119–139
10. Diffenbaugh et al., 2005 N.S. Diffenbaugh, J.S. Pal, R.J. Trapp, F. Giorgi Fine-scale processes regulate the response of extreme events to global climate change Proc. Natl. Acad. Sci. U.S.A., 102 (2005), pp. 15774–15778
11. Fang and Stefan, 2009 X. Fang, H.G. Stefan Simulations of climate effects on water temperature, dissolved oxygen, and ice and snow covers in lakes of the contiguous United States under past and future climate scenarios Limnol. Oceanogr., 54 (2009), pp. 2359–2370
12. Fang et al., 2012 X. Fang, S.R. Alam, H.G. Stefan, L. Jiang, P.C. Jacobson, D.L. Pereira Simulations of water quality and oxythermal cisco habitat in Minnesota lakes under past and future climate scenarios Water Qual. Res. J. Can., 47 (2012), pp. 375–388
13. Hadley et al., 2013 Hadley, K.R., Paterson, A.M., Stainsby, E.A., Michelutti, N., Yao, H., Rusak, J.A., Ingram, R., McConnell, C., Smol, J.P., 2013. Climate warming alters thermal stability but not stratification phenology in a small north-temperate lake. Hydrol. Process. Hamilton and Schladow, 1997 D.P. Hamilton, S.G. Schladow Prediction of water quality in lakes and reservoirs, Part I. Model description Ecol. Model., 96 (1997), pp. 91–110
14. Hampton et al., 2013 S.E. Hampton, C.A. Strasser, J.J. Tewksbury, W.K. Gram, A.E. Budden, A.L. Batcheller, C.S. Duke, J.H. Porter Big data and the future of ecology Front. Ecol. Environ., 11 (2013), pp. 156–162
15. Hanson et al., 2011
16. P.C. Hanson, D.P. Hamilton, E.H. Stanley, N. Preston, O.C. Langman, E.L. Kara Fate of allochthonous dissolved organic carbon in lakes: a quantitative approach PLoS One, 6 (2011), p. e21884
17. Heffernan et al., 2014 J.B. Heffernan, P.A. Soranno, M.J. Angilletta Jr., L.B. Buckley, D.S. Gruner, T.H. Keitt, J.R. Kellner, J.S. Kominoski, A.V. Rocha, J. Xiao Macrosystems ecology: understanding ecological patterns and processes at continental scales Front. Ecol. Environ., 12 (2014), pp. 5–14
18. Hipsey et al., 2013 M.R. Hipsey, L.C. Bruce, D.P. Hamilton GLM General Lake Model. Model Overview and User Information The University of Western Australia Technical Manual, Perth, Australia (2013)
19. Hondzo and Stefan, 1993 M. Hondzo, H.G. Stefan Lake water temperature simulation model J. Hydraulic Eng., 119 (1993), pp. 1251–1273
20. Hostetler and Small, 1999 S.W. Hostetler, E.E. Small Response of north American freshwater lakes to simulated future climates1 JAWRA J. Am. Water Resour. Assoc., 35 (1999), pp. 1625–1637
21. Hostetler et al., 2011 Hostetler, S.W., Alder, J.R., Allan, A.M., 2011. Dynamically downscaled climate simulations over North America: Methods, evaluation and supporting documentation for users. U.S. Geological Survey, Reston, Virginia, OFR 2011–1238, 64 pp., 2011.
22. Keller et al., 2008 M. Keller, D.S. Schimel, W.W. Hargrove, F.M. Hoffman A continental strategy for the National Ecological Observatory Network Front. Ecol. Environ., 6 (2008), pp. 282–284
23. Kent and Wong, 1982 C. Kent, J. Wong An index of littoral zone complexity and its measurement Can. J. Fish. Aquat. Sci., 39 (1982), pp. 847–853
24. Keppel and Wardell-Johnson, 2012 G. Keppel, G.W. Wardell-Johnson Refugia: keys to climate change management Global Change Biol., 18 (2012), pp. 2389–2391
25. Killworth and Carmack, 1979 P.D. Killworth, E.C. Carmack A filling-box model of river-dominated lakes Limnol. Oceanogr., 24 (1979), pp. 201–217
26. Kratz et al., 1998 T.K. Kratz, P.A. Soranno, S.B. Baines, B.J. Benson, J.J. Magnuson, T.M. Frost, R.C. Lathrop Interannual synchronous dynamics in north temperate lakes in Wisconsin, USA Manage. Lakes Reservoirs During Global Climate Change, 42 (1998), pp. 273–287
27. Kucharik et al., 2000 C.J. Kucharik, J.A. Foley, C. Delire, V.A. Fisher, M.T. Coe, J.D. Lenters, C. Young-Molling, N. Ramankutty, J.M. Norman, S.T. Gower Testing the performance of a dynamic global ecosystem model: water balance, carbon balance, and vegetation structure Global Biogeochem. Cycles, 14 (2000), pp. 795–825
28. Leonard and Duffy, 2013 L. Leonard, C.J. Duffy Essential terrestrial variable data workflows for distributed water resources modeling Environ. Model. Software, 50 (2013), pp. 85–96
29. Livingstone, 2003 D.M. Livingstone Impact of secular climate change on the thermal structure of a large temperate central European lake Climatic Change, 57 (2003), pp. 205–225
30. Livingstone, 2008 D.M. Livingstone A change of climate provokes a change of paradigm: taking leave of two tacit assumptions about physical lake forcing Int. Rev. Hydrobiol., 93 (2008), pp. 404–414
31. Livingstone and Dokulil, 2001 D.M. Livingstone, M.T. Dokulil Eighty years of spatially coherent Austrian lake surface temperatures and their relationship to regional air temperature and the North Atlantic Oscillation Limnol. Oceanogr., 46 (2001), pp. 1220–1227
32. Luoto and Heikkinen, 2008 M. Luoto, R.K. Heikkinen Disregarding topographical heterogeneity biases species turnover assessments based on bioclimatic models Global Change Biol., 14 (2008), pp. 483–494

33. Magnuson et al., 1979 J.J. Magnuson, L.B. Crowder, P.A. Medvick Temperature as an ecological resource Am. Zool., 19 (1979), pp. 331–343

34. Magnuson et al., 2000 J.J. Magnuson, D.M. Robertson, B.J. Benson, R.H. Wynne, D.M. Livingstone, T. Arai, R.A. Assel, R.G. Barry, V. Card, E. Kuusisto Historical trends in lake and river ice cover in the Northern Hemisphere Science, 289 (2000), pp. 1743–1746

35. Markfort et al., 2010 C.D. Markfort, A.L. Perez, J.W. Thill, D.A. Jaster, F. Porté-Agel, H.G. Stefan Wind sheltering of a lake by a tree canopy or bluff topography Water Resour. Res., 46 (2010)

36. Mesinger et al., 2006 F. Mesinger, G. DiMego, E. Kalnay, K. Mitchell, P.C. Shafran, W. Ebisuzaki, D. Jović, J. Woollen, E. Rogers, E.H. Berbery North American regional reanalysis Bull. Am. Meteorol. Soc., 87 (2006)

37. Mitchell et al., 2004 K.E. Mitchell, D. Lohmann, P.R. Houser, E.F. Wood, J.C. Schaake, A. Robock, B.A. Cosgrove, J. Sheffield, Q. Duan, L. Luo The multi-institution North American Land Data Assimilation System (NLDAS): utilizing multiple GCIP products and partners in a continental distributed hydrological modeling system J. Geophys. Res.: Atmos. (1984–2012) (2004), p. 109

38. Mooij et al., 2010 W.M. Mooij, D. Trolle, E. Jeppesen, G. Arhonditsis, P.V. Belolipetsky, D.B. Chitamwebwa, A.G. Degermendzhy, D.L. DeAngelis, L.N.D.S. Domis, A.S. Downing Challenges and opportunities for integrating lake ecosystem modelling approaches Aquat. Ecol., 44 (2010), pp. 633–667

39. O'Reilly et al., 2003 C.M. O'Reilly, S.R. Alin, P.-D. Plisnier, A.S. Cohen, B.A. McKee Climate change decreases aquatic ecosystem productivity of Lake Tanganyika, Africa Nature, 424 (2003), pp. 766–768

40. Paerl and Huisman, 2008 H.W. Paerl, J. Huisman Blooms like it hot Science, 320 (2008), p. 57 Palmer et al., 2014 M.E. Palmer, N.D. Yan, K.M. Somers Climate change drives coherent trends in physics and oxygen content in North American lakes Climatic Change (2014), pp. 1–15

41. Poole and Atkins, 1929 Poole, H.H., Atkins, W.R.G., 1929. Photo-electric measurements of submarine illumination throughout the year. J. Mar. Biol. Assoc. U.K. (New Series) 16, 297–324.

42. R Development Core Team, 2013 R Development Core Team, 2013. R: A Language and Environment for Statistical Computing. R Foundation for Statistical Computing.

43. Read and Rose, 2013 J.S. Read, K.C. Rose Physical responses of small temperate lakes to variation in dissolved organic carbon concentrations Limnol. Oceanogr., 58 (2013), pp. 921–931

44. Read et al., 2011 J.S. Read, D.P. Hamilton, I.D. Jones, K. Muraoka, L.A. Winslow, R. Kroiss, C.H. Wu, E. Gaiser Derivation of lake mixing and stratification indices from high-resolution lake buoy data Environ. Model. Software, 26 (2011), pp. 1325–1336

45. Roach et al., 2013 J.K. Roach, B. Griffith, D. Verbyla Landscape influences on climate-related lake shrinkage at high latitudes Global Change Biol., 19 (2013), pp. 2276–2284

46. Schindler et al., 1990 D.W. Schindler, K.G. Beaty, E.J. Fee, D.R. Cruikshank, E.R. DeBruyn, D.L. Findlay, G.A. Linsey, J.A. Shearer, M.P. Stainton, M.A. Turner Effects of climatic warming on lakes of the central boreal forest Science, 250 (1990), pp. 967–970

47. Schneider and Hook, 2010 Schneider, P., Hook, S.J., 2010. Space observations of inland water bodies show rapid surface warming since 1985. Geophys. Res. Lett. 37. Sharma et al., 2007

48. S. Sharma, D.A. Jackson, C.K. Minns, B.J. Shuter Will northern fish populations be in hot water because of climate change? Global Change Biol., 13 (2007), pp. 2052–2064

49. Shuter et al., 2013 B.J. Shuter, C.K. Minns, S.R. Fung Empirical models for forecasting changes in the phenology of ice cover for Canadian lakes Can. J. Fish. Aquat. Sci., 70 (2013), pp. 982–991

50. Smol et al., 2005 J.P. Smol, A.P. Wolfe, H.J.B. Birks, M.S. Douglas, V.J. Jones, A. Korhola, R. Pienitz, K. Rühland, S. Sorvari, D. Antoniades Climate-driven regime shifts in the biological communities of arctic lakes Proc. Natl. Acad. Sci. U.S.A., 102 (2005), pp. 4397–4402

51. Soranno et al., 1999 P.A. Soranno, K.E. Webster, J.L. Riera, T.K. Kratz, J.S. Baron, P.A. Bukaveckas, G.W. Kling, D.S. White, N. Caine, R.C. Lathrop Spatial variation among lakes within landscapes: ecological organization along lake chains, Ecosystems, 2 (1999), pp. 395–410

52. Soranno et al., 2010 P.A. Soranno, K.S. Cheruvelil, K.E. Webster, M.T. Bremigan, T. Wagner, C.A. Stow Using landscape limnology to classify freshwater ecosystems for multi-ecosystem management and conservation BioScience, 60 (2010), pp. 440–454

53. Stefan et al., 1996 H.G. Stefan, M. Hondzo, X. Fang, J.G. Eaton, J.H. McCormick Simulated long-term temperature and dissolved oxygen characteristics of lakes in the north-central United States and associated fish habitat limits Limnol. Oceanogr., 41 (1996), pp. 1124–1135

54. Stefan et al., 2001 H.G. Stefan, X. Fang, J.G. Eaton Simulated fish habitat changes in North American lakes in response to projected climate warming Trans. Am. Fish. Soc., 130 (2001), pp. 459–477

55. Suggitt et al., 2011 A.J. Suggitt, P.K. Gillingham, J.K. Hill, B. Huntley, W.E. Kunin, D.B. Roy, C.D. Thomas Habitat microclimates drive fine-scale variation in extreme temperatures Oikos, 120 (2011), pp. 1–8

56. Tanentzap et al., 2008 A.J. Tanentzap, N.D. Yan, B. Keller, R. Girard, J. Heneberry, J.M. Gunn, D.P. Hamilton, P.A. Taylor Cooling lakes while the world warms: effects of forest regrowth and increased dissolved organic matter on the thermal regime of a temperate, urban lake Limnol. Oceanogr., 53 (2008), p. 404

57. Thain et al., 2005 D. Thain, T. Tannenbaum, M. Livny Distributed computing in practice: the Condor experience Concurrency Comput.: Pract. Exp., 17 (2005), pp. 323–356

58. Torbick et al., 2013 N. Torbick, S. Hession, S. Hagen, N. Wiangwang, B. Becker, J. Qi Mapping inland lake water quality across the Lower Peninsula of Michigan using Landsat TM imagery Int. J. Remote Sens., 34 (2013), pp. 7607–7624

59. Weinberger and Vetter, 2012 S. Weinberger, M. Vetter Using the hydrodynamic model DYRESM based on results of a regional climate model to estimate water temperature changes at Lake Ammersee Ecol. Model., 244 (2012), pp. 38–48

60. Wetzel and Likens, 2000 Wetzel, R.G., Likens, G.E., 2000. The heat budget of lakes, in: Limnological Analyses. Springer, pp. 45–56.

61. Weyhenmeyer et al., 1999 G.A. Weyhenmeyer, T. Blenckner, K. Petterson Changes of the plankton spring outburst related to the North Atlantic Oscillation Limnol. Oceanogr., 44 (1999), pp. 1788–1792

62. Wilby et al., 2000 R.L. Wilby, L.E. Hay, W.J. Gutowski, R.W. Arritt, E.S. Takle, Z. Pan, G.H. Leavesley, M.P. Clark Hydrological responses to dynamically and statistically downscaled climate model output Geophys. Res. Lett., 27 (2000), pp. 1199–1202

63. Winder and Schindler, 2004 M. Winder, D.E. Schindler Climatic effects on the phenology of lake processes Global Change Biol., 10 (2004), pp. 1844–1856

64. Wismer and Christie, 1987 Wismer, D.A., Christie, A.E., 1987. Temperature relationships of Great Lakes fishes. Spec. publ./Great Lakes fishery commiss.

65. Woolway et al., 2014 R.I. Woolway, S.C. Maberly, I.D. Jones, H. Feuchtmayr A novel method for estimating the onset of thermal stratification in lakes from surface water measurements Water Resour. Res. (2014)

CHAPTER 5

ASSESSMENT OF COASTAL HAZARDS DUE TO FUTURE CHANGES OF TROPICAL CYCLONES

Nobuhito Mori, Tetsuya Takemi

Keywords: Climate change, Tropical cyclones, Downscaling, Impact assessment, Coastal hazards

INTRODUCTION

Tropical cyclones (TCs) are one of the major meteorological hazards as a cause of flooding, landslides, damaging winds, high waves, and storm surges. Depending on their size, intensity, track, and translation speed, TCs affect wide regions in the tropical, the sub-tropical, and the mid-latitude regions. After reaching a certain intensity, TC is specifically called a typhoon in the western North Pacific, a hurricane in the northeastern Pacific and in the North Atlantic, and a cyclone in the north Indian Ocean. Such intense TCs have spawned devastating damages from the past up to the present; recent examples include Hurricane Katrina in 2005, Cyclone Nargis in 2008, Typhoon Morakot in 2009, Hurricane Sandy in 2012, Typhoon Haiyan in 2013, and Tropical Cyclone Pam in 2015. For the purpose of mitigating and preventing disasters due to such intense storms, timely and proper warning based on accurate numerical weather prediction is prerequisite. Although numerical weather prediction of such destructive TCs is still a scientific challenge, some studies have recently shown that extended-range predictions of TCs beyond two weeks can be conducted under certain synoptic-scale conditions (Nakano et al., 2015 and Xiang et al., 2015). Given the progress in quantitative weather forecasting of the track and intensity of TCs, the accuracy of warning for anticipated disasters will be improved.

From the perspective of mitigating and preventing disasters, long-term projections of TCs under climate change are also important. Global warming in the future is expected to affect the characteristics of TCs, in particular their frequency, intensity, and track. Such future changes of TCs would give significant impacts on natural disasters. Therefore, in order to mitigate and prevent TC-related disasters in the long term we need to assess the characteristics of TCs under global warming and the resulting impacts of future TCs on natural disasters.

Among the intense TCs in recent years, Cyclone Nargis (2008), Hurricane Sandy (2012), Typhoon Haiyan (2013), and Tropical Cyclone Pam (2015) caused significant coastal disasters. Because most of the major urban areas in the world are located in the coastal regions, intense TCs that make landfall or pass near the coastal regions affect a large number of people and social infrastructures. With the continuous growth of populations and social infrastructures in coastal urban regions in the future, intense TCs will increasingly be a threat to the human society especially in coastal regions. The impact assessment of coastal hazards due to anticipated changes of intense TCs under future climate change is of great concern. Because of the rapidly growing population in Asia (United Nations, 2012), TCs in the western North Pacific and in the Indian Ocean more significantly affect the economic and industrial activities as well as human lives than those in other ocean basins. Here we focus on TCs in the western North Pacific, i.e., typhoons.

The direct consequence of sea-level rise is inundation of low-lying coastal areas, which is a long-term concern and a gradual process of climate change effects. The Intergovernmental Panel on Climate Change (IPCC) the Fifth Assessment Report (AR5) (2013) states that climate change exacerbates the vulnerability of coastal regions to other extreme and impulsive physical processes, such as storm surges and storm waves. Future changes in storm surge and ocean wave climates are a dynamic side issue of how climate change influences coastal regions (e.g. Mori et al., 2010; Hemer et al., 2013). If extreme weather events become stronger in the future, it is necessary to seriously consider the

Citation: Nobuhito Mori, Tetsuya Takemi, Assessment of coastal hazards due to future changes of tropical cyclones, http://dx.doi.org/10.1016/j.wace.2015.09.002.

effects of these dynamic phenomena to prevent and reduce the impact of coastal disasters. The wave hindcasts in the Atlantic Ocean show more significant wave height increase in the region off the Canadian coast and the northwest of Ireland but less significant change in the North Sea and in the region off the Scandinavian coast. In addition, the long-term changes to storm surge and storm waves are important considerations for structures near the coastal zone. For example, a coastal breakwater is designed with a maximum storm surge level and pressure from a maximum wave condition, a so-called design wave, for a predetermined design lifetime. Generally, coastal areas that will be most affected by increases in storm surge are the heavily populated mega-delta regions of East Asia, South-East Asia, South Asia and the Gulf of Mexico.

Tropical cyclone (TC) activity has major impacts on the delta regions located in the subtropical to mid-latitude zones. In these regions, 1-hPa decrease in atmospheric pressure corresponds to 1-m change in the barotropic effect to storm surge, and wind-induced surge is proportional to the square of wind speed (e.g. Nielsen, 2009). Also, wind induced surge is related to the moving speed and duration of a storm, and to the bathymetry of the bay. Therefore, storm surge is not only sensitive to TC intensity, but also to the track of the cyclone. Moreover, a certain number of TC tracks is necessary to make an assessment of storm surge risk at a particular location.

Future storm surge projections have been conducted for particular places such as Europe and East Asia. For example, Lowe and Gregory (2005) and Woth (2005) have analyzed future storm surges around the UK and in the North Sea regions, and they showed significant increases in storm surge elevations for the UK coast and the continental North Sea coast resulting from increases in the intensities of mid-latitude storms. Yasuda et al. (2014) and Zhao et al. (2014) show the impact of future storm surge changes in East Asia. For the subtropical to mid-latitude zones, however, future storm surge projections are limited due to the lack of knowledge of how TCs in these regions will change. furthermore, in these latitudes the majority of coastal regions measuring a few hundred kilometers in area have limited historical records of TC landfall. Therefore, assessing the TC risk in a particular region is difficult even considering the present climate alone. One way to overcome the limited amount of actual events is to use statistical modeling of TC events and another way is pseudo global warming experiments by a regional climate model.

Combining dynamic and statistical climate models gives a useful tool to assess the risks of future TCs and related coastal hazard. This study describes a brief review of future projections of TC characteristics due to global climate change. Then, we provide a review and ongoing research on the impact assessment of tropical cyclones for both dynamic downscaling and statistical models. Finally, we describe how TC characteristics change in the future climate changes coastal hazard based on the different approaches. A schematic flow chart of projection and impact assessment of climate change is shown inFig. 1 (the numbers in Fig. 1 indicate the sections subsections in the manuscript).

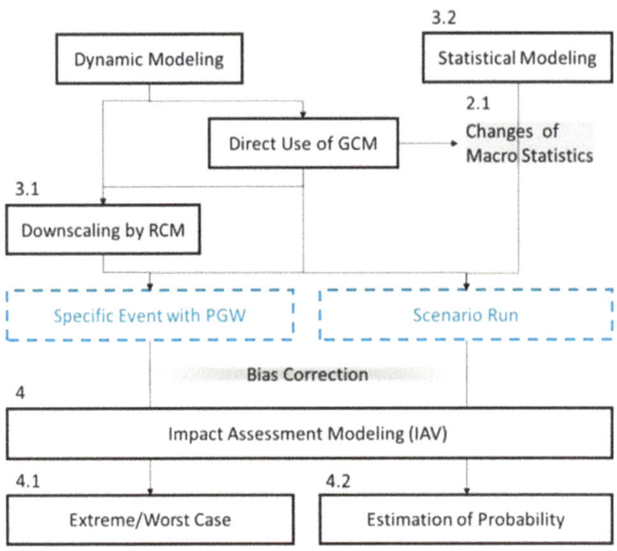

Figure 1: Schematic flow chart of projection and impact assessment of climate change. Numbers in figure indicate the sections.

PAST AND FUTURE CHANGES OF TROPICAL CYCLONES

In this section, we briefly overview the current understanding on the changes in the characteristics of TCs in the past and in the future. The current status of the understanding which has gained consensus among the broad scientific community is found in IPCC AR5 (2013). Based on the results included in IPCC (2013) and from other recent studies, we describe the characteristics on TCs in the western North Pacific from an impact assessment perspective.

Past Changes of Tropical Cyclones

First, we give an overview on the changes of TCs in the past. For the assessment of the long-term changes in the TC activities, quantitative information on the intensity and track of TCs is necessary. Such reliable data have been available since the beginning of the satellite era starting from around the 1970s. Weather surveillance geostationary satellites give reliable data on TCs over each ocean basin, although the starting point of such observations depends on each ocean basin. For example, the observations by the Japanese Geostationary Meteorological Satellite "Himawari" over the western half of the Pacific started in 1977. Before the satellite era, aircraft observations played a role in quantifying the activity of TCs. However, without aircraft or satellite observations it is quite difficult to gain quantitative information on TCs. Therefore, data availability imposes a strong limitation on the assessment of the long-term changes in past TC activity. Although efforts have been made to archive long-term data on the TC activity back to the middle of the 19th century (International Best Track Archive for Climate Stewardship (IBTrACS), which is maintained by The National Oceanic and Atmospheric Administration (NOAA)), data that have uniformity in time and space are not available.

Because of the limited data availability, the confidence level in assessing the centennial-scale, long-term changes in TC activity in AR5 is overall low. An exception is for the activity of TCs in the North Atlantic; it is assessed in AR5 as virtually certain (i.e., the probability of 99% or greater) that the frequency of strong hurricanes in the North Atlantic has increased since the 1970s. However, firm conclusions on the changes in the frequency and intensity of TCs in the western North Pacific have not been obtained. InKunkel et al. (2013), it is concluded that robust detection of trends in the TC activity in the Atlantic and the western North Pacific is significantly constrained by data heterogeneity and deficient quantification of internal variability, and that attribution of the past changes is challenged by a lack of consensus on the physical linkages between climate forcing and TC activity.

As far as the TCs in the western North Pacific are concerned, Fig. 2 indicates the temporal variation of the number of named TCs (which attained typhoon intensity) in the western North Pacific, based on the best-track data archived at the Japan Meteorological Agency (JMA)/ Regional Specialized Meteorological Center (RSMC) Tokyo. There is no discernible trend in the frequency of TCs that attained a typhoon intensity. If we focus only on the strongest typhoons, the trend in the frequency of such strongest storms is also not clear. For example, the number of the strongest typhoons which attained a lifetime maximum intensity of 920 hPa or lower, corresponding roughly to the Category-5 storms by the Saffir–Simpson scale (Knaff and Zehr 2007), does not indicate a long-term trend for the past 60 years (Fig. 3). From both Fig. 2 and Fig. 3, it is seen that there is a large interannual variability of the number of typhoons.

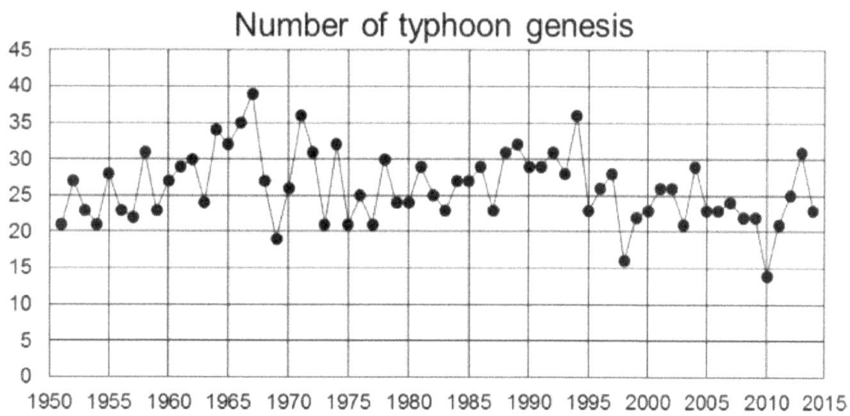

Figure 2: The historical variation of the annual number of typhoons in the western North Pacific from 1951 to 2014.

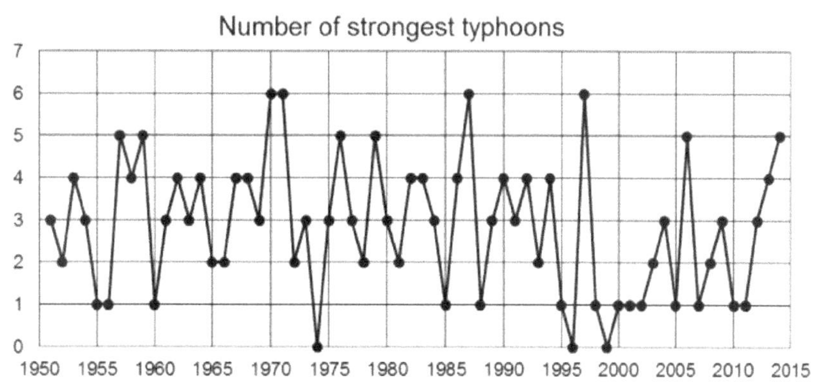

Figure 3: The same as Fig. 2, except for typhoon with the lifetime maximum intensity of 920 hPa or lower.

In the western North Pacific regions (and in the Indian Ocean), the influences of the coupled atmospheric and oceanic system play an important role in controlling the interannual variability of weather and climate phenomena. The activity of TCs in the western North Pacific as well as in the Indian Ocean is related to the interannual variability such as El Niño/La Niña and Southern Oscillation (ENSO) (Chia and Ropelewski, 2002, Wang and Chan, 2002, Camargo et al., 2007 and Li and Zhou, 2012) and Indian Ocean Dipole Mode (Yuan and Cao, 2013 and Mahala et al., 2015) and to the intraseasonal variability such as the Madden–Julian Oscillation (Liebmann et al., 1994,Frank and Roundy, 2006, Camargo et al., 2009, Huang et al., 2011, Tsuboi and Takemi, 2014 and Tsuboi et al., 2015) and other intraseasonal phenomena (Yoshida et al.

2014). Furthermore, it is suggested that intraseasonal variation would provide conditions for the increase in the number of land-falling typhoons over Japan (Nakazawa, 2006). Because the long-term trend in the TC activity over the western North Pacific is affected not only by climate change but also by interannual and intraseasonal variability, long-term data with high quality and reliability are indispensable to detect the long-term TC trend in the past.

Future changes of Tropical Cyclones

Compared with the past trend of the TC activity, the statement of AR5 on future changes in the TC activity is more advanced. It is concluded in AR5 that the increase in the activity of intense TCs in the western North Pacific and the North Atlantic has a likelihood of 50% probability or greater, although the confidence level depends on the ocean basins.

Future projections on the changes in TC activity are conducted by performing numerical experiments with general circulation models (GCMs) or regional climate models (RCMs) nested in GCMs. ITo represent TCs in climate models, proper settings on computational domain, spatial resolution, and physics parameterization are required. Because TCs travels over a vast region from the tropical oceans to mid-latitude areas under the influence of large-scale atmospheric circulations, higher-resolution GCMs are more appropriate than RCMs for future projections of TC activity. With the development of a high-resolution atmospheric GCM (Mizuta et al., 2006 and Mizuta et al., 2012), the assessment of future changes of TC activity has been significantly advanced since an earlier study by Oouchi et al. (2006).

Murakami and Sugi (2010) investigated the effects of model resolution on projected climatological features of TCs by changing the resolution from 180 km down to 20 km, and showed that a resolution of 60 km or finer is required for projecting future changes in the frequency of intense TCs. The horizontal resolution of 20 km of the Meteorological Research Institute (MRI)-AGCM captures not only the TC activity (Murakami et al., 2012a) but also climatic characteristics in general (Mizuta et al., 2012). In addition to the resolution issue, prescribing sea surface temperature also affects the representations of projected future climate (Mizuta et al., 2014). In the context of TCs, Murakami et al. (2012b) investigated uncertainties for projected future changes in TC activity by performing multi-physics and multi-SST ensemble experiments, and suggested that the differences in SST spatial patterns prescribed in GCMs would cause substantial variations and uncertainties in projected future changes of TG frequency at ocean-basin scales. The sensitivity of the projected TC activity to different climate models is summarized in Knutson et al. (2010). Murakami et al. (2014) further investigated influences of model biases on projected future changes in the frequency of TC occurrence using MRI-AGCM under various emission scenarios including the Special Report on Emissions Scenarios (SRES) A1B emission scenario and the representative concentration pathway (RCP) 4.5 and 8.5 scenarios and showed that model biases introduce an uncertainty of approximately 10% in the total future changes. It was also indicated that the influence of biases depends on the model physics rather than on model resolutions or emission scenarios. For TCs in the western North Pacific, the biases result of TC frequency occurrence in the underestimate in the western North Pacific.

Owing to the remaining uncertainties in model simulations, the confidence level in the future projections of TC activity is not high. Despite these deficiencies, the global frequency of TC occurrence is projected to decrease (Oouchi et al., 2006 and Sugi et al., 2009), while the activity of intense TCs is projected to increase in some basins. Based on the study by Knutson et al. (2010) and previous studies, AR5 concludes that under the SRES A1B emission scenario the global frequency of TCs will either decrease or remain unchanged while the mean intensity of TCs will increase. In contrast to such changes in the global-scale characteristics of TCs under global warming, regional-scale assessments of TC activity have a lower confidence level.

Using the climate simulations with the 20-km mesh MRI-AGCM under the SRES A1B scenario, Murakami et al. (2011) investigated the changes in TC activity in a projected future at the end of the 21st century. They indicated that there is a significant reduction by about 23% in the numbers of TC genesis and occurrence and that fewer TCs will form in the western portion of the western North Pacific while more TCs will form in the southeastern part of the western North Pacific. Owing to the changes in the genesis location of TCs and in addition to changes in large-scale steering flows for the TC movement, an eastward shift in the prevailing northward-recurving TC tracks during the peak TC season, i.e., from July to October, was found. With the recurving TC tracks shifted eastward, there is a significant reduction by 44% in the frequency of TCs that approach coastal regions of Southeast Asia. However, it was indicated that future changes in the number of TC landfalls were not statistically significant. Murakami et al. (2011) noted that the changes in large-scale circulation would critically depend on the spatial pattern of future sea surface temperature.

In the study of Murakami et al. (2011), the analyses were made for the outputs from a single simulation case. Therefore, Murakami et al. (2012b) further investigated the effects of SST spatial patterns as well as cumulus parameterizations on the future climate simulations with the use of AGCM at the 60-km resolution. It was shown that all the ensemble experiments consistently project reductions in the frequency of TC genesis and occurrence in the western North Pacific. The genesis and occurrence of TCs were shown to increase over the central Pacific; this was also found in Li et al. (2010). By taking into account the result indicated in Murakami et al. (2011), the decrease of TCs in the western Pacific and the increase in the central Pacific seem to correspond to the eastward shift of recurving TCs. Murakami et al. (2012b) indicated that the spatial contrast in projected future changes in the TC frequency between the western North Pacific and the central Pacific is primarily due to the global warming.

By using an updated version of MRI-AGCM (Mizuta et al., 2012), Murakami et al. (2012a)demonstrated that the new MRI-AGCM successfully simulates a more realistic present-day global climate of TCs and that it is able to reproduce extremely intense TCs corresponding to Categories 4 and 5. The projected frequency of TC occurrence was shown to consistently decrease in the western part of the western North Pacific while increase in the central Pacific. Furthermore, it was indicated that the global frequency of intense TCs is projected to increase. From a viewpoint of impact assessment, it is important to note that in Murakami et al. (2012a) all coastal regions were indicated to suffer from an increasing intensity by 1–7% in terms of maximum surface winds due to TCs and it was suggested that damages due to TCs may become more severe under global warming, particularly in the northern part of the western North Pacific, despite the projected decrease in the frequency of TCs that approach coastal regions.

Future projections of typhoons were conducted also with a coupled atmosphere–ocean model. Mori et al. (2013) performed ensemble numerical experiments of near-future projections targeted for the period of 2016-2035 by using a number of the coupled atmosphere–ocean global climate model, the Model for Interdisciplinary Research on Climate (MIROC) (Nozawa et al., 2005, Watanabe et al., 2010 and Sakamoto et al., 2012), and indicated that significant reductions (approximately 14%) in TC number are found especially in the western part of the western North Pacific. Although their results are based on the near-future climate scenarios when global warming signal is less pronounced than at the end of the 21st century, the results are consistent with the long-term future projections on typhoons.

From these recent studies the understanding on the future projections of tropical cyclones over the western North Pacific has advanced. With such advanced projections, meteorological hazards due to typhoons have been studied. In Table 14.3 of IPCC (2013), it is summarized that extreme precipitation is projected to increase near the center of tropical cyclones that make landfall in Japan and along the coasts of the East China Sea, Sea of Japan, South China Sea, Gulf of Thailand, and Andaman Sea.

Mechanisms for the decrease in the frequency of TCs are attributed to the weakening of tropical circulation and convective activity with the stabilization of the atmospheric stratification under global warming (Sugi et al., 2002 and Sugi et al., 2012). Satoh et al. (in press) further investigated the mechanisms for the reduction in the global frequency of TCs under global warming and suggested it is a result of future intensification of TCs under the constraint that the contribution of convective mass flux of TCs remains the same or smaller.

Although it is still difficult to make quantitative estimates of the strengths of future TCs, other macroscopic characteristics of TCs should be known for impact assessment. As described above, the track of TCs is projected to shift due to changes in the genesis location and to changes in large-scale steering flows for the TC movement (Murakami et al., 2011 and Mori, 2012). Here we show the results of analysis of TC track shift based on two MRI models, MRI-AGCM-3.1S and MRI-AGCM-3.2S in CMIP5 and six AOGCMs (CCCMA CGCM3, CSIRO 3.0 and 3.5, GFDL CM2.0, MIROC3.2hires, ECHAM5) in CMIP3. The TC-detection method employed six criteria to identify TCs and compare the simulated annual global-mean TC genesis number with observed data (see Murakami and Sugi, 2010). The TC genesis locations were counted for each 5×5° grid area over the global domain. This detection method identifies observed TCs reasonably well. The TC positions were counted over 2×2° grid points every 6 h. The total count is defined as the frequency of TC occurrence, which indicates the probability of a TC occurring in a designated basin.

The TC processes examined in this study include cyclogenesis, cyclolysis, translation speeds, directions, and central pressure changes. To identify how these processes change from present-day climates to future projected climates, the mean and standard deviation were analyzed to identify signatures in TC changes. Significant changes were observed in the cyclogenesis number, cyclogenesis location, cyclolysis location and average central pressure but here we only focus on track changes for the sake of brevity.Fig. 4 illustrates the weighted ensemble-averaged centroid shifts for cyclogenesis and their corresponding deviations for each ocean basin. The upper and lower panels indicate shifts in the cyclogenesis centroids in the longitudinal and latitudinal directions, respectively. The wider vertical column illustrates the ensemble-averaged shift between present and future climates, and the narrower vertical column indicates the relative difference in the cyclogenesis centroid between the IBTrACS and the ensemble-averaged centroids of the present climate as an expected error for model reproducibility in the present climate. Additionally, the two vertical bars illustrate the standard deviation for the present climate (left) and future climate (right), respectively. The simple ensemble-averaged centroid shift was estimated as the mean of the centroid of each model. A clear trend the cyclogenesis centroid shifts in future projections is observed inFig. 4. Future cyclogenesis centroids move toward the center of the ocean basin in the longitudinal direction and toward the poles in the latitudinal direction. The future centroid shift in the longitudinal direction is several times greater than the centroid shift in the latitudinal direction. For example, the cyclogenesis centroid in the WNP moves toward the east-northeast direction, and the centroid in the ENP moves toward the west-northwest direction. Second, the future cyclogenesis centroid shift has the same order of relative error which defines the difference between the observed (IBTrACS) centroid and the model ensemble averaged centroid for the present climate condition. The relative errors are higher for the future cyclogenesis centroid shifts in the NIO, WNP and ENP. Specifically, the future centroid in the WNP is expected to shift eastward but the relative error shows a westward bias. The uncertainty associated with these cyclogenesis characteristics in these basins is significant and nontrivial. Therefore, the changes in the expected future cyclogenesis locations has sufficient uncertainty to warrant careful projections of future tropical cyclones, and discussions should be conducted quantitatively basin by basin.

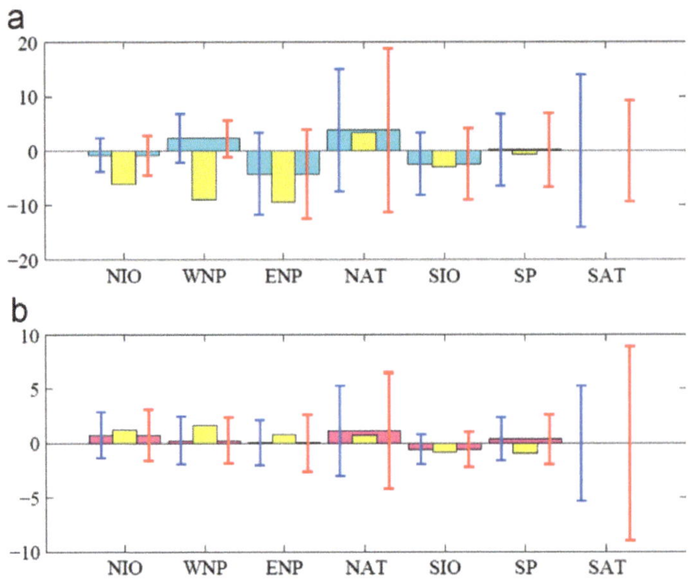

Figure 4: Weighted ensemble averaged future cyclogenesis centroid shifts, standard deviation, and error relative to IBTrACS for each ocean basin (wider vertical column: ensemble averaged shift, narrower column: difference relative to IBTrACS, left side bar: model standard deviation of present climate, right side bar: model standard deviation of future climate; abbreviations in the figure indicate as NIO: North Indian Ocean, WNP: Western North Pacific, ENP: Eastern North Pacific, NAT: North Atlantic Ocean, SIO: Southern Indian Ocean, SP: Southern Pacific Ocean, SAT: Southern Atlantic Ocean).

IMPACT ASSESSMENT OF TROPICAL CYCLONES

Dynamical Downscaling Approach

A methodology of Downscaling Typhoon Hazards for impact Assessment Studies

For the purpose of assessing the impacts of typhoons on natural disasters, high-resolution data are necessary. The horizontal resolutions of GCMs are getting higher and higher on the order of 10 km. With the advances in resolutions of GCM simulations, the representations of TCs in GCMs have been significantly improved, as described inSection 2.2. However, cumulus clouds and cumulus-scale processes play important roles in generating and forming TCs, which mean that a resolution of higher than 10 km is required in representing quantitatively the intensity and evolution of TCs. By conducting sensitivity experiments of TCs to resolutions, Bryan and Rotunno (2009) indicated that convergence in the numerical solutions of TCs is achieved when radial grid spacing is on the order of 1 km.

Furthermore, from the meteorological hazard perspective, quantitative estimations of rain and wind are important. Kanada et al., (2010) clearly demonstrated that the representation of rainfall in a higher-resolution (i.e., 5 km) regional climate model (RCM) outperforms that in the 20-km-mesh MRI-AGCM. Jimenez et al. (2008) and Takemi, 2009 and Takemi, 2013 indicated a benefit of higher-resolution simulations in quantitatively representing wind speed in complex terrain. Oku et al. (2010) conducted 1-km resolution, regional-meteorological simulations of a severe typhoon that spawned damaging strong winds in the western part of Japan, where complex terrain features are evident, and compared the simulated results with those obtained by a different regional meteorological model. The comparison of simulated surface winds in the two regional models indicated that slight differences in the model topography significantly affect the representations of the wind speed extremes in each model; a sharper representation of topography leads to higher wind extremes in the simulation. They insisted that differences in the reproduction of topography in the model induce differences in the representations of extreme weather.

From these studies, it is considered that a higher-resolution simulation will benefit not only from a better representation of convective-scale processes but also from a better reproduction of topography. Therefore, RCM simulations at higher resolutions on the order of 1 km are quite useful for the impact assessment studies on TCs. A high-resolution RCM developed by MRI, called Non-Hydrostatic Regional Climate Model (NHRCM) (Sasaki et al., 2008, Sasaki et al., 2012, Nakano et al., 2012 and Murata et al., 2015), is one example. Nakano et al. (2010) conducted 5-km-resolution regional simulations using NHRCM for a warm-season climate of Japan from June to October in 2002–2006 to simulate 57 typhoons during the period and examined precipitation characteristics directly and in-directly influenced by those typhoon; direct (indirect) precipitation was defined as that occurred within (outside) the 300-km radius from the centers of typhoons. It was shown that for precipitation of intensity exceeding 5 mm h^{-1}, direct precipitation is more frequent than indirect precipitation. The effects of topography on the precipitation characteristics due to TCs were investigated in detail for the case of Typhoon Morakot (2009) that spawned devastating landslides in complex terrain (Zhang et al., 2010, Fang et al., 2011, Xie and Zhang, 2012, Hall et al., 2013 and Wu, 2013).Minamide and Yoshimura (2014) examined the effects of complex terrain in the Philippines on the distribution of precipitation due to a severe typhoon that caused significant disasters in the region and indicated orographic enhancement of precipitation.

Another point to be considered in assessing the impacts of TCs on meteorological hazards is that rain and wind extremes are quite dependent on the track, intensity, and translation speed of TCs. The characteristics of rain and wind due to TCs would significantly vary if the track, intensity, and translation speed of TCs are different. This imposes a severe limitation on the impact assessment for TCs, because the number of damaging TCs in the history is limited. In other words, we are not able to assess the impacts of TC hazards on natural disasters in a certain area if the area has not suffered from significant TC disasters in the past, or even if the number of the historical events is only a few. A sufficient amount of samples is required for the assessment of TC impacts.

To increase the number of severe TC samples, numerical experiments capable of controlling the track and intensity of TCs are useful. By examining the sensitivity of rain and wind caused by TCs to TC track and intensity from a large number of simulated TCs in the numerical experiments, we are able to use the simulated data to investigate the spatial and temporal features of rain and wind extremes in response to a certain track and intensity of a simulated TC.

In order to control the track and intensity of TCs, a TC bogus scheme that can extract a TC vortex and relocate it to a different position has been developed (Yoshino et al., 2008,Ishikawa et al., 2013 and Shimokawa et al., 2014). Potential vorticity (PV) is a useful meteorological parameter for extracting and relocating a TC vortex, because PV values within the core of a TC are significantly higher than those in the tropospheric surroundings. A TC signature is easily separated from the environmental troposphere with the use of PV. In the procedure, all the meteorological variables such as pressure, temperature, and wind are first converted to PV. After extracting a high PV region associated with a TC, the high PV region is relocated to a different position. Relocating the TC position, meteorological variables are then obtained by inverting the PV, conducted using an approach by Davis and Emanuel (1991).

With this PV inversion technique, Ishikawa et al. (2013) examined the impacts of a severe typhoon reproduced in the future climate simulation under the SRES A1B scenario by conducting downscaling numerical experiments through controlling the TC track, and they investigated the influences of rainfall induced by different typhoons on river discharge to assess a possible maximum flood event in the Tone River basin near Tokyo. Oku et al. (2014) used the PV inversion technique of Ishikawa et al. (2013) to investigate sensitivities of orographic rainfall induced by Typhoon Talas (2011) to the track and intensity of the typhoon. Typhoon Talas (2011) was a slowly moving storm that affected the Kii Peninsula, a mountainous region in Japan, for a long period of time causing a significant rainfall amount estimated to be over 2000 mm for this specific event (JMA, 2011). By conducting a large number of downscaling experiments, they successfully identified the worst-case scenarios on the track and intensity of the typhoon that spawned the maximum amount of rainfall over the Kii Peninsula. With the present typhoon bogus scheme, we are able to develop an idea of worst-case scenarios on typhoon hazards.

In the following, we describe the results of the downscaling simulations for two severe typhoon cases which caused significant coastal disasters. One is a historical severe event: Typhoon Vera (1959) which caused disastrous storm surges in the coastal region of the Ise Bay, Japan in September 1959. The other example is a recent event: Typhoon Haiyan which spawned devastating disasters due to storm surges in the Leyte Gulf in November 2013.

A Case Study of Typhoon Vera (1959)

Typhoon Vera (1959) was one of the most disastrous meteorological phenomena in the modern history of Japan. Typhoon Vera gained its maximum intensity, in terms of central pressure, of 895 hPa on 23 September 1959, and made landfall on 26 September. The central pressure at the time of the landfall was 929 hPa, the 2nd lowest on record since 1951. The typhoon caused storm surges, strong winds, heavy rainfall, and flooding, which killed more than 5000 people. Storm surges spawned devastating damages in the Ise Bay, which is the reason why the typhoon is also named the Isewan Typhoon.

In order to better reproduce Typhoon Vera in numerical simulations, better reanalysis data for use as initial and boundary conditions of a regional model are needed. Recently JMA released a long-term reanalysis data for the period starting from 1958, i.e., the Japanese 55-year Reanalysis (JRA-55) (Ebita et al., 2011 and Kobayashi et al., 2015), which enables us to conduct downscaling numerical simulations of meteorological events dating back to 1958. The JRA-55 data used here have a spatial resolution of 1.25 degrees and a temporal interval of 6 h. The overall performance of JRA-55 was described in Kobayashi et al. (2015). As far as TCs are concerned, Murakami (2014)demonstrated that the characteristics of TC frequency and structure are reproduced better in the JRA-55 dataset than in other reanalysis datasets.

Using JRA-55 as the initial and boundary conditions, we perform regional-scale simulations of Typhoon Vera in September 1959. The model used here is the Weather Research and Forecasting (WRF)/Advanced Research WRF model, version 3.3.1 (Skamarock et al., 2008). The computational domains on the Lambert conformal map projection are set with the use of the two-way nesting capability; the outer domain (Domain 1) has an area of 4875 km by 4150 km at the 5-km grid spacing, and the inner domain (Domain 2) has an area of 400 km by 400 km at the 1-km grid (Fig. 5). The center of Domain 1, which covers the Japanese Archipelago and most of the western North Pacific, is set to be at 140°E and 27°N, while Domain 2 covers a central part of the Japanese main island including the Ise Bay. The model top is set to the 20-hPa level, and there are 56 vertical levels, whose intervals are stretched with height. Physics parameterizations employed in the simulations are a single-moment 6-class scheme for cloud microphysics, the Kain–Fritsch scheme for cumulus clouds (applied only for Domain 1), the Yonsei University scheme for boundary-layer turbulent mixing, a Monin–Obukhov similarity scheme for surface fluxes, a 5-layer thermal diffusion scheme for land-surface processes, and a short-wave and long-wave radiative transfer scheme. The details of these schemes can be found in Skamarock et al. (2008).

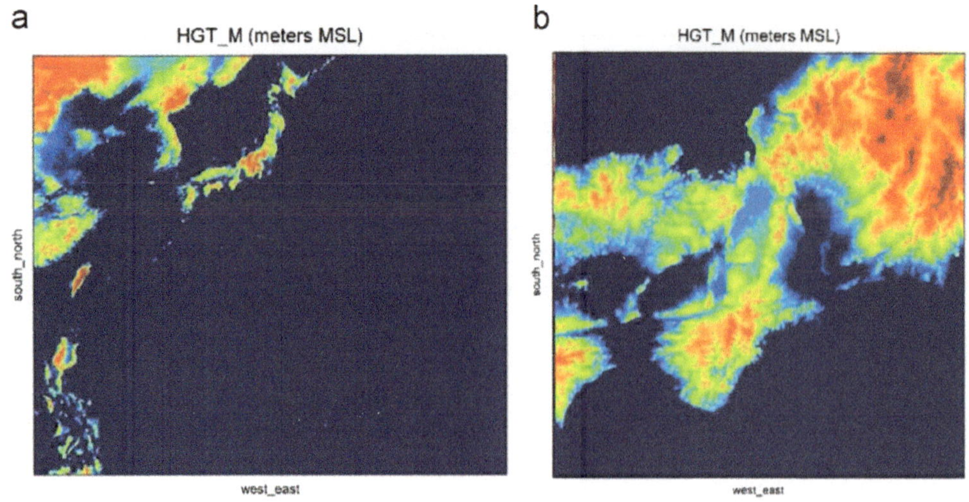

Figure 5: The computational domains for the simulations of Typhoon Vera (1959): (a) Domain 1, and (b) Domain 2.

For the purpose of the simulated fields mainly holds the JRA-55 analysis fields, we impose nudging toward the analysis fields during the time integration. The nudging terms with a time constant of 0.00028 s^{-1} are applied to the east–west and the north–south component of winds above the 700-hPa level.

With these settings described above, we perform numerical simulations of Typhoon Vera under real conditions in September 1959. In the simulations, we examine the uncertainties of the simulated typhoon to the initial condition by changing the initial time of the time integration. The initial times examined here are 1200 UTC 22 September (CASE001), 0000 UTC 22 September (CASE002), 1200 UTC 21 September (CASE003), 0000 UTC 21 September (CASE004), and 1200 UTC 20 September (CASE005).

In addition, we perform simulations of Typhoon Vera under an assumed warming climate. The idea for this is to consider how the natural hazards caused by a typhoon with a Vera-class intensity would change under a warming climate condition. Because Typhoon Vera caused one of the worst natural disasters in Japan, the assessment of natural hazards due to such a worst-class typhoon under global warming is important from a viewpoint of disaster prevention and mitigation. If such a worst-class typhoon is simulated in GCMs, direct downscaling from the GCM outputs should be promising. However, it is not expected that the future-climate simulations with a GCM can reproduce a severe typhoon with track and intensity very similar to Typhoon Vera in 1959, because the genesis and movement of TCs are sensitively affected by slight changes in atmospheric circulation and vertical stability. Therefore, direct downscaling from future-climate simulations of GCM is quite difficult.

An effective approach for assessing the impacts of future warming on regional-scale weather and climate was developed by Sato et al. (2007), referred to as a pseudo global warming experiment. The pseudo global warming experiment is designed to add climate change components to the analysis field for the real atmosphere. Climate change components here are defined as the deficits of the future climate states to the present climate states represented in GCM simulations. We employ this pseudo global warming experiment to assess the impacts of global warming on a Vera-like typhoon.

The global warming deficits are computed from the present- and future-climate simulations of the 20-km-mesh MRI-GCM under the RCP8.5 scenario (Mizuta et al., 2014). In the present study, we compute the mean global warming deficits averaged for the month of September under the present and the future climate. We add the warming deficits for mean temperature, pressure, and geopotential height to the JRA-55 analysis fields during the Typhoon Vera event in September 1959. We do not add the warming deficits for relative humidity and horizontal wind components; the reasons being that there is no significant difference in relative humidity between the present and the future climate (e.g., Takemi et al., 2012), and the differences in winds would significantly change the tracks of the simulated typhoons in the future climate from those in the present climate, which will make the future assessment for the specific regions difficult. In the present pseudo global warming experiments, the model settings are exactly the same with those for Typhoon Vera under the real condition of September 1959. The simulations initialized at 1200 UTC 22 September, 0000 UTC 22 September, 1200 UTC 21 September, 0000 UTC 21 September, and 1200 UTC 20 September are referred to as CASE101, CASE102, CASE103, CASE104, and CASE105.

The intensity of the simulated typhoons under the September 1959 condition and the future September condition are compared in terms of the central surface pressure and the maximum surface wind within the TC core in Fig. 6 and Fig. 7, respectively. The differences of the typhoon intensity among the cases with different initial times are also shown in Fig. 6 and Fig. 7. In Fig. 6, the best estimates on the central pressure according to the best-track data of JMA are also exhibited. Compared with the best-track central pressure, the simulations for the 1959 condition well reproduce the temporal change of the typhoon intensity especially after 1200 UTC 24 September. Although CASE005 seems to fail to capture the deepening of the central pressure, it captures the intensity evolution after 1200 UTC 25 September. Based on the favorable performance of the simulations under the 1959 condition, we can assess the typhoon changes under the assumed future climate change.

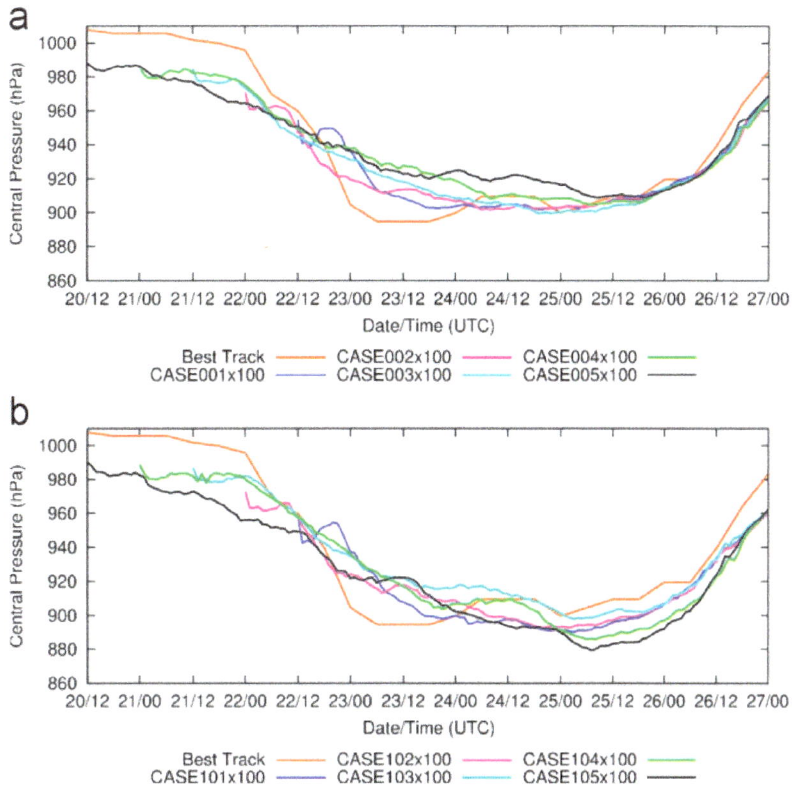

Figure 6: Time series of the central surface pressure of the simulated typhoons in (a) September 1959 and in (b) the future climate condition. The red line indicates the best-track data of JMA.

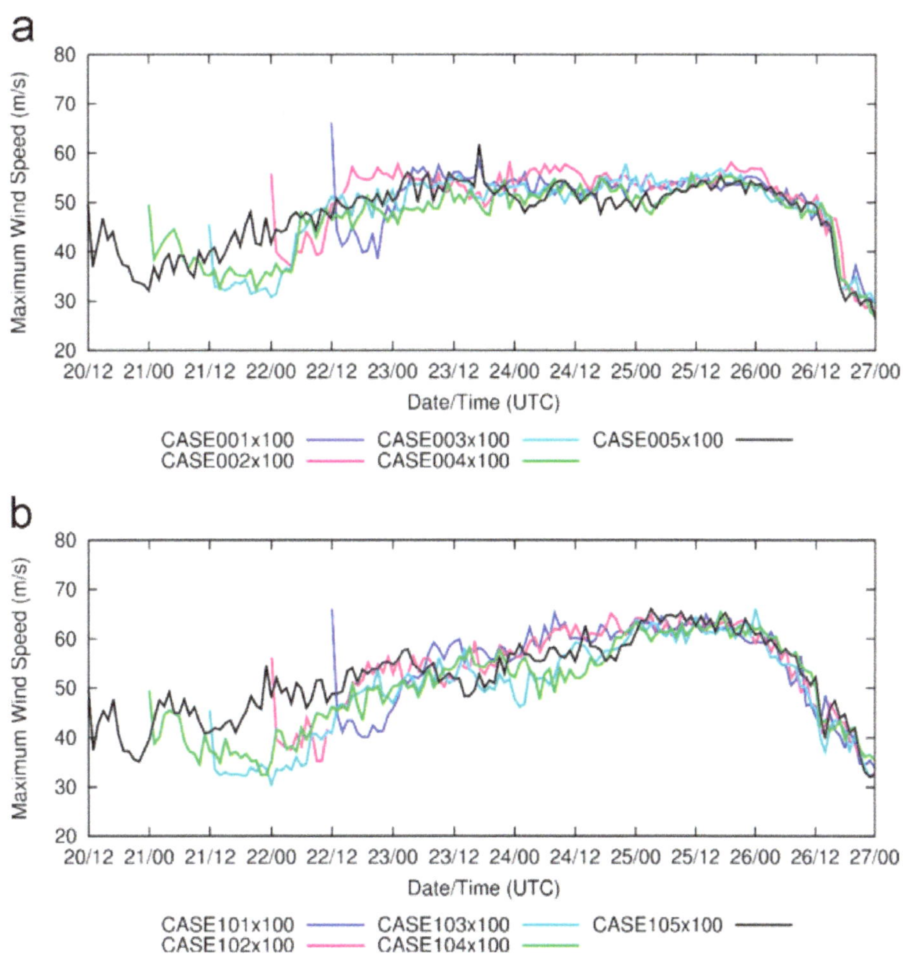

Figure 7: The same as Fig. 6, except for the maximum surface wind. Note that there are no wind data in the JMA's best-track data.

Comparing Figs. 6 and 7, it is clearly seen that the typhoon intensity during the simulated time period in all the cases with different initial times is overall stronger in the future climate condition than in the 1959 condition. It is also found that there are some sensitivities of the evolution of the typhoon intensity to the initial times; this suggests that the initial conditions for the numerical simulations are one of the uncertainties that should be taken into account in order to assess the impacts of climate change on typhoon evolution and intensity.

Table 1 further compares the differences of the simulated typhoon intensity between the 1959 and the future condition. Both the central surface pressure and the maximum surface wind indicate that the simulated intensity is unanimously stronger in the future condition than in the 1959 condition, despite the sensitivities to the initial times. In other words, it is anticipated that the intensity of a typhoon which has features similar to Typhoon Vera (1959) will be stronger in an assumed future September condition than in the actual September 1959 condition. This result seems to be robust irrespective of the initial conditions. This further suggests that the resulting natural disasters would be worse in a future climate condition than in the past if a similar extreme typhoon develops.

Table 1: The maximum intensity of the simulated typhoons initialized at the five times for the conditions of September 1959 and the future September climate. The maximum intensity refers to the central surface pressure (in hPa) and maximum surface wind (in m s^{-1}) of the typhoons

Initial time	September 1959		Future September	
	Pressure	Wind	Pressure	Wind
1200 UTC 22 Sep	901.9	58.3	890.2	65.3
0000 UTC 22 Sep	901.8	58.4	893.0	65.5
1200 UTC 21 Sep	899.5	57.9	898.1	66.2
0000 UTC 21 Sep	904.9	55.4	886.0	65.6
1200 UTC 21 Sep	909.0	61.8	879.4	66.1

A case Study of Typhoon Haiyan (2013)

Typhoon Haiyan (2013) developed in November 2013 and caused devastating damages over the Philippines. In particular, severe storm surges swept over the coastal areas of the Leyte Gulf. The central surface pressure of Typhoon Haiyan deepened to 895 hPa at its mature stage (according to the JMA's best-track data) and was a Category-5 storm. For its extreme intensity, Lin et al. (2014) proposed that Category-6 should be added to the existing Saffir-Simpson scale to respond to anticipated intensification of TCs in a warming climate.

Mori et al. (2014) quantitatively evaluated the storm surge level over the Leyte Gulf by conducting numerical simulations. To compute the storm surge, simulated outputs of the WRF model downscaled from the National Centers for Environmental Prediction (NCEP) Final Analysis data were used. In the WRF simulations, they examined a number of model settings to investigate the impacts of the uncertainties due to meteorological model simulations on quantitative assessment of storm surges and found that small changes in the simulated track and intensity of the typhoon in the meteorological model significantly affect the storm surge simulations.

Similar to the WRF model settings of Mori et al. (2014), we conducted another numerical simulation of Typhoon Haiyan. In this experiment, the time integration for both the 3-km-mesh domain (Domain 1, 4000 km by 2000 km area) and the nested 1-km-mesh domain (Domain 2, 2000 km by 700 km area) was initiated at 0000 UTC 5 November 2013. This case reproduced the minimum central surface pressure of 896.6 hPa during the lifetime of the simulated typhoon, which is regarded as the best performance among all the cases examined. Fig. 8 demonstrates the spatial distribution of the maximum surface wind speed diagnosed at the 10-m height during the simulated time period of the WRF model for this simulation case. It is seen that the regions with high wind speeds are very concentrated along the typhoon track, corresponding to a compact structure of the typhoon core. Such a concentrated region of high wind speeds is considered to significantly influence the storm surge assessment using the meteorological model outputs, because the uncertainties in representing the typhoon track and intensity in the meteorological simulations cannot be ignored in assessing the storm surge at a regional coastal scale. The uncertainties in meteorological model simulations are still large in comparison to the spatial scales of the regions for impact assessments.

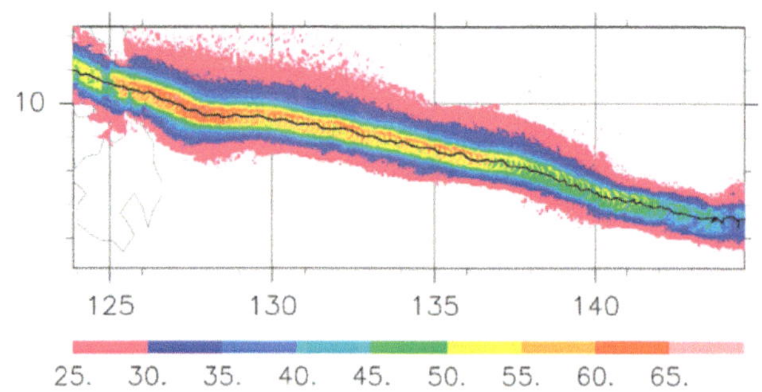

Figure 8: The spatial distribution of the maximum surface wind speed (color shading, in m s^{-1}) during the simulated time period of the WRF model at each grid point. The solid line indicates the observed track according to the JMA's best-track data.

Takayabu et al. (in press) investigate the effects of climate change on a worst-case storm surge by examining the case of Typhoon Haiyan and indicated that the worst-case storm surge would be worse owing to the effects of global warming on the intensification of an extreme typhoon. Similar to the study of Mori et al. (2014), downscaling experiments at the 1-km resolution were conducted with the use of the WRF model in Takayabu et al. In their study, the 1-km resolution simulations were performed for 16 cases each in the actual 2013 condition and in an assumed natural condition without global warming. Excluding the 6 sensitivity experiments to the initial time, Fig. 9 shows the track and intensity of the 10 simulated typhoons in the actual condition and the natural condition. It is seen that the typhoons are stronger in the actual case than in the natural case particularly in the longitudinal region between 125 and 130°E. These stronger TCs simulated in the actual environments led to more severe storm surge under the influences of climate change that evolved since the preindustrial era. Developing the idea of worst-case scenarios as natural hazards due to extreme weather events is an important issue for the assessment of climate change impact on natural disasters.

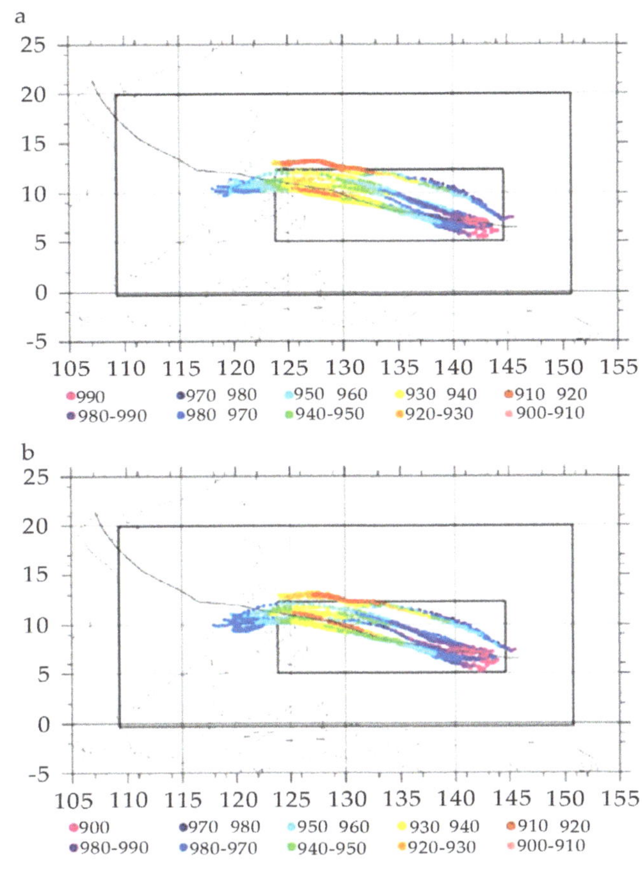

Figure 9: The track and the central surface pressure of simulated Typhoon Haiyan in different environmental conditions under (a) the actual 2013 condition and (b) an assumed natural condition without global warming. The boxes indicate the computational domains (Domain 1 for the outer box and Domain 2 for the inner box), and the solid line denotes the Haiyan's track of the JMA's best-track data.

Recently, Huang et al. (2015) suggested that a suppressive effect of subsurface oceans on the intensification of future TCs projected by a large number of state-of-the-art climate models and indicates that future subsurface ocean environments may be more suppressive than the existing subsurface ocean environments, which varies from region to region. Atmospheric and oceanic coupling is a key to further enhance our understanding on the changes of TCs under climate change.

Statistical Downscaling Approach

Statistical projections of TCs have been examined in engineering and insurance industries. There are two different approaches to design structures and assess the risks of TCs. The primitive engineering approach estimates landfall rates along the coastline based on historical records. This is called a site-specific approach or a local model that provides TC statistics for a particular location of interest (Russell, 1971). The statistics or probability density functions (pdfs) of TC parameters include central pressure, radius of maximum winds, direction, moving speed, and distance of closest approach; these are analyzed in the site-specific approach. Applying a site-specific approach to climate change is difficult because all TC parameters at all locations need to consider the influences of climate change.

The second approach to statistical projection of TCs involves a stochastic tropical cyclone model/storm track method (Vickery et al., 2000, James and Mason, 2005,Emanuel et al., 2006 and Nakajo et al., 2014) which is a type of Monte Carlo simulation. The stochastic tropical cyclone model (denoted STM hereafter) simulates a TC track by random draw from a temporal–spatial pdf of cyclogenesis locations based on a historical data set of TCs; it is possible to expand this approach to a basin-wide scale. The STM simulates the full track of a tropical cyclone, from cyclogenesis to cyclolysis, where cyclolysis includes the transition from a TC to an extra-tropical system. The cyclogenesis and cyclolysis indicate generation and weakening of low pressure circulation physically. However, these are simply used for generation and dissipation of TC in the study of statistical projection of TC. The STM models TC translation as a function of TC location and moving speed at each time step.

Fig. 10 shows one such historical TC data set, IBTrACS, and simulated TC data using the STM by Nakajo et al. (2014). The changing ratios of TC direction, speed and central pressure are provided in pdf format based on the historical record of TCs as shown in Fig. 10. The STM has biases due to an inadequate or missing depiction of physical processes, but it shows good agreement with the local model if the target area has relatively few landfalls. Qualitatively, the overview of the TC tracks in each basin is reasonably distributed. The simulated peak in the WNP shows good agreement with observed data, and simulated distributions in other basins also show reasonable reproducibility. The global grid averaged error against the historical data is 0.06 #/year (# indicates number of TC per 3° by 3° grid) and indicates reasonable accuracy. The direction, moving speed and intensity were verified against both spatial patterns and local characteristics, respectively (Nakajo et al. 2014).

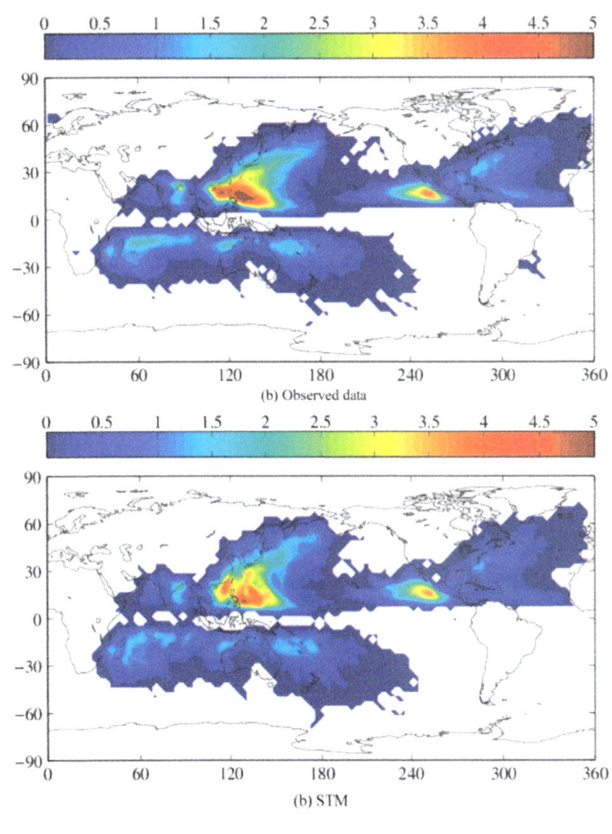

Figure 10: Observed and simulated mean TC number by IBTrACS and stochastic tropical cyclone model in 5×5° grid per year.

Using an STM for future climate projections is one option to assess tropical cyclone risks. If the STM satisfies the TC characteristics in the present climate condition, then a sufficient number of TC events can be obtained at arbitrary locations. If appropriate future TC conditions are provided, the future projection of TCs using an STM provides a tool to quantitatively evaluate risk along a given coast line. The future projection of TC track and intensity can be estimated by the STM. The statistical properties of TCs in the future climate are summarized as follows.

1. Cyclogenesis frequency decreases (Section 2.2).
2. Cyclogenesis centroids shift toward the center of a basin (Fig.4 in Section 2.2).
3. Cyclolysis centroids shift toward the center of a basin (Fig.4 in Section 2.2).
4. Central pressures of TCs decrease quadratically in the polar direction (Section 2.2).
5. Remaining TC characteristics are unchanged in the future climate.

The TC track shifts are assumed to be changed linearly from cyclogenesis to cyclolysis centroid changes over the ocean. All of these statistical changes are assumed to be the results of future TCs based on the GCM model ensemble averaged values analyzed based on CMIP3 and CMIP5 (Mori, 2012).

The changes listed above were computed by the STM over a time-frame of 10,000 years on a global scale. The difference between the future condition and the present condition amounts to significant changes in both TC frequency and intensity. However, the extreme change in central pressure in the future climate shows different characteristics. The extreme minimum TC central pressure with a 100 year return period is calculated by extreme value analysis using 10,000 years simulated data as shown in Fig. 11. In the future climate condition the minimum central pressure with a 100 year return value shows both positive and negative changes, and the spatial distribution of extreme minimum central pressure is different from the averaged central pressure. Overall, the extreme minimum central pressure will be smaller (i.e., stronger TCs) at latitudes above 30°N and 30°S. The extreme value of minimum central pressure shows positive change from subtropical to mid-latitudes around 30°, and it shows negative change in the polar direction. These complex changes in the extreme values result from a combination of assumed future changes described above. Although the cyclogenesis and cyclolysis shifts and the mean central pressure changes in the latitudinal direction reduce the central pressure of TCs in the middle latitudes, the cyclogenesis and cyclolysis shifts in the longitudinal direction either enhance or reduce the extreme central pressure change of TCs in the future climate.

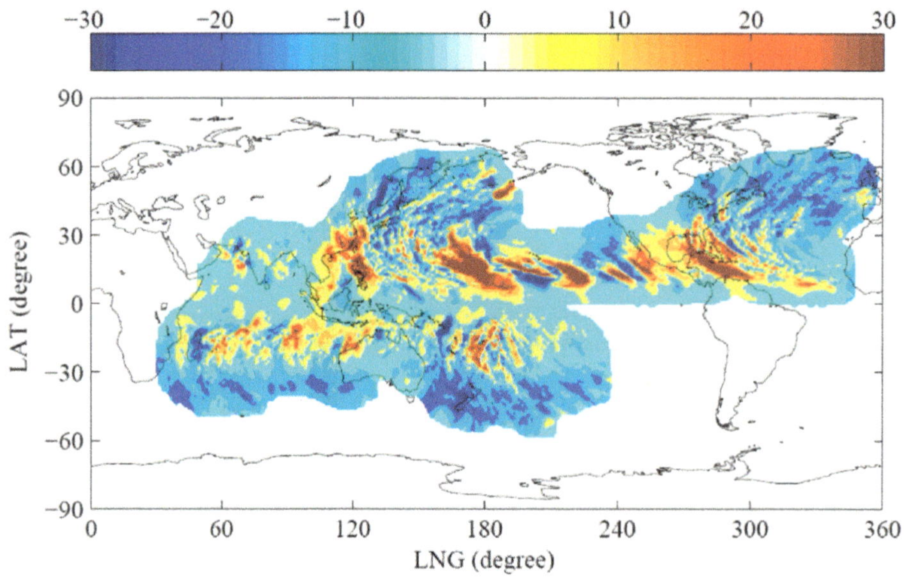

Figure 11: Simulated future change of maximum typhoon central pressure in 100 years return period (DP [hPa]/year).

The statistical projection presented here is limited to the numerical values of the mean and extreme central pressures of the TC. We showed the averaged results but it is possible to evaluate standard deviation as an indicator of uncertainty of protection, which is another advantage of using a stochastic model. The STM can increase the number of TCs, and thus it is possible to apply it to a regional study of storm surge projection due to climate change. This will be presented in the next section.

EXPECTED CHANGES OF COASTAL DISASTERS

Storm Surges

A case Study of Typhoon Vera (1959)

As described in Section 3.1.2, the Ise Bay including Nagoya, Japan, suffered from the most severe storm surge disaster in 1959. A storm surge projection with inundation over the land was conducted targeting the Ise Bay considering projected climate change. Future changes of typhoon characteristics are analyzed to estimate possible changes in intensity, generation numbers, and tracks in the Pacific Ocean, and the influence of climate change on storm surges in the Ise Bay are simulated as a case study.

Storm surge simulation is conducted using fully-coupled Surge–Wave–Tide model (SuWAT) developed by Kim et al. (2008). The SuWAT model consists of a storm surge module using the nonlinear shallow water equations and a wave module using the spectral wave model SWAN considering tidal effects at the lateral boundary. The details of the model are described by Kim et al. (2008). In order to simulate inundation, the model was modified to take into account overflow into the land with bottom roughness based on land use data. In this study the wave model was basically not used and the atmospheric forcing was given by an empirical typhoon model to reduce computational cost. The atmospheric pressure field is estimated using the Mitsuta and Fujii (1987) model, and the wind field is simulated by Myers and Malkin (1961).

The storm surge model employs four domain nesting for the historical analysis. The spatial resolution of the coarsest domain is 12,150 m and the finest domain resolution is 450 m. The maximum storm surge result and the surge height time series show good agreement with observation (not shown). It is confirmed that the maximum leading edge of flooding into the land, and maximum surge height of 3.24 m at Nagoya port, respectively. The storm surge and inundation area by the hindcast show good agreement with historical record. Furthermore, the storm surge in the pseudo global warming scenario with the worst course of typhoon Vera was simulated. A series of simulations was carried out using the intensified central pressure obtained from analysis of CMIP5 and the changed typhoon track considering the analyzed typhoon characteristics changes as the pseudo global warming experiments. The worst course of typhoon Vera was estimated by shifting and rotating the track.

First, Vera was simulated under the present depth and sea wall condition and compared with the historical record. Fig. 12(a) shows the numerical results of maximum surge height and inundation area. The storm surge ran up along the river and was source of flooding upstream. A series of storm surges by extreme typhoons considering the track changed worst course was simulated by pseudo warming experiments. The results of these worst course extreme typhoons show that the maximum surge height becomes larger and inundation area increases significantly as shown in Fig. 12(b). The result indicates that the storm surge disaster increases when the typhoon is intensified in the future climate condition. When the typhoon intensifies, the maximum surge heights will increase in the range of 1.3–1.0 m. The future change of inundation area due to storm surge will increase significantly by about 30%.

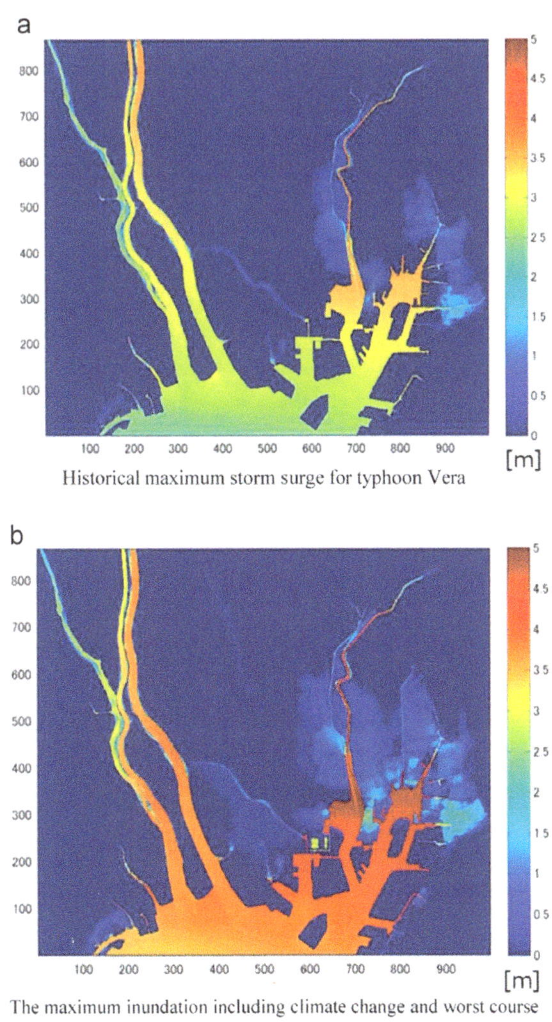

Figure 12: The maximum storm surge for a case study of typhoon Vera.

A case Study of Typhoon Haiyan (2013)

As described in Section 3.1.2, Typhoon Haiyan is the most severe landfall typhoon in the last 30 years. The combination of several storm surge factors causes extreme changes in water surface level at particular locations. To understand these characteristics for Super Typhoon Haiyan in detail, to evaluate the current performance of typhoon and storm surge prediction systems, and to guide reconstruction efforts of devastated areas, it is necessary to validate models and determine the local properties of storm surge.

Based on WRF model hindcast for Haiyan in Section 3.1.3, storm surge was simulated using the surge–wave coupling model (SuWAT). Three-domain two-way nesting was performed with spatial resolutions from 0.1° (D1) to 0.0067° (D3). The final domain (D3) size was 2.0°×2.0° with a resolution of 670 m. Fig. 13 shows the maximum surge forced by WRF with land inundation survey data (Tajima et al., 2014). The pressure-driven storm surge was approximately 1.1 m and was limited to within 50 km of the eye of the TC. However, the area affected by wind-driven storm surge was several times larger than that affected by the pressure-driven surge. Areas where maximum storm surge exceeded 3 m were mainly limited to the Tacloban area. The maximum water level at Tacloban was estimated to be 5.15 m and is in good agreement with the post-survey data measured from watermarks remaining on the land. Although the local increasing of water level due to local topographic effects, the simulation gave reasonably accurate results of the increase and decrease of maximum water surface level near the Tacloban area. However, the maximum water surface level is very sensitive to NWP results and the mean error of maximum water surface level between the numerical results and the post-survey data ranged from 1.08 to 4.15 m depending on the NWP model configuration. The WRF series indicate that the sensitivity of hindcast maximum water surface level, the error for which was 38–219%, depends on the configuration of the numerical setup. These factors make it difficult to quantitatively forecast or hindcast the storm surge during typhoon Haiyan.

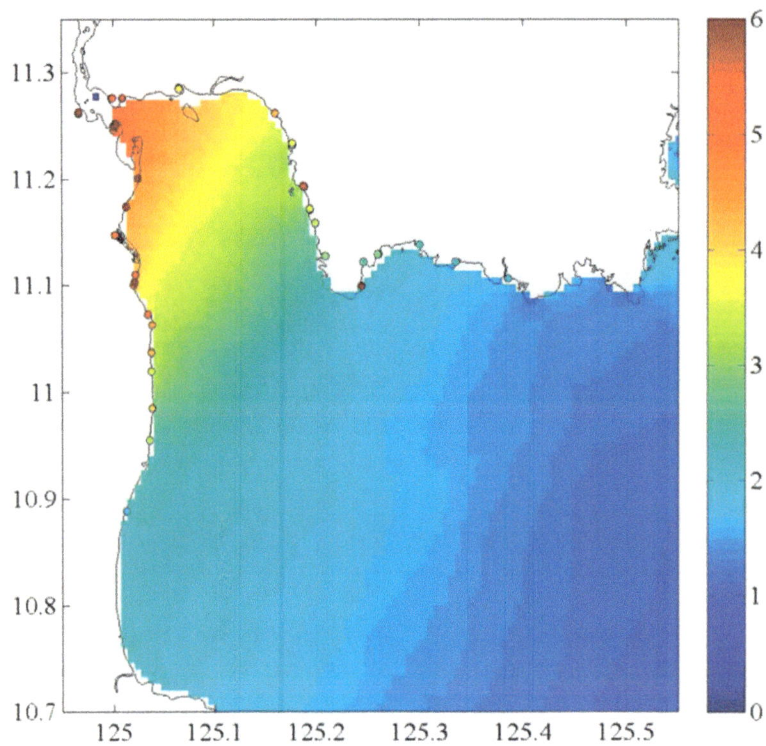

Figure 13: Simulated maximum water surface elevation in Tacloban and surrounding area with survey data (circles).

Application of Stochastic Tropical Cyclone model with Neural network for Storm Surge

The pseudo global warming experiments and related application to storm surge and others are powerful tools for regional impact assessment under the assumed conditions. However, it is important to estimate probabilistic information of extreme events considering uncertainty of tracks, speed, etc. Therefore, the application of stochastic tropical cyclone for storm surge provides future changes of storm surges at particular location.

As an example application, we selected Osaka Bay which faces the Pacific Ocean in the western part of Japan. Synthetic TCs were generated over 5000 years using a stochastic tropical cyclone model with historical conditions. The total number of TC is greater than 2000, thus we selected 100 potentially high impact TCs for storm surges based on their tracks, minimum central pressures, and speeds. The selected tracks were selected to minimize distance from bay center and an angle of incidence to bay center, respectively.Fig. 14 shows selected simulated typhoon tracks of the stochastic tropical cyclone model for Osaka bay. All selected TC tracks are basically passing from southwest to northeast over Osaka bay. The dynamical storm surge simulation was conducted for the 100 selected TCs by surge–wave–tide coupled model (SuWAT) as shown in Section 4.1.1.

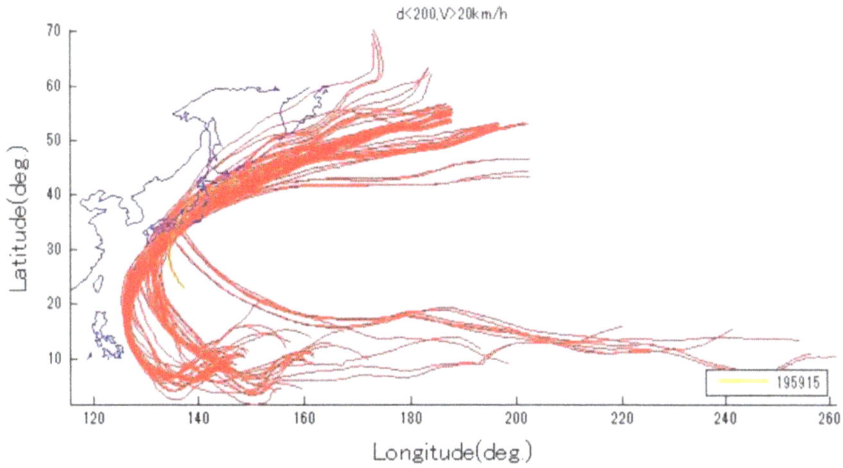

Figure 14: Selected simulated typhoon tracks by stochastic tropical cyclone model for Osaka bay.

Using the dynamically simulated storm surge based on the selected synthetic TC data, a neural network model was applied as a function of minimum central pressure, distance to bay, angle of incidence to bay and moving speed. Fig. 15 shows training, validation and test of storm surge height based on the stochastic tropical cyclone model and neural network model. The blue bar is input training data and sky-blue bar is validation of storm surge height. The root-mean-square error of validation phase is less than 0.2 m, and therefore, the evaluated neural network model works well for the estimation of maximum storm surge using TC information without required a dynamical storm surge model. Given the cheap computational cost of the neural network model, it is possible to estimate storm surge height over this 5000 year period based on the synthetic TCs from the stochastic tropical cyclone model. The estimated return periods of storm surge heights are 60, 100, and 300 years for 3, 4, and 5 m surge height, respectively. Although it is difficult to validate these values due to a lack of historical data, it is important to estimate long-term return value of extreme storm surge for impact assessment and adaptation strategies. In addition, the synthetic TC data can estimate the long-term return period of pseudo global warming experiments estimation of joint probability of similar TC tracks, intensity and speed. The combination of synthetic TC data and pseudo global warming experiments are a powerful tool to estimate probabilities and worst case scenarios of extreme storm surges.

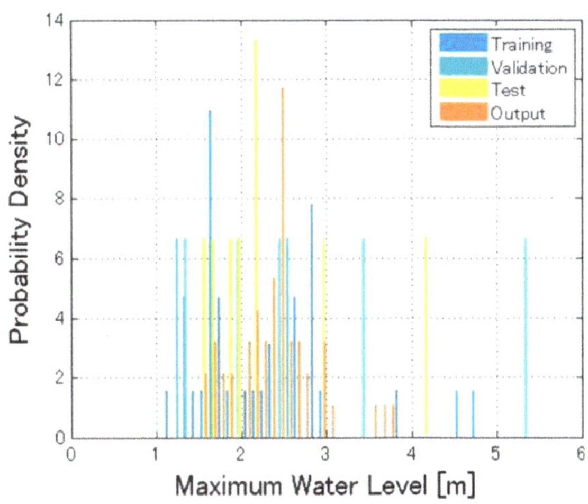

Figure 15: Training, validation and test of storm surge height based on stochastic tropical cyclone model and neural network model.

Extreme Ocean WAVES

Understanding future changes of storm waves and storm surges is important for assessing and mitigating the impact of climate change on coastal, marine and ocean environments and on engineering problems. For example, coastal breakwaters are designed for a maximum storm surge level and dynamic pressure from a maximum wave condition over a predetermined design lifetime. Therefore, it is important to discuss future wave climate change from the present to the future (Mori et al., 2010, Hemer et al., 2013). As a case study, future extreme wave climate is projected based on an AGCM and spectral wave model. The future extreme wind and wave values are analyzed using extreme value analysis, and the characteristics of extreme values and the influence of TCs on the future wave climate around the NWP are discussed in detail.

The ocean wave projection uses the product of global climate projections from a high-resolution AGCM, MRI-AGCM-3.1S. The MRI-AGCM-3.1S has a horizontal resolution of T959L60 (corresponding to about 20 km), and the future climate is simulated under the A1B scenario (Kitoh et al., 2009). As shown in the previous section, one major target of developing MRI-AGCM-3.1S (denoted hereafter as GCM) was to quantitatively resolve TCs in the Pacific Ocean. Time-slice experiments were conducted for three computational periods for 1979–2004 (present climate), 2015–2031 (near-future climate), and 2075–2100 (future climate), respectively. External forcing of the GCM is from mean SST of CMIP3 AOGCM projections and observed SST is used in the present period. Global wave climate projections were simulated by the spectral wave model SWAN (Booij et al., 1999), using sea surface winds at 10 m height U_{10} of the GCM. Global wave computations were carried out using spherical coordinates in the latitude range 80N–80S with 1.25° resolution.

Extreme values of significant wave height H_s were analyzed by fitting candidate distribution functions to the data of present and future climates respectively. The Mann-Kendall rank test was applied and extreme values is analyzed by the peak over threshold approach (POT) and the General Pareto Distribution (GPD) following the standard methodology of extreme value analysis. Fig. 16 shows differences in the 50-year return values of H_s between the present and future climate. There are clear increases in 50-year return values in the TC-dominant regions around the middle latitudes of the western North Pacific, the eastern North Pacific, the North Atlantic, and off the coast of Madagascar. These results indicate that future TCs will become stronger, and the range of their tracks in the low to middle latitudes will be larger than in the present climates. This is consistent with other climate projection studies (e.g., Knutson et al., 2010; Murakami and Sugi, 2010). The spatial patterns of H_s in both the present and future climates are characterized by larger values off the south and east coasts of Japan. The extreme values of H_s are caused by TC passing through this area. Similar peak extreme values ofH_s can be observed in the west North Pacific, the North Atlantic, and off the coast of Madagascar. As Fig. 16 shows, a significant increase in extreme values can be observed in the future climate around Japan. Remarkable increases of U_{10} are found to the south and west off the coasts of Japan. This area is related to TC generation and propagation in the summer season. The significance of increased H_s in this region is larger than that of wind speed due to characteristics of wind wave generation. Namely momentum transfer from wind to ocean is proportional to U_{10} so the impact of TC to H_s is larger than the wind field.

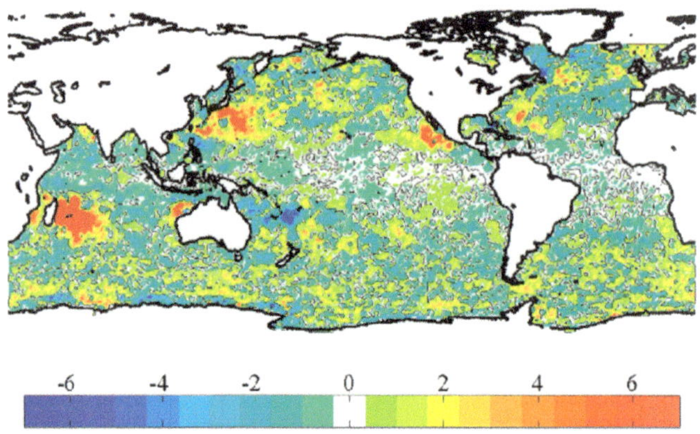

Figure 16: Future change of the 50-yr return value of H_s (2075–2099 minus 1979–2004).

A series of individual TC tracks is extracted from the source GCM data set. The future TC number is decreased significantly around the western part of the west North Pacific at the lower latitudes and is increased slightly in the eastern part of the mid-latitude west North Pacific (off the coast of Japan). These futures include two changes in TC characteristics, namely that TC genesis decreases and that TC tracks shift to the east side of the west North Pacific. This is consistent with previous studies as described in the previous section. The combination of reduced cyclogenesis and shifted tracks means that in the future more TCs will pass off the coast of Japan in the future climate depending on the region. As discussed in the previous section, future change in TC intensity is a key issue pertaining to extreme wind and wave climate change. Fig. 17 shows future change of the minimum central pressure, ΔP_{min}, both for the NWP at a global scale. There is a deeper striped pattern for ΔP_{min} at 15–20N latitude. The spatial distribution of ΔP_{min} in the NWP is different in the lower latitudes between 15–20N. TC wind speed depends on radius and intensity, and projected ΔP_{min} exceeds −30 hPa, which roughly corresponds to a 10–25 m/s wind speed increase. Future change of extreme TC intensity has a remarkable impact on extreme wave conditions.

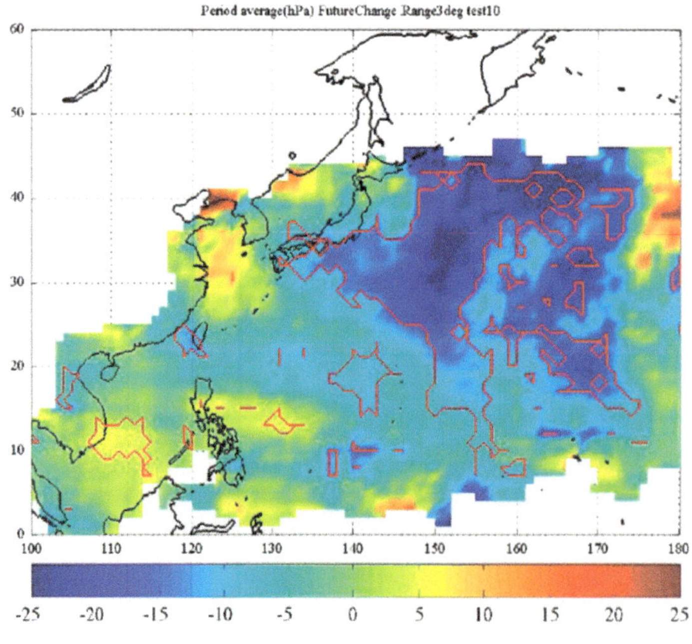

Figure 17: Future change of minimum of central pressure over computational period (2075–2099 minus 1979–2004, unit: hPa).

A similar significant decrease of ΔP_{min} is observed in the Atlantic Ocean, the Southern Indian Ocean and the Western South Pacific Ocean, where significant increases in future extreme wave height are expected. It can be concluded that projecting future extreme waves requires an accurate description of TC by GCM. This is one weakness of the climate model, whose uncertainties make future changes of middle latitude extreme wave climate difficult to project at regional scales, quantitatively.

CONCLUSION

Impact assessment of future coastal hazard due to climate change requires accurate descriptions of tropical cyclones (TCs) using GCM in the TC influenced regions, but this is difficult task. The performance of GCM and future changes in the dynamic and thermodynamic conditions in tropical oceans will influence TC genesis frequency and a locational shift toward the center of the basin. This study reviews and shows the latest related research and original results of impact assessment of TCs and related coastal hazards.

First, we briefly overviewed the current understanding on the changes in the characteristics of TCs in the past and in the future. We summarized the characteristics on TCs in the western North Pacific in the future climate from an impact assessment perspective. Second, we presented a review and ongoing research of impact assessment of tropical cyclones using both dynamical downscaling and statistical models. The dynamical downscaling simulations for two severe typhoon cases which caused significant coastal disasters were demonstrated as an example of dynamical downscaling of a super typhoon with a regional meteorological model. One example was Typhoon Vera (1959), known as Isewan Typhoon, that hit the coastal regions in central Japan, i.e., the Ise bay, in September 1959. The other example was Typhoon Haiyan in 2013, intended to show an example of event attribution like technique. Furthermore, a stochastic tropical cyclone model was introduced to increase the number of TCs which is useful for regional impact assessment. Assessing the impacts of such extreme TCs on coastal hazards emphasizes the importance of considering worst-case typhoons as a scenario basis in order to investigate the consequences of extreme weather events to natural hazards and disasters under a changing climate. This worst-case approach should be promising for impact assessment studies.

Finally, several examples of impact assessment of storm surge and extreme wave changes were presented. The future changes of storm surge are not only sensitive to TC intensity but are also needed to consider future changes of cyclogenesis and tracks. Both statistical and dynamical downscaling are important for the projection of storm surge height on regional scales. Future increases in extreme waves were also detected at the middle latitudes, the west North Pacific Ocean, the east North Pacific Ocean, the North Atlantic Ocean and the South Indian Ocean, regions mainly influenced by tropical cyclone activity. Changes in the return values of H_s in the Northwest Pacific Ocean over the next 50 years are significantly increased present climate values.

Changes in both TC intensity and track are linked to future changes in extreme storm surge and wave climate in middle latitude. The combination of reduced cyclogenesis and shifts in TC track will increase the number of changes and uncertainties in quantitatively assessing the impacts of future TC changes on storm surge and extreme characteristics. Further study is required to reduce the uncertainty of extreme coastal hazard projections due to climate change, with the natural variability of wave climate for both present and future climates being especially important for assessment, and these remain as topics for future study.

ACKNOWLEDGMENTS

The authors express our appreciation for Dr. Sota Nakajo of Kumamoto University, Dr. Yoko Shibutani of Tottori University, Dr. Tomoya Shimura of Kyoto University and Dr. Hiroyuki Murakami of the Geophysical Fluid Dynamics Laboratory for their kind support. We would also like to acknowledge Dr. Osamu Arakawa and Dr. Sachie Kanada for producing the global warming deficit data simulated by MRI-AGCM. This work was conducted under the framework of the "Precise Impact Assessments on Climate Change" of the Program for Risk Information on Climate Change (SOUSEI Program) supported by the Ministry of Education, Culture, Sports, Science, and Technology—Japan (MEXT). This work was also supported by a Kakenhi Grant-in-Aid of MEXT.

REFERENCES

1. Booij et al., 1999N. Booij, R.C. Ris, L.H. HolthuijsenA third-generation wave model for coastal regions: 1. Model description and validationJ. Geophys. Res.: Oceans (1978-2012), 104(C4) (1999), pp. 7649–7666
2. Bryan and Rotunno, 2009G.J. Bryan, R. RotunnoThe maximum intensity of tropical cyclones in axisymmetric numerical model simulationsMon. Weather Rev., 137 (2009), pp. 1770–1789 http://dx.doi.org/10.1175/2008MWR2709.1
3. Camargo et al., 2007S.J. Camargo, K.A. Emanuel, A.H. SobelUse of a genesis potential index to diagnose ENSO effects on tropical cyclone genesisJ. Clim., 20 (2007), pp. 4819–4834
4. Camargo et al., 2009S.J. Camargo, M.C. Wheeler, A.H. SobelDiagnosis of the MJO modulation of tropical cyclogenesis using an empirical indexJ. Atmos. Sci., 66 (2009), pp. 3061–3074
5. Chia and Ropelewski, 2002H.H. Chia, C.F. RopelewskiThe interannual variability in the genesis location of tropical cyclones in the northwest PacificJ. Clim., 15 (2002), pp. 2934–2944
6. Davis and Emanuel, 1991C.A. Davis, K.A. EmanuelPotential vorticity diagnostics of cyclogenesisMon. Weaher Rev., 119 (1991), pp. 1929–1953
7. Ebita et al., 2011A. Ebita, S. Kobayashi, Y. Ota, M. Moriya, R. Kumabe, K. Onogi, Y. Harada, S. Yasui, K. Miyaoka, K. Takahashi, H. Kamahori, C. Kobayashi, H. Endo, M. Soma, Y. Oikawa, T. IshimizuThe Japanese 55-year Reanalysis "JRA-55": An interim reportSci. Online Lett. Atmos., 7 (2011), pp. 149–152 http://dx.doi.org/10.2151/sola.2011-038
8. Emanuel et al., 2006K. Emanuel, S. Ravela, E. Vivant, C. RisiA statistical deterministic approach to hurricane

risk assessmentBull. Am. Meteor. Soc., 87 (2006), pp. 299–314

9. Fang et al., 2011X. Fang, Y.-H. Kuo, A. WangThe impacts of Taiwan topography on the predictability of Typhoon Morakot's record-breaking rainfall: a high-resolution ensemble simulationWeather Forecast., 26 (2011), pp. 613–633 http://dx.doi.org/10.1175/WAF-D-10-05020.1

10. Frank and Roundy, 2006W. Frank, P. RoundyThe role of tropical waves in tropical cyclogenesisMon. Weather Rev., 134 (2006), pp. 2397–2417

11. Hall et al., 2013J.D. Hall, M. Xue, L. Ran, L.M. LeslieHigh-resolution modeling of Typhoon Morakot (2009): Vortex Rossby waves and their role in extreme precipitation over TaiwanJ. Atmos. Sci., 70 (2013), pp. 163–186 http://dx.doi.org/10.1175/JAS-D-11-0338.1

12. Hemer et al., 2013M.A. Hemer, Y. Fan, N. Mori, A. Semedo, X.L. WangProjected changes in wave climate from a multi-model ensembleNat. Clim. Change (2013) http://dx.doi.org/10.1038/nclimate1791

13. Huang et al., 2011P. Huang, C. Chou, R. HuangSeasonal modulation of tropical intraseasonal oscillations on tropical cyclone geneses in the western North PacificJ. Clim., 24 (2011), pp. 6339–6352

14. Huang et al., 2015P. Huang, I.-I. Lin, C. Chou, R.-H. HuangChange in ocean subsurface environment to suppress tropical cyclone intensification under global warmingNat. Commun., 6 (2015), p. 7188 http://dx.doi.org/10.1038/ncomms8188

15. IPCC, 2013IPCC, 2013. In: Stocker, T.F., D. Qin, G.-K. Plattner, M. Tignor, S.K. Allen, J. Boschung, A. Nauels, Y. Xia, V. Bex and P.M. Midgley (eds.), Climate Change 2013: The Physical Science Basis. Contribution of Working Group I to the Fifth Assessment Report of the Intergovernmental Panel on Climate Change. Cambridge University Press, Cambridge, United Kingdom and New York, NY, USA.1535 pp.

16. Ishikawa et al., 2013H. Ishikawa, Y. Oku, S. Kim, T. Takemi, J. YoshinoEstimation of a possible maximum flood event in the Tone River basin, Japan caused by a tropical cycloneHydrol. Process., 27 (2013), pp. 3292–3300 http://dx.doi.org/10.1002/hyp.9830

17. James and Mason, 2005M. James, L. MasonSynthetic Tropical Cyclone DatabaseJ. Waterw., Port, Coast., Ocean Eng., 131 (4) (2005), pp. 181–192

18. Japan Meteorological Agency, 2011Japan Meteorological Agency, 2011. Annual Report on the Activities of the RSMC Tokyo – Typhoon Center 2011. Available online at: 〈http://www.jma.go.jp/jma/jma-eng/jma-center/rsmc-hp-pub-eg/AnnualReport/2011/Text/Text2011.pdf〉.

19. Jimenez et al., 2008P.A. Jimenez, J.F. Gonzalez-Rouco, J.P. Montavez, J. Navarro, E. Garcia-Bustamante, F. ValeroSurface wind regionalization in complex terrainJ. Appl. Meteor. Climatol., 47 (2008), pp. 308–325

20. Kanada et al., 2010S. Kanada, M. Nakano, T. KatoClimatological characteristics of daily precipitation over Japan in the Kakushin regional climate experiments using a non-hydrostatic 5-km-mesh model: Comparison with an outer global 20-km-mesh atmospheric climate modelSci. Online Lett. Atmos., 6 (2010), pp. 117–120 http://dx.doi.org/10.2151/sola.2010-030

21. Kitoh et al., 2009A. Kitoh, T. Ose, K. Kurihara, S. Kusunoki, M. SugiProjection of changes in future weather extremes using super-high-resolution global and regional atmospheric models in the KAKUSHIN Program: Results of preliminary experimentsHydrol. Res. Lett., 3 (2009), pp. 49–53

22. Knaff and Zehr, 2007J.A. Knaff, R.M. ZehrReexamination of tropical cyclone wind–pressure relationshipsWeather Forecast., 22 (2007), pp. 71–88 http://dx.doi.org/10.1175/WAF965.

23. Knutson et al., 2010T.R. Knutson, J.L. McBride, J. Chan, K. Emanuel, G. Holland, C. Landsea, I. Held, J.P. Kossin, A.K. Srivastava, M. SugiTropical cyclones and climate changeNat. Geosci., 3 (2010), pp. 157–163

24. Kobayashi et al., 2015S. Kobayashi, Y. Ota, Y. Harada, A. Ebita, M. Moriya, H. Onoda, K. Onogi, H. Kamahori, C. Kobayashi, H. Endo, K. Miyaoka, K. TakahashiThe JRA-55 Reanalysis: general specifications and basic characteristicsJ. Meteor. Soc. Jpn., 93 (2015), pp. 5–48 http://dx.doi.org/10.2151/jmsj.2015-001

25. Kim et al., 2008S. Kim, T. Yasuda, H. MaseNumerical analysis of effects of tidal variations on storm surges and wavesAppl. Ocean Res., 30 (4) (2008), pp. 311–322

26. Kunkel et al., 2013K.E. Kunkel, T.R. Karl, H. Brooks, J. Kossin, J.H. Lawrimore, D. Arndt, L. Bosart, D. Changnon, S.L. Cutter, N. Doesken, K. Emanuel, P.Y. Groisman, R.W. Katz, T. Knutson, J. O'Brien, C.J. Paciorek, T.C. Peterson, K. Redmond, D. Robinson, J. Trapp, R. Vose, S. Weaver, M. Wehner, K. Wolter, D. WuebblesMonitoring and understanding trends in extreme storms: state of knowledgeBull. Am. Meteor. Soc., 94 (2013), pp. 499–514 http://dx.doi.org/10.1175/BAMS-D-11-00262.1

27. Li and Zhou, 2012R.C.Y. Li, W. ZhouChanges in western Pacific tropical cyclones associated with the Niño–Southern oscillation cycleJ. Clim., 25 (2012), pp. 5864–5878

28. Li et al., 2010T. Li, M. Kwon, M. Zhao, J.-S. Kug, J.-J. Luo, W. YuGlobal warming shifts Pacific tropical cyclone locationGeophys. Res. Lett., 37 (2010), p. L21804

29. Lin et al., 2014I.-I. Lin, I.-F. Pun, C.-C. Lien"Category-6" supertyphoon Haiyan in global warming hiatus: Contribution from subsurface ocean warmingGeophys. Res. Lett., 41 (2014) http://dx.doi.org/10.1002/2014GL061281

30. Lowe and Gregory, 2005J.A. Lowe, J. GregoryThe effects of climate change on storm surges around the United KingdomPhilos. Trans. R. Soc. Lond. A: Math. Phys. Eng. Sci., 363 (2005), pp. 1313–1328

31. Liebmann et al., 1994B.B. Liebmann, H.H. Hendon, J.D. GlickThe relationship between tropical cyclones of the western Pacific and Indian Oceans and the Madden–Julian oscillationJ. Meteor. Soc. Japan, 72 (1994), pp. 401–412

32. Mahala et al., 2015B.K. Mahala, B.K. Nayak, P.K. MohantyImpacts of ENSO and IOD on tropical cyclone

activity in the Bay of BengalNat. Hazards, 75 (2015), pp. 1105–1125

33. Myers and Malkin, 1961V.A. Myers, W. MalkinSome Properties of Hurricane Wind Fields as Deduced From Trajectories, 49, US Department of Commerce, Weather Bureau (1961), p. 45

34. Mitsuta and Fujii, 1987Y. Mitsuta, T. FujiiAnalysis and Synthesis of Typhoon Wind Pattern Over JapanDisaster Prevention Research Institute, Kyoto University (1987), pp. 169–185

35. Minamide and Yoshimura, 2014M. Minamide, K. YoshimuraOrographic effect on the precipitation with Typhoon Washi in the Mindanao Island of the PhilippinesSci. Online. Lett. Atmos., 10 (2014), pp. 67–71 http://dx.doi.org/10.2151/sola.2014-014

36. Mizuta et al., 2006R. Mizuta, K. Oouchi, H. Yoshimura, A. Noda, K. Katayama, S. Yukimoto, M. Hosaka, S. Kusunoki, H. Kawai, M. Nakagawa20 km-Mesh global climate simulations using JMA-GSM modelJ. Meteor. Soc. Jpn., 84 (2006), pp. 165–185

37. Mizuta et al., 2012R. Mizuta, H. Yoshimura, H. Murakami, M. Matsueda, H. Endo, T. Ose, K. Kamiguchi, M. Hosaka, M. Sugi, S. Yukimoto, S. Kusunoki, A. KitohClimate simulations using MRI-AGCM3.2 with 20-km gridJ. Meteor. Soc. Jpn., 90A (2012), pp. 233–258

38. Mizuta et al., 2014R. Mizuta, O. Arakawa, T. Ose, S. Kusunoki, H. Endo, A. KitohClassification of CMIP5 future climate responses by the tropical sea surface temperature changesSci. Online. Lett. Atmos., 10 (2014), pp. 167–171 http://dx.doi.org/10.2151/sola.2014-035

39. Mori et al., 2013M. Mori, M. Kimoto, M. Ishii, S. Yokoi, T. Mochizuki, Y. Chikamoto, M. Watanabe, T. Nozawa, H. Tatebe, T.T. Sakamoto, Y. Komuro, Y. Imada, H. KoyamaHindcast prediction and near-future projection of tropical cyclone activity over the western North Pacific using CMIP5 near-term experiments with MIROCJ. Meteor. Soc. Jpn., 91 (2013), pp. 431–452

40. Mori, 2012Mori, N., 2012. Projection of Future Tropical Cyclone Characteristics based on Statistical Model, In Cyclones Formation, Triggers and Control. Nova Science Publishers, Inc. pp. 249–270.Mori et al., 2010N. Mori, T. Yasuda, H. Mase, T. Tom, Y. OkuProjection of extreme wave climate change under the global warmingHydrol. Res. Lett., 4 (2010), pp. 15–19

41. Mori et al., 2014N. Mori, M. Kato, S. Kim, H. Mase, Y. Shibutani, T. Takemi, K. Tsuboki, T. YasudaLocal amplification of storm surge by Super Typhoon Haiyan in Leyte GulfGeophys. Res. Lett., 41 (2014), pp. 5106–5113 http://dx.doi.org/10.1002/2014GL060689

42. Murakami and Sugi, 2010H. Murakami, M. SugiEffect of model resolution on tropical cyclone climate projectionsSola, 6 (2010), pp. 73–76

43. Murakami, 2014H. MurakamiTropical cyclones in reanalysis data setsGeophys. Res. Lett., 41 (2014), pp. 2133–2141 http://dx.doi.org/10.1002/2014GL059519

44. Murakami and Sugi, 2010H. Murakami, M. SugiEffect of model resolution on tropical cyclone climate projectionsSci. Online. Lett. Atmos., 6 (2010), pp. 73–76 http://dx.doi.org/10.2151/sola.2010-019

45. Murakami et al., 2011H. Murakami, B. Wang, A. KitohFuture change of western North Pacific typhoons: Projections by a 20-km-mesh global atmospheric modelJ. Clim., 24 (2011), pp. 1154–1169

46. Murakami et al., 2012aH. Murakami, Y. Wang, H. Yoshimura, R. Mizuta, M. Sugi, E. Shindo, Y. Adachi, S. Yukimoto, M. Hosaka, S. Kusunoki, T. Ose, A. KitohFuture changes in tropical cyclone activity projected by the new high-resolution MRI-AGCMJ. Clim., 25 (2012), pp. 3237–3260

47. Murakami et al., 2012bH. Murakami, R. Mizuta, E. ShindoFuture changes in tropical cyclone activity projected by multi-physics and multi-SST ensemble experiments using the 60 km-mesh MRI-AGCMClim. Dyn., 39 (2012), pp. 2569–2584

48. Murakami et al., 2014H. Murakami, P.-C. Hsu, O. Arakawa, T. LiInfluence of model biases on projected future changes in tropical cyclone frequency of occurrenceJ. Clim., 27 (2014), pp. 2159–2181 http://dx.doi.org/10.1175/JCLI-D-13-00436.1

49. Murata et al., 2015A. Murata, H. Sasaki, H. Kawase, M. Nosaka, M. Oh'izumi, T. Kato, T. Aoyagi, F. Shido, K. Hibino, S. Kanada, A. Suzuki-Parker, T. NagatomoProjection of future climate change over Japan in ensemble simulations with a high-resolution regional climate modelSci. Online Lett. Atmos., 11 (2015), pp. 90–93 http://dx.doi.org/10.2151/sola.2015-022

50. Nakano et al., 2010M. Nakano, S. Kanada, T. KatoStatistical analysis of simulated direct and indirect precipitation associated with typhoons around Japan using a cloud-system resolving modelHydrol. Res. Lett., 4 (2010), pp. 6–10 http://dx.doi.org/10.3178/hrl.4.6

51. Nakano et al., 2012M. Nakano, T. Kato, S. Hayashi, S. Kanada, Y. Yamada, K. KuriharaDevelopment of a 5-km-mesh cloud-system-resolving regional climate model at the Meteorological Research InstituteJ. Meteor. Soc. Jpn., 90A (2012), pp. 339–350

52. Nakano et al., 2015M. Nakano, M. Sawada, T. Nasuno, M. SatohIntraseasonal variability and tropical cyclogenesis in the western North Pacific simulated by a global nonhydrostatic atmospheric modelGeophys. Res. Lett., 42 (2015), pp. 565–571 http://dx.doi.org/10.1002/2014GL062479

53. Nakajo et al., 2014S. Nakajo, N. Mori, T. Yasuda, H. MaseGlobal stochastic tropical cyclone model based on principal component analysis with cluster analysisJ. Appl. Meteorol. Climatol., Am. Meteorol. Soc., 53 (2014), pp. 1547–1577 http://dx.doi.org/10.1175/JAMC-D-13-08.1

54. Nakazawa, 2006T. NakazawaMadden-Julian Oscillation activity and typhoon landfall on Japan in 2004Sci. Online Lett. Atmos., 2 (2006), pp. 136–139 http://dx.doi.org/10.2151/sola.2006-035

55. Nilsen, 2009P. NilsenCoastal Estuarine Processes, Advanced Series on Ocean EngineeringWorld Scientific Publishing Company (2009), p. 360

56. Nozawa et al., 2005T. Nozawa, T. Nagashima, H. Shiogama, S.A. CrooksDetecting natural influence on surface air temperature change in the early twentieth

centuryGeophys. Res. Lett., 32 (2005) http://dx.doi.org/10.1029/2005GL023540 L20719

57. Oku et al., 2010Y. Oku, T. Takemi, H. Ishikawa, S. Kanada, M. NakanoRepresentation of extreme weather during a typhoon landfall in regional meteorological simulations: a model intercomparison study for Typhoon Songda (2004)Hydrolo. Res. Lett., 4 (2010), pp. 1–5 http://dx.doi.org/10.3178/hrl.4.1

58. Oku et al., 2014Y. Oku, J. Yoshino, T. Takemi, H. IshikawaAssessment of heavy rainfall-induced disaster potential based on an ensemble simulation of Typhoon Talas (2011) with controlled track and intensityNat. Hazards Earth Syst. Sci., 14 (2014), pp. 2699–2709 http://dx.doi.org/10.5194/nhess-14-2699-2014

59. Oouchi et al., 2006K. Oouchi, J. Yoshimura, H. Yoshimura, R. Mizuta, S. Kusunoki, A. NodaTropical cyclone climatology in a global warming climate as simulated in a 20-km-mesh global atmospheric model: frequency and wind intensity analysisJ. Meteor. Soc. Jpn., 84 (2006), pp. 259–276

60. Russell, 1971L.R. RussellProbability distributions for hurricane effects, Journal of the WaterwaysHarb. Coast. Eng. Div., 97 (1) (1971), pp. 139–154

61. Sakamoto et al., 2012T.T. Sakamoto, Y. Komuro, T. Nishimura, M. Ishii, H. Tatebe, H. Shiogama, A. Hasegawa, T. Toyoda, M. Mori, T. Suzuki, Y. Imada, T. Nozawa, K. Takata, T. Mochizuki, K. Ogochi, S. Emori, H. Hasumi, M. KimotoMIROC4h: a new high-resolution atmosphere-ocean coupled general circulation modelJ. Meteor. Soc. Jpn., 90 (2012), pp. 325–359

62. Sasaki et al., 2008H. Sasaki, K. Kurihara, I. Takayabu, T. UchiyamaPreliminary experiments of reproducing the present climate using the non-hydrostatic regional climate modelSci. Online Lett. Atmos., 4 (2008), pp. 25–28

63. Sasaki et al., 2012 H. Sasaki, A. Murata, M. Hanafusa, M. Oh'izumi, K. KuriharaProjection of future climate change in a non-hydrostatic regional climate model nested within an atmospheric general circulation modelSci. Online Lett. Atmos., 8 (2012), pp. 53–56

64. Sato et al., 2007T. Sato, F. Kimura, A. KitohProjection of global warming onto regional precipitation over Mongolia using a regional climate modelJ. Hydrol., 333 (2007), pp. 144–154

65. Satoh et al., 2015M. Satoh, Y. Yamada, M. Sugi, C. Kodama, A.T. NodaConstraint on future change in global frequency of tropical cyclones due to global warmingJ. Meteor. Soc. Jpn., 93 (2015), pp. 489–500

66. Skamarock et al., 2008Skamarock, W.C., Klemp, J.B., Dudhia, J., Gill, D.O., Barker, D.M., Duda, M.G., Huang, X.Y., Wang, W., Powers, J.G., 2008. A description of the Advanced Research WRF Version 3. NCAR Tech. Note, NCAR/TN-47+STR. 113 pp.Shimokawa et al., 2014

67. S. Shimokawa, T. Murakami, S. Iizuka, J. Yoshino, T. YasudaA new typhoon bogussing scheme to obtain the possible maximum typhoon and its application for assessment of impacts of the possible maximum storm surges in Ise and Tokyo Bays in JapanNat. Hazards, 74 (2014), pp. 2037–2052

68. Sugi et al., 2002M. Sugi, A. Noda, N. SatoInfluence of the global warming on tropical cyclone climatology: An experiment with the JMA Global ModelJ. Meteor. Soc. Jpn., 80 (2002), pp. 249–272

69. Sugi et al., 2009M. Sugi, H. Murakami, J. YoshimuraA reduction in global tropical cyclone frequency due to global warmingSci. Online Lett. Atmos., 5 (2009), pp. 164–167

70. Sugi et al., 2012M. Sugi, H. Murakami, J. YoshimuraOn the mechanism of tropical cyclone frequency changes due to global warmingJ. Meteor. Soc. Jpn., 90A (2012), pp. 397–408

71. Tajima et al., 2014Y. Tajima, T. Yasuda, B.M. Pacheco, E.C Cruz, K. Kawasaki, H. Nobuoka, F. BrionesInitial report of JSCE-PICE joint survey on the storm surge disaster caused by Typhoon HaiyanCoast. Eng., 56(01) (2014), p. 1450006

72. Takayabu et al., 2015I. Takayabu, K. Hibino, H. Sasaki, H. Shiogama, N. Mori, Y. Shibutani, T. TakemiClimate change effects on the worst-case storm surge: a case study of Typhoon HaiyanEnviron. Res. Lett., 10 (2015), p. 064011 http://dx.doi.org/10.1088/1748-9326/10/6/064011

73. Takemi, 2009T. TakemiHigh-resolution numerical simulations of surface wind variability by resolving small-scale terrain features Theor. Appl. Mech. Jpn., 57 (2009), pp. 421–428 http://dx.doi.org/10.11345/nctam.57.421

74.

75. Takemi, 2013T. TakemiHigh-resolution meteorological simulations of local-scale wind fields over complex terrain: a case study for the eastern area of Fukushima in March 2011Theor. Appl. Mech. Jpn., 61 (2013), pp. 3–10 http://dx.doi.org/10.11345/nctam.61.3

76. Takemi et al., 2012T. Takemi, S. Nomura, Y. Oku, H. IshikawaA regional-scale evaluation of changes in environmental stability for summertime afternoon precipitation under global warming from super-high-resolution GCM simulations: a study for the case in the Kanto PlainJ. Meteor. Soc. Jpn., 90A (2012), pp. 189–212 http://dx.doi.org/10.2151/jmsj.2012-A10

77. Tsuboi and Takemi, 2014A. Tsuboi, T. TakemiThe interannual relationship between MJO activity and tropical cyclone genesis in the Indian OceanGeosci. Lett., 1 (2014), p. 9 http://dx.doi.org/10.1186/2196-4092-1-9

78. Tsuboi et al., 2015A. Tsuboi, T. Takemi, K. Yoneyama2015: Environmental characteristics for the tropical cyclone genesis in the Indian Ocean during the CINDY2011/DYNAMO Field ExperimentJ. Meteor. Soc. Jpn. (2015) (in preparation)

79. United Nations, 2012United Nations, 2012. World Urbanization Prospects: The 2011 revision. United Nations, New York.Vickery et al., 2000P. Vickery, P. Skerlj, L. TwisdaleSimulation of Hurricane Risk in the U.S. using Empirical Track ModelJ. Struct. Eng., 126 (10) (2000), pp. 1222–1237

80. Wang and Chan, 2002B. Wang, J.C.L. ChanHow strong ENSO events affect tropical storm activity over the

Western North PacificJ. Clim., 15 (2002), pp. 1643–1658

81. Watanabe et al., 2010M. Watanabe, T. Suzuki, R. O□ishi, Y. Komuro, S. Watanabe, S. Emori, T. Takemura, M. Chikira, T. Ogura, M. Sekiguchi, K. Takata, D. Yamazaki, T. Yokohata, T. Nozawa, H. Hasumi, H. Tatebe, M. KimotoImproved climate simulation by MIROC5: mean states, variability, and climate sensitivityJ. Clim., 23 (2010), pp. 6312–6335
82. Woth, 2005K. WothNorth Sea storm surge statistics based on projections in a warmer climate: How important are the driving GCM and the chosen emission scenario?Geophys. Res. Lett., 32 (2005) http://dx.doi.org/10.1029/2005GL023762
83. Wu, 2013C.-C. WuTyphoon Morakot: key findings from the Journal TAO for improving prediction of extreme rains at landfallBull. Am. Meteor. Soc., 94 (2013), pp. 155–160 http://dx.doi.org/10.1175/BAMS-D-11-00155.1
84. Xiang et al., 2015B. Xiang, S.-J. Lin, M. Zhao, S. Zhang, G. Vecchi, T. Li, X. Jiang, L. Harris, J.-H. ChenBeyond weather time-scale prediction for Hurricane Sandy and Super Typhoon Haiyan in a global climate modelMon. Weather Rev., 143 (2015), pp. 524–535 http://dx.doi.org/10.1175/MWR-D-14-00227.1
85. Xie and Zhang, 2012B. Xie, F. ZhangImpacts of typhoon track and island topography on the heavy rainfalls in Taiwan associated with Morakot (2009)Mon. Weather Rev., 140 (2012), pp. 3379–3394 http://dx.doi.org/10.1175/MWR-D-11-00240.1
86. Yasuda et al., 2014T. Yasuda, S. Nakajo, S. Kim, H. Mase, N. Mori, K. HorsburghEvaluation of future storm surge risk in East Asia based on state-of-the-art climate change projectionCoast. Eng., 83 (2014), pp. 65–71 http://dx.doi.org/10.1016/j.coastaleng.2013.10.003
87. Yoshida et al., 2014R. Yoshida, Y. Kajikawa, H. IshikawaImpact of boreal summer intraseasonal oscillation on environment of tropical cyclone genesis over the western North PacificSci. Onlline. Lett. Atmos., 10 (2014), pp. 15–18 http://dx.doi.org/10.2151/sola.2014-004
88. Yoshino et al., 2008J. Yoshino, T. Murakami, K. Kobayashi, T. YasudaAn estimation method for potential maximum storm surge heights using a new tropical cyclone initialization scheme and a coupled atmosphere–ocean-wave modelSolut. Coast. Disasters Conf., 2008 (2008), pp. 256–267
89. Yuan and Cao, 2013J.P. Yuan, J. CaoNorth Indian Ocean tropical cyclone activities influenced by the Indian Ocean Dipole modeSci. China: Earth Sci., 56 (2013), pp. 855–865
90. Zhang et al., 2010 F. Zhang, Y. Weng, Y.-H. Kuo, J.S. Whitaker, B. XiePredicting Typhoon Morakot's catastrophic rainfall with a convection-permitting mesoscale ensemble systemWeather Forecast., 25 (2010), pp. 1816–1825 http://dx.doi.org/10.1175/2010WAF2222414.1
91. Zhao et al., 2014C. Zhao, G. Jianzhong, P. DingImpact of sea level rise on storm surge around the Changjiang estuaryJ. Coast. Res., 68 (2014), pp. 27–34 http://dx.doi.org/10.2112/SI68-004.1

CHAPTER 6

IMPACT OF WEATHER EXTREMES ON GLOBAL FOOD SECURITY

Sika Gbegbelegbe, Uran Chung, Bekele Shiferaw, Siwa Msangi, Kindie Tesfaye

Keywords: Spatial bio-economic modelling, Weather extreme, IMPACT, Food security, Developing world

INTRODUCTION

The world food system is more vulnerable to weather extremes for various reasons, the key one being that crop yield changes have not been keeping up with rising food demand worldwide (Ortiz et al., 2008 and Boyer et al., 2013). Yield growth rates have collapsed or are stagnating in about 30%, 38% and 39% of global harvested area for maize, rice and wheat, respectively (Ray et al., 2012). In addition, wheat yield gains across the developing world have remained below population growth in recent years (Ortiz et al., 2008). Other factors that make the global food system more vulnerable to weather extremes include the globalization of grain markets and rising fuel prices which increase food transportation costs (Brown and Funk, 2008).

The globalization of grain markets implies that weather extremes that reduce food harvests in key breadbaskets influence world food prices and can have negative effects on food security in poor net maize-importing countries (Chung et al., submitted for publication). However, there is no empirical study on the subject. Most studies that aim to estimate the socio-economic impact of climate extremes use statistical approaches to quantify the direct economic losses brought by climate extremes (Pielke and Landsea, 1998, Changnon, 2003a, Hallegatte, 2007 and Pielke, 2007). Such direct losses usually consist of estimated financial losses (e.g. property or crop losses) closely linked to the climate extreme (Changnon, 2003a, Changnon, 2003b, Hallegatte, 2007 and Hallegatte et al., 2007). Other studies have attempted to quantify the effect of climate extremes on human mortality (Kunkel et al., 1999 and Deschenes and Moretti, 2009). Studies that have estimated the primary and secondary effects of climate extremes used process-based models to analyse the impact of extreme weather events on poverty vulnerability in developing countries (Adger, 1999 and Ahmed et al., 2009), producer and consumer welfare in Spain (Mechler et al., 2010), and food security in China (Nelson et al., 2010). Unlike statistical models, process-based models have the ability to consider trade effects in the socio-economic analysis of climate extremes.

Mechler et al. (2010) used a process-based model, to quantify the economic effect of a combined heat wave and drought in Spain. Similarly, Nelson et al. (2010) used a process-based model to estimate the impact of an extended drought in China on food security in the nation. These two studies highlighted the need to consider trade effects in the economic analysis of extreme events. Since the USA is the leading maize producer and exporter globally, it is very likely that the extreme weather event of 2012 in the USA would have indirect secondary effects on food security in other parts of the world where maize is a staple food.

This study assesses the impact of the 2012 weather extreme in the USA on food security across the developing world. The study also estimates the plausible effects on food security of a similar weather extreme occurring in 2050.

MATERIALS AND METHODS

Methodological Framework: spatial Bio-economic Modelling

A spatial bio-economic framework which is described in Nelson et al. (2010) and illustrated in Fig. 1 was used to estimate the global impact of the climate extreme occurring in the USA. Crop and spatial modelling were combined to estimate the biophysical impact of the climate extreme on

Citation: Sika Gbegbelegbe, Uran Chung, Bekele Shiferaw, Siwa Msangi, Kindie Tesfaye, Impact of weather extremes on global food security, http://dx.doi.org/10.1016/j.wace.2014.05.005.

maize yields in the USA under various climate models. The analysis involves a high performance computing cluster where the Geographic Resources Analysis Support System (GRASS, http://grass.osgeo.org/) is used to handle spatial data defined and the Decision Support System for Agrotechnology Transfer (DSSAT) (Hoogenboom et al., 1999 and Jones et al., 2003) was used to simulate crop yield and other biophysical variables. The biophysical results were defined at the 5′ spatial resolution were averaged to the level of Food Production Units (FPUs), which are production zones defined within countries (Rosegrant and et al., 2012). These results were then inputted into the International Model for Policy Analysis of Agricultural Commodities and Trade (IMPACT) to derive socio-economic results. IMPACT is a multi-market, multi-country model that focuses on the agricultural sector and projects trends in national and global food security under alternative scenarios on population growth, income growth and future climates (Rosegrant et al., 2008 and Rosegrant et al., 2012).

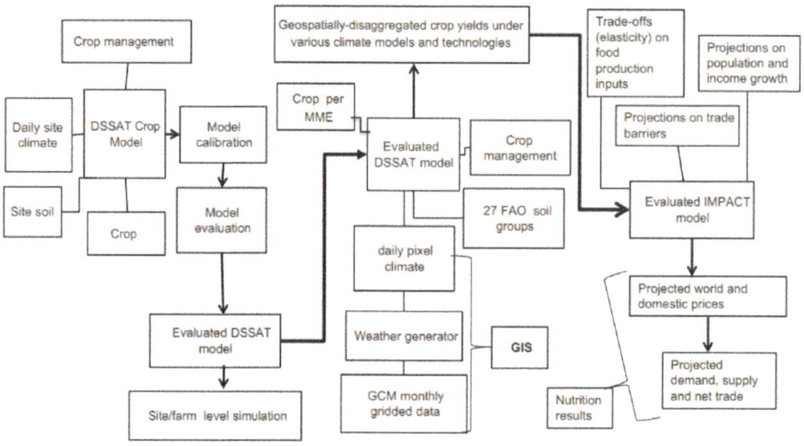

Figure 1: Illustration of the bio-economic modelling framework; sources consist of authors and adaptation from (Rosegrant et al., 2008).

Incorporation of Biophysical Results into IMPACT

Chung et al. (submitted for publication) used geo-spatial crop modelling at the 5′ spatial resolution to estimate the impact of the weather extreme on maize yields across the USA. The biophysical results were then aggregated to the levels of Food Production Units and inputted into IMPACT. More specifically, the biophysical maize yield change brought by the extreme weather event is used as maize area change (represented as "al" in eq. 1) for irrigated and rainfed maize in the USA. For example, the biophysical analysis byChung et al. (submitted for publication) implies a reduction of 46% in rainfed maize produced in Ohio. This loss which is represented by "al" in Eq. (1) is inputted as maize area loss in 2012 due to the weather extreme in IMPACT. After 2012, the maize area in the USA is brought back to trend by re-adjusting area growth rates. More specifically, the area growth rate is adjusted as follows in 2012:

$$ar = \text{area recovery rate} = \frac{1}{1+al} \tag{1}$$

Since IMPACT involves a partial equilibrium economic model, simulated per capita income would not change from one socio-economic scenario to the next. In reality, the weather extreme, by reducing maize production in the USA, would affect the wages of maize farmers and other agents involved in the maize value chain in the USA. In turn, agricultural GDP and hence national income within the USA would be affected. Since the USA is a major maize producer and exporter across the world, the climate extreme would most likely affect wages in other countries. The effect of the climate extreme on wages would ultimately affect the ability of households at cushioning themselves against increased food insecurity. A multi-country Computable General Equilibrium (CGE) model would be able to capture the effect of the climate extreme on GDP across world countries, as agricultural and non-agricultural income are endogenous variables in CGE models. IMPACT, a multi-market, multi-country partial equilibrium which focuses on the agricultural sector, estimates the impact of the climate extreme on food availability per capita without assessing the secondary effects of the climate extreme on wages and hence on the ability of households to cushion themselves against increased food insecurity.

IMPACT is well calibrated for baseline food production and consumption across the world: simulated annual production and consumption of selected commodities from IMPACT is similar to FAO's averaged annual production between 1999 and 2001 (Fig. 2).

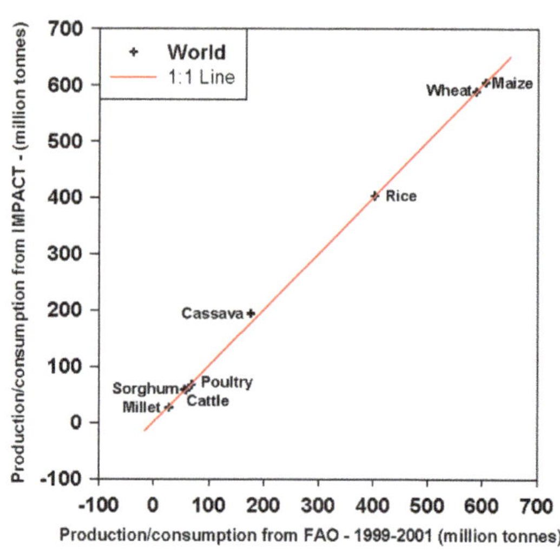

Figure 2: Validation of IMPACT – simulated versus observed world production/consumption values for selected commodities.

Weather Extreme Scenarios

Six scenarios are analysed in this study. They all involve one socio-economic scenario (Base) which consists of moderate growth in population and world economies between 2000 and 2050. Agricultural technologies are assumed to remain unchanged over the years, although yield improvements from enhanced agronomic practices are considered in the model. Consumers' diets are assumed to change with income growth and higher income is associated with increased consumption of animal-sourced foods.

In the first scenario (Base-2012), the emission of greenhouse gases stops in 2000 such that the projected climate around 2050 is identical to the baseline climate (1950–2000). In the second scenario (Base-2012-EW), the climate extreme occurs in 2012 under the baseline climate (Table 1).

Table 1: Bio-economic scenarios involving a weather extreme

Scenario name	Socio-economic scenario	Climate model	Emission scenario	Weather extreme (yes/no)
Scenarios for 2012				
Base-2012	Base	Baseline	None	No
Base-2012-WE	Base	Baseline	None	Yes
Scenarios for 2050				
Base-2050	Base	Baseline	None	No
CSI2050-B1-EW	Base	CSIRO-Mk3.0	B1	Yes
MIR2050-A1-EW	Base	MIROC 3.2	A1	Yes

The three other scenarios relate to the 2050s (Table 1). The third scenario (Base-2050) projects the world food system in 2050, under the baseline climate scenario and the base socio-economic scenario. In the fourth and fifth scenarios, namely 'CSI2050-B1-EW' and 'MIR2050-A1-EW', the weather extreme occurs in the USA in 2050 under the CSIRO and MIROC GCMs, respectively. Previous analysis has shown that the combination of the CSIRO-Mk3.0 climate model for 2050 and the B1 emission scenario leads to the mildest changes in the key variables affecting crop growth, namely precipitation and temperature, compared to the baseline climate (Nelson et al., 2010). Similarly, the combination of the MIROC 3.2 climate model with the A1B emission scenario around 2050 leads to the largest changes in mean precipitation and temperature compared to the baseline climate (Nelson et al., 2010). Hence, the range of maize yields in the USA when the weather extreme occurs under climate change and the yield range of other crops on a global scale should all be encompassed by the yields generated under the 'CSI2050-B1-EW' and 'MIR2050-A1-EW' scenarios.

RESULTS: IMPACT OF THE WEATHER EXTREME ON FOOD SECURITY

Impact of the 2012 Weather Extreme on Food Security Across Developing World

Under the 'Base-2012' scenario, maize production in the USA would reach 334 million metric tons in 2012. However, the extreme weather event consisting of combined heat wave and drought would decrease maize production in the USA by 29% compared to the 'Base-2012' scenario. On the other hand, maize consumption in the USA, which would have stood at 244 million metric tons under the 'Base-2012' scenario, would decrease by 5% only due to the weather extreme. Among all maize uses, the consumption of maize as animal feed would experience the largest decrease. The decrease in total maize consumption would be half of that of total maize production in volume terms and this implies that the country will have to reduce its net maize exports to meet its demand requirements. More specifically, net maize exports in the USA would decrease by 83 million tons under the 'Base-2012-EW' scenario compared to the 'Base-2012' scenario (Fig. 3).

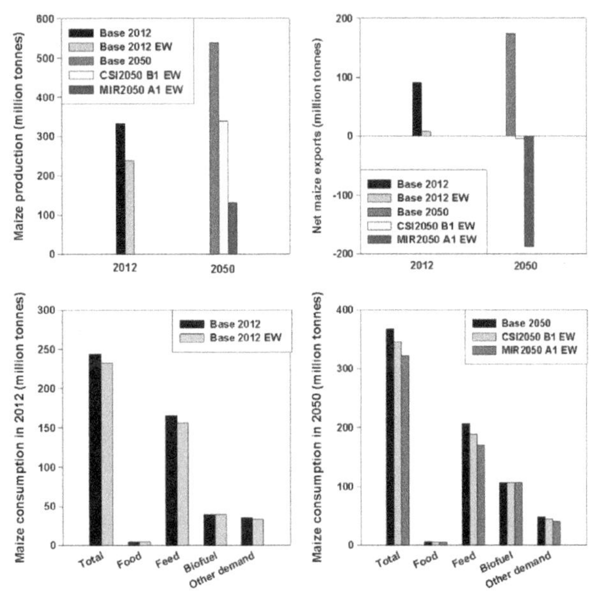

Figure 3: Projected impact of extreme weather event on maize production consumption and net exports for the USA – results from IMPACT.

The decrease in maize production in the USA due to the weather extreme would act as a negative supply shock across world maize markets. Hence, world maize prices would increase to reflect increased maize scarcity on a global scale. The increased world maize prices would make maize production attractive in other regions not affected by the extreme weather event: they would increase their maize production (Table 2). Across the developing world, E & SE Asia and the LAC region would lead the increase in maize production in volume terms (Table 2). However, the increased maize production in other regions of the world would not be enough to compensate the decrease in maize production in the USA. Hence, global maize production would fall by about 48 million tons compared to the 'Base-2012' scenario; this drop would represent about 6% of global maize production under the 'Base-2012'; scenario. The projected increase in maize production in areas not affected by the weather extreme (Table 2) is likely to be overstated. Some large maize-producing countries, including China, plant maize around the same time as the USA. These countries would not be able to adjust their maize production the year when the weather extreme hits the USA.

Table 2: projected impact of the weather extreme on global maize production and net maize trade – results from IMPACT

	Production change in million tonnes (%)			Change in net exports in million tonnes (%)		
Region	Base-2012-EW vs Base-2012	CSI2050-B1-EW vs Base-2050	MIR2050-A1-EW vs Base-2050	Base-2012-EW vs Base-2012	CSI2050-B1-EW vs Base-2050	MIR2050-A1-EW vs Base-2050
SSA	4.5 (11)	8.4 (13)	21.1 (32)	9.3 (−94)	23.5 (−47)	51.2 (−102)
CWANA	1.4 (7)	−2.9 (−8)	−5.5 (−16)	2.9 (−18)	1.8 (−5)	4.2 (−10)
South Asia	2.1 (11)	2.2 (5)	11.2 (25)	3.6 (−151)	10.4 (−36)	28.0 (−96)
E & SE Asia	15.4 (8)	18.0 (6)	48.3 (16)	32.9 (−58)	64.1 (−63)	139.9 (−138)
LAC	13.5 (13)	33.6 (16)	42.1 (20)	19.7 (−167)	49.1 (253)	73.3 (377)
USA	−95.4 (−29)	−201.1 (−37)	−407.8 (−76)	−83.8 (−92)	−178.7 (−103)	−362.2 (−208)
ROW	10.4 (11)	21.2 (16)	48.1 (37)	15.4 (288)	29.8 (108)	65.6 (238)
World	−48.1 (−6)	−120.7 (−9)	−242.5 (−18)	0.0 (0)	0.0 (0)	0.0 (0)
	Major net maize exporters in world (million tonnes)					
Country	Base-2012	Base-2012-EW	Country	Base-2050	CSI2050-B1-EW	MIR2050-A1-EW
United States	91.6	7.8	United States[c]	174.0	−4.7	−188.2
Argentina	10.3	13.5	Argentina	20.9	30.7	37.1
Ctr. Eur.[d]	8.1	11.6	Brazil	19.5	48.6	57.8
France[a]	6.0	8.1	Ctr. Eur.[d]	11.7	19.0	22.7
Ukraine[a]	4.7	5.6	Ukraine[a]	11.7	13.7	14.8
Brazil[a]	−4.4	5.3	Russia[a]	8.4	9.9	12.7
Russia[a]	1.8	2.3	France[a]	9.4	9.1	20.6
Vietnam[a]	1.4	2.0	Adriatic[e],[a]	4.1	7.7	8.7
India[a]	−1.5	1.7	China[b]	−55.7	−1.6	60.1
Global exports	127.8	71.9	Global exports	271.0	168.0	291.9

a. becomes major exporter when climate extreme occurs
b. become major exporter when climate extreme occurs under MIR-2050-A1-EW only
c. no longer major exporter when climate extreme occurs in 2050 under climate change
d. region includes Bulgaria, Czech Republic, Hungary, Moldova, Romania, Slovak Republic, and Slovenia
e. region includes Albania, Bosnia and Herzegovina, Croatia, Greece, Montenegro, Serbia, and Macedonia

The decrease in global maize production would affect maize trade across the globe. All regions across the developing world would reduce their net maize imports. South Asia and the LAC region would even become net maize exporters. Under the 'Base-2012' scenario, the leading maize exporters globally in 2012 would be the USA and Argentina. These two countries would account for 80% of global maize exports in 2012, with the USA alone accounting for 72% of global maize exports. With the climate extreme, the USA would become the fourth leading maize exporter by accounting for 8% of global maize exports. India and Brazil, two countries that would have been net maize importers under the 'Base-2012' scenario, would join the list of leading maize exporters. Vietnam would be the only country from E & SE Asia to join the list of major maize exporters (Table 2).

The decrease in global maize production means that global maize consumption would also have to decrease. E & SE Asia would experience the largest decrease in maize consumption in volume terms. However, Africa would experience the largest relative decrease. More specifically, maize consumption in SSA would decrease by 9% compared to the 'Base-2012' scenario. By contrast, E & SE Asia would experience a decrease of 7% (Table 3). The bulk of the reduction in maize consumption would relate to food in SSA. In all other regions of the developing world, the bulk of the reduction in maize consumption would relate to animal feed. The only exception would be South Asia where the consumption of maize as food and feed would decrease by roughly 40% each (Table 3).

Table 3: Projected impact of climate extreme on global maize consumption – results from IMPACT

	Food	Feed	Biofuels	Other use	Total	Calorie - base	Caloric change (food)	Caloric change (maize)
Region	Base-2012-EW vs Base-2012 in million tonnes (%)					2012	2012	2012
SSA	−3.7 (−10)	−0.4 (−5)	0.0 (0)	−0.7 (−10)	−4.8 (−9)	2,220.2	−35.3	−39.1
CWANA	−0.5 (−4)	−0.9 (−4)	0.0 (0)	−0.1 (−4)	−1.5 (−4)	2,776.6	−3.9	−5.7
South Asia	−0.7 (−7)	−0.6 (−7)	0.0 (0)	−0.2 (−7)	−1.5 (−7)	2,389.8	−4.8	−3.2
E & SE Asia	−2.2 (−6)	−14.0 (−8)	0.0 (0)	−1.3 (−5)	−17.5 (−7)	3,034.1	−9.5	−7.5
LAC	−1.8 (−7)	−3.6 (−5)	0.0 (0)	−0.8 (−6)	−6.2 (−6)	2,812.9	−28.1	−26.1
USA	−0.3 (−6)	−9.2 (−6)	0.0 (0)	−2.1 (−6)	−11.6 (−5)	3,687.6	−4.4	−6.1
ROW	−0.3 (−5)	−3.9 (−6)	0.0 (0)	−0.8 (−6)	−5.0 (−6)	3,287.0	−1.9	−2.7
Region	CSI2050-B1-EW vs Base-2050 in million tonnes (%)					2050	2050	2050
SSA	−11.3 (−15)	−1.9 (−7)	0.0 (0)	−2.0 (−15)	−15.1 (−13)	2,650.5	−70.3	−54.1
CWANA	−1.1 (−6)	−3.3 (−6)	0.0 (0)	−0.3 (−6)	−4.7 (−6)	3,091.7	−24.9	−8.0
South Asia	−1.3 (−10)	−6.5 (−12)	0.0 (0)	−0.4 (−9)	−8.2 (−11)	2,649.0	−33.3	−4.8
E & SE Asia	−3.7 (−9)	−40.5 (−12)	0.0 (0)	−1.9 (−7)	−46.1 (−11)	3,560.0	−40.5	−12.0
LAC	−3.1 (−10)	−11.2 (−8)	0.0 (0)	−1.3 (−9)	−15.5 (−8)	3,044.5	−59.5	−35.0
USA	−0.5 (−8)	−18.0 (−9)	0.0 (0)	−4.0 (−8)	−22.4 (−6)	4,009.0	−20.3	−9.2
ROW	−0.5 (−8)	−6.9 (−9)	0.0 (0)	−1.3 (−9)	-8.6 (−8)	3,562.5	−23.2	−4.3
Region	MIR2050−A1−EW vs Base-2050 in million tonnes (%)					2050	2050	2050
SSA	−22.4 (−29)	−3.7 (−14)	0.0 (0)	−4.0 (−30)	−30.1 (−26)	2,650.5	−108.0	−107.5
CWANA	−2.4 (−14)	−6.7 (−13)	0.0 (0)	−0.6 (−12)	−9.7 (−13)	3,091.7	−24.9	−17.3
South Asia	−2.8 (−20)	−13.2 (−24)	0.0 (0)	−0.9 (−20)	−16.8 (−23)	2,649.0	−32.0	−10.2
E & SE Asia	−7.8 (−18)	−79.6 (−24)	0.0 (0)	−4.2 (−15)	−91.7 (−23)	3,560.0	−58.8	−25.6
LAC	−6.4 (−20)	−22.1 (−16)	0.0 (0)	−2.7 (−19)	−31.2 (−17)	3,044.5	−121.8	−73.0
USA	−1.0 (−17)	−36.3 (−18)	0.0 (0)	−8.3 (−17)	−45.6 (−12)	4,009.0	−41.4	−19.3
ROW	−0.9 (−16)	−13.8 (−18)	0.0 (0)	−2.7 (−18)	−17.4 (−17)	3,562.5	−13.6	−9.0

One indirect effect of the weather extreme consists of the changes in the demand for maize substitutes and complements. The negative impact of the climate extreme on global maize production would lead to higher maize prices and hence would incite consumers to substitute maize with other products. Another indirect effect of the weather extreme consists of changes in the production of other crops. Maize competes with other crops for the allocation of land and other agricultural inputs. Hence, the increased maize production recorded in countries not affected by the climate extreme should affect the production of other agricultural crops. These two indirect effects should lead to some changes in the production and consumption of other major cereals, namely rice and wheat; and other important staple foods, including cassava in Sub-Saharan Africa. Overall, the production of rice and wheat would increase globally by 2 million tons (Table 4). Such increase reflects a global shift away from the consumption of maize, which has become scarcer and hence more expensive, towards the consumption of other major cereals. CWANA, South Asia and E & SE Asia would increase their production of rice and wheat by a small margin; In SSA and LAC, the production of wheat and rice would be taxed by the increase in maize production (Table 2) and hence would decrease slightly (Table 4). The consumption of rice and wheat would substantially increase everywhere except in South Asia where their consumption would decrease by 89,000 tons compared to the 'Base-2012' scenario. In all other regions of the developing world, the consumption of wheat and rice would increase by 1.2 million tons (Table 4).

Table 4: Projected impact of the climate extreme on the production and consumption of rice, wheat and cassava – results from IMPACT.

	Base-2012-EW vs Base-2012	CSI2050-B1-EW vs Base-2050	MIR2050-A1-EW vs Base-2050
Region	Change in production of rice and wheat in thousand tons (%)		
SSA	−8.0 (−0.0)	732.8 (1.7)	−975.2 (−2.2)
CWANA	149.9 (0.1)	−6,018.3 (−3.6)	3,044.9 (1.8)
South Asia	565.5 (0.3)	−16,914.6 (−6.3)	8,746.3 (3.2)
E & SE Asia	555.2 (0.2)	−6,989.6 (−1.8)	17,529.2 (4.6)
LAC	−6.0 (−0.0)	−1,008.3 (−1.6)	−4,266.8 (−6.6)
USA	−16.2 (−0.0)	9,843.7 (10.4)	−7,661.1 (−8.1)
ROW	747.3 (0.3)	9,877.1 (2.3)	−2,362.4 (−0.6)
World	1,987.6 (0.2)	−10,477.2 (−0.7)	14,055.0 (1.0)
	Change in consumption of rice and wheat in thousand tons (%)		
SSA	302.0 (0.8)	−1,660.8 (−1.4)	2,721.6 (2.2)
CWANA	339.8 (0.2)	−363.0 (−0.2)	3,520.7 (1.5)
South Asia	−88.8 (−0.0)	−4,033.0 (−1.4)	−1,034.1 (−0.4)
E & SE Asia	406.6 (0.1)	−3,870.1 (−1.0)	1,874.3 (0.5)
LAC	176.4 (0.3)	−365.4 (−0.6)	748.5 (1.2)
USA	246.9 (0.6)	28.4 (0.0)	1,764.0 (2.8)
ROW	604.8 (0.3)	−213.3 (−0.1)	4,460.0 (1.7)
World	1,987.6 (0.2)	−10,477.2 (−0.7)	14,055.0 (1.0)
	Change in production of cassava in thousand tons (%)		
SSA	356.3 (0.2)	−2,049.6 (−0.8)	1,015.1 (0.4)
CWANA	1.7 (0.3)	7.6 (0.7)	13.5 (1.3)
South Asia	25.4 (0.3)	46.8 (0.4)	531.3 (4.0)
E & SE Asia	8.9 (0.0)	−307.1 (−0.5)	1,380.6 (2.2)
LAC	64.5 (0.2)	1,144.9 (2.1)	−2,152.5 (−3.9)
USA	0.0 (−0.0)	0.0 (0.2)	−0.8 (−9.9)
ROW	0.4 (0.1)	147.3 (23.0)	271.0 (42.4)
World	457.1 (0.2)	−1,010.1 (−0.3)	1,058.3 (0.3)
	Change in consumption of cassava in thousand tons (%)		
SSA	535.2 (0.4)	−180.3 (−0.1)	2,074.8 (0.8)
CWANA	2.0 (0.1)	−14.5 (−0.6)	−4.0 (−0.2)

South Asia	−15.3 (−0.2)	−69.0 (−0.7)	−110.5 (−1.1)
E & SE Asia	14.3 (0.0)	−74.4 (−0.1)	−38.1 (−0.1)
LAC	−88.8 (−0.2)	−615.8 (−1.3)	−827.1 (−1.7)
USA	0.2 (0.1)	−1.7 (−0.7)	−2.2 (−0.9)
ROW	9.5 (0.1)	−54.2 (−0.8)	−34.7 (−0.5)
World	457.1 (0.2)	−1,010.1 (−0.3)	1,058.3 (0.3)

The importance of cassava in African diets is illustrated by the fact that Sub-Saharan Africa would experience the highest increase in cassava production and consumption after the weather extreme. More specifically, the sub-region would account for 78% of the increase of 457 thousand tons in world cassava production in 2012 under the 'Base-2012-WE' scenario (Table 4). The sub-region would also increase its cassava consumption by 535 thousand tonnes and would have to rely on imports to meet its consumption requirements (Table 4).

The drop in global maize production coupled with some adjustments in food consumption across the globe would lead to world maize prices increasing by 17% in 2012 if the climate extreme occurs under the baseline climate (Fig. 4). World prices of wheat and rice would also increase, albeit slightly: about 1%, each (Fig. 4).

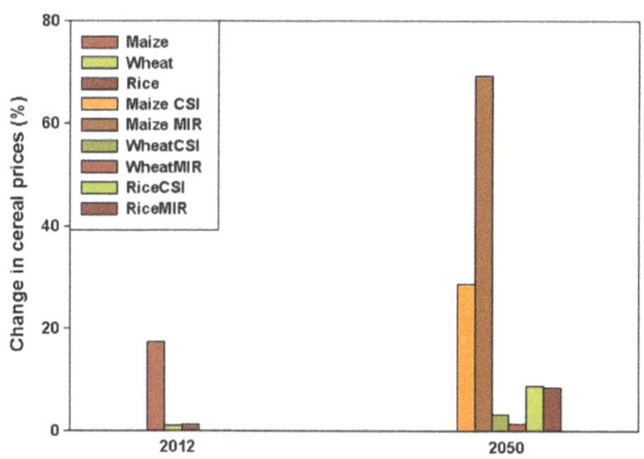

Figure 4: Projected impact of the climate extreme on key cereal prices.

Daily caloric intake derived from maize would decrease most in SSA compared to other regions of the developing world when the weather extreme affects the USA (Table 3). This, coupled with SSA having the lowest average per capita caloric intake in 2012 (Table 3), suggests that consumers in SSA would find it more difficult to substitute maize for other food products. Hence, the effects of the weather extreme on food security would likely be worst in SSA. Another region that would experience a substantial decrease in caloric intake from maize is the LAC region; however, its average per capita daily caloric intake, under the 'Base-2012' scenario, would be much higher than that of SSA (Table 3).

The number of people at risk of hunger would increase most in SSA, the region that would also experience the largest decrease in per capita caloric availability due to the weather extreme. Within SSA, countries that would experience an increase of 1 million or more in the number of people at risk of hunger would mainly be from Eastern Africa and would consist of Kenya, Tanzania, Ethiopia, Malawi, Zambia, Zimbabwe and the Democratic Republic of Congo (Fig. 5).

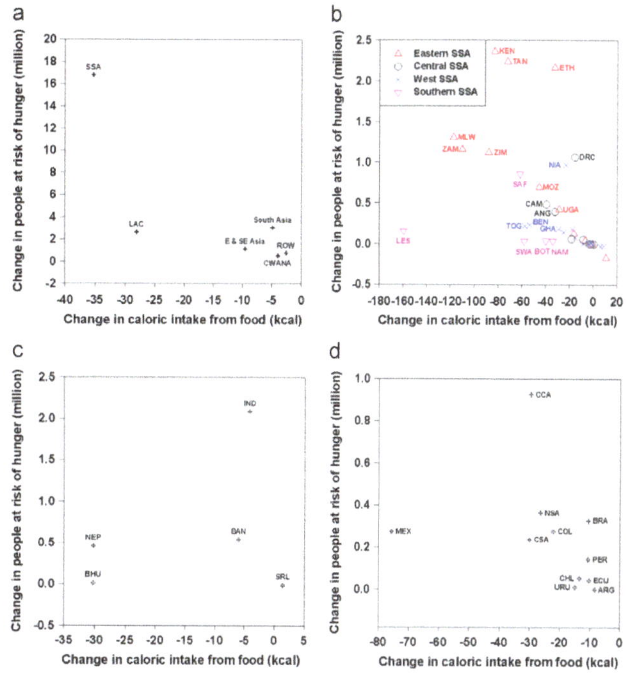

Figure 5: Projected impact of extreme weather event on number of people at risk of hunger in 2012 – results from IMPACT. CCA region includes Belize, Costa Rica, Cuba, Dominican Republic, El Salvador, Guatemala, Haiti, Honduras, Nicaragua, and Panama; CSA includes Bolivia and Paraguay; NSA includes French Guiana, Guyana, Suriname and Venezuela. (a) World, (b) SSA, (c) south asia and (d) LAC.

Despite experiencing larger reductions in per capita caloric availability, Malawi would have a smaller increase in the number of people at risk of hunger compared to Kenya and Tanzania (Fig. 5). The higher population levels in Kenya and Tanzania would explain this result. Kenya, Tanzania and Malawi would have similar per capita caloric intake in 2012 under the 'Base-2012' scenario with 2070, 2081 and 2018 calories, respectively. Hence, their shares of people at risk of hunger would be similar: they would stand at 30, 27 and 34 for Kenya, Tanzania and Malawi, respectively. However, the population in each of Kenya and Tanzania would be twice as high as that of Malawi. More specifically, Kenya, Tanzania and Malawi would have about 43, 48 and 16 million people, respectively. The combination of similar per capita caloric intake across the 3 countries and a much smaller population in Malawi implies that the weather extreme would lead to a smaller increase in the number of at-risk people in Malawi, even if the country experiences a larger reduction in caloric intake, compared to Kenya and Tanzania. However, in relative terms, the effect of the weather extreme would be more severe in Malawi compared to Kenya or Tanzania. The relative increase in the number of people at risk of hunger would be 24% in Malawi compared to the 'Base-2012' scenario; it would stand at 19 and 17% in Kenya and Tanzania, respectively. Relative values do away with the effect that population levels can have on estimating the impact of the weather extreme.

Ethiopia and Zambia would also have similar daily per capita caloric intake in 2012, under the 'Base-2012' scenario; however, the Ethiopian population would be 6 times higher than that of Zambia. Hence, the weather extreme would lead to Ethiopia having a higher increase in the number of people at risk of hunger (Fig. 4), although its relative effect would be more severe in Zambia which would experience a higher reduction in caloric intake due to the weather extreme. In relative terms, the increase in the number of people at risk of hunger would be 20% in Zambia and 5% in Ethiopia.

Lesotho, which would experience the largest reduction in caloric intake in SSA would also account for the largest relative increase in the number of people at risk of hunger: 46%. Lesotho would be followed by South Africa which would experience a relative increase of 41% in the number of people at risk of hunger. South Africa and Tanzania would have similar population sizes in 2012. They would also experience a similar reduction in per capita caloric intake due to the weather extreme (Fig. 5). However, South Africa would have a higher per capita caloric intake which would ensure that a smaller portion of its population would be at risk of hunger in 2012 under the 'Base-2012' scenario. The higher caloric intake in South Africa would also dampen the negative effect of the weather extreme on the number of people at risk of hunger (Fig. 5). In relative terms, South Africa would experience a larger increase in the number of people at risk of hunger compared to Tanzania, due to its smaller population at risk of hunger under the 'Base-2012' scenario.

The LAC region, which would experience the second largest decrease in per capita caloric availability due to the weather extreme, would see an increase of 2.6 million in the number of people at risk of hunger. Within the LAC region, countries from the Caribbean and Central America (CCA) would be hardest-hit, despite experiencing a smaller decrease in per capita caloric intake compared to Mexico (Fig. 5). In 2012, prior to the weather extreme, Mexico would have a higher daily per capita caloric intake compared to CCA; the caloric intake levels would stand at 3124 and 2458 in Mexico and CCA, respectively. Hence, CCA would have a higher share of its population at risk of hunger: 21% in CCA versus 5% in Mexico. These would correspond to 16 million people at risk of hunger in CCA versus 5.8 million in Mexico whose population would be 1.5 times higher than that of CCA in 2012. The weather extreme would lead to a higher reduction in caloric intake in Mexico; however, its effect of food security would not be worse in Mexico compared to CCA, because of the less precarious food security situation in Mexico, prior to the weather extreme. In relative terms, the increase in the number of people at risk of hunger caused by the weather extreme would be higher in CCA compared to Mexico: 5.8 versus 4.8%.

Across the LAC region, few other countries CSA, NSA, Chile, Uruguay, and Colombia would experience a higher relative increase in the number of people at risk of hunger compared to CCA (Fig. 5). CSA and NSA would have the highest relative increase in the number of people at risk of hunger with a value of 8.5%, each. Chile, Uruguay, and Columbia would follow with increases of 7.2%, 7%, and 6%, respectively.

In the CSA and NSA regions, caloric intake in 2012 prior to the weather extreme would be smaller than that of CCA. The decrease in food caloric intake brought by the weather extreme in the CSA and NSA regions would be similar to that of the CCA region. Hence, the proportional increase in the number of people at risk of hunger would be lower in CCA compared to CSA and NSA.

Chile, Colombia and Uruguay would have a higher caloric intake compared to CCA prior to the weather extreme. Hence, compared to the CCA region, these three countries would have a smaller proportion of their population that would be at risk of hunger prior to the weather extreme. In addition, each of Chile, Colombia and Uruguay would have a much smaller population compared to the CCA region. Hence, the proportional increase in the number of people at risk of hunger due to the weather extreme would be higher in these three countries compared to CCA.

Interestingly, South Asia, where the reduction in caloric intake would be less than one-sixth of that of SSA, would be the second hardest-hit region in terms of food insecurity; the number of people at risk of hunger in this region would increase by 3 million due to the weather extreme. India alone would account for more than two-thirds of the increase in the number of at-risk people in South Asia, despite experiencing a reduction in caloric intake that would be substantially less than that of Nepal or Bhutan (Fig. 5).

The very high population in India explains why the increase in the number of people at risk of hunger would be much higher in value terms compared to Nepal or Bhutan. By 2012, India's population would stand at around 1.2 billion; in Nepal and Bhutan, the population numbers would stand at 31 million and 749 thousand, respectively. Prior to the weather extreme, daily per capita caloric intake would be higher in India than in Nepal and Bhutan. The values would be 2423, 2296 and 2229 in India, Nepal and Bhutan, respectively. Hence, the share of the population at risk of hunger would be lower in India; these shares would be 17%, 18% and 28% in India, Nepal and Bhutan, respectively. In relative terms, the impact of the weather extreme on the number of people at risk of hunger would be slightly higher in Nepal than in Bhutan: 8% and 6%, respectively. By contrast, India would experience a 1% increase in the number of people at risk of hunger due to the weather extreme. Apart from Bhutan and Nepal, Bangladesh would also have a higher relative increase in the number of people at risk of hunger compared to India.

Surprisingly, despite experiencing the largest reduction in maize consumption across the developing world (Table 3), E & SE Asia would not experience a substantial reduction in caloric intake compared to SSA and the LAC region (Fig. 5). This is mainly caused by the fact the bulk of the reduction in maize consumption in E & SE Asia would consist of a reduction in animal feed. In addition, the region would have the highest base caloric intake in 2012 without the weather extreme, among all developing regions (Table 3). Hence, E & SE Asia would experience the fourth largest increase in the number of people at risk of hunger, in both value and relative terms. The region would be ranking behind SSA, LAC and South Asia.

Results – Impact of a 2050 Weather Extreme on Global Food Security

Under the 'Base-2050' scenario, the world economies are projected to experience moderate growth in per capita income between 2000 and 2050. Under such scenario, countries which were ranked as developing countries in 2012 might no longer be considered as such in 2050. Hence, in this section, they are referred as developing countries/regions of 2012.

Under the 'Base-2050' scenario, global maize production would almost double by 2050 compared to 2012. Just like in the 2000s and 2010s, the USA would remain the leading maize producer and exporter in the world in 2050: the country would account for 40% and 64% of global maize production and exports, respectively (Table 2). Hence, an extreme weather event that would affect maize production in the USA in 2050 could still have serious repercussions for food security in vulnerable regions.

If the weather extreme occurs in 2050 instead of 2012, its effects on maize production in the USA would be more pronounced. More specifically, the reduction in maize production caused by the weather extreme would range between 37% under the 'CSI2050-B1-EW' climate scenario and 76% under the 'MIR2050-A1-EW' climate scenario. However, maize consumption in the USA would barely decrease. In this case, the country would have to become a net maize importer to meet its consumption requirements: its net maize imports would range between 4.6 and 19 million tons (Fig. 3).

The substantial decrease in maize production in the USA in 2050 coupled with sustained maize consumption within the country would translate into high maize prices that would incite other countries not affected by the climate extreme, to increase their maize production (Table 2). Maize production would increase everywhere, except in the CWANA region (Table 2) where maize would still not be a key food product in 2050 (Table 3). However, despite the increased maize production in countries not affected by the climate extreme, global maize production would still decrease by 9% to 18%, compared to the 'Base-2050' scenario.

By 2050, under the 'base-2050' scenario, the USA, Argentina, Brazil and Central Europe would lead global maize exports and would account together for 80% of global maize exports. However, the weather extreme would force the USA out of the list of the major maize exporters in 2050; other countries that would become major maize exporters globally would mainly come from the European continent (Table 2).

The weather extreme in 2050 would reduce global maize consumption. E & SE Asia would experience by far the largest reduction in maize consumption in volume terms (Table 3). In addition, the reduction in maize consumption in this region would mainly affect animal feed. SSA and the LAC region would be next, relative to the reduction in maize consumption. In each of these regions, maize consumption would decrease by 15–30 million tonnes due to the weather extreme. In SSA, the bulk of the reduction would relate to human food whereas in the LAC region, it would relate to animal feed (Table 3).

If the weather extreme occurs in 2050 under the milder climate change scenario, the decrease in global maize production would be accompanied by a decrease in the production of wheat and rice. This implies that the increased demand for rice and wheat brought by the weather extreme would not be strong enough to increase the production of these two commodities. Overall, the consumption of rice and wheat would decrease across all world regions, except the USA, where it would remain stagnant (Table 4). Similarly, the production of rice and wheat would decrease across most regions of the developing world of 2012, except in SSA, which would experience an increase in the production of maize, rice and wheat under the weather extreme (Table 4). In South Asia, E & SE Asia and LAC, the increase in maize production brought by the weather extreme would be accompanied by a decrease in the production of rice and wheat (Table 4).

The weather extreme would see SSA increase its production of key cereals in 2050 under the milder climate change scenario. This increase would come at the expense of reducing the production of other crops, including cassava. More specifically, the sub-continent would lead all regions relative to the reduction in cassava production (Table 4). SSA would also experience the second-largest reduction in cassava consumption if the weather extreme occurs in 2050 under the milder climate change scenario (Table 4). The LAC region would experience the largest increase in cassava production and the largest reduction in cassava consumption if the weather extreme occurs under the milder climate change scenario (Table 4). Since the region was a net cassava exporter prior to the weather extreme (data not shown), the change in its cassava production and consumption would be used as exports to other regions.

If the weather extreme occurs in 2050 under the harsher climate change scenario, the decrease in global maize production would be accompanied by an increase in the global consumption of rice and wheat (Table 4). In addition, the changes in the production and consumption of rice and wheat across the developing world would be similar to those that would occur, if the weather extreme occurred in 2012.

If the weather extreme occurs under the harsher climate change scenario, SSA would increase its consumption of not only rice and wheat, but also cassava. The sub-continent would increase its cassava consumption by more than 2 million tons; it would also be the only region that would experience an increase in cassava consumption due to the weather extreme (Table 4). Cassava production would also increase in the SSA (Table 4), but the sub-continent would still need to rely on imports to meet its consumption requirements. In the LAC region, the increase of 42 million tons in maize production would come at the cost of reducing the production of other crops, including rice, wheat and cassava. Cassava production would decrease by more than 2 million tons whereas the production of combined rice and wheat would decrease by 7.6 million tons (Table 4).

If the climate extreme occurs in 2050 under climate change, maize prices in the long run would increase by 30% to 70%, compared to the 'Base-2050' scenario (Fig. 4). Long-run prices for wheat would increase from 1% to 3% whereas long run prices for rice would increase from 8% to 9% (Fig. 4). Interestingly, the prices of rice and wheat would be higher if the weather extreme occurred under the milder climate change scenario compared to the harsher climate change scenario. The higher prices, related to the milder climate change scenario, are partly fuelled by the decrease in the global production of rice and wheat when the weather extreme occurs under the milder climate change scenario (Table 4).

Despite experiencing the largest reduction in maize consumption under the weather extreme, E & SE Asia would rank below SSA, South Asia and the LAC region, in the number of people at risk of hunger (Fig. 6 and Fig. 7). The combination of the highest base caloric intake in 2050 and relatively small reductions in caloric intake due to the weather extreme in E & SE Asia would explain this result. The reduction in maize consumption in E & SE Asia would mainly consist of a reduction in animal feed (Table 3). This would translate into a smaller reduction in caloric intake compared to other regions such as SSA where the bulk of the reduction in maize consumption would consist of a reduction in food (Table 3).

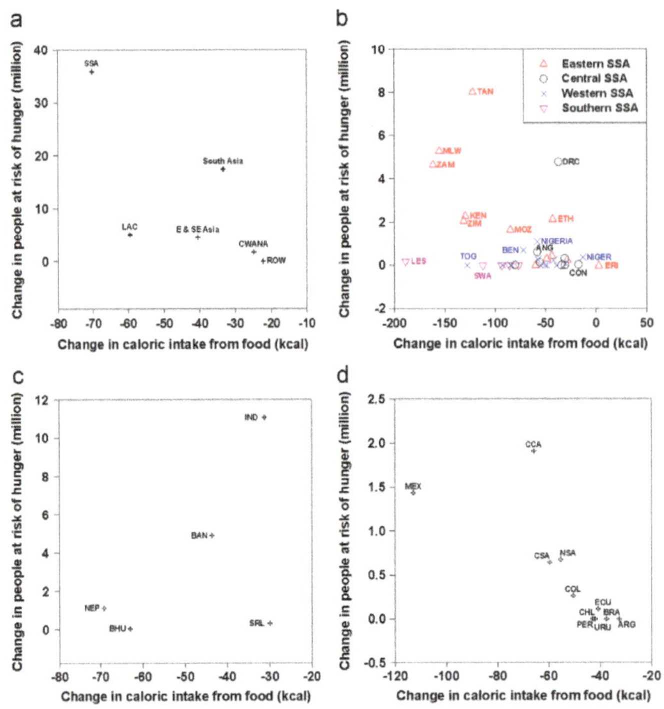

Figure 6: Projected impact of the extreme weather event on number of people at risk of hunger under the CSI2050 A1 EW scenario – results from IMPACT. CCA region includes Belize, Costa Rica, Cuba, Dominican Republic, El Salvador, Guatemala, Haiti, Honduras, Nicaragua, and Panama; CSA includes Bolivia and Paraguay; NSA includes French Guiana, Guyana, Suriname and Venezuela. (a) World, (b) SSA, (c) south asia and (d) LAC.

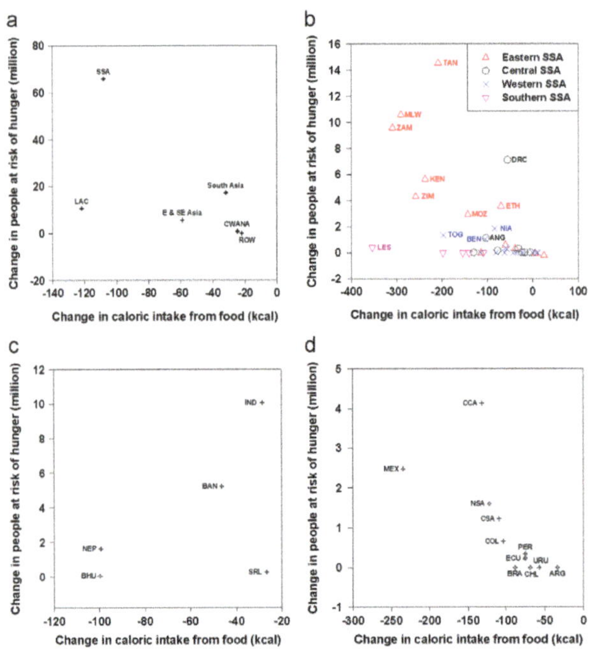

Figure 7: Projected impact of the extreme weather event on number of people at risk of hunger under the MIR2050 B1 EW scenario – results from IMPACT. CCA region includes Belize, Costa Rica, Cuba, Dominican Republic, El Salvador, Guatemala, Haiti, Honduras, Nicaragua, and Panama; CSA includes Bolivia and Paraguay; NSA includes French Guiana, Guyana, Suriname and Venezuela. (a) World, (b) SSA, (c) south asia and (d) LAC.

SSA would experience the largest increase in the number of people at risk of hunger, if the weather extreme occurs in 2050 under climate change (Fig. 6 and Fig. 7). The region would be unique in combining one of the lowest caloric intakes by 2050 with one of the highest reductions in caloric intake due to the weather extreme (Table 3). The increase in the number of people at risk of hunger in SSA would range between 36 and 66 million people. In relative terms, the number of people at risk of hunger in SSA would rise by 14% to 26% compared to the 'Base-2050' scenario.

If the weather extreme occurs under the milder climate change scenario, the countries within SSA that would experience an increase of 1 million or more in the number of people at risk of hunger are Tanzania, Malawi, Zambia, the Democratic Republic of Congo, Kenya, Zimbabwe, Ethiopia, Mozambique and Nigeria (Fig. 6); most of them would be in East Africa. However, if the weather extreme occurs under the harsher climate change scenario, more countries would experience an increase of at least 1 million in the number of people at risk of hunger. Additional countries that would join the countries listed above are Angola in Central Africa; and Benin and Togo in West Africa (Fig. 7).

In relative terms, the countries in SSA that would experience the largest increases in the number of people at risk of hunger under the weather extreme would also be located in eastern Africa. If the weather extreme occurs under the milder climate change scenario, countries with the largest relative increase in the number of people at risk of hunger would be Gambia in West Africa with an increase of 163% and Lesotho in southern Africa with an increase of 68%. These two countries would be followed by Kenya, Tanzania, Zambia, Malawi, Mozambique and Zimbabwe with 59%, 45%, 36%, 34%, 29%, and 23%, respectively. If the weather extreme occurs under the harsher climate change scenario, Togo in West Africa would experience the highest relative increase with 305%; it would be followed by Gambia in West Africa with 163%, Lesotho in southern Africa with 148%. Kenya, Tanzania, Zambia, Malawi, Mozambique and Zimbabwe would follow with an increase of 146%, 82%, 76%, 68%, 52% and 47%, respectively.

In southern Africa, only Lesotho would experience an increase in the number of people at risk of hunger, if the weather extreme occurs in 2050 under climate change (Fig. 5 and Fig. 6). More specifically, the number of people at risk of hunger would rise by 200 thousand under the milder climate change scenario to 421 thousand under the harsher climate change scenario (Fig. 6). If the weather extreme occurs under the milder climate change scenario (CSI2050-A1-EW), daily caloric intake would decrease by 112, 93, 86 and 77, in Swaziland, South Africa, Botswana and Namibia, respectively (Fig. 6). The reductions in caloric intake would be larger under the harsher climate change scenario. However, none of these countries would experience an increase in the number of at-risk people (Fig. 6 and Fig. 7). This implies that by 2050, daily caloric intake in all of southern Africa, Lesotho excluded, would have risen enough to offset the negative effect of the weather extreme on national food security.

In Central Africa, 3 countries, namely Cameroon, Equatorial Guinea and Gabon, would experience no increase in the number of people at risk of hunger is the weather extreme occurs under the milder climate change scenario (Fig. 6). Among these three countries, Cameroon would experience the largest decrease in caloric intake whereas Equatorial Guinea would experience the lowest decrease. In these three countries too, average per capita caloric intake would have risen enough by 2050 to counter the negative effects of the weather extreme on national food security. Among all 3 countries, oil-producing Gabon would be the richest by 2050 with a per capita income that would be 4 times higher than that of Equatorial Guinea and 7 times higher than that of Cameroon. Gabon would also be having the highest caloric intake prior to the weather extreme: 3411 calories per capita. If the weather extreme occurred in 2050 under the harsher climate change scenario, the same three countries would see no increase in their number of people at risk of hunger (Fig. 7). In all other countries in Central Africa, the weather extreme would worsen national food security under any of the climate change scenarios. These countries are Angola, Central African Republic, Chad, Congo and the Democratic Republic of Congo. Moreover, in all these countries, except Congo, the reduction in caloric intake brought by the weather extreme would be higher under the harsher climate change scenario compared to the milder scenario (Fig. 6 and Fig. 7).

In half of all western African countries, food security would not worsen if the climate extreme occurs in 2050 under the milder climate change scenario. These countries include Ghana, Guinea, Guinea Bissau, Côte d'Ivoire, Mali, Senegal, and Sierra Leone. Benin would experience the second largest increase in the number of people at risk of hunger in West Africa. The country would experience an increase of 700,000 people and it would be followed by Niger with 360,000, Gambia with 260,000 and Burkina Faso with 240,000 (Fig. 7).

Apart from Togo, all other countries in West Africa that did not experience an increase in the number of people at risk of hunger under the milder climate change would have the same outcome under the harsher climate change scenario (Fig. 7). In Togo, the reduction in caloric intake would rise from 127 under the milder climate change scenario to 200 under the harsher climate change scenario. As a result, the number of people at risk of hunger would rise from 0 under the milder scenario to 1.4 million under the harsher scenario. In Benin, Burkina Faso and Nigeria, the increase in number of people at risk of hunger would be worse if the weather extreme occurs under the harsher climate change scenario compared to the milder scenario. More specifically, under the harsher climate change scenario, the number of people at risk of hunger would rise to 1 million, 360,000 and 1.8 million in Benin, Burkina Faso and Nigeria, respectively (Fig. 7). In the other countries, namely Gambia, Liberia and Niger, the increase in the number of people at risk of hunger would be smaller if the weather extreme occurs under the harsher climate change scenario compared to the milder scenario. In Gambia and Niger, under the harsher climate change scenario, the number of people at risk of hunger would rise to 260,000 and 4000 people, respectively; it would decrease by 31,000 people in Liberia (Fig. 7).

Within the LAC region, the number of people at risk of hunger would rise from 5 million under the milder climate change scenario to 11 million under the harsher climate change scenario (Fig. 6 and Fig. 7). This result suggests that, as the momentum of climate change worsens, the negative effect of the weather extreme on food security in the LAC region would also worsen.

Within the LAC region, the Caribbean Central America and Mexico would be hardest-hit: the number of people at risk of hunger would increase by 2–4 million in CCA and by 1.4–2.5 million in Mexico (Fig. 6 and Fig. 7). In relative terms, CCA would experience an increase of 13–28% in the number of people at risk of hunger compared to the 'Base-2050' scenario. In Mexico, the change in the number of people at risk of hunger would range from 25% to 43%. The higher at-risk population in CCA, under the 'Base-2050' scenario, explains why the weather extreme would lead to a higher increase in the number of people at risk of hunger, despite CCA experiencing smaller reductions in per capita caloric intake compared to Mexico.

By 2050, under the 'Base-2050' scenario, Mexico would have a substantially higher daily per capita caloric intake compared to CCA, even if the latter would have seen an improvement in food security between 2012 and 2050. Indeed, average per capita daily caloric intake under the baseline climate model would be 2700 by the year 2050, compared to 2460 in the year 2012 for the CCA. In Mexico, per capita caloric intake would remain high by 2050 and would barely change between 2012 and 2050; it would increase from 3100 to 3200. By 2050, Mexico would also have a slightly larger population compared to CCA, under the baseline climate model: 144 versus 109 million people. The substantially higher caloric intake in Mexico coupled with a slightly higher population leads to Mexico having a lower at-risk population compared to CCA by 2050, under the baseline climate model. Indeed, by 2050, under the 'Base-2050' scenario, the number of people at risk of hunger would reach 14 million in CCA; it would reach 5.8 million in Mexico.

In other regions not as populous as CCA, the relative increase in the number of people at risk of hunger would be very high, despite these regions experiencing reductions in caloric intake similar to those of CCA and also sharing similar per capita caloric intake in 2050 under the 'Base-2050' scenario. NSA and CCA would share similar daily per capita caloric intake by 2050, under the 'Base-2050' scenario: 2700. Moreover, the reductions in caloric intake caused by the weather extreme would be similar across the two regions (Fig. 6 and Fig. 7). Yet, CCA would experience a higher increase in the number of people at risk of hunger (Fig. 6 and Fig. 7). However, the relative increase in the number of people at risk of hunger in NSA would range between 34% and 80% and hence, would be much higher than that of CCA. The higher population in CCA explains this result: by 2050, the population would be 44 million in NSA compared to 108 million in CCA.

South Asia would also be the second hardest-hit region after SSA, if the weather extreme occurred in 2050, instead of 2012: the number of people at risk of hunger would rise by 17 million people (Fig. 6 and Fig. 7). The largest increase in the number of people at risk of hunger would occur in India. Interestingly, the reduction in caloric intake in India would be less than half that of Nepal and yet, India would experience an increase of 10–11 million in the number of at-risk people, unlike Nepal which would experience an increase of 1.1–1.6 million (Fig. 6 and Fig. 7). The higher population in India explains these results. The two countries would have similar average per capita caloric intake by 2050, under the 'Base-2050' scenario. In India, daily caloric intake would stand at 2700; in Nepal, it would stand at 2500. However, India's population would reach 1.7 billion by 2050, whereas Nepal's population would reach 46 million. In relative terms, the increase in the number of at-risk people in India would range between 8% and 7% compared to a scenario involving perfect climate change. In Nepal, the increase would range between 28 and 41%.

DISCUSSION

USDA's estimates are in agreement with the simulated results on annual maize production and consumption in the USA in 2012, after the weather extreme. Based on USDA's estimates, annual maize production in the USA was estimated at 314 million tons in the 2011/12 season; world maize production stood at around 881 million tons over the same period (Foreign Agricultural Service, 2012). Hence, the share of the USA in global maize production was around 36%. Based on the simulated results from IMPACT, the projected share of the USA in global maize production after the weather extreme in 2012 would be 32%. The simulated results from IMPACT imply that US maize production in 2012, after the weather extreme, would be about 238 million tons and world maize production would stand at 741 million tons.

Similarly, the estimates from USDA suggest that, with a consumption volume of 279 million tons in 2012, the USA accounted for 32% of global maize consumption (Foreign Agricultural Service, 2012). The simulated results from IMPACT imply that, with the weather extreme in 2012, US maize consumption would stand at 244 million and would account for 33% of global maize consumption.

However, the estimates from USDA differ substantially from those of IMPACT on maize trade volumes. The simulated results from IMPACT imply that, with net maize exports amounting to 8 million tons under the weather extreme in 2012, the US would account for 11% of world maize exports. By contrast, USDA's estimates imply that net maize exports in the USA amounted to 14 million tons and the country accounted for 20% of global maize exports in 2012 (Foreign Agricultural Service, 2012). In addition, maize exports from the USA were much lower in 2012 than in earlier years (Chung et al., submitted for publication); they were also accompanied by a substantial decrease in US maize stocks, compared to earlier years (Chung et al., submitted for publication).

The simulated results from IMPACT imply that the change in US maize stocks remains the same over the years, with or without the weather extreme. Such discrepancy between USDA's estimates and the simulated results from IMPACT reflect the limitations of IMPACT, a partial equilibrium model that assumes that changes in population, income and food stocks are exogenous. The projected substantial decrease in net maize exports from the USA would lead to substantial but unrealistic effects on food security across the globe. Hence, the magnitude of the results from IMPACT should be used with caution; however, the trends observed in the results are likely to be valid.

The study results demonstrate that climate extremes that negatively affect crop production among major world exporters can indeed have negative effects on food security in other regions. The study results imply that across the developing world, Sub-Saharan Africa, South Asia and the LAC region are likely to suffer most from the climate extreme that occurred in the USA, in 2012. Within SSA, the countries that would experience the highest increase in the number of people at risk of hunger due to the weather extreme would mainly be from eastern Africa; however, the countries that would experience the highest relative increase in the number of at-risk people would include countries from eastern and southern Africa; these countries, which would experience an increase of 10% or more in the number of at-risk people, would be Lesotho, South Africa, Malawi, Zambia, Kenya, Tanzania, Swaziland, Zimbabwe, Uganda and Bostwana. Within the LAC region, the CCA would experience the largest increase in the number of people at risk of hunger due to the weather extreme; however, in relative terms, CSA, NSA, Chile, Uruguay and Colombia, would also see their food security worsen due to the weather extreme. Within South Asia, India alone would account for two-thirds of the increase in the number of people at risk of hunger. Other countries that would experience substantial increases in their number of at-risk people include Nepal and Bhutan.

The relative effects of the weather extreme on long-run maize production would be worse if it occurred in 2050 instead of 2012. The negative impact of the weather extreme on maize production in the USA and hence on global maize production would be much worse in 2050 than in 2012. With the non-implementation of climate change adaptation strategies across the world food baskets between the 2000s and 2050s, climate change would be eroding crop productivity gains over the years. Hence, the weather extreme in 2050 would further weaken maize productivity in the USA, which is projected to remain the leading maize producer and exporter in 2050.

Similarly, the negative effects of the weather extreme on food security would be worse if it occurred in 2050 compared to 2012. Globally, the relative increase in the number of people at risk of hunger would be 1.4% in 2012; if the weather extreme occurred in 2050, the relative increase in the number of at-risk people would range between 8% and 13%. However, the hardest-hit regions would remain the same, whether the weather extreme occurs in 2012 instead of 2050: SSA, South Asia and the LAC region. Across these regions, food security would improve between 2012 and 2050, under the baseline climate model and moderate growth in per capita income across world economies. In few countries located in southern, western and central Africa, and in the LAC region, per capita caloric intake would have risen enough to offset the negative effects of the weather extreme. However, in the other countries of SSA, South Asia and the LAC region, the weather extreme would erode much of the gains made in food security.

The effect of the weather extreme on food security in vulnerable countries can be mitigated through social protection programs including cash transfers and food aid (Chung et al., submitted for publication). By 2012, the LAC region had strong social protection programs (Ferreira and Robalino, 2010) which likely mitigated the negative effect of the weather extreme on food security in the region. Another strategy for enhancing the adaptive capacity of countries to weather extremes would consist of changing policies that favour the use of maize or other food crops for biofuel production. The proportion of US maize production used for ethanol production increased from 1% in the 1980s to 25% in 2007/2008 (Piesse and Thirtle, 2009 and Capehart, 2014). The increase was fuelled in part by favourable US policies (Yano et al., 2010 and Capehart, 2014). Some of the maize used in ethanol production could be diverted into replenishing US maize stocks which could be used to buffer the production loss brought by weather extremes.

Our results illustrate the considerable utility of the process-based spatial bio-economic framework, as a tool for assessing the impacts of extreme weather events. The socio-economic model (IMPACT) is calibrated for the year 2000 and assumes moderate growth in per capita income followed by the 2012 weather extreme. The results from the model were close to the reality despite the fact that it is unlikely that moderate economic growth was observed across the whole world between 2000 and 2012. Two major events, the global food price crisis in 2008 which was followed by the global financial crisis, were not captured in model. These events had large influence on global food prices and farmers' decisions. Similarly, speculation was not captured in model.

CONCLUSION

One of the objectives of this study was to quantify the potential effects of the 2012 weather extreme in the USA on food security across the developing world. Study results suggests that the extreme climate of 2012 that occurred in the USA is likely to increase food insecurity among poor communities where maize provides a substantial portion of daily caloric intake and where households cannot easily adjust their food consumption patterns in the face of increased maize scarcity. Our results indicate that countries where food security would worsen due to the weather extreme are located in eastern and southern Africa; South Asia; and the LAC region.

If a similar weather extreme were to occur in the USA in 2050 under climate change, its effects on global food production and security would be worse, assuming no adaptation to climate change over the years. In addition, the hardest-hit regions would remain the same, whether the weather extreme occurs in 2012 instead of 2050: SSA, South Asia and the LAC region. However, for few countries in Sub-Saharan Africa and LAC, per capita caloric intake would have risen enough to completely offset the negative effect of the weather extreme on food insecurity.

Future research would involve developing a CGE model that can consider the spillover effects of weather extremes. Similarly, frequency distribution on weather extremes could be used in combination with a CGE model to project the impact of climate extremes in the future.

ACKNOWLEDGEMENTS

This research was supported by the CGIAR research programme on Climate Change, Agriculture and Food Security (CCAFS).

REFERENCES

1. Adger, 1999W.N. AdgerSocial Vulnerability to Climate Change and Extremes in Coastal VietnamWorld Dev., 27 (2) (1999), pp. 249–269
2. Ahmed et al., 2009

3. A.S. Ahmed, S.N Diffenbaugh, W.T. HertelClimate volatility deepens poverty vulnerability in developing countriesEnviron. Res. Lett., 4 (3) (2009), p. 8

4. Boyer et al., 2013J.S. Boyer, P. Byrne, K.G. Cassman, et al.The U.S. drought of 2012 in perspective: a call to actionGlob. Food Secur., 2 (3) (2013), pp. 139–143

5. Brown and Funk, 2008M.E Brown, C.C. FunkFood security under climate changeScience, 319 (5863) (2008), pp. 580–581

6. Capehart, 2014Capehart, T., 2014. Corn Policy. 2014. From 〈http://www.ers.usda.gov/topics/crops/corn/policy.aspx#.U23shygr00G〉 .Changnon, 2003a

7. S. ChangnonPresent and future economic impacts of climate extremes in the United StatesGlob. Environ. Change Part B: Environ. Hazards, 5 (3–4) (2003), pp. 47–50

8. Changnon, 2003bS. ChangnonMeasures of economic impacts of weather extremes.Bull. Am. Meteorol. Soc., 84 (2003), pp. 1231–1235

9. Chung et al., (submitted)Chung, U., Gbegbelegbe, S., Shiferaw, B., et al., 2014. Modeling the effect of a heat wave on maize production in the USA and its implications on food security in the developing world. (submitted for publication).

10. Deschenes and Moretti, 2009O. Deschenes, E. MorettiExtreme Weather Events, Mortality, and MigrationThe Rev. Econ. Stat., 91 (4) (2009), pp. 659–681

11. Ferreira and Robalino, 2010F.H. G. Ferreira, D. RobalinoSocial Protection in Latin America: Achievements and LimitationsWorld Bank (2010)

12. Foreign Agricultural Service, 2012Foreign Agricultural Service, U., 2012. Grain: World Markets and Trade. Washington DC, p. 55.

13. Hallegatte, 2007S. HallegatteThe use of synthetic hurricane tracks in risk analysis and climate change damage assessmentJ. Appl. Meteorol. Climatol., 46 (11) (2007), pp. 1956–1966

14. Hallegatte et al., 2007S. Hallegatte, J.-C Hourcade, P. DumasWhy economic dynamics matter in assessing climate change damages: illustration on extreme eventsEcol. Econ., 62 (2) (2007), pp. 330–340

15. Hoogenboom et al., 1999G. Hoogenboom, P.W. Wilkens, G.Y. Tsuji, et al.DSSAT Version 3University of Hawai, Honolulu, Hawaii (1999)

16. Jones et al., 2003J.W. Jones, G. Hoogenboom, C.H. Porter, et al.The DSSAT cropping system modelEur. J. Agron., 18 (2003), pp. 235–265

17. Kunkel et al., 1999K.E. Kunkel, R.A. Pielke, S.A. ChangnonTemporal fluctuations in weather and climate extremes that cause economic and human health impacts: a reviewBull. Am. Meteorol. Soc., 80 (6) (1999), pp. 1077–1098

18. Mechler et al., 2010R. Mechler, S. Hochrainer, A. Aaheim, et al.Modelling economic impacts and adaptation to extreme events: insights from European case studiesMitig. Adapt. Strateg. Glob. Change, 15 (7) (2010), pp. 737–762

19. Nelson et al., 2010G.C. Nelson, M.W. Rosegrant, A. Palazzo, et al.Food security, Farming, and Climate Change to 2050: Scenarios, Results, Policy Options. IFPRI Research MonographIFPRI, Washington DC (2010), p. 131

20. Ortiz et al., 2008R. Ortiz, H.-J. Braun, J. Crossa, et al.Wheat genetic resources enhancement by the International Maize and Wheat Improvement Center (CIMMYT)Genet. Resour. Crop Evol., 55 (7) (2008), pp. 1095–1140

21. Pielke, 2007R.A. PielkeFuture economic damage from tropical cyclones: sensitivities to societal and climate changesPhilos. Trans. R. Soc. A, 365 (1860) (2007), pp. 2717–2729

22. Pielke and Landsea, 1998R.A. Pielke, C.W. LandseaNormalized hurricane damages in the United States: 1925–95Weather Forecast., 13 (3) (1998), pp. 621–631

23. Piesse and Thirtle, 2009J. Piesse, C. ThirtleThree bubbles and a panic: an explanatory review of recent food commodity price eventsFood Policy, 34 (2) (2009), pp. 119–129

24. Ray et al., 2012D.K. Ray, N. Ramankutty, N.D. Mueller, et al.Recent patterns of crop yield growth and stagnationNat. Commun., 3 (2012), p. 1293

25. Rosegrant et al., 2012M. Rosegrant, et al.International Model for Policy Analysis of Agricultural Commodities and Trade (IMPACT): Model DescriptionIFPRI, Washington DC (2012)

26. Rosegrant et al., 2008M.W. Rosegrant, C. Ringler, S. Msangi, et al.International Model for Policy Analysis of Agricultural Commodities and Trade (IMPACT): Model DescriptionIFPRI, Washington DC (2008)

27. Yano et al., 2010Y. Yano, D. Blandford, Y.R. SurryDo current US ethanol policies make sense?Policy Issues-Agricultural and Applied Economics Association, (10) (2010), pp. 1–4

CHAPTER 7

DUST STORM FREQUENCY AND IMPACT OVER EASTERN AUSTRALIA DETERMINED BY STATE OF PACIFIC CLIMATE SYSTEM

Milton S. Speer

Keywords: Dust, Eastern Australia, ENSO, SAM, Tropospheric circulation

INTRODUCTION

Dust storms are an important part of the global dust cycle. For example, atmospheric dust potentially affects the radiative balance and chemical composition of the atmosphere and may provide nutrients to terrestrial and marine ecosystems (Kohfeld and Harrison, 2001). Dust can also contribute to land degradation, resulting in economic loss, reduced biological diversity and social hardship while immediate effects include adverse impacts on human health and local disruption of transport through reduced visibility. The first spatial study of dust storm occurrence over Australia was by Middleton (1984), who found the highest frequency of dust storms occur in the centre of Australia, where there was an average of 10.8 at Alice Springs and a maximum of 65 events per year. Previous studies have investigated wind regimes associated with Australian dust storms. For example, Ekström et al. (2004) identified September to February dust storms over inland Australia as being most likely controlled by the timing of non-precipitating cold fronts over southern and central Australia. They concluded that the dust storm occurrence is related to the extent to which ridging from the subtropical Indian Ocean high dominates the Great Australian Bight (GAB). A more westward displacement of high pressure in the GAB allows cold fronts to enter the continent and increase the potential for dust storm activity. Lamb et al. (2009) noted a decrease in September–February dust storm frequency over central eastern Australia from the mid-1970s in the context of the Pacific-wide climate system. They showed that the dust oscillation was associated locally with a strengthening and then weakening of the southerly component of the low-level tropospheric wind over the dust-prone region. Leslie and Speer (2006) found that from a total of 55 dust storm occurrences over central eastern Australia between January 1995 and February 2005, there were 43 associated with cold fronts. Of those 43, 24 fronts were embedded in the zonal westerlies and the remaining 19 resulted from cold fronts over eastern Australia where high-pressure systems over the GAB generated strong post-frontal south to southeast winds. This questions the larger proportion of westerly wind-induced dust occurrences than southerly wind dust occurrences from 1995 to 2005, given that prior to the mid-1970s southerly wind occurrences dominate. The main aim of this study is to highlight the strong influence that tropospheric circulation changes of the Pacific climate system and resulting synoptic characteristics had on dust storm frequency and impact over central eastern Australia prior to the mid-1970s compared to the post mid-1970s.

DATA AND METHODOLOGY

Dust Observations

The inclusion of dust observations from the Australian Bureau of Meteorology (BoM) in this study covers the September–February period from 1957/58 to 2010/11. There were some important changes in BoM dust observing practice over the period. Centred on 1957, the International Geophysical Year, there was a gradual increase in the number of BoM stations Australia-wide. The 12 BoM stations in this study

Citation: Milton S. Speer, Dust storm frequency and impact over Eastern Australia determined by state of Pacific climate system, http://dx.doi.org/10.1016/j.wace.2013.10.004.

have complete records for the whole study period, except for Dubbo after 1999 and after 2001 for Boulia. There was a Pacific-wide climate shift in the mid-1970s, which has been reported by several authors (Lamb et al., 2009, Trenberth and Hurrell, 1994 and Meehl et al., 2009). In order to address a hypothesised decrease in dust days between the two periods 1957/58 to 1975/76 and 1976/77 to 2010/11 (hereafter referred to as P1 and P2, respectively), the change in the way dust was reported from 1973 is described in the following section. Also presented in the following section is an analysis of the dust frequency differences between P1 and P2 in terms of the mean 3 pm daily wind speed, wind direction and mean daily wind speed at four of the 12 synoptic stations.

Methodology

The area of interest is defined as the Lake Eyre drainage basin of central eastern Australia (Fig. 1), which is the main dust source region that affects eastern Australia (McTainsh and Pitblado, 1987). In this study, a dust day is defined when any one of the 12 BoM stations covering central eastern Australia within the Lake Eyre basin source region reported a synoptic observation with a dust code. In other words, if at least one station reported dust at one of the routine daily synoptic reporting times then it was included as an occurrence. Each dust day observation of the 12 reporting stations from September to February 1957–58 to September to February 2010–11 were then classified synoptically from the BoM MSLP archives as *southerly*, if the frontal wind changed to a southerly owing to high pressure in the GAB, *westerly* if the frontal system in the GAB was embedded either in the zonal westerly winds or low pressure itself was located in the GAB or *local,* if dust was reported in locally strong winds not associated with a frontal system.

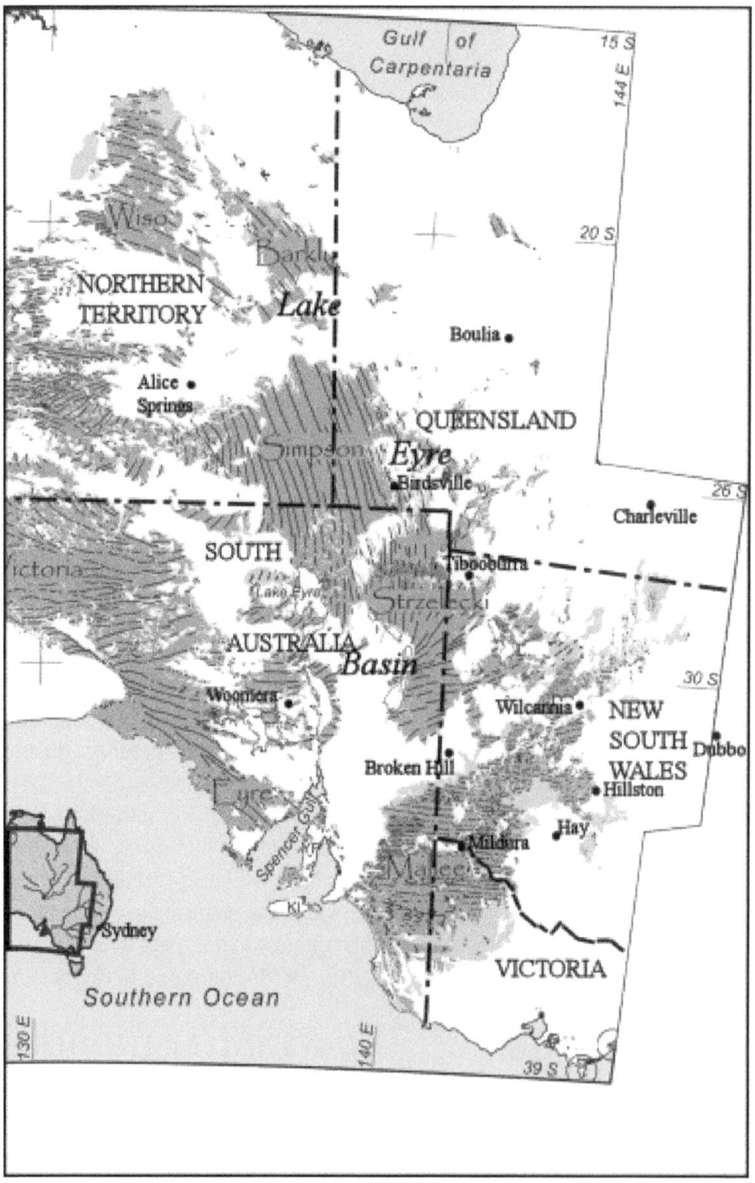

Figure 1: Map of central eastern Australia centred on the Lake Eyre basin showing 12 representative BoM dust reporting stations. All 12 stations recorded at least 9 am and 3 pm synoptic observations of dust from 1957 to 2011 except after 1999 for Dubbo and after 2001 for Boulia when both became automatic weather stations. Also shown are central Australian dune fields mapped from Landsat TM imagery (after Hesse (2010), Fig.1) with principal dune fields named. Light grey areas are sand plains; medium grey areas are dune fields. Representative longitudinal dune orientations are shown by black lines. State names and borders have been added.

Prior to 1973, there were several codes for reporting dust at synoptic times relating to the timing of observed dust. From 1973 these codes were changed with a resulting increase in use of an existing code. Similarly, several codes describing dust severity were changed to include visibility thresholds, which also increased the use of the existing code. The end result was that even though the number of codes for reporting dust reduced from 1973 one would expect neither a decrease nor an increase in dust reporting after 1973. However, an analysis is presented of four individual September–February station dust data to support this proposition. The four stations (Alice Springs, Boulia, Charleville and Hay) have purposely been chosen to represent a sample covering an arc from the north through northeast, east and southeast, which maximises the area covered by the influence of the Lake Eyre basin (Fig. 1). There certainly is a decrease in dust after 1973 (Fig. 2a), but extreme rainfall occurred over eastern Australia between 1973 and 1976. This La Niña period was arguably the second most pronounced back-to-back La Niña periods over the last century (Record-breaking, 2010) and was very likely the major cause of reduced dust storm activity in other La Niña periods (see Fig. 2a). Therefore, dust day frequency for the four individual stations of Alice Springs, Boulia, Charleville and Hay shown in Fig. 2(b) confirms the decrease in dust in La Niña periods found for the total 12 stations in Fig. 2(a). Table 1 shows for September–February the mean 3 pm daily wind speeds and mean daily wind speeds for all days and dust days only of the four stations divided into the two periods 1957/58 to 1975/76 and 1976/77 to 2010/11 (P1 and P2). In both periods, the 3 pm mean wind speeds of dust days only for the four stations (including the eight not shown) are significantly higher statistically (1% level) from those of all days. Similarly, their mean daily wind speeds of dust days only are significantly higher than their mean daily wind speeds of all days for both periods, confirming that days on which dust are recorded tend to be much windier than days when no dust is recorded, as expected.

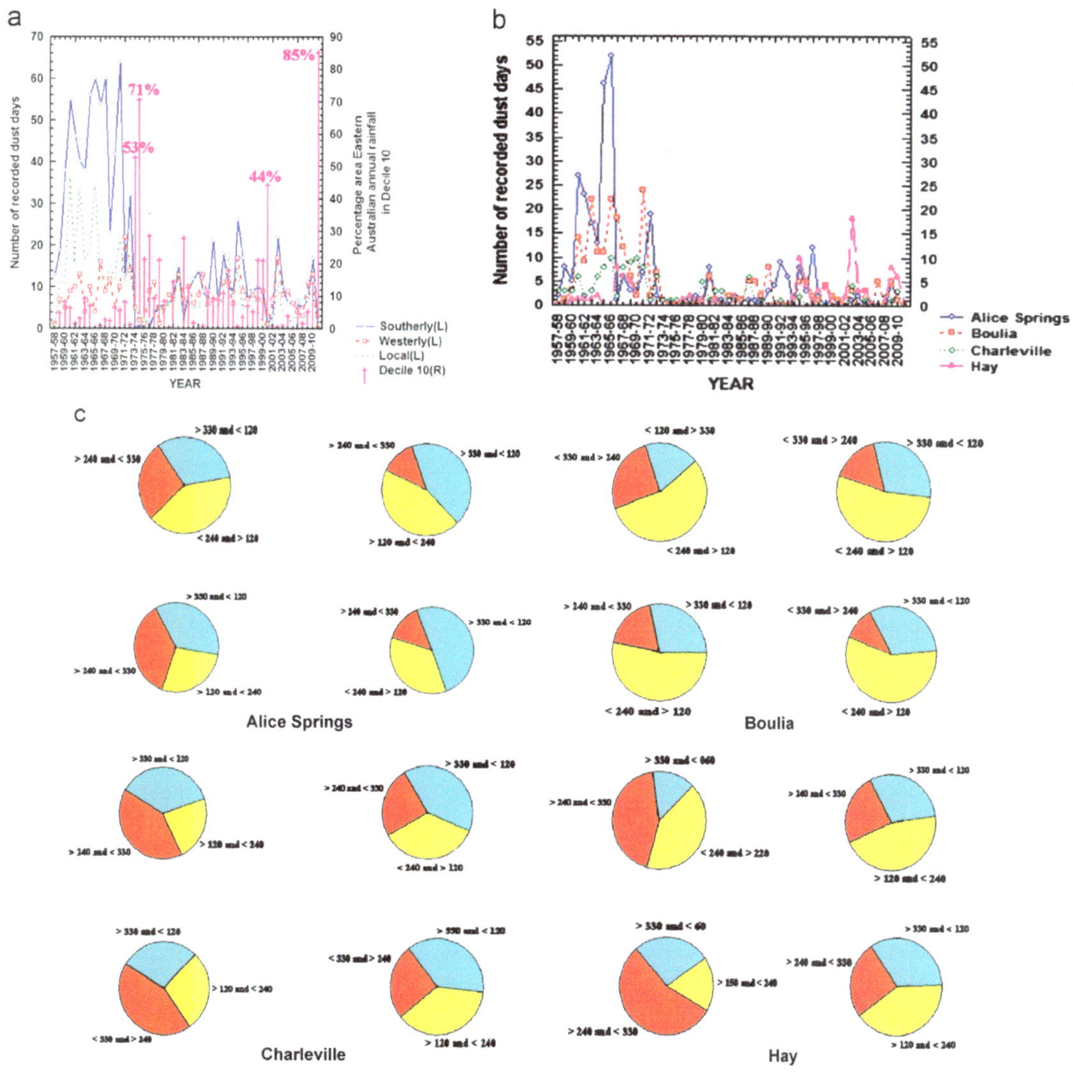

Figure 2: (a) Total number of *southerly, westerly* and *local* dust days for September to February 1957/58 to 2010/11 covering central eastern Australia based on 12 synoptic reporting stations. Also plotted is the percentage area of eastern Australia (defined by the states of QLD, NSW, Victoria and Tasmania) covered by decile 10 annual rainfall. Years above 40% area covered by decile 10 are marked individually. (b) Total number of*southerly, westerly* and *local* dust days for September to February 1957/58 to 2010/11 for the four stations of Alice Springs, Boulia, Charleville and Hay. (c) Pie diagrams relating to the four stations, Alice Springs, Boulia, Charleville and Hay, showing proportions of wind direction (deg.), frequency of dust days only (top left panel of each station for P1; bottom left panel of each station for P2) and all days (top right panel of station for P1; bottom right panel of each station for P2). Yellow signifies predominantly southerly component winds, red signifies predominantly westerly component winds and light blue represents mostly northeast component winds.

Table 1: Mean 3 pm daily wind speed (km/h) and number of observations (*N*), mean daily wind speed (km/h) and number of observations (*N*) for both P1 and P2 for the 4 stations: Alice Springs, Boulia, Charleville and Hay. Also shown is the number of daily synoptic observations for each station

Station	Period *(Sep to Feb)*	Mean 3 pm daily wind speed (km/h) and number of observations (*N*)				Mean daily wind speed (km/h) and number of observations (*N*)				Number of daily synoptic observations
		Dust days	*N*	All days	*N*	Dust days	*N*	All days	*N*	
Alice Springs	1957–1975	22.5	230	15.0	3421	18.5	233	10.9	3444	7 to 8
	1976–2011	24.8	76	16.1	6321	17.8	76	11.9	6343	8
Boulia	1957–1975	27.0	164	12.6	3330	25.8	165	13.1	3380	7
	1976–2011	19.5	68	11.5	6190	19.5	68	10.8	6317	7
Charleville	1957–1975	19.7	91	12.7	3363	13.9	91	9.2	3384	7 to 8
	1976–2011	22.0	36	15.1	6321	15.8	36	11.8	6342	7 to 8
Hay	1957–1975	30.2	26	15.2	3125	28.1	27	15.5	3376	2 to 4
	1976–2011	22.5	79	9.0	6088	15.6	79	7.4	6285	2 to 4

The orientation of the longitudinal sand dunes shaped by the wind in Fig. 1 should be reflected in the wind direction such that those stations to the north and northeast (Alice Springs and Boulia, respectively) should have predominantly southerly winds while further south, particularly Hay, an increasingly westerly wind component should be evident. This is indeed the case in the pie diagrams of wind direction for three of the four stations in Fig. 2(c). There is a predominant southerly component wind compared to westerly winds for all days at Alice Springs in P1, whereas for dust days southerly wind events slightly decrease at the expense of westerly events but still dominate. However, in P2 the proportion of westerly dust events compared to southerly events becomes larger. At Boulia, southerly winds dominate all days and dust days for both periods with just a slight increase in westerly winds for dust events compared to all days for both periods. Charleville is interesting because while southerly winds dominate all days for both periods compared to westerly winds, dust events occur more frequently as westerly events than southerly events for both periods. A possible explanation is that the Charleville area in central southern Queensland, in addition to being influenced by anticyclonic winds from high-pressure in the GAB, is also influenced by the oscillatory nature of the well-known quasi-permanent northeast Australian surface heat trough in the warmer spring/summer months with westerly winds converging into the trough on its western side (Skinner and Leslie, 1999). Southerly component winds dominate for all days at Hay in both periods. Southerly dust events occur in about the same proportion as westerly events in P1 (note, however, that they are all southwesterly with no southeasterly events in the top left panel for Hay). However, southerly dust events decreased markedly in P2, reflecting the rise in westerly dust events generally during P2 for the 12 stations shown in Fig. 2(a).

The September–February dust data for P1 and P2, including the seasonal break-up (SON and DJF), were then related to surface (MSLP) vector wind anomalies in order to assess changes in the main synoptic characteristics of weather systems responsible. The resulting images were provided by the NOAA/ESRL Physical Sciences Division, Boulder Colorado (http://www.esrl.noaa.gov/psd/), which were originally developed from the NCEP/NCAR reanalysis project by Kalnay et al. (1996). The southern annular mode (SAM), alternatively referred to as the Antarctic oscillation (AAO), is the leading mode of variability in the extra-tropical southern hemispheric circulation and is characterised by deep, zonally symmetric or "annular" structures with geopotential height perturbations of opposing signs in the polar cap region and in the surrounding zonal ring centred near 45°S (Thompson and Wallace,2000). The AAO (SAM) index originally developed by Gong and Wang (1999) is slightly different from the modified Nan and Li index (Nan and Li, 2003) that is used here and is defined as the difference in the normalised monthly zonal MSLP between 40°S and 70°S. The seasonal SAM data values were sourced from: http://web.lasg.ac.cn/staff/ljp/data-NAM-SAM-NAO/SAM-AAO.htm.

RESULTS AND DISCUSSION

Analysis of Dust day Occurrence in Terms of Extreme Rainfall

As shown in Fig. 2(a), a noticeable feature is the large number of dust occurrences prior to the mid-1970s' climate shift (see Lamb et al., 2009). Also plotted on Fig. 2(a) are the years showing the percentage area of eastern Australia covered by decile 10 rainfall. For rainfall purposes, the BoM defines eastern Australia by the area that includes Queensland, NSW, Victoria and Tasmania at: http://www.bom.gov.au/climate/change/about/temp_timeseries.shtml.

In such high-rainfall years, vegetation is likely to be a prominent feature over the normally sparsely vegetated Lake Eyre drainage basin and hence hinder dust storm generation. Clearly, there is a drop in dust day occurrence in each of the decile 10 rainfall years marked. Particularly noticeable are the very few dust day observations from September to February during the La Niña periods of 1973–76, 2000–01 and 2011. La Niña periods are defined by the BoM at: http://www.bom.gov.au/climate/enso/lnlist/.

Dust Frequency Related to Tropospheric Circulation Changes

In both SON (Fig. 3a) and DJF (Fig. 3b) prior to the mid-1970s, dust storms associated with *southerly* wind anomalies dominate. In SON, dust storms with *westerly* frontal systems are more frequent than in DJF. This is not surprising since by summer the seasonal poleward movement of the mid-latitude westerlies starts occurring. The number of dust days classified as *local* increased in DJF compared to SON. This is because thunderstorm-scale downdrafts and gusts are likely to contribute more to local dust occurrences in DJF than in SON.

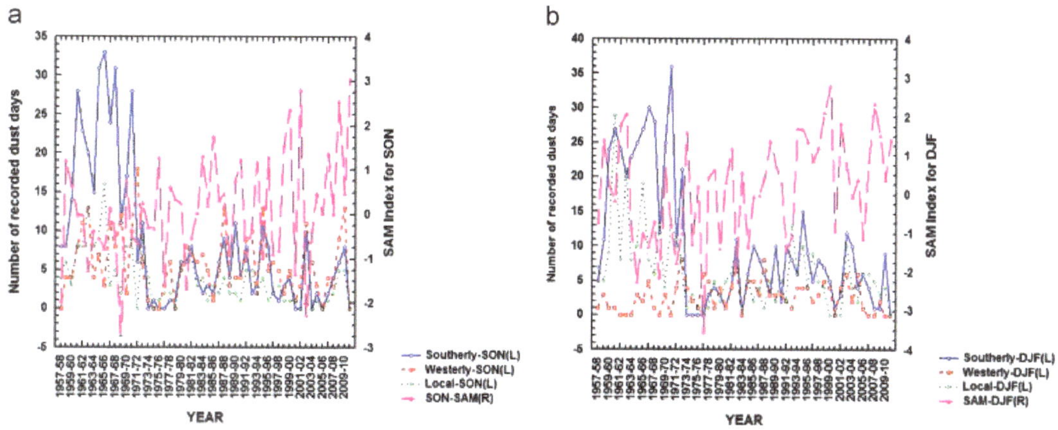

Figure 3: (a) Total number of southerly, westerly and local dust days for SON 1957 to 2010 Also shown is the SAM index for SON, (b) Total number of southerly, westerly and local dust days for DJF 1957/58 to 2010/11. Also shown is the SAM index for DJF.

There was a major change to the above relationships among *southerly*, *westerly* and*local* dust storms after the mid-1970s, in that *southerly* occurrences decreased markedly down to a level approximately equal to the number of *westerly* occurrences. As a result the total number of annual, SON and DJF dust frequencies have remained much less than prior to the mid-1970s. However, the strong southerly wind anomalies prior to the mid-1970s over eastern Australia emanate from the state of the Pacific climate system then, rather than from an increase in GAB frontal systems, as Ekström et al. (2004) concluded. The south to southeasterly wind anomalies are located on the western side of a cyclonic pressure anomaly adjacent to the east Australian coast at the surface and through the depth of the troposphere over northeast Australia. The surface (MSLP) vector wind anomalies are shown in Fig. 4 for P1 and P2. A similar feature in both figures is the anticyclonic vector wind anomalies emanating from the GAB and over South Australia, even though the vector wind anomalies are much weaker (an order of magnitude less) and variable in direction over central eastern Australia after 1976 (Fig. 4b). Nevertheless, this implies that southerly vector wind anomalies are a regular feature influencing wind-blown dust over central eastern Australia. This is exemplified on larger, geological time scales by the orientation of longitudinal sand dunes in the Lake Eyre basin (Hess, 2010). Longitudinal dunes are elongated ridges of aeolian sediments that run parallel to the resulting vector winds creating the sand movement (King, 1960). Fig. 1shows these longitudinal dunes as having been shaped by westerly winds in the south of the basin with south to southeast-shaped longitudinal dunes over northwest NSW, northeast South Australia, western Queensland and the Northern Territory.

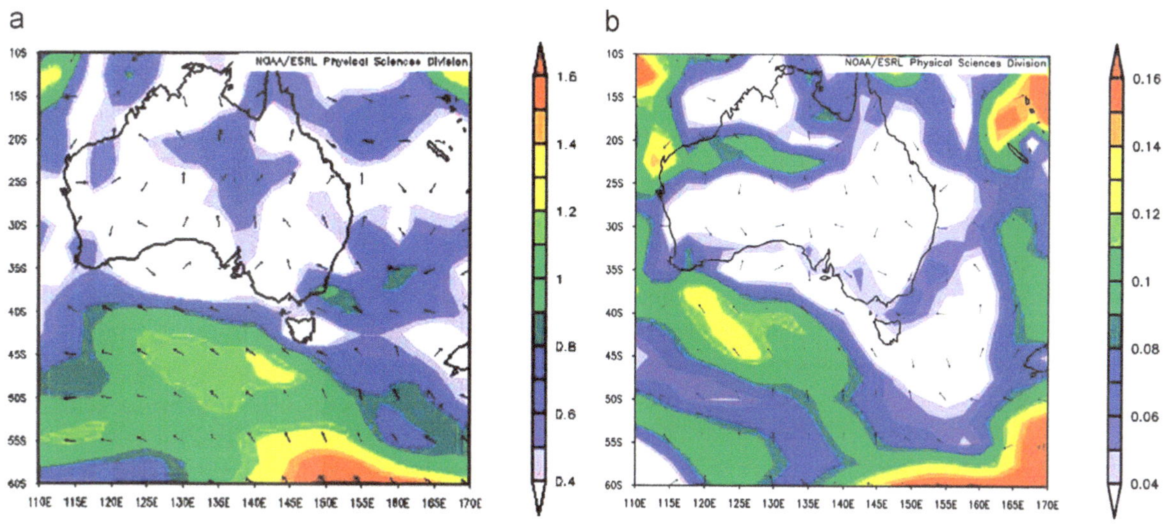

Figure 4: (a) NCEP/NCAR reanalysis MSLP vector wind (m/s) composite anomaly for September–February 1957/58 to 1975/76. Contour interval=0.1 m/s, Fig. 4(b) NCEP/NCAR reanalysis MSLP vector wind (m/s) composite anomaly for September–February 1976/77 to 2010/11. Contour interval=0.01 m/s.

DUST FREQUENCY AND SAM

Even though more frontal systems penetrated inland southern and central Australia prior to the mid-1970s owing then to large-amplitude mid-tropospheric troughs in a negative phase of the SAM (see Speer et al., 2009), it was the Pacific climate system influence of producing enhanced south to southeast wind anomalies that resulted in the larger frequency of dust storms. Over the study period from 1957 to 2011 the SAM index has trended from a negative phase in the 1950s and the 1960s to increasingly positive. This has resulted in a poleward contraction of frontal systems affecting southern and central Australia (e.g. Hendon et al., 2007 and Meneghini et al., 2007) and yet total *westerly* dust storm frequency emanating from the Lake Eyre basin has changed little. The most likely explanation is that in the resulting lower-rainfall regime across southern Australia, transient strong, but nevertheless, infrequent westerly frontal systems (particularly in SON) have been slightly more efficient in producing dust storms than earlier in the period. This follows from the fact that the SAM acts on shorter, synoptic time scales of up to a few weeks (Hartmann and Lo, 1998; Lorenz and Hartmann, 2001 and Barnes and Hartmann, 2010) and in its negative phase this is manifested as brief excursions of frontal systems/zonal westerlies crossing eastern Australia at more equatorward latitudes than when SAM is in a positive phase. Moreover, the zonal westerlies in SON have not yet retreated to the more poleward latitudes of their DJF position. Fig. 3(a) and (b) highlights this lack of association between increasing positive SAM values after the 1970s and the stable overall frequency of *westerly* dust storms.

Dust storms reaching Australia›s east coast

When westerly winds result in dust storms reaching the Australian east coast they affect a much more densely populated area than over inland eastern Australia. However, these events are very rare within a relatively small total number of *westerly* dust occurrences. For example, Table 2 shows that there were 6 days in the last 70 years from September to February when dust storms were reported in Sydney on Australia's east coast. Furthermore, all these occurred after prolonged dry periods over the dust source region of the Lake Eyre basin, including 5 of the 6 days when dust was observed during or towards the end of El Niño episodes. The main point here is that even in westerly dust occurrences, the extent to which dust will affect a larger area extending to the east coast is determined by the state of the Pacific climate system.

Table 2: September–February dust storms that have reached Sydney on Australia's east coast over the last 70 years (source: BoM significant weather summaries). Also shown are the antecedent rainfall conditions over the dust source region and whether during an El Niño episode a outside period of study.

Date of dust storm	Antecedent rainfall	El Niño
4 and 8 Jan 1942[a]	Decile 2–4 Apr. 1941 to Jan. 1942	✓
2 Dec 1957	Below decile 4–7 from Mar. to Nov. 1957	✓
24 and 29 Sep 1968	Mostly decile 2–3 from Lake Eyre to Sydney Aug. to Oct. 1968	X
25 Sep 2009	Below decile 3, May to Oct. 2009	✓

SUMMARY AND CONCLUSIONS

The pie diagrams of wind direction from the four stations show that southerly dust events over eastern Australia (September to February) have decreased after the mid-1970s' climate shift and that the corresponding frequencies of dust for the four individual stations do reflect those frequencies for the 12 stations shown. Changes in dust storm observing codes from 1973 are unlikely to have been a major factor in the decrease after the mid-1970s. Rather, extreme rainfall years are more likely to have been a major contributing factor to the large decreases in dust occurrence from 1973 to 1976, in addition to the other La Niña periods of 2000–01 and 2010/2011.

Dust storms resulting from synoptic-scale mid-latitude frontal systems affect inland, eastern Australia predominantly through spring/early summer but extend to the east coast only rarely when strong, westerly winds crossing the continent are sustained over the coast. These extreme *westerly* wind dust events are anomalous to the more pervasive *southerly* wind dust events that shape the sand dunes over central inland eastern Australia. Therefore, a change in dust storm frequency affecting the densely populated east coast is subject to changes in a small number of extreme events. With typically low annual rainfall totals over the dust source region of the Lake Eyre basin, the main climate drivers affecting widespread eastern Australian dust storms are antecedent dry conditions, predominantly associated with El Niño episodes, as already noted in previous studies. It was shown, however, that although antecedent dry conditions are very important, the synoptic-scale frontal systems that result in the transport of dust over the east coast occur during both negative and positive phases of the southern annular mode. This finding is consistent with the fact that the mid-latitude westerlies are usually strongest at their most equatorward position in Australian longitudes in late winter/spring, regardless of the SAM phase. It was also shown that the state of the Pacific climate system is important in determining the larger frequency of dust storms in the La Niña-dominated period prior to the mid-1970s as a result of the stronger southerly wind vector anomalies then present.

The stable total frequency of westerly wind-induced dust storms during spring/summer over central eastern Australia during the two periods suggests little change would be expected regarding their frequency over time even though the seasonal westerly winds are expected to contract further poleward under increased global warming. However, a complicating factor is that the increasing likelihood of longer dry spells under climate change would imply reduced vegetation for longer periods over the dust-prone source region of Lake Eyre basin, which could worsen the impact of dust storm events. Furthermore, longer dry spells throughout the year would increase the risk of more frequent westerly dust storm events in the other seasons (autumn/winter) when strong, westerly frontal systems may also occur.

REFERENCES

1. Barnes and Hartmann, 2010E.A. Barnes, D.L. HartmannDynamical feedbacks of the Southern Annular Mode in winter and summerJ. Atmos. Sci., 67 (2010), pp. 2320–2330
2. Ekström et al., 2004M. Ekström, G.H. McTainsh, A. ChappellAustralian dust storms: temporal trends and relationships with synoptic pressure distributions (1960–99)Int. J. Climatol., 24 (2004), pp. 1581–1599
3. Gong and Wang, 1999D. Gong, S. WangDefinition of Antarctic oscillation indexGeophys. Res. Lett., 26 (1999) (1999), pp. 459–462
4. Hartmann and Lo, 1998D.L. Hartmann, F. LoWave-driven zonal flow vacillation in the Southern HemisphereJ. Atmos. Sci., 55 (1998) (1998), pp. 1303–1315
5. Hendon et al., 2007H.H. Hendon, D.W.J. Thompson, M.C. WheelerAustralian rainfall and surface temperature variations associated with the Southern Hemisphere annular modeJ. Clim., 20 (2007), pp. 2452–2467
6. Hess, 2010P.P. HessThe Australian desert dune fields: formation and evolution in an old, flat, dry continentP. Bishop, B. Pillans (Eds.), Australian Landscapes, 346, Geological Society, London (2010), pp. 141–163 (Special Publications)
7. Kalnay et al., 1996E. Kalnay, et al.The NCEP/NCAR reanalysis 40-year projectBull. Am. Meteorol. Soc., 77 (1996), pp. 437–471
8. King, 1960D. KingThe sand ridge deserts of South Australia and related aeolian landforms of the quaternary arid cycles,T. Roy. Soc. South Aust., 83 (1960), pp. 99–108
9. Kohfeld and Harrison, 2001K.E. Kohfeld, S.P. HarrisonDIRTMAP: the geological record of dustEarth-Sci. Rev., 54 (1–3) (2001), pp. 81–114
10. Lamb et al., 2009P.J. Lamb, L.M. Leslie, R.P. Timmer, M.S. SpeerMultidecadal variability of eastern Australian dust and Northern New Zealand sunshine: associations with Pacific climate systemJ. Geophys. Res., 114 (2009) http://dx.doi.org/10.1029/2008JDO11184
11. Leslie and Speer, 2006L.M. Leslie, M.S. SpeerModelling dust transport over central eastern AustraliaMeteorol. Appl., 13 (2006), pp. 1–27
12. Lorenz and Hartmann, 2001D.J. Lorenz, D.L. HartmannEddy-zonal flow feedback in the Southern HemisphereJ. Atmos. Sci., 58 (2001), pp. 3312–3327
13. McTainsh and Pitblado, 1987G.H. McTainsh, J.R. PitbladoDust storms and related phenomena measured from meteorological records in AustraliaEarth Surf. Processes Landforms, 12 (1987), pp. 415–424
14. Meehl et al., 2009G.A. Meehl, A. Hu, B.D. SanterThe mid-1970s climate shift in the Pacific and the relative roles of forced versus inherent decadal variabilityJ. Clim., 22 (D09106) (2009), pp. 780–792
15. Meneghini et al., 2007B. Meneghini, I. Simmonds, N. SmithAssociation between Australian rainfall and the Southern annular modeInt. J. Climatol., 27 (2007), pp. 109–121
16. Middleton, 1984N.J. MiddletonDust storms in Australia: frequency, distribution and seasonalitySearch, 15 (1984), pp. 46–47
17. Nan and Li, 2003S. Nan, J. LiThe relationship between summer precipitation in the Yangtze River valley and the previous Southern Hemisphere annular modeGeophys. Res. Lett., 30 (24) (2003), p. 2266 http://dx.doi.org/10.1029/2003GL018381
18. Record-breaking, 2010Record-breaking La Nina events – Analysis of the La Nina life-cycle and the impacts and significance of the 2010–11 and 2011–12 La Nina events in Australia. Published by the Bureau of Meteorology, GPO Box 1299, Melbourne, VIC. 3001. www.bom.gov.au
19. Skinner and Leslie, 1999T.C.L. Skinner, L.M. LeslieNumerical prediction of the summertime ridge-trough system over northeastern AustraliaWeather Forecast., 14 (1999), pp. 306–325
20. Speer et al., 2009M.S. Speer, L.M. Leslie, A.O. FierroAustralian east coast rainfall decline related to large scale climate driversClim. Dyn., 36 (2009), pp. 1419–1429 http://dx.doi.org/10.1007/s00382-009-0726-1
21. Thompson and Wallace, 2000
22. D.W.J. Thompson, J.M. WallaceAnnular modes in the extratropical circulation. Part I: Month-to-month variabilityJ. Clim., 13 (2000), pp. 1000–1016
23. Trenberth and Hurrell, 1994K.E. Trenberth, J.W. HurrellDecadal atmosphere-ocean variations in the PacificClim. Dyn., 9 (1994), pp. 303–319

CHAPTER 8

LONG TERM SPATIAL AND TEMPORAL RAINFALL TRENDS AND HOMOGENEITY ANALYSIS

Arun Kumar Taxak, A.R. Murumkar and D.S. Arya

Keywords: Trend, Homogeneity Mann–Kendall test, Sen's slope estimator, Rainfall, Wainganga

INTRODUCTION

Rainfall is one of the key climatic variables that affect both the spatial and temporal patterns of water availability. One of the challenges posed by climate change/climate variability is ascertainment, identification and quantification of trends in rainfall and their implications on river flows in order to assist in formulation of adaptation measures through appropriate strategies for water resources management. It is also recognised that rainfall is one of the key climatic variables that affect both the spatial and temporal patterns on water availability (De Luis et al., 2000).

In the analysis of trends of rainfall in the Indian Himalayas, Basistha et al. (2009)observed that rainfall has decreased in the Indian Himalayas during last century as a sudden shift, rather than gradual trend. Kumar and Jain (2011) found decreasing trend in the annual rainfall and rainy days in 15 basins out of 22 basins in India. Even, consolidation of recent studies that have been carried out on analysis of rainfall which is the key input into the hydrologic system, there is conclusive evidence that rainfall is decreasing in Asia (Sinha Ray and Srivastava,, Khan et al., 2000, Shrestha et al., 2000,Mirza, 2002, Lal, 2003, Min et al., 2003, Goswami et al., 2006 and Dash et al., 2007). Some investigators (i.e. Cayan and Peterson, 1989, Lins and Slack, 1999 and Jain and Lall, 2000) have reported evidence of trends (possibly due to anthropogenic influences) and long-term variability of climate. Studies suggest South Asia most vulnerable to climate change.

Citation: Arun Kumar Taxak , A.R. Murumkar , D.S. Arya, Long term spatial and temporal rainfall trends and homogeneity analysis, http://dx.doi.org/10.1016/j.wace.2014.04.005.

Analysis of rainfall trends is important in studying the impacts of climate change for water resources planning and management (Haigh, 2004). It has been recognised that global or continental scale observations of historical climate are less than useful for local or regional scale planning (Barsugli et al., 2009, Brekke et al., and Raucher, 2011). Thus, the evaluation of historical trends or future projections on a regional or local scale is needed. In this study, an attempt has therefore been made to study the trends in the annual rainfall series in the Wainganga basin, India to find if there have been any significant changes in the rainfall trends during 1901–2012.

STUDY AREA AND DATA USED

Study Area

Wainganga basin is a sub-basin of the Godavari River basin which is located from 78°00′ to 81°00′ East longitudes and 19°60′–22°07′ North latitudes as shown in Fig. 1. The total catchment area of the basin is 51,421 km^2 with an elevation ranges from 144 to 1208 m (Fig. 1) above mean sea level. The basin is bounded in the North by Central India hills, in the South and East by the Eastern Ghats and in the West by Maikala hill range. The Chiroli Hills form the watershed dividing the Wainganga basin from the Narmada basin. It is a typical basin considered from geographical and geological point of view covering major parts of the states Maharashtra (27,350 km^2), Madhya Pradesh (23,109 km^2)and small portions of Chhattisgarh states (962 km^2). The river in its initial reaches flows westwards and thereafter turns southwards in Madhya Pradesh and continues to flow Southwards through the Maharashtra State.

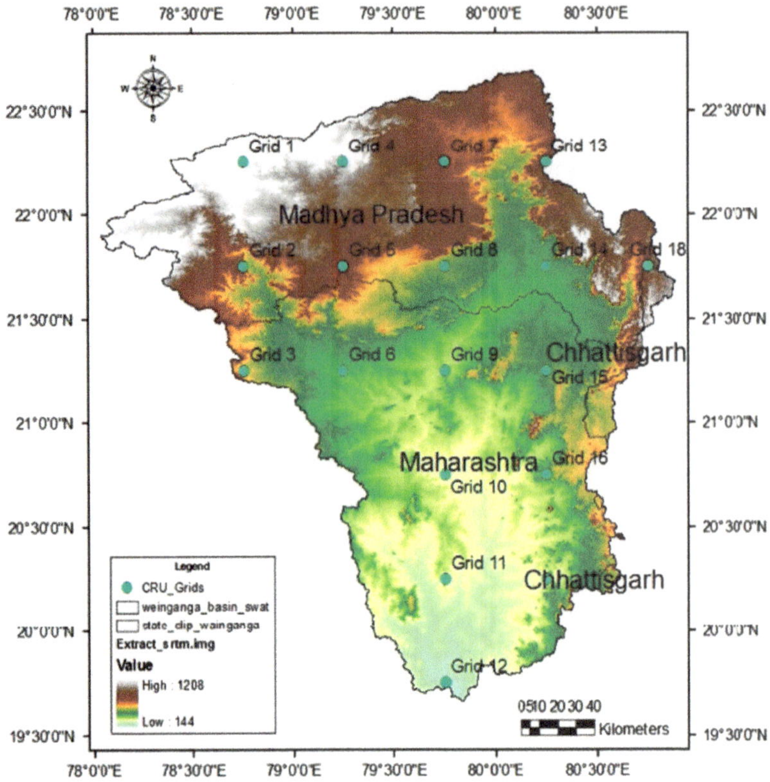

Figure 1: Location of Wainganga basin and the grid points with DEM.

The climate of the basin is characterised as summer from March to May with monsoon season from June to September having some rains in post monsoon season too. Mean daily maximum temperature varies from 26–30 °C in July to 32–33 °C in October. The minimum temperatures are observed in the month of January in the range of 10–15 °C witnessing the winter season. The water users in Wainganga sub basin face many challenges to manage their water resources. Flooding has been a major problem in the past. Though most parts of the basin receives a rainfall of about 150 cm during the monsoon months, there are very few water conservation structures on the main river, and therefore only a few irrigation schemes exists in the basin.

Data Used

CRU TS 3.21 dataset comprises more than 4000 weather stations distributed around the world. The data set contain nine climatic variables: cloud cover, diurnal temperature range, frost day frequency, rainfall, daily mean temperature, monthly average daily maximum temperature, vapour pressure, Potential Evapo-transpiration and wet day frequency. However in this study only rainfall data for the period 1901–2012 at 0.5×0.5° resolution was used. The data was downloaded from Centre of Environmental Data Archival (http://badc.nerc.ac.uk). Eighteen grid points of CRU TS 3.21 dataset cover the whole Wainganaga basin as seen in Fig. 1.

METHODOLOGY

Annual series was prepared for each grid and annual scale trend analysis was performed. The monthly rainfall data was divided into four seasons: Monsoon season (June to September), post-monsoon season (October and November), pre-monsoon season (March to May) and cold winter season (December–February) for seasonal analysis. After fixing the most probable change point, the trend analysis was performed on two partial duration series before and after the change point.

The following tests were used for carrying out the rainfall trend analysis:

- Significance of autocorrelation was detected by using student's t test at lag-1 in annual and seasonal rainfall series.
- Mann–Kendall (MK)/Modified Mann–Kendall (MMK) tests were applied to the non-autocorrelated/autocorrelated series to detect the presence of trend in annual and seasonal rainfall.
- Magnitude of a trend was estimated by Theil and Sen's slope estimator test.
- Changes were calculated as percentage change over the period of 112 year in annual and seasonal rainfall series.
- Pettitt Mann–Whitney (PMW) test was applied to detect most probable change year in the annual and monsoonal rainfall series.
- Percentage change in rainfall over a mean was calculated on the basis of change year in annual and monsoonal rainfall series.
- Inverse distance weighted interpolation technique was used to explore spatial temporal rainfall trend over a basin with the help of Arc GIS.
- Homogeneity in annual series was tested by Pettitt's test, SNHT test and Buishand's test .

- Homogeneity in trends was tested by van Belle and Hughes' homogeneity trend test to obtain global trend over a basin.

RESULTS

Preliminary analysis

Preliminary analysis of the rainfall data shows that mean annual rainfall varies from 1097 mm in western part (Grid #1) to 1541 mm in the eastern part (Grid #16).

The standard deviation varies from 186.03 mm to 282.2 mm for Grid #12 and #9, respectively. The skewness varies between 0.09 and 0.36, predominantly positive skewness with average around 0.21 indicating that annual rainfall during the period is asymmetric and it lies to the right of the mean over all the stations. Kurtosis varies from 1.40 to 4.88 with an average around 3.26. The coefficient of variation (CV) varies between 16.04% (Grid #13) and 21.46% (Grid #9) with an average coefficient of variation 20.97% in the entire basin. Preliminary analysis of the rainfall concluded that the zones of usually heavy rainfall are the zones of least variability and zones of lower rainfall are the zone of higher variability.

Fig. 2a and b represents the spatial mean annual and mean monsoonal rainfall variation over Wainganga basin. Spatial distribution of annual rainfall map shows annual decrease in rainfall from Grid #14 (North East) to Grid #1 (North West) and Grid #12 (South West). In monsoon season, Grid #14 receives more rainfall while Grid #17 less.

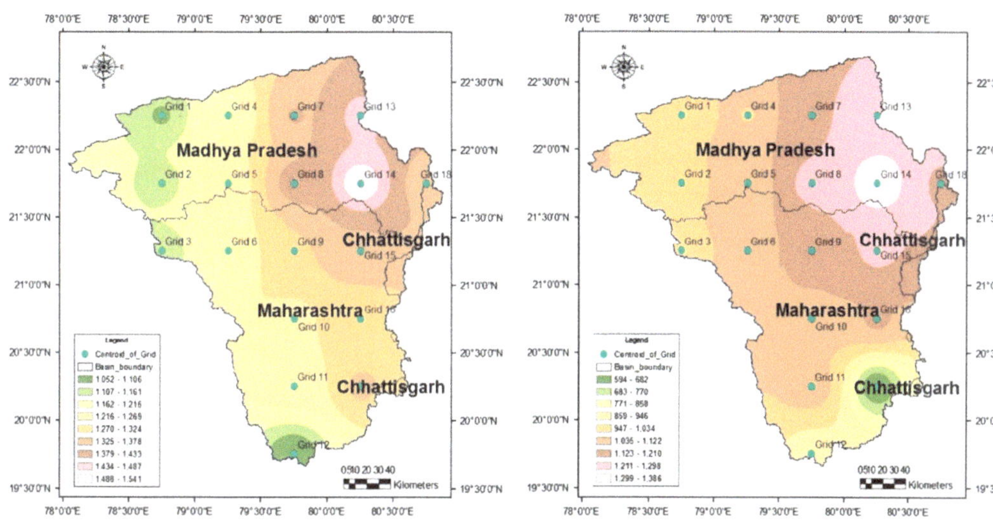

Figure 2: Rainfall Map in Wainganga basin (a) annual (b) monsoon.

Exploratory Data Analysis

The running mean is not resistant to (robust) to local fluctuations. Therefore, to reduce the local fluctuations, the annual and seasonal data series are fitted with LOWESS (Cleveland, 1979, Cleveland, 1984 and Helsel and Hirsch,) regression curves to identify patterns over time. The LOWESS regression curve, as seen in Fig. 3, indicates increasing pattern in the rainfall upto 1930 in annual series and a decreasing trend thereafter. Fig. 4 shows the LOWESS regression curves for Seasonal rainfall series depicting more or less the same patterns. From Fig. 4 it can be seen that monsoon rainfall series has followed similar trend to annual rainfall series. Post monsoon rainfall has shown steep increasing trend up to 1940 after which it decreases gradually. Pre monsoon rainfall series has shown no significant trend, increasing marginally throughout the century. Winter rainfall series has shown no significant tend in the first half but steep decreasing after 1980s.

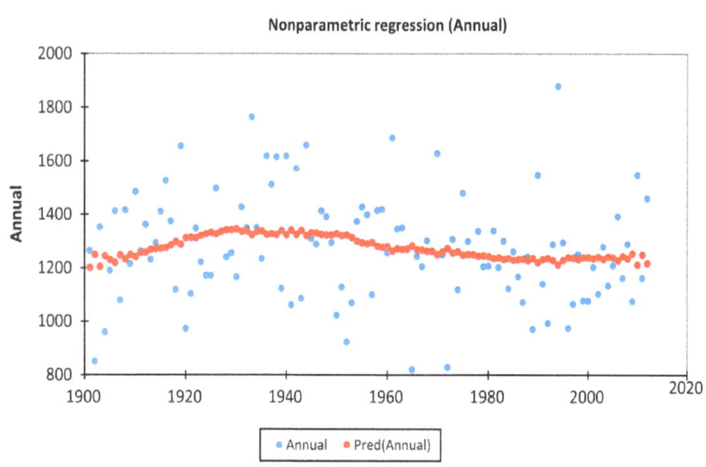

Figure 3: LOWESS regression lines for annual rainfall in Wainganga basin (Pred means prediction).

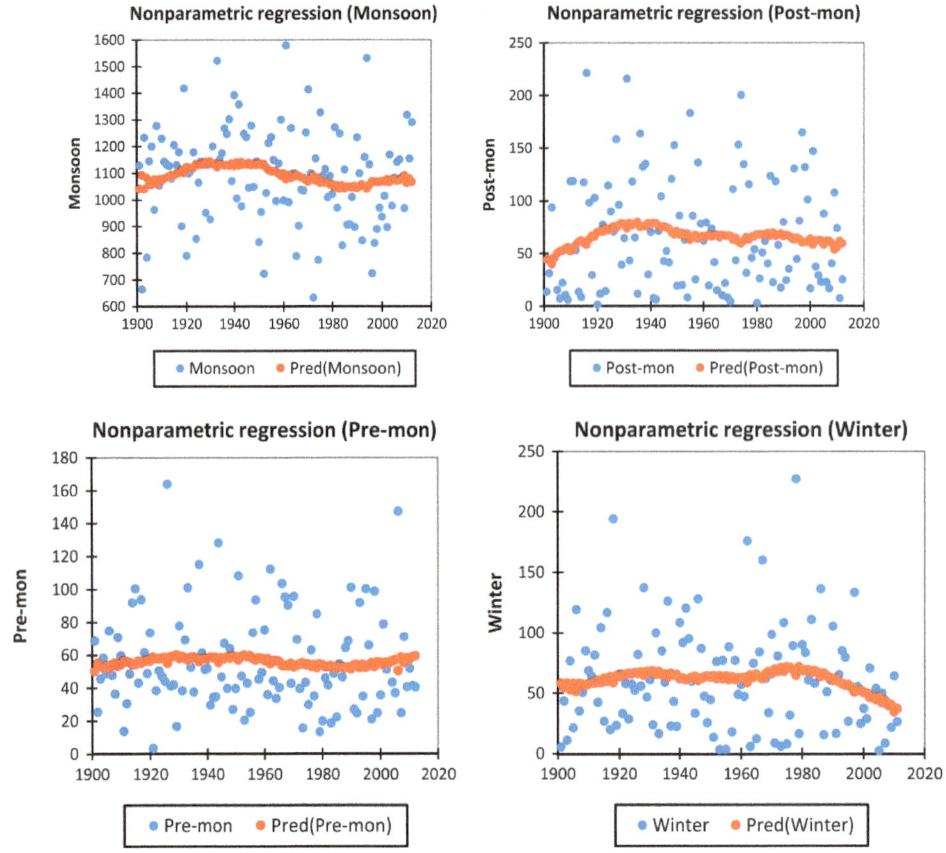

Figure 4: LOWESS regression lines for seasonal rainfall in Wainganga basin (Pred means prediction).

Annual and Seasonal Scale Rainfall Trend Analysis During 1901–2012

The results of the autocorrelation analysis for 1901–2012 are presented in Table 1 along with Z statistics of MK/MMK test. Statistically, only two annual series were autocorrelated at 5% significance level and five seasonal rainfall series: Grid #17 data is autocorelated on annual as well as seasonal scale. Most of the annual series exhibits a decreasing trend, out of which only seven has significant decreasing trend except Grid #1, 4 and 12 with non-significant increasing trend. On the basis of seasonal trend analysis, non-significant increasing trend is observed only in post monsoon season while seven grid points show significant decreasing trend in monsoon rainfall and non-significant decreasing trends in pre-monsoonal and winter rainfall over 112 years (at the 5% level).Table 1 also shows the percentage changes over mean values in annual and seasonal rainfall. Shaded grids shows more than 10% changes in two series and darker shades means more number of series. Grids #9, #10, #15, #16 and #18 shows maximum series (i.e. 5) showing 10% changes. All grids shows non-significantly decrease in magnitude of change in rainfall during the winter season.

Table 1: Result of MK (MMK) test (at 5% level) and percentage change over 1901–2012 with autocorrelation

Grids	Seasonal									
	Annual		Monsoon		Post-Monsoon		Pre-Monsoon		Winter	
	Z-value	% change	Z-value	% change	Z-val-ue	% change	Z-val-ue	% change	Z-val-ue	% change
1	1.33	9.35	1.80	12.79	−0.39	−9.54	0.17	3.35	−1.78	−35.86
2	−0.03	−0.13	0.29	2.01	**−0.05**	−1.93	−0.07	−1.48	−1.85	−37.95
3	−1.82	−12.70	−1.77	−12.98	0.26	4.38	−0.02	−0.36	−1.38	−26.97
4	0.44	2.21	0.93	6.11	−0.65	−12.63	0.08	1.48	−1.26	−24.06
5	−0.58	−3.56	−0.41	−2.23	−0.34	−6.20	−0.18	−3.01	−1.27	−23.51
6	−1.97*	−13.39	−2.20*	−14.59	0.41	7.99	−0.66	−14.77	−0.98	−19.39
7	−0.74	−3.77	−0.37	−2.29	−0.90	−17.43	0.05	0.60	−0.53	−10.97

8	−2.00*	−9.98	−1.51	−8.32	−0.01	−0.13	−1.11	−23.76	−1.28	−26.01
9	−1.97*	−11.78	−1.66	−10.87	0.90	16.01	−1.46	−34.45	−1.34	−27.41
10	−1.76	−10.43	−1.52	−9.04	0.85	16.98	−1.07	−22.88	−1.60	−28.78
11	−1.23	−5.54	−0.98	−4.62	0.88	18.07	−0.11	−2.52	−0.92	−19.55
12	**0.02**	0.43	−0.15	−0.79	0.80	17.42	1.11	18.76	−0.83	−18.01
13	−1.86	−8.56	−1.30	−7.24	−0.58	−12.54	−0.52	−9.09	−1.23	−21.59
14	−2.63*	−13.09	−2.24*	−12.14	0.42	7.58	−1.36	−31.49	−1.53	−32.23
15	−2.34*	−13.72	−2.12*	−12.98	0.88	15.51	−1.41	−31.45	−1.49	−28.71
16	−2.19*	−12.63	−1.98*	−12.10	0.80	16.93	−1.42	−30.96	−1.33	−26.17
17	**−0.08**	−1.03	**0.00**	0.00	**1.49**	30.55	**0.37**	7.97	**−0.46**	−2.48
18	−2.94*	−15.64	−2.71*	−15.58	0.83	16.06	−1.51	−32.69	−1.69	−36.32

MMK values in bold represents presence of autocorrelation in rainfall series. Negative/positive Z value indicates decreasing/increasing trend. Shaded cells show that a grid has more than 2 series showing 10% changes.

Represents significant trend.

Fig. 5(a and b) shows Grid #1 and #18 with highest rising and falling trends in annual and monsoon rainfall at 5% significance respectively have been identified as the two poles in the wainganga basin with respcect to trend. Spatially, the falling trend observed at Grid #18 is found to slowely transform to rising trend in Grid #1and #12 through the internediate grids lying between two poles.

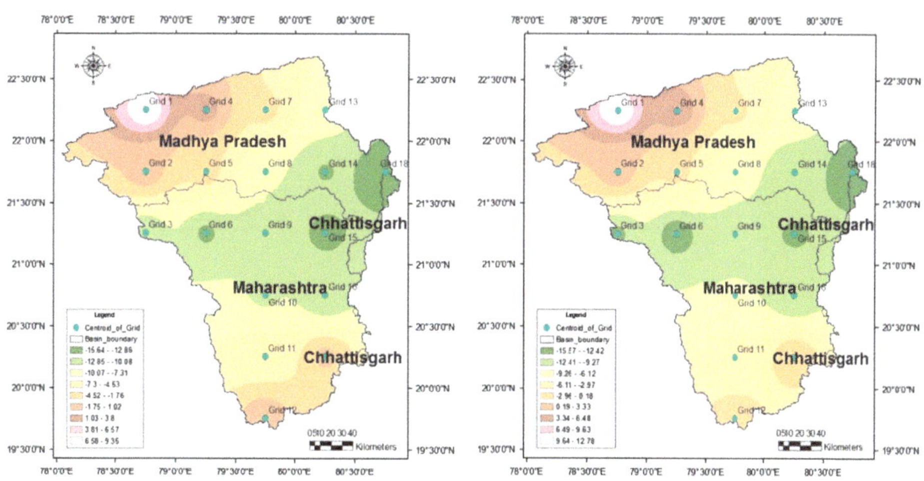

Figure 5: Spatial distribution of change magnitude as percentage of mean in (a) annual rainfall and (b) monsoon rainfall in Wainganga basin during 1901–2012.

Grid-wise Change Point Analysis

Table 2 shows result of change point probability in annual rainfall is computed using PWM test. Pettitt's test, SNHT test and Buishand's test were used to test the homogeneity of a rainfall series. The table depicts that year 1948 is found to be the most probable change year using PWM test. Homogeneity analysis also shows that 8 annual rainfall series are heterogeneous around the year 1948 – it means there is a significant change in the mean before and after the detected change point. SNHT test is known to find change point towards the beginning and end of the series whereas Buishand's and Pettitt's tests are sensitive to find the changes in the middle of a series (Mart´ınez et al., 2009). The results given in Table 2 are in agreement with Mart´ınez et al. (2009)conclusion showing most of the series homogenous.

Table 2: Most probable change year by PMW test, Pettitt's test, SNHT test and Buishand's test

Grid	PWM		Pettitt's test			SNHT			Buishand's test		
	Year	Prob.	K value	Year	Trend	To value	Year	Trend	Q value	Year	Trend
1	1930	91.27	759	1930	Ho	6.75	1930	Ho	12.23	1930	Ho
2	1930	53.28	424	1930	Ho	2.85	1904	Ho	7.37	1930	Ho

3	**1949**	**97.89**	955	**1949**	**Ha**	7.71	1949	Ho	**14.64**	**1949**	**Ha**
4	1925	77.88	597	1925	Ho	6.22	1909	Ho	9.043	1925	Ho
5	1964	73.48	560	1964	Ho	3.57	1902	Ho	7.53	1964	Ho
6	**1949**	**97.24**	921	**1949**	**Ha**	6.74	1949	Ho	13.69	1949	Ho
7	1970	92.54	783	1970	Ho	4.72	1971	Ho	11.14	1970	Ho
8	**1971**	**98.18**	973	**1971**	**Ha**	5.76	1963	Ho	12.65	1963	Ho
9	**1947**	**95.66**	561	1947	Ho	7.10	1942	Ho	13.71	1944	Ho
10	1947	92.34	779	1947	Ho	4.64	1963	Ho	11.36	1963	Ho
11	1963	89.06	723	1963	Ho	2.99	1963	Ho	9.12	1963	Ho
12	1964	56.91	446	1964	Ho	2.93	1926	Ho	7.68	1926	Ho
13	**1971**	**99.51**	1121	**1971**	**Ha**	7.70	1971	Ho	**14.21**	**1971**	**Ha**
14	**1948**	**99.36**	1092	**1948**	**Ha**	8.63	1948	Ho	**15.50**	**1948**	**Ha**
15	**1947**	**98.66**	1009	**1947**	**Ha**	8.63	1947	Ho	**15.42**	**1947**	**Ha**
16	**1947**	**98.21**	975	**1947**	**Ha**	7.73	1947	Ho	**14.59**	**1947**	**Ha**
17	1994	41.52	356	1994	Ho	1.10	1909	Ho	3.20	1994	Ho
18	**1948**	**99.82**	1224	**1948**	**Ha**	**11.57**	**1947**	**Ha**	**17.88**	**1948**	**Ha**

Ha- Heterogeneous series, Ho-Homogenous series.

An analysis across the tests shows that there are 6 common grids where all the tests are in agreement about the change point year. The changes in the mean of these series along with the rainfall plots are shown in Fig. 6(a) Grid #3, (b) Grid #13, (c) Grid #14, (d) Grid #15, (e) Grid #16 and (f) Grid #18. Of these six grids, five grids shows the change point as 1948 except grid #13 where it is 1971. The location of this grid might be playing a significant role as this grid has the highest elevation sloping towards south if compared to other grids having south east slope direction. These geographical settings will promote more orographic precipitation. The figures clearly show that there is a noticeable change in the rainfall mean of the series before and after the change point in all the grids. Table 3shows the percentage change in annual and monsoon rainfall between two partial series around 1948. It demonstrates that almost all the grid show a decrease in annual and monsoonal rainfall except. The highlighted rows in Table 3 are those grids where all the tests are in agreement about the change point. These grids show that the changes in the mean are more than 10% except in Grid #3. The lower changes in this grid may be because of its geographical setting leading to a different change point.

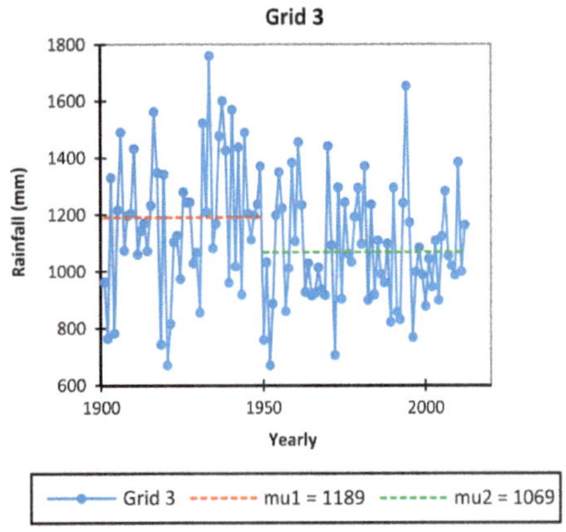

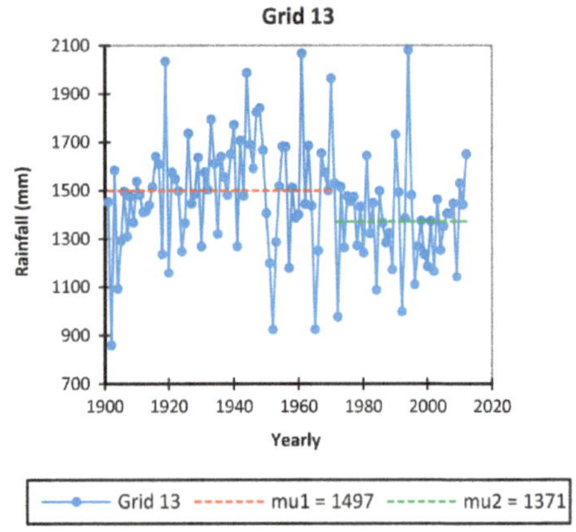

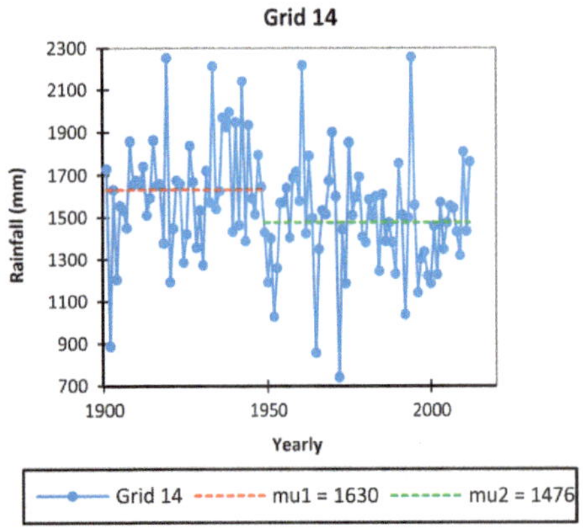

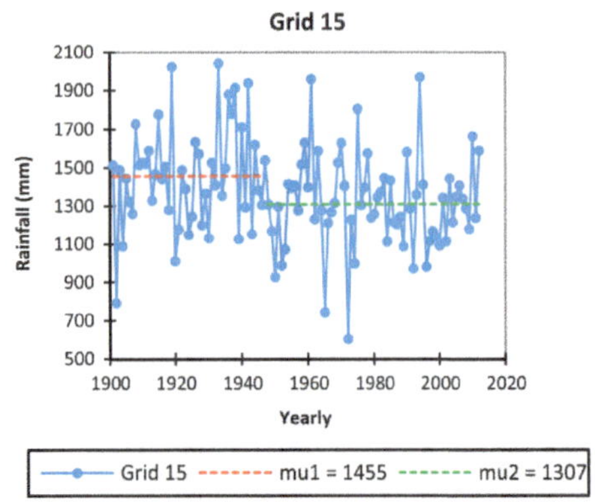

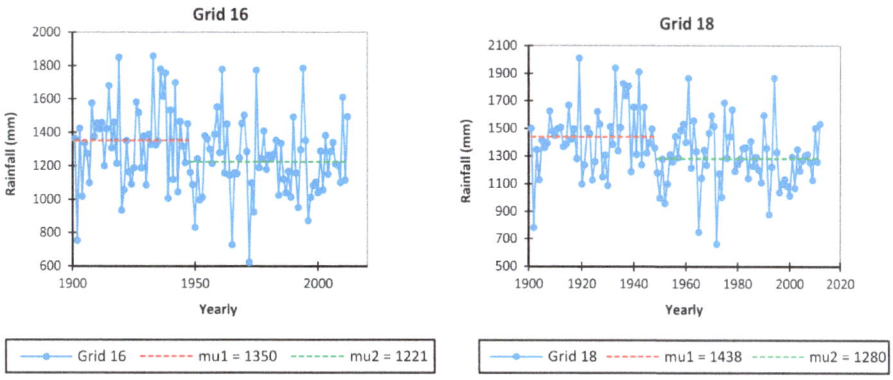

Figure 6: Change year in annual rainfall series (a) Grid #3, (b) Grid #13, (c) Grid #14, (d) Grid #15, (e) Grid #16 and (f) Grid #18 (mu1 and mu2 represents the mean rainfall before and after the change point).

Table 3: Change in percentage in the 1949–2012 mean over the 1901–48. The shaded rows are the grids where series are found heterogeneous using all the tests. Grids in bold face shows the change in the mean are more than 10%

	Annual	Monsoon
Grid 1	2.52	5.12
Grid 2	−2.49	−0.79
Grid 3	**−10.39**	**−10.34**
Grid 4	−0.32	1.83
Grid 5	−4.15	−2.96
Grid 6	−9.64	**−10.22**
Grid 7	−4.67	−3.32
Grid 8	−7.92	−6.99
Grid 9	**−11.01**	**−10.88**
Grid 10	−7.93	−7.80
Grid 11	−3.91	−3.86
Grid 12	−0.73	−0.61
Grid 13	−7.27	−5.94
Grid 14	**−10.45**	**−9.99**
Grid 15	**−11.08**	**−11.15**
Grid 16	**−10.09**	**−10.26**
Grid 17	2.70	5.51
Grid 18	**−12.35**	**−12.10**

Rainfall Trend Analysis using two Partial Series for Periods 1901–1948 and 1949–2012

The results of the autocorrelation analysis for 1901–1948 and 1949–2012 are presented in Table 4 and Table 5 along with Z statistics of MK/MMK test, respectively. From table it is concluded that annual and seasonal series show significant increasing trend in the basin during the period 1901–1948 (Table 4), which turns decreasing during the period 1949–2012 (Table 5) showing an overall decreasing trend in the basin. Maximum % increasing change in annual rainfall is found at Grid #7 (28.18% during 1901–1948) and maximum decreasing % change in annual rainfall is found at Grid #17 (−11.05 during 1949–2012). Seasonal change in monsoon rainfall, approximately 80% of annual rainfall, is 24.44% at Grid #7 which is in agreement with the changes found in annual rainfall during 1901 to 1948. However, maximum seasonal changes during 1949–2012 in monsoon season are −5.55% at Grid #1, it is −4.92% at Grid #17 which are also in agreement with the annual rainfall data of that duration. All partial series, annual and seasonal during 1901–1948, at Grid #4, 5, 7 and #13 show that the changes are more than 10% in all the seasons shown in the darkest shade in Table 4. More than 10% changes are observed in post-monsoon, pre-monsoon and winter rainfall at these grid points (except Grid #12) during 1949–2012 shown in dark colour in Table 5.

Table 4: Result of MK (MMK) test (at 5% level) and percentage over 1901–48 with autocorrelation

Grids	Seasonal									
	Annual		Monsoon		Post-Monsoon		Pre-Monsoon		Winter	
	Z-value	% change	Z-value	% change	Z - value	% change	Z-value	% change	Z-value	% change
1	2.50*	27.14	2.44*	24.16	1.44	50.73	0.21	8.52	1.31	37.03
2	1.80	20.79	**0.74**	20.02	1.37	50.82	0.12	5.32	1.08	30.70
3	1.27	14.24	1.34	14.85	1.08	35.19	0.02	1.04	1.06	34.51
4	2.74*	26.71	**1.04**	23.54	1.46	41.26	0.54	14.02	1.33	40.64

5	2.05*	18.80	**0.82**	16.60	1.42	47.28	0.33	13.73	1.24	35.31
6	1.00	10.12	0.73	7.25	1.27	30.59	−0.25	−8.62	0.71	25.85
7	3.30*	28.18	2.53*	24.44	1.39	52.45	0.68	22.39	1.65	44.51
8	2.07*	17.59	**0.75**	13.26	1.41	49.75	−0.10	−2.03	1.93	50.22
9	0.45	4.88	0.09	1.40	0.95	33.48	−0.28	−11.27	1.50	56.48
10	0.56	6.18	0.26	2.22	1.24	43.92	−0.21	−7.85	1.24	41.59
11	**0.44**	7.93	**0.28**	6.25	1.43	45.89	−0.10	−4.77	0.96	30.79
12	1.27	14.45	1.00	11.73	1.58	53.63	−0.13	−4.36	0.57	22.27
13	3.72*	24.59	2.91*	20.65	1.52	55.75	0.60	22.26	1.64	41.55
14	1.25	11.07	1.08	7.37	1.20	39.24	−0.73	−19.43	2.00*	63.49
15	0.70	5.69	0.47	3.48	1.09	40.50	−0.35	−8.05	1.64	59.65
16	0.63	5.50	0.52	4.99	1.32	46.29	0.02	0.35	0.86	30.34
17	**0.38**	9.65	**0.05**	0.27	0.50	9.40	**0.66**	12.90	**0.11**	0.57
18	1.06	9.24	0.60	5.32	1.41	49.97	−0.54	−22.81	1.81	55.06

MMK values in bold represents presence of autocorrelation in rainfall series. Negative/positive Z value indicates decreasing/increasing trend. Shaded cells show that a grid has more than 2 series showing 10% changes.

*Represents significant trend.

Table 5: Result of MK (MMK) test (at 5% level) and percentage over 1949–2012 with autocorrelation

Grids	Seasonal									
	Annual		Monsoon		Post-Monsoon		Pre-Monsoon		Winter	
	Z-value	% change	Z-value	% change	Z-value	% change	Z-value	% change	Z-value	% change
1	0.12	1.18	0.28	2.91	−1.01	−31.43	0.27	5.67	−0.65	−15.83
2	0.03	0.27	0.43	4.13	−0.46	−14.34	−0.13	−3.79	−0.32	-9.88
3	0.04	0.21	0.26	3.13	0.10	5.38	0.51	14.94	−0.09	-1.16
4	−0.53	−4.29	0.04	0.81	−0.98	−26.01	−0.45	−10.42	−0.82	−19.19
5	−0.16	−1.38	0.04	0.42	−0.64	−19.07	−0.59	−16.66	−0.45	−15.18
6	−0.25	−1.46	−0.04	−0.28	−0.06	−0.92	−0.39	−12.30	−0.21	−4.65
7	−0.57	−4.09	−0.25	−1.94	−0.63	−17.51	−0.43	−11.23	−0.53	−11.67
8	−0.64	−4.09	0.09	0.58	−0.27	−8.10	−0.65	−20.20	−0.52	−14.05
9	0.08	1.36	1.04	8.54	0.43	11.89	−0.33	−7.76	0.07	1.34
10	−0.11	−0.83	0.59	4.37	0.21	5.11	−0.21	−5.08	−0.05	−1.26
11	−0.57	−2.81	−0.08	−0.58	0.32	9.40	0.23	5.97	0.48	10.94
12	−0.39	−3.77	−0.47	−4.27	**0.17**	11.72	0.53	12.18	0.60	20.08
13	−1.33	−8.16	−0.68	−5.55	−0.62	−17.21	−0.55	−13.42	−0.53	−12.53
14	−0.54	−4.05	0.06	0.71	−0.02	−0.70	−0.42	−12.82	−0.42	−9.29
15	−0.09	−0.58	0.53	3.86	0.23	5.65	−0.41	−13.50	−0.20	−4.52
16	−0.02	−0.15	0.61	4.84	0.04	2.30	−0.64	−17.46	0.00	0.00
17	−0.68	−11.05	−0.65	−4.92	−0.52	−12.58	0.42	6.72	−0.84	−4.50
18	−0.37	−2.72	0.00	0.00	0.16	4.53	−0.62	−17.78	−0.16	−5.63

MMK values in bold represents presence of autocorrelation in rainfall series. Negative/positive Z value indicates decreasing/increasing trend. Shaded cells shows that a grid has more than 2 series showing 10% changes.

Estimation of Manitude of Trend slope in Annual and Sesaonal Rainfall Series

The box-plots of Theil-Sen's slopes for annual and seasonal rainfall series over the basin are shown in Fig. 7 for three time periods during 1901–2012, 1901–1948 and 1949–2012. At the annual scale, the median of slopes are located below the zero line, implying a decreasing trend over two spans i.e. 1901–2012 and 1949–2012, while monsoon season shows an increaing trend in second partial series. During 1901–1948, median of slopes of annual and seasonal series are located above zero line, indicated a increasing trend.

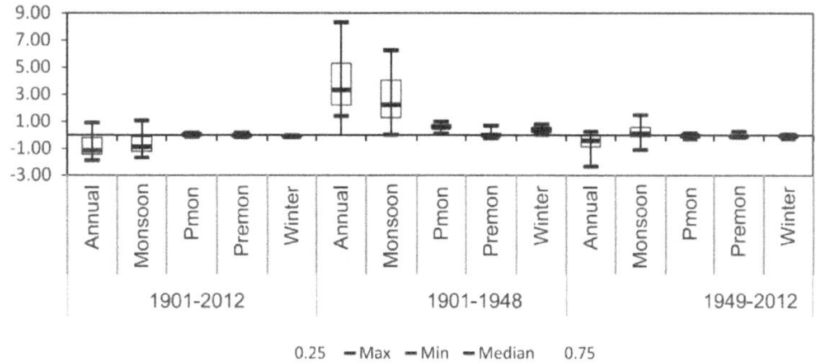

Figure 7: Box-and-Whisker plot of magnitude of trend slope in rainfall series using Theil and Sen's Slope over three time periods.

Analysis of Homogeneity of Trends

To test homogeneity of trends across annual and seasonal rainfall, van Belle and Hughes' homogeneity trend test was applied on all the 18 grids in the basin. Results of the test are given in Table 6. With p=18, $\chi^2_{homogenous}$ is found to be 4.11, 3.40, 0.22, 1.33 and 3.0 for annual, monsoon, post-monsoon, pre-monsoon and winter series respectively. At 5% significance level, null hypothesis of homogeneity of trends is accepted for annual and seasonal trends during 1901–2012, 1901–1948 and 1949–2012. It means the overall the trends of annual series for entire duration are decreasing in the entire Wainganga basin.

Table 6: Results of homogeneity of trends between grid points on annual and seasonal basis

Series	χ2homogenous		
	1901–2012	1901–1948	1949–2012
Annual	4.11*	1.76*	0.57*
Monsoon	3.40*	0.64*	0.10*
Post-Monsoon	0.22*	0.42*	0.43*
Pre-Monsoon	1.33*	0.13*	0.42*
Winter	3.00*	0.56*	0.47*

Indicates significant homogeneous trend in annual seasonal rainfall across grids over a basin.

Trend analysis of Mean Rainfall Data over entire Wainganga Basin

Mean rainfall series of annual rainfall data and seasonal rainfall data using all grids' data were prepared for three time periods during 1901–2012, 1901–1948 and 1949–2012. The result of mean, autocorrelation, Z statistics of Mann–Kendall (MMK) test, Sen's slope estimator test and % change for annual and seasonal rainfall series are given inTable 7(a–c). There is statistically non-significant decreasing trend found in annual and seasonal series in the entire basin during 1901–2012. There is statistically non-significant increasing trend in annual and seasonal rainfall over the basin during the period 1901–1948, which becomes decreasing during the period 1949–2012. There is an overall decrease (−8.45%) in the annual rainfall during the period 1901–2012 over a basin.

Table 7: Results of MK (MMK) test statistic, Theil and Sen's slope and % change over Wainganga basin from

	Mean	MK (MMK)	Sen's slope	% change
(a) 1901–2012				
Annual	1275	−1.75	−0.96	−8.45
Monsoon	1090	−1.33	−0.67	−6.85
Post	66	0.19	0.02	4.17
Pre	56	−0.79	−0.05	−10.68

Winter	61	**−0.38**	−0.11	−20.09
As per Table 1				
(b) 1901–48				
Annual	1319	1.86	4.52	16.43
Monsoon	1125	**0.64**	2.58	10.99
Post-mon	69	1.32	0.63	44.07
Pre-mon	57	0.12	0.05	4.41
Winter	65	**0.93**	0.60	44.31
(c) 1949-2012				
Annual	1243	−1.00	−1.32	−6.80
Monsoon	1064	−0.04	−0.03	−0.19
Post-mon	65	−0.37	−0.09	−8.95
Pre-mon	54	−0.10	−0.01	−1.59
Winter	59	−0.30	10.10	−11.28

CONCLUSIONS

In the present study, trend and homogeneity of trends for annual and seasonal rainfall series was analysed for Wainganga basin during the period 1901–2012 using CRU 0.5×0.5° gridded rainfall data. From the study, it is concluded that annual and monsoon rainfall is decreased in the Wainganga basin during the period 1901–2012. The most probable year of change in the basin is 1948. There is an increasing trend in the basin during the period 1901–1948, which got reversed during the period 1949–2012 resulting in an overall decreasing trend in the basin. There is an overall decrease (−8.45%) in annual rainfall in the basin during the period 1901–2012. It is also found that the trends are homogeneous in the entire basin. Overall, the results of this study are in agreement with other studies conducted by many researcher in central India (Duhan and Pandey, 2013, Kumar and Jain, 2011 and Singh et al., 2005).

Global climate shift (Baines, 2006) or weakening global monsoon circulation (Pant, 2003), reduction in forest cover (Nair et al., 2003, Ray et al., 2003 and Ray et al., 2006) and increasing aerosol due to anthropogenic activities (Ackerman et al., 2000,Ramanathan et al., 2001, Ramanathan et al., 2005 and Sarkar and Kafatos, 2004) may be the probable causes of change in rainfall which were reported by Basistha et al. (2009).

REFERENCES

1. Alexandersson and Moberg, 1997H. Alexandersson, A. MobergHomogenization of Swedish temperature data. Part I: a homogeneity test for linear trendsInt. J. Climatol., 17 (1997), pp. 25–34
2. Alexandersson, 1986H. AlexanderssonA homogeneity test applied to precipitation dataInt. J. Climatol., 6 (1986), pp. 661–675Ackerman et al., 2000
3. A.S. Ackerman, O.B. Toon, D.E. Stevens, D.E. Stevens, A.J. Heyms field, V. Ramanathan, E.J. WeltonReduction of tropical cloudiness by sootScience, 288 (2000), pp. 1042–1047
4. Anderson, 1941R.L. AndersonDistribution of the serial correlation coefficientsAnn. Math. Stat., 8 (1) (1941), pp. 1–13
5. Baines, 2006Baines, P.G., 2006. The late 1960s Global Climate Shift and its influence on the Southern Hemisphere. In: Proceedings of 8 ICSHMO, Brazil, INPE, pp. 1477–1482.
6. Barsugli et al., 2009Barsugli, J., Anderson, C., Smith, J.B., Vogel, J.M., 2009. Options for Improving Climate Modeling to Assist Water Utility Planning for Climate Change. White Paper, Water Utility Climate Alliance.
7. Basistha et al., 2009Ashoke Basistha, D.S. Arya, N.K. GoelAnalysis of historical changes in the Indian HimalayasInt. J. Climatol., 29 (2009), pp. 555–572
8. Brekke et al.,Brekke, L.D., Kiang, J.E., Olsen, J.R., Pulwarty, R.S., Raff, D.A., Turnipseed, D.P., Webb, R.S., White, K.D., 2009. Climate Change and Water Resources Management: A Federal Perspective. US Geological Survey Circular, 1331, 65p. 〈http://pubs.usgs.gov/circ/1331/〉 .
9. Buishand, 1982T.A. BuishandSome methods for testing the homogeneity of precipitation recordsJ. Hydrol., 58 (1982), pp. 11–27
10. Cayan and Peterson, 1989D.R. Cayan, D.H. PetersonThe influence of north Pacific atmospheric circulation on streamflows in the westGeophys. Monogram Ser., 55 (1989), pp. 375–397
11. Cleveland, 1979W.S. ClevelandRobust locally weighted regression and smoothing scatter plotsJ. Am. Stat. Assoc., 74 (1979), pp. 829–836
12. Cleveland, 1984W.S. ClevelandGraphs in scientific publicationsAm. Stat., 38 (4) (1984), pp. 261–269
13. Cunderlik and Burn, 2004J.M. Cunderlik, D.H. BurnLinkages between regional trends in monthly maximum flows and selected climatic variablesASCE J. Hydrol. Eng., 9 (4) (2004), pp. 246–256

14. Dash et al., 2007S.K. Dash, R.K. Jenamani, S.R. Kalsi, S.K. PandaSome evidence of climate change in twentieth-century IndiaClim. Change, 85 (2007), pp. 299–321
15. de la Casa and Nasello, 2010A. de la Casa, O. NaselloBreakpoints in annual rainfall trends in Córdoba, ArgentinaAtmos. Res., 95 (2010), pp. 419–427
16. de la Casa and Nasello, 2012A. de la Casa, O.B. NaselloLow frequency oscillation of rainfall in Córdoba, Argentina, and its relation with solar cycles and cosmic raysAtmos. Res., 113 (2012), pp. 140–146
17. De Luis et al., 2000M. De Luis, J. Raventós, J.C. González-Hidalgo, J.R. Sánchez, J. CortinaSpatial analysis of rainfall trends in the region of Valencia (east Spain), 2000Int. J. Climatol., 20 (12) (2000), pp. 1451–1469
18. Duhan and Pandey, 2013D. Duhan, A. PandeyStatistical analysis of long term spatial and temporal trends of precipitation during 1901–2002 at Madhya Pradesh, IndiaAtmos. Res., 122 (2013), pp. 136–149
19. Goswami et al., 2006B.N. Goswami, V. Venugopal, D. Sengupta, M.S. Madhusoodanam, P.K. XavierIncreasing trends of extreme rain events over India in a warming environmentScience, 314 (2006), pp. 1442–1445
20. Haigh, 2004M.J. HaighSustainable management of headwater resources: the Nairobi headwater declaration (2002) and beyondAsian J. Water, Environ. Pollut., 1 (1–2) (2004), pp. 17–28
21. Hamed and Rao, 1998K.H. Hamed, A.R. RaoA modified Mann–Kendall trend test for auto correlated dataJ. Hydrol., 204 (1998), pp. 182–196
22. Helsel and Hirsch,Helsel, D.R., Hirsch, R.M., 2002. Statistical Methods in Water Resources. United States Geological Survey, 524.Jain and Lall, 2000S. Jain, U. LallMagintude and timing of annual maximum floods, trends and large scale climatic associations for the Blacksmith Fork River, UtahWater Resour. Res., 36 (12) (2000), pp. 3641–3651
23. Kendall, 1975M.G. KendallRank Correlation MethodsCharles Graffin, London (1975)
24. Khan et al., 2000T.M.A. Khan, O.P. Singh, M.D. Sazedur RahmanRecent sea level and sea surface temperature trends along the Bangladesh coast in relation to the frequency of intense cyclonesMar. Geo-desy, 23 (2000), pp. 103–116
25. Krishnakumar et al., 2009K.N. Krishnakumar, G.S.L.H.V. Prasada Rao, C.S. GopakumarRainfall trends in twentieth century over Kerala, IndiaAtmos. Environ., 43 (2009), pp. 1940–1944
26. Kumar and Jain, 2011V. Kumar, S.K. JainTrends in rainfall amount and number of rainy days in river basins of India (1951–2004)Hydrol. Res., 42 (2011), p. 4
27. Lal, 2003M. LalGlobal climate change: India's monsoon and its variabilityJ. Environ. Stud. Policy, 6 (2003), pp. 1–34
28. Lins and Slack, 1999H.F. Lins, J.R. SlackStream flow trends in the United StatesGeophys. Res. Lett., 26 (2) (1999), pp. 227–230
29. Mann, 1945H.B. MannNonparametric tests against trendEconometrica, 13 (1945), pp. 245–259
30. Mart´ınez et al., 2009M.D. Mart´ınez, C. Serra, A. Burgue˜no, X. LanaTime trends of daily maximum and minimum temperatures in Catatonia (NE Spain) for the period 1975–2004Int. J. Climatol., 30 (2) (2009), pp. 267–290 http://dx.doi.org/10.1002/joc.1884
31. Min et al., 2003S.K. Min, W.T. Kwon, E.H. Parkand, Y. ChoiSpatial and temporal comparisons of droughts over Korea with East AsiaInt. J. Climatol., 23 (2003), pp. 223–233
32. Mirza, 2002M.Q. MirzaGlobal warming and changes in the probability of occurrence of floods in Bangladesh and implicationsGlob. Environ. Chang., 12 (2002), pp. 127–138
33. Nair et al., 2003U.S. Nair, R.O. Lawton, R.M. Welch, R.A. PielkeImpact of land use on Costa Rican tropical montane cloud forests: sensitivity of cumulus cloud field characteristics to low land deforestationJ. Geophys. Res., 108 (D7) (2003), pp. 4206–4219
34. Pant, 2003G.B. PantLong-term climate variability and change over monsoon AsiaJ. Indian Geophys. Union, 7 (3) (2003), pp. 125–134
35. Pettitt, 1979A.N. PettittA non-parametric approach to the change-point detectionAppl. Stat., 28 (2) (1979), pp. 126–135
36. Ramanathan et al., 2001V. Ramanathan, P.J. Crutzen, J.T. Keihl, D. RosenfeldAerosols, climate and the hydrological cycleScience, 294 (2001), pp. 2119–2124
37. Ramanathan et al., 2005V. Ramanathan, C. Chung, D. Kim, T. Bettge, L. Buja, J.T. Keihl, W.M. Washington, Q. Fu, D.R. SikkaAtmospheric brown clouds: impacts on South Asian climate and hydrological cycleProc. Natl. Acad. Sci. USA, 102 (15) (2005), pp. 5326–5333
38. Rao et al., 2003A.R. Rao, K.H. Hamed, H.L. ChenNonstationarities in Hydrologic and Environmental Time SeriesKluwer Academic Publishers, The Netherlands (2003), p. 362
39. Raucher, 2011Raucher, R.S., 2011. The Future of Research on Climate Change Impacts on Water: A Workshop Focused on Adaption Strategies and Information Needs. Water Research Foundation. 〈http://www.waterrf.org/projectsreports/publicreportlibrary/4340.pdf〉.
40. Ray et al., 2003 D.K. Ray, U.S. Nair, R.M. Welch, Q. Han, J. Zeng, W. Su, T. Kikuchi, T.J. LyonsEffects of land use in Southwest Australia: 1. Observations of cumulus cloudiness and energy fluxesJ. Geophys. Res., 108 (D14) (2003), p. 4414 http://dx.doi.org/10.1029/2002JD002654
41. Ray et al., 2006D.K. Ray, U.S. Nair, R.O. Lawton, R.M. Welch, R.A. PielkeImpact of land use on Costa Rican tropical montane cloud forests: sensitivity of orographic cloud formation to deforestation in the plainsJ. Geophys. Res., 111 (2006), p. D02108 http://dx.doi.org/10.1029/2005JD006096
42. Sarkar and Kafatos, 2004S. Sarkar, M. KafatosInterannual variability of vegetation over the Indian sub-continent and its relation to the different meteorological

parametersRemote Sens. Environ., 90 (2004), pp. 268–280

43. Sen, 1968P.K. SenEstimates of the regression coefficient based on Kendall's tauJ. Am. Stat. Assoc., 63 (1968), pp. 1379–1389

44. Shifteh Somee et al., 2012B. Shifteh Somee, A. Ezani, H. TabariSpatiotemporal trends and change point of precipitation in IranAtmos. Res., 113 (2012), pp. 1–12

45. Shrestha et al., 2000A.B. Shrestha, C.P. Wake, J.E. Dibb, P.A. MayewskiPrecipitation fluctuations in the Nepal Himalaya and its vicinity and relationship with some large scale climatological parametersInt. J. Climatol., 20 (2000), pp. 317–327

46. Singh et al., 2005Singh, O.P., Sontakke, N.A., Singh, N.H., Pandey Kumar, A., 2005. Recent trend in spatio temporal variation of precipitation over India – an investigation into basin-scale rainfall fluctuations. In:Proceedings of Symposium S6 Held During the Seventh IAHS Scientific Assembly at Foz do Iguacu, Brazil, April 2005. IAHS Publication 296. pp. 273–282.

47. Sinha Ray and Srivastava,Sinha Ray, K.C., Srivastava, A.K., 1999. Is there any change in extreme events like droughts and heavy rainfall? INTROPMET- 97, 2–5 December 1997, IIT New Delhi, 1999.

48. Subash et al., 2011N. Subash, S.S. Singh, N. PriyaVariability of rainfall and effective onset and length of the monsoon season over a sub-humid climatic environmentAtmos. Res., 99 (2011), pp. 479–487

49. Tabari et al., 2011H. Tabari, B. Shifteh Somee, M.R. ZadehTesting for long-term trends in climatic variables in IranAtmos. Res., 100 (2011), pp. 132–140

50. Theil, 1950H. TheilA rank invariant method of linear and polynomial regression analysis, Part 3Nederl. Akad. Wetensch. Proc., 53 (1950), pp. 1397–1412

51. van Belle and Hughes,van Belle, G., Hughes, J., 1984. Nonparametric Tests for Trend in Water Quality. Water Resources Research.Yue et al., 2003S. Yue, P. Pilon, B. PhinneyCanadian streamflow trend detection: impacts of serial and cross-correlationHydrol. Sci. J., 48 (1) (2003), pp. 51–63

52. Yue and Hashino, 2003S. Yue, M. HashinoLong term trends of annual and monthly precipitation in JapanJ. Am. Water Resour. Assoc., 39 (3) (2003), pp. 587–596

53. Yue et al., 2002bS. Yue, P. Pilonand, G. CavadiasPower of the Mann–Kendall and Spearman's rho tests for detecting monotonic trends in hydrological seriesJ. Hydrol., 259 (2002), pp. 254–271

CHAPTER 9 AFFECT ON VEGETATION CYCLE FROM CLIMATE CHANGE

Nabil Laanaia, Dominique Carrer, Jean-Christophe Calvet, Christian Pagé

Keywords: Climate change, Adaptation, Modeling, Agriculture, Forestry, Phenology, Soil moisture

INTRODUCTION

Adapting agriculture and forestry to climate change in a global context of rising demand for food and energy will be a key objective in the next decades. The basic information needed to build adaptation strategies is provided by impact models able to simulate vegetation growth in response to climate variables (Tubiello et al., 2007). The latter are provided by climate models. Both climate and impact models are affected by uncertainties, which have to be described and quantified as much as possible (e.g.Déqué and Somot, 2010 and Soussana et al., 2010). In many European countries, agricultural and forestry practices present local differences related to soil characteristics and, also, often related to local climates. Therefore, downscaling the large scale information given by global climate models is key. Moreover, an important component of adaptation is the development of new ground networks able to monitor biophysical indicators such as soil moisture (Calvet et al., 2007) and methods permitting the identification of vulnerable areas where the added value of such networks would be high.

In this study, we focus on metropolitan France, and we address local aspects of the impact of climate change on vegetation growth using the SURFEX ("Surface Externalisée") modeling platform (Masson et al., 2013). The IPCC climate simulations, either CMIP3 A1B or CMIP5 RCP, show significant trends towards more precipitation in northern Europe and less precipitation in southern Europe and the Mediterranean basin (Jacob et al., 2014). From this point of view, France lies in an area between northern and southern Europe, where trends in precipitation are less pronounced. However, the marked increase in air temperature (T_{air}) may enhance evapotranspiration in spite of the antitranspirant action of CO_2 on plants. This may trigger a soil water deficit at springtime, earlier than under the present climate, and increase the irrigation demand (Calvet et al., 2008).

In order to simulate the impact of climate change on cultivated plant species, Brisson and Levrault (2010) used specific crop and forestry models over 12 sites representative of the main climatic regions in France. Their simulations suggested that wheat yields and the forage production could slightly increase in the future. On the other hand, the viability of forests, vineyards, and summer crops such as maize could be impacted by more frequent droughts in many regions. Loustau et al. (2005) used several forestry models over 5 locations in French plains. Their simulations indicated "a slight increase in potential forest yield until 2013–2050, followed by a plateau or a decline around 2070–2100". These trends were more or less pronounced from one location to another and from one tree species to another.

As past studies have focused on a limited number of sites in France, there is a need for more extensive evaluation of the plant phenology response across species. Models able to represent various vegetation types have been developed in the last two decades. They rely on generic approaches permitting the simulation of land surface water, energy and CO_2 fluxes, together with soil moisture and vegetation biomass. In this study, we used the Interactions between Soil, Biosphere, and Atmosphere model (ISBA) developed by Meteo-France for

Citation: Nabil Laanaia, Dominique Carrer, Jean-Christophe Calvet, Christian Pagé, Affect on vegetation cycle from climate change, Climate Risk Management, Available online 15 June 2016, ISSN 2212-0963, http://dx.doi.org/10.1016/j.crm.2016.06.001.

meteorological, hydrological and climatic applications. The CO_2-responsive version of this model (Calvet and Soussana, 2001 and Gibelin et al., 2008) is able to simulate plant growth and carbon storage. The model uses relatively few parameters while representing key processes. In particular, drought-avoiding and drought-tolerant plants can be distinguished thanks to a refined representation of the effects of soil water deficit on photosynthesis parameters (e.g. mesophyll conductance). Phenology is directly driven by photosynthesis and, using satellite leaf area index (LAI) products, Szczypta et al. (2014) showed that realistic LAI simulations can be obtained over Europe.

This work aims at assessing uncertainties of future trends on biophysical variables (leaf onset, leaf senescence, maximum LAI) for four vegetation types (rainfed straw cereals and grasslands, broadleaf forests, needleleaf forests) over a large number of sites representative of the agricultural and forest regions in France. Maize and vineyards are not considered as they are not explicitly represented in our simulations. As a key objective of this study is to quantify the impact of uncertainties in climate simulations, we use an ensemble of eleven climate models (Fig. 1) to drive the ISBA model. We apply classical statistical methods to assess: (1) trends in time series of downscaled simulations of biophysical variables and plant phenology indicators, (2) the consistency of the various climate simulations at a given location. Finally, the analysis of the spatial variability of the response of plant growth to climate change is investigated and the main spatial patterns identified in this study are summarized.

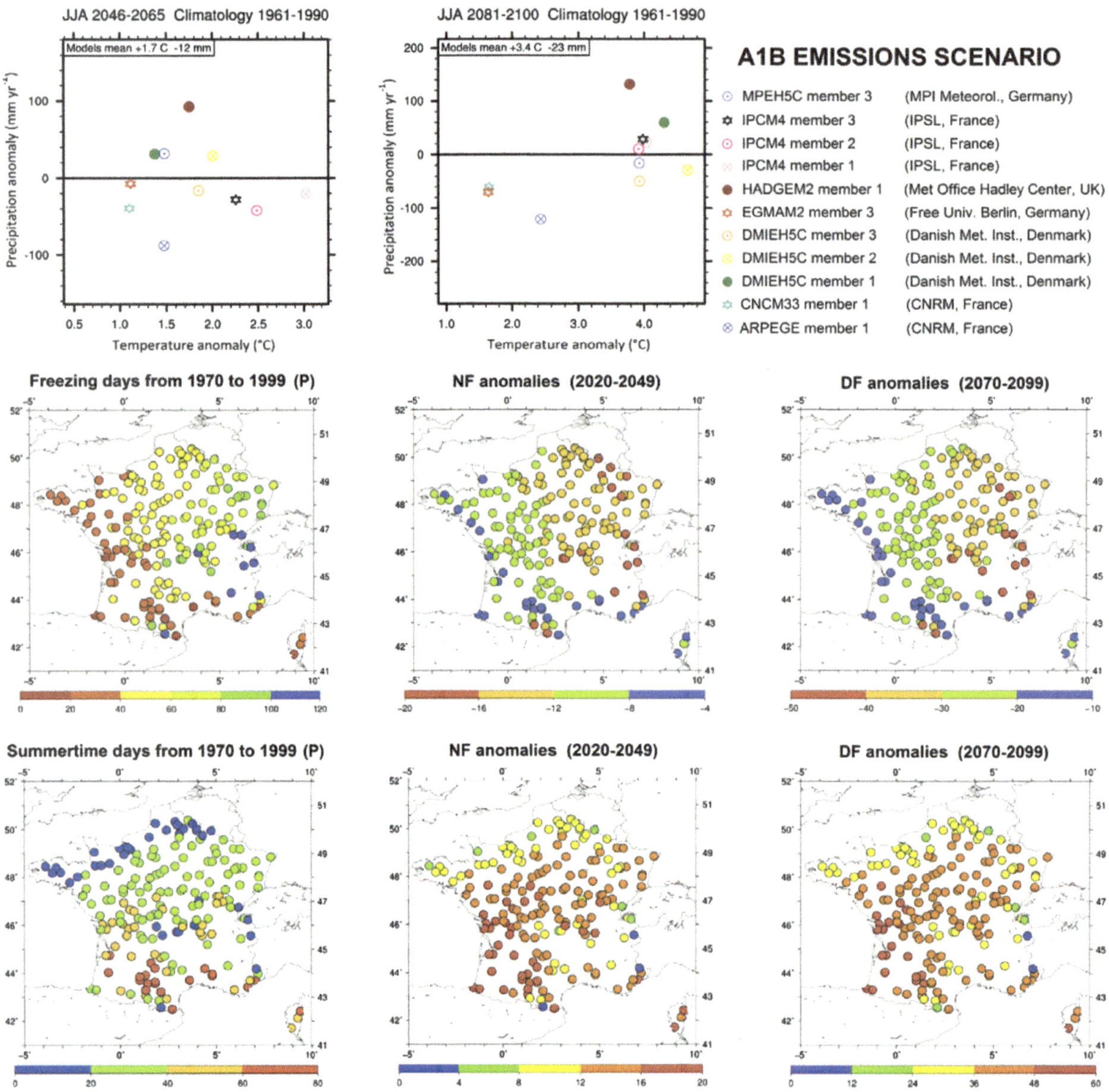

Figure 1: The eleven downscaled A1B climate simulations used in this study. (Top, from left to right) average summertime (June–July–August) cumulated precipitation anomalies vs. air temperature anomalies for 2046–2065 and 2081–2100, with respect to the 1961–1990 period. (Middle, from left to right) average number of freezing days for the past ("P", from 1970 to 1999), and anomalies (in days) for near-future ("NF", from 2020 to 2049) and distant future ("DF", from 2070 to 2099). (Bottom, from left to right) average number of summertime days for P, and anomalies (in days) for NF and DF. Freezing and summertime days are defined as days with at least one hourly air temperature value below 0° and above 25°, respectively, at 191 locations.

The climate simulations and the ISBA land surface model are presented in Section 2, together with the design of the simulations of biophysical variables. The results are shown in Section 3 and discussed in Section 4. The conclusions of this study are summed up in Section 5.

MATERIAL AND METHODS

Climate Simulations

The downscaled climate simulations used in this study are derived from the SCRATCH2010 database (http://www.cerfacs.fr/~page/scratch/), which has been used in various climate change impact studies in France (e.g. Gouache et al., 2012 and Habets et al., 2013). SCRATCH2010 climate simulations are generated using a statistical downscaling method applied to global large-scale climate simulations. The methodology is described in Boé et al. (2006), and in Boé and Terray (2008). It is based on weather-typing and builds on the physical links between the large-scale circulation and the local scale climate, from which statistical relationships can be derived. The spatial scale of representation is changed but the temporal distribution of weather variables is close to the parent global simulation. This statistical downscaling method is used to produce climate projections at a fine scale over France (8 km by 8 km), at a hourly time step. It is first trained on a reference period using observations to define the statistical relationships, and then applied to the climate simulations. The reference atmospheric reanalysis used to train the method is "Système d'Analyse Fournissant des Renseignements A la Neige" (SAFRAN) (Durand et al., 1993 and Quintana-Segui et al., 2008). SAFRAN provides hourly values of liquid and solid precipitation, incoming solar radiation, incoming longwave radiation, air temperature, air humidity, atmospheric pressure, and wind speed over France. Boé et al. (2006) showed that the climatology derived from the downscaled climate simulations for past decades is very close to the observed climatology. The original large scale climate simulations correspond to the CMIP3 Special Report on Emission Scenario (SRES) A1B scenario (IPCC, 2000). In order to produce an estimate of potential uncertainty through a broad sampling of future possibilities, sampling both chaotic variability and uncertainty in climate modeling (Harding et al., 2015), we used eleven climate simulations, listed in Fig. 1. They cover a 150-yr period, from 1950 to 2099. Three periods are considered: near future (NF) (2020–2049), distant future (DF) (2070–2099) and a reference past period P (1970–1999). The average cumulated precipitation and air temperature anomalies of the climate simulations over France are shown in Fig. 1 for summertime. At summertime, the sign of the precipitation anomaly varies from one climate simulation to another and the same results are found for the other seasons (not shown). This illustrates the difficulty to predict the precipitation anomalies for this part of Europe. More details about the downscaled climate simulations can be found in Pagé and Terray (2010) and Pagé et al. (2010). We used a single climatic scenario (A1B), which is restrictive. In future works, the use of other scenarios such as RCP (Jacob et al., 2014) will help characterize the uncertainties, especially in DF conditions.

Selection of Representative Sites

The downscaled climate model inputs are available on a regular grid of 8 km by 8 km, representing 8602 grid-cells. Since we took the option to use eleven climate models we conducted our study for a limited number of sites in France, due to computational cost constrains. The ISBA simulations were made for a subset of 191 grid-cells representing the main French agricultural and forest regions. This subset of 191 grid-cells consists of the 45 cropland and 48 grassland homogeneous areas used in Calvet et al. (2012), of the 12 sites considered by Brisson and Levrault (2010), and of 86 forested areas (one in each eco-regions in France as defined by IFN, 2011). Fig. 1 shows the spatial distribution of the 191 sites.

The ISBA land Surface Model

In this study, we use these latest improvements of the model within the SURFEX version 7.3 software. This version of the model is similar to the one validated using satellite LAI products by Lafont et al. (2012) and Szczypta et al. (2014), except for the addition of a subroot-zone soil layer (Boone et al., 1999) and an enhanced multilayer canopy radiative transfer model (Carrer et al., 2013). The values of the ten key parameters of the model are listed in Table 1 for the vegetation types considered in our simulations. Key parameters of the photosynthesis model respond to the soil water stress permitting the representation of drought-avoiding and drought-tolerant responses to drought. For low vegetation and for trees, the response to drought is based on the meta-analysis of Calvet (2000) and Calvet et al. (2004), respectively. The intrinsic leaf-level water use efficiency (WUE) depends on the photosynthesis parameters and on [CO_2]. WUE responds to soil moisture: in the drought-avoiding (-tolerant) approach, WUE increases (decreases) in response to drought, provided the fraction of the extractable soil moisture content fraction is higher than a critical Soil Wetness Index value (SWI_C).

Table 1: Standard values of ISBA-A-gs parameters (Gibelin et al., 2008) for 4 vegetation types (straw cereals and C3 grasslands). The mesophyll conductance at a leaf temperature of 25 °C, in well-watered conditions, g_m, is in units of mm s^{-1}, g_c is the cuticular conductance, in mm s^{-1}, SWI_C is the critical extractable soil moisture content fraction, dimensionless, τ_{max} is the maximum leaf span time, in days, LAI_{min} is the minimum leaf area index, in m^2 m^{-2}, N_L is the leaf nitrogen concentration in% of dry mass for an atmospheric CO_2 concentration of 371 ppm, e is the SLA (specific leaf area) sensitivity to N_L, in m^2 kg^{-1} $\%^{-1}$, f is SLA at N_L = 0%, in m^2 kg^{-1}. The nitrogen dilution parameter D (derived from Calvet et al., 2008) is dimensionless

Vegetation type	g_m	g_c	SWI_C	Response to drought	τ_{max}	LAI_{min}	N_L	e	f	D	SLA response to +50% [CO_2]
Straw cereals	1	0.25	0.3	Avoiding	150	0.3	1.3	3.79	9.84	0.17	−2.2%
Grasslands	1	0.25	0.3	Tolerant	150	0.3	1.3	5.56	6.73	0.24	−4.8%

Broadleaf forests	3	0.15	0.3	Tolerant	230	0.3	2.0	4.83	2.53	0.45	−13.3%
Needleleaf forests	2	0.00	0.3	Avoiding	365	1.0	2.8	4.85	−0.24	0.19	-7.5%

The surface water, energy, and CO_2 fluxes are computed separately for all the vegetation types together with separate soil water storage. Different soil water storage capacities are used to differentiate forests from low vegetation, in relation to contrasting rooting depths. The root-zone depth and the soil depth values for low vegetation are 1.5 m and 2.0 m, 2.0 m and 3.0 m for high vegetation, respectively. Relatively low uniform values of root-zone soil water holding capacity are obtained: 129 mm for low vegetation and 172 mm for high vegetation, on average. However, additional water can be used by the plants through capillary rises from base-flow soil layers. The water budget is calculated, which allows the simulation of the surface and of the root zone moisture. The latter is used to characterize the soil water stress.

The ISBA model was extensively validated and compared to a large variety of observations such as surfaces fluxes (Carrer et al., 2012, Balzarolo et al., 2014 and Garrigues et al., 2015), and satellite-derived LAI and soil moisture (Lafont et al., 2012 and Szczypta et al., 2014). In order to make sure that the ISBA model captures known phenomena, the meta-analysis of Yin (2002), based on 170 data cases of 62 species compiled from the literature, is used to ensure credibility for future estimates of the SLA response to CO_2. Using in situ agricultural statistics, Calvet et al. (2012) andCanal et al. (2014) have shown that this approach is able to represent to a large extent the interannual variability of the straw cereal yields and of the forage dry matter production over France. However, extremely wet years (e.g. 2007 in Canal et al., 2014) can be interpreted by the model as favorable to cereal yield, while in the real world they may turn out unfavorable. This is caused by shortcomings in the representation of constraints on field operations, such as the lack of field accessibility (Trnka et al., 2015).

The Specific Leaf Area (SLA) response to a 50% increase in the atmospheric concentration of CO_2 (hereafter referred to as $[CO_2]$) is shown in Table 1 in order to illustrate the nitrogen dilution approach used in the model:

The ratio of future to past leaf mass-based nitrogen concentration is related to the ratio in $[CO_2]$ as:

$$\frac{Future\ N_L}{Past\ N_L} = \left(\frac{Future\ [CO_2]}{Past\ [CO_2]}\right)^{-D} \quad (1)$$

where D is a nitrogen dilution parameter derived by Calvet et al. (2008) from the meta-analysis of Yin (2002) for various plant functional types. The past N_L values given inTable 1 correspond to a past $[CO_2]$ value of 371 ppm. Values of D are given in Table 1for the four vegetation types considered in this study (straw cereals, grasslands, broadleaf forests, and needleleaf forests). The SLA value is derived from N_L using leaf trait plasticity parameters e and f (Gibelin et al., 2006):

$$SLA = e \times N_L + f \quad (2)$$

DESIGN OF THE SIMULATIONS

The ISBA model is forced by the hourly atmospheric variables derived from the downscaling of the eleven climate simulations listed in Fig. 1. The simulations cover a 150-yr period, from 1950 to 2099 and three time periods are used in our analysis: P (1970–1999), NF (2020–2049), and DF (2070–2099). In addition to the downscaled climate simulations, mean annual values of $[CO_2]$ consistent with the A1B scenario are used. For example: 311 ppm in 1950, 367 ppm in 2000, 522 ppm in 2050, and 700 ppm in 2099. The past $[CO_2]$ value of 371 ppm used in Eq. (1) corresponds to year 2002.

The ISBA simulations are made for 191 grid-cells (Section 2.2). The simulations obtained for each of the four vegetation types and for each of the eleven climate models consist of 191 daily time series of LAI and root-zone soil moisture, covering a 150-yr period. In order to initialize the carbon pools, the same number of simulations is performed before starting the final simulations through a spin-up procedure using the 1950–1959 period (i.e. the 10 first years) repeated 15 times.

In order to evaluate the impact of the nitrogen dilution Eq. (1), an additional set of simulations is performed with D = 0, for four climate simulations: HADGEM2_1, IPCM4_1, CNCM33_1, and DMIEH5C_3.

Biophysical Indicators

Four biophysical indicators derived from the ISBA simulations are used in this study. Three of them correspond to phenological dates (in units of days within each considered year): leaf onset, leaf senescence, date of maximum LAI. Annual dates of leaf onset and leaf senescence are retrieved from the LAI simulations, following Gibelin et al. (2006)and Szczypta et al. (2014). Leaf onset (senescence) corresponds to the date when the departure of LAI from its minimum annual value between 1 September of the previous year and the date of maximum LAI becomes higher (respectively, lower) than 40% of the amplitude of the annual cycle. It may happen that LAI values are higher than the 40% threshold before or after the date of maximum LAI. In this case, the minimum LAI is used to determine the date.

The fourth biophysical indicator, hereafter referred as drought duration, is the mean yearly cumulated duration of dry soil conditions. The latter are defined as mean Soil Wetness Index (SWI, dimensionless) lower than $SWI_C = 0.3$. The SWI can be written as:

$$SWI = (\theta - \theta_{WILT})/(\theta_{FC} - \theta_{WILT}) \quad (3)$$

where θ is the root-zone volumetric water content (in $m^3\ m^{-3}$), and the subscript "FC" and "WILT" indicate soil moisture at field capacity and at wilting point, respectively.

Statistical Analysis

A statistical analysis is applied to the time series of the biophysical indicators in order to assess trends and uncertainties:

- The trend analysis aims at determining whether the impact of climate change on a given variable is significant or not. The Mann-Kendall trend test is applied to the 1970–2049 period for the near future (NF) conditions, and to the 1970–2099 period for the distant future (DF) conditions. The test null hypothesis is “no atrend” and the null hypothesis is rejected if the test p-value is lower than 0.01.

- The uncertainty analysis evaluates the consistency of the impact of the various climate models. The ANOVA variance test is applied to NF vs. P anomalies and to DF vs. P anomalies. The test null hypothesis is “identical anomaly means” and the alternative hypothesis is “at least one climate model differs from the other models”. The null hypothesis is rejected if the test p-value is lower than 0.05.

Trends and uncertainties are presented in Section 3.

Spatial Classification Method

The spatial patterns are identified through a clustering method, the k-means classification technique. This method is applied to anomalies of leaf onset and leaf senescence dates, in either NF or DF conditions, for the four vegetation types. This represents 16 configurations. Thirty anomaly values are considered at each site, corresponding to the 30 years of the NF or DF time period (Fig. 1). Each anomaly value consists of the mean across the 11 climate simulations listed in Fig. 1. A number of tests (not shown) demonstrated that four clusters per configuration are enough to represent the spatial variability of the impact of climate change. In order to facilitate the analysis of the results, only four configurations (among 16) are used, all in DF conditions: straw cereals leaf onset and leaf senescence, broadleaf forests leaf onset and leaf senescence. Common cluster grid-cells are used for straw cereals and grasslands (low vegetation), and for broadleaf and needleleaf forests (high vegetation). Hereafter, these two sets of cluster grid-cells are referred to as cluster ‘LV’ (for low vegetation) and cluster ‘HV’ (for high vegetation). The same cluster grid-cells are used for NF conditions, as preliminary tests showed that similar results are obtained.

This clustering method, combined with the Mann-Kendall and ANOVA statistical tests, permits the identification of five areas where specific leaf onset or leaf senescence anomaly attributes are simulated. Fig. 3 presents the correspondence between the areas and the clusters. The analysis of the results focusses on these areas and on clusters presenting the lowest uncertainty range in DF conditions.

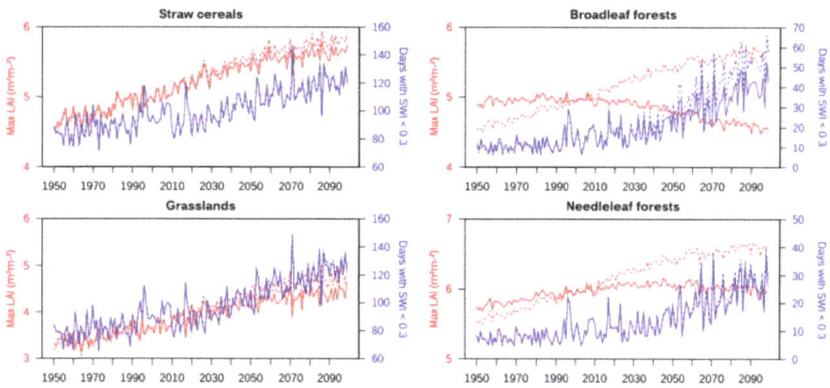

Figure 2: Impact of climate change on (red) mean annual maximum LAI, (blue) drought duration defined as days with SWI < 0.3: (left, from top to bottom) straw cereals and grasslands, (right, from top to bottom) broadleaf and needleleaf forests, with and without nitrogen dilution in response to [CO_2] increase (solid and dashed lines, respectively). LAI and drought duration are mean values across 191 locations and four climate simulations (HADGEM2_1, IPCM4_1, CNCM33_1, DMIEH5C_3). (For interpretation of the references to color in this figure legend, the reader is referred to the web version of this article.)

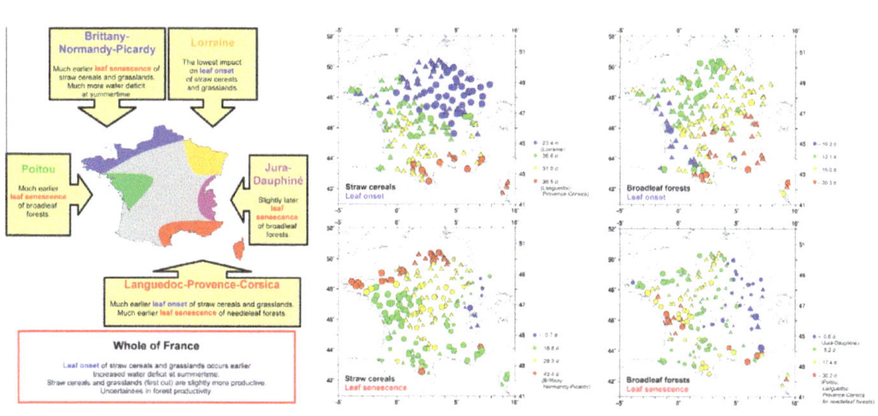

Figure 3: Main trends in France on the phenology and on the biomass production simulated by the ISBA model related to climate change in the 21st century. (Left) overview of results. (Right) detailed regional impact of climate change at 191 locations for distant future (2070–2099) on straw cereals and broadleaf forests: (top) leaf onset anomalies, (bottom) leaf senescence anomalies (in days). The size of the symbols is proportional to the number of climatic simulations (11 as a maximum) presenting a significant trend (Mann-Kendall test). Circles denote consistent trends, triangles denote divergent trends (ANOVA test). Colors differentiate clusters: the blue, green, yellow, and red colors of the symbols are for the lowest (blue) to the highest (red) impact, i.e. cluster 1 and cluster 4, respectively. (For interpretation of the references to color in this figure legend, the reader is referred to the web version of this article.)

RESULTS

Mean trends of Air Temperature

In order to illustrate the local climate attributes derived from the spatial downscaling of the climate simulations, Fig. 1 shows the average annual number of days with at least one hourly air temperature (T_{air}) value below freezing level (freezing days), and above 25 °C (summertime days), respectively, together with the NF and DF anomalies, at 191 locations (see Section 2.2). While the number of freezing days tends to decrease, the number of summertime days tends to increase. Fig. 1 shows that in past conditions, the average number of freezing days is larger than 100 d in the Jura-Dauphiné area, about 80 d or more in Lorraine, and lower than 40 d in Langedoc-Provence-Corsica and Poitou (see geographical information in Fig. 3). In the latter two areas (Langedoc-Provence-Corsica and Poitou), the future anomalies in units of days are less pronounced. However, expressing the anomalies as a fraction of the past values, these areas present the largest impact of climate change on freezing days: from −20% to −50% in NF conditions, and −60% to −100% in DF conditions, to be compared with −15% to −20% in NF conditions, and −40% to −50% in DF conditions in the Lorraine and Jura-Dauphiné areas.

As far as summertime days are concerned, they are less frequent in the Brittany-Normandy-Picardy area (less than 20 d in P conditions), and more frequent in the Langedoc-Provence-Corsica area (more than 60 d in P conditions). For the Brittany-Normandy-Picardy area, their number increases less than in other regions: less than 12 d in NF conditions, less than 36 d in DF conditions. However, as for freezing days, the relative change in summertime days is more pronounced in areas presenting small numbers, such as Brittany-Normandy-Picardy: more than +50% in NF, and more than +150% in DF conditions for the area.

Mean Trends of Peak LAI and Drought Duration

The ISBA model is able to account for the antitranspirant and fertilizing effects of CO_2. In particular, the decrease in SLA related to nitrogen dilution (Section 2.3) tends to limit the increase in LAI caused by the fertilizing effect of CO_2 as shown by Fig. 2 for low vegetation (straw cereals and grasslands) and for high vegetation (broadleaf and needleleaf forests). This figure show time series of the average maximum LAI for the 150-yr period considered in this study, with and without activating the nitrogen dilution (Eq. (1)) process in the model. The same maximum LAI value is obtained in 2002, with and without activating the nitrogen dilution, as the past $[CO_2]$ value of 371 ppm used in Eq. (1) corresponds to year 2002 (Section 2.4). After 2002, the impact of nitrogen dilution is noticeable (LAI difference higher than 0.2 m^2 m^{-2}) after 2020 for broadleaf forests, 2030 for needleleaf forests, 2040 for grasslands, and 2060 for straw cereals. The impact of nitrogen dilution is more pronounced for the forests, especially for broadleaf forests. While the maximum LAI of low vegetation tends to increase, the trend observed for forests is much lower is relation to the more pronounced nitrogen dilution. In nitrogen dilution simulations, maximum LAI of broadleaf forests even tends to decrease after 2010, and after 2050 for needleleaf forests.

Fig. 2 also shows the mean drought duration, with and without activating the nitrogen dilution option. The lower LAI values obtained with the activation of nitrogen dilution generate less plant transpiration and limit the increase in drought duration. While the impact of maximum LAI on droughts is limited for low vegetation and for needleleaf forests, it is quite marked for broadleaf forests. In all cases, drought duration markedly increases after 2020.

The simulated impact of climate change and CO_2 effects on photosynthesis (or Gross Primary Production, GPP), Net Primary Production (the difference between GPP and autotrophic respiration), and the ratio of autotrophic respiration to GPP (R/GPP). These three quantities increase through time, especially after 2010. However, in response to the sharp increase in the R/GPP ratio, NPP tends to stabilize after 2060. The impact of nitrogen dilution on these trends is only noticeable for trees.

Trend and Uncertainty Mapping

The results of the statistical analysis and of the spatial classification are shown in Fig. 3for straw cereals and for broadleaf forests in DF conditions, for both leaf onset and leaf senescence. In these maps, the size of the symbols is proportional to the number of climate models for which a significant trend in the considered biophysical variable is found. When at least one climate model differs from the other models circles are replaced by triangles. The four sorted clusters can be distinguished using colors: from blue for the lowest impact (cluster LV-1 or HV-1) to red for the largest impact (cluster LV-4 or HV-4).

For straw cereals, the leaf onset cluster LV-4 covers the Languedoc-Provence-Corsica area, and occurs 36 days earlier than in past conditions, on average. The LV-1 cluster covers the Lorraine area with a leaf onset occurring 23 days earlier than in past conditions. The leaf senescence cluster LV-4 covers the Brittany-Normandy-Picardy area and occurs 43 days earlier than in past conditions. Similar results are found for grasslands (not shown).

For high vegetation, the trends are less marked than for low vegetation and leaf onset and leaf senescence uncertainties are both more pronounced (more triangles and smaller symbols, respectively). Also, the spatial distribution of the clusters is more complex than for low vegetation.

The results of the statistical trend and uncertainty tests for leaf onset and leaf senescence. The fraction of significant trends throughout sites and climate models (Mann-Kendall test) and the fraction of sites for which at least one climate model differs from the other models (ANOVA test) are used as metrics to characterize trends and uncertainties, respectively. The following interpretations can be proposed:

- Maximum LAI in DF conditions: for all the clusters, resulting from either leaf onset or leaf senescence classifications, the largest impact and the smallest uncertainties are observed for grasslands; the largest uncertainties are observed for broadleaf forests, which present a systematic decrease in peak LAI, while the other plant types present an increase in peak LAI. This is consistent with Fig. 2.

- Maximum LAI in NF conditions: needleleaf forests present the largest uncertainties, in relation to the relatively small response of peak LAI (Fig. 2).
- Leaf onset of straw cereals systematically displays the largest impact but uncertainty is larger than for grasslands in both DF and NF conditions.
- Leaf onset of forests presents the highest uncertainty values; the trend towards earlier leaf onset is more pronounced for needleleaf forests than for broadleaf forests in both DF and NF conditions.
- Leaf senescence presents smaller trends than leaf onset in both DF and NF conditions; needleleaf forests display the lowest uncertainty values; grasslands are often more impacted than other vegetation types; a good level of confidence is obtained for the leaf senescence cluster HV-4 that covers the Languedoc-Provence-Corsica (especially needleleaf forests in this area) and Poitou (especially broadleaf forests in this area) and cluster HV-1 over the Jura-Dauphiné area (Fig. 3); the cluster HV-4 leaf senescence occurs −30 days and −22 days earlier for broadleaf and needleleaf forests, respectively; in contrast to other areas, the cluster HV-1 leaf senescence occurs +0.6 day latter for broadleaf forests in Jura-Dauphiné.

Impact of Climate Change on the Vegetation Cycle

Fig. 4 illustrates the main findings from the trend analysis (Section 3.3), for straw cereals and broadleaf forests. The average annual cycle of LAI is shown across three 30-yr periods (P, NF, DF), for straw cereals and broadleaf forests. Each subfigure presents four clusters: leaf onset cluster 1 and cluster 4, and leaf senescence cluster 1 and cluster 4. The earlier leaf onset and the increase in peak LAI values triggered by climate change are particularly evident for straw cereals. Slightly later leaf senescence dates are observed for leaf senescence cluster LV-1 LAI cycles. Apart from these clusters, the future LAI presents lower values than past LAI at summertime (July–August) for straw cereals. For broadleaf forests, a decrease in LAI is observed during the Autumn (September–October). While leaf senescence cluster LV-1 presents systematically higher LAI values in the future, the reverse is observed for leaf senescence cluster HV-1 in September–October. Results for grasslands and needleleaf forests.

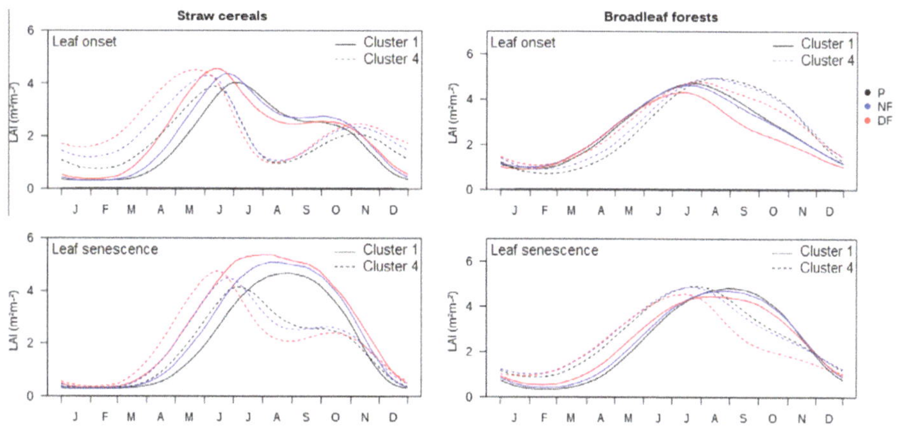

Figure 4: Mean seasonal time series of (left) "straw cereals" and (right) "broadleaf forests" vegetation type LAI across climate models and sites within clusters presenting the (solid lines) lowest and (dashed lines) highest anomalies of (top) leaf onset and (bottom) leaf senescence, for past, near future and distant future ("P", "NF", "DF", respectively).

DISCUSSION

To what extent is Plant Phenology Driven by soil Moisture?

This is a key question as today in situ networks measuring soil moisture are much less developed than weather station networks measuring T_{air}. The drought duration indicator derived from our simulations can be used to assess the past and future impact of soil moisture on phenology. Fig. 5 presents leaf onset and leaf senescence responses of straw cereals and broadleaf forests to the mean annual T_{air} and to the annual drought duration. It appears that leaf senescence correlates better with drought duration ($r^2 > 0.9$) than with T_{air} ($r^2 < 0.8$). On the other hand, leaf onset correlates better with T_{air} than with drought duration. For grasslands and for winter crops such as straw cereals, leaf onset occurs at the beginning of spring in conditions where soil moisture content is generally close to field capacity. Leaf senescence generally occurs in dry conditions where soil moisture is the main limiting factor of plant growth. This is a consistent result, as leaf senescence occurring in dry soil conditions is the normal plant development process of cereals and grasslands.

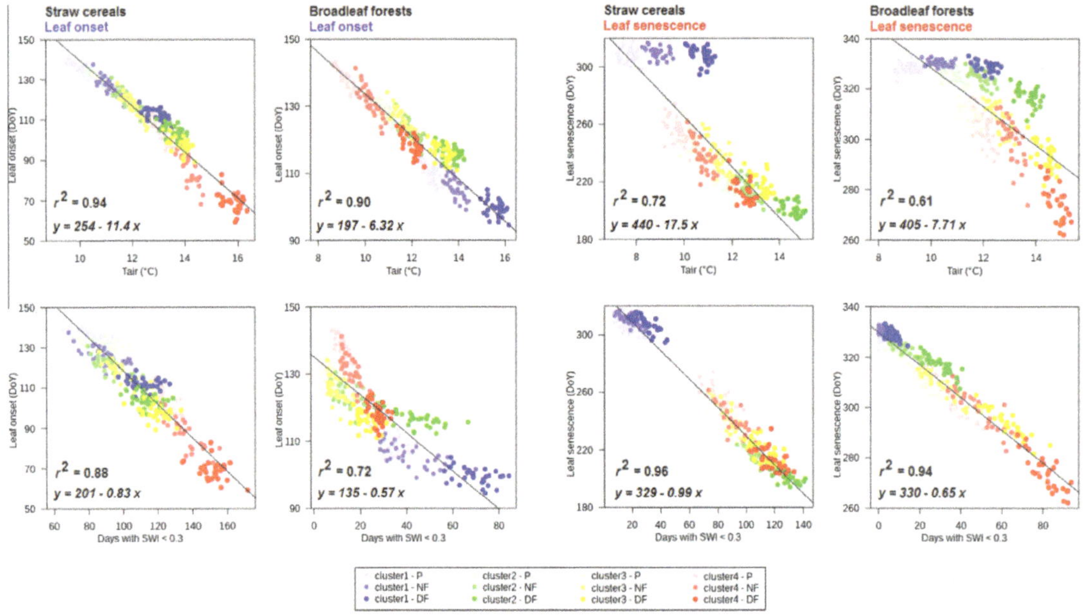

Figure 5: Phenology indicators (in days within each considered year, or Day of Year (DoY)) vs. climatic indicators: (top) the annual mean of hourly air temperature (T_{air}) and (bottom) the mean cumulated duration of droughts defined as mean SWI < 0.3, across the 11 climate models, for the three reference 30-yr periods (P, NF, or DF). From left to right: leaf onset of straw cereals, leaf senescence of straw cereals, leaf onset of broadleaf forests, leaf senescence of broadleaf forests. Each dot corresponds to one year within a reference period, for a cluster presenting the (blue) lowest or the (red) highest anomalies. (For interpretation of the references to color in this figure legend, the reader is referred to the web version of this article.)

Therefore, drought duration is a key variable to monitor the impacts of climate change on vegetation, especially in the leaf senescence cluster 4 areas in Fig. 3: Languedoc-Provence-Corsica, Brittany-Normandy-Picardy, and Poitou. This shows that developingin situ soil moisture networks in these areas could help monitoring the long-term impacts of climate change.

The type of evidence for vegetation responses shown here is also available via studies using agrometeorological indicators. While indicators based on air temperature such as growing-degree-days can be used to characterize the beginning of the growing season, the length of the growing season "is defined by soil water availability (...) in low-latitude arid regions where precipitation is primary in determining vegetation and crop growth" (Feng and Hu, 2004). Fig. 5 shows that in mid-latitude regions where soil moisture-based indicators are not relevant today, such indicators may be highly relevant in the future. For example, when all the straw cereal simulations are considered together, the leaf senescence date correlates with drought duration ($r^2 = 0.96$). On the other hand a rather poor result is obtained for Cluster 1, with $r^2 = 0.22$, showing that in this region (in eastern France), drought duration is not a very good indicator of leaf senescence, even in the future, in relation to a small change in the leaf senescence date (+0.6 d in DF conditions). Since leaf onset occurs earlier, the length of the growing season increases (see theSupplement). A similar result was obtained by Harding et al. (2015) over the UK: as drought duration does not increase, "over most of the UK the growing season length increases almost two months allowing a near year-long growing season over much of Southern England".

Uncertainties Related to the CO_2 Effect

The effect of atmospheric CO_2 concentration on plant growth and on nitrogen dilution has a noticeable impact on our simulations, as illustrated by Fig. 2. Eq. (1) predicts a decrease of N_L in response to rising [CO_2], in relation to nitrogen dilution. As a consequence, SLA also decreases following the leaf trait relationship Eq. (2). The combination of Eqs. (1) allows the prediction of the future SLA values. Table 1 shows that given the parameter values used for D, e, f, and N_L, the SLA decreases from about 2% for straw cereals, to 13% for broadleaf forests, in response to a rise of 50% in [CO_2]. The marked decrease of maximum LAI of broadleaf forests after 2010 shown by Fig. 2 is consistent with the higher SLA sensitivity to a rise in [CO_2] reported in Table 1. Simple calculations using Eqs. (1) show that the SLA sensitivity to [CO_2] does not vary much in response to changes in N_L. The larger sensitivity obtained for broadleaf forests is triggered by the larger value of the D nitrogen dilution parameter (D = 0.45) derived from the meta-analysis of Yin (2002) for this vegetation type. Following Calvet et al. (2008), we use constant values of D. However, it must be noted that Yin (2002) showed that D could be influenced to some extent by air temperature and by the incoming solar radiation. Also, studies reporting observations of the plant response to CO_2 are not always consistent (Leakey et al., 2009). For example, Gutiérrez et al. (2009) found an increase in SLA for wheat in response to [CO_2] enrichment, instead of the decrease in SLA reported by other studies. Further research is needed to progress in the simulation of the plant response to [CO_2].

Advantages and Limitations of a

The generic ISBA model is neither a crop model nor a forestry model. The generic modeling approach allows the representation of the main vegetation types across the globe at various scales (from the local scale to the global scale). The differences in the simulated response of vegetation and soil variables can be easily analyzed. A limitation of this approach is that agricultural practices and forest management cannot be represented in detail. Adaptation strategies such as crop type swapping or under-planting of more suitable tree species in forests are not represented. From this point of view, our simulations provide biophysical indicators that can be used to identify climate-driven risks at given locations. But they cannot be used to provide detailed recommendations to help farmers or foresters adapt their practices. In order to fulfill this objective, specialized crop and forestry models must be used (Brisson and Levrault, 2010).

However, it must be noticed that the results summarized in Fig. 3 are consistent with the findings of Brisson and Levrault (2010) for wheat, as shown by Table 2. Quantitative information on the response of soft wheat as simulated by the CERES agricultural model (Gabrielle et al., 2002) forced by the ARPEGE A1B simulation can be found in Brisson and Levrault (2010). In particular, changes in phenological dates (wheat flowering date in particular) and in the annual coefficient of variation of yield are reported for the Mons site located in an area in northern France corresponding to the leaf onset cluster 1. For wheat, flowering occurs before maximum LAI is reached (Song et al., 2014), and changes in leaf onset dates as defined in this study can be considered as an approximation of changes in the flowering date. As shown by Table 2, changes in phenological dates are similar. Overall, trends in production increase and towards a decrease of the interannual variability are consistent. However, our large scale simulations tend to generate a lower interannual variability of the biomass production (i.e. a smaller annual coefficient of variation (ACV)) and a larger increase of the biomass production in the future. ACV values smaller than 0.1 are typical of large scale straw cereal yield statistics in France (Canal et al., 2014). Our results do not show an increase in adverse weather events able to reduce wheat yield. As shown in Table 2, our simulations indicate a reduction in the interannual variability of the maximum above-ground biomass of straw cereals: −20% in DF conditions for cluster 1. Table 2 also shows that this result is consistent with Brisson and Levrault (2010). We find a similar result for the whole of France (−16%). A different result was reported by Trnka et al. (2015), with a trend towards a larger frequency of events detrimental to wheat yield under both RCP4.5 and RCP8.5 scenarios. Over northern and western France, they found that the main cause of this trend was a decrease in field accessibility during key field operations. This difference can be explained by the fact that we do not simulate field accessibility.

Table 2: Simulated response of straw cereals to climate change (this study) vs. soft wheat CERES simulations inBrisson and Levrault (2010).

	Production increase		Leaf onset date change		Change in growing season length		ACV (*)	Change in ACV (*)	
Period	NF	DF	NF	DF	NF	DF	P	NF	DF
Brisson and Levrault (2010) (soft wheat)	+13%	+14%	−9 d	−17 d	−10 d	−20 d	0.330	−36%	−19%
	(Yield, France average)		(Flowering at Mons)		(At Mons)		(Yield, at Mons)		
This study (straw cereals)	+17%	+29%	−9.5 d	−23 d	−7 d	−22 d	0.065	−4%	−20%
	(Above-ground biomass, average)		(Leaf onset cluster 1 mean)		(Leaf onset cluster 1 mean)		(Maximum above-ground biomass, Leaf onset cluster 1 mean)		

Annual Coefficient of Variation (ACV, i.e. ratio of the standard deviation to the mean).

Regarding forests, we checked that our simulations of Gross Primary Production (GPP) and Net Primary Production (NPP) (see the Supplement) are consistent with time series produced by the CASTANEA model in Loustau et al. (2005) for beech and oak. However, more quantitative comparisons are difficult since Loustau et al. (2005) used the B2 emissions scenario.

Another limitation of this study is the definition of the contour of specific areas with the clustering method. The spatial determination of these areas could be improved by using more grid-cells.

CONCLUSIONS

In spite of the increased water deficit at summertime in France, winter straw cereals and grasslands (first cut) should be slightly more productive at the end of the century than today, in relation to earlier leaf onset dates, with greatest changes in leaf onset occurring in Languedoc-Provence-Corsica. Brittany-Normandy-Picardy would experience a much more pronounced soil water deficit at summertime, resulting in much earlier leaf senescence and shorter growing season length of straw cereals and grasslands. On the other hand, we might expect smaller changes in Lorraine. Future forest productivity as a whole is affected by large uncertainties. However, it seems likely that leaf senescence will occur much earlier in Poitou for broadleaf forests, and slightly later in Jura-Dauphiné. Leaf senescence should occur much earlier in Languedoc-Provence-Corsica for needleleaf forests.

Although we used a generic approach unable to simulate agricultural and forestry practices, our results are consistent with the findings of Brisson and Levrault (2010), who used specialized crop and forestry models. With respect to previous studies, we performed a more extensive spatial analysis of the impact of climate change, at 191 locations. Trends were analyzed in NF and DF conditions, together with the consistency of the biophysical variables across climate simulations. We analyzed the uncertainties related to climate simulations. The largest uncertainties on peak LAI values are observed for needleleaf forests in NF conditions and for broadleaf forests in DF conditions. Forests also present the largest uncertainties for leaf onset. Although leaf onset of straw cereals is more impacted than for grasslands, uncertainties are larger.

Using the plant trait meta-analysis of Yin (2002), we investigated the impact of CO_2 on plant morphology and we found a larger sensitivity for broadleaf forests triggered by a larger nitrogen dilution. Further research is needed to progress in the simulation of the plant response to [CO_2] as this is a key process.

Since the SURFEX modeling platform is able to work at a global scale, the approach used in this study could be applied to other regions in future works provided downscaled climate simulations are available. In particular, new CMIP5 RCP climate simulations could be used.

ACKNOWLEDGEMENTS

This work was supported by ANR in the framework of the ORACLE project (ANR-10-CEPL-011). We thank Nathalie de Noblet (LSCE) and Iñaki Garcia de Cortazar Atauri (INRA) for their helpful comments and advices, and Catherine Meurey (CNRM) for technical support. We want to pay tribute to the memory of Nadine Brisson (INRA), who played a key role in the initiation of this project.

REFERENCES

1. Balzarolo et al., 2014M. Balzarolo, S. Boussetta, G. Balsamo, A. Beljaars, F. Maignan, J.-C. Calvet, S. Lafont, A. Barbu, B. Poulter, F. Chevallier, C. Szczypta, D. PapaleEvaluating the potential of large scale simulations to predict carbon fluxes of terrestrial ecosystems over a European Eddy Covariance network Biogeosciences, 11 (2014), pp. 2661–2678 http://dx.doi.org/10.5194/bg-11-2661-2014
2. Boé and Terray, 2008J. Boé, L. TerrayA weather-type approach to analyzing winter precipitation in France: twentieth-century trends and the role of anthropogenic forcingJ. Clim., 21 (2008), pp. 3118–3133 http://dx.doi.org/10.1175/2007JCLI1796.1
3. Boé et al., 2006J. Boé, L. Terray, F. Habets, E. MartinA simple statistical-dynamical downscaling scheme based on weather types and conditional resamplingJ. Geophys. Res., 111 (2006), p. D23106 http://dx.doi.org/10.1029/2005JD006889
4. Boone et al., 1999A. Boone, J.-C. Calvet, J. NoilhanInclusion of a third soil layer in a land-surface scheme using the force-restore methodJ. Appl. Meteorol., 38 (11) (1999), pp. 1611–1630
5. Brisson and Levrault, 2010N. Brisson, F. Levrault (Eds.), Livre vert du projet Climator, Changement climatique, agriculture et forêt en France: simulations d'impacts sur les principales espèces. ANR, INRA, ADEME, ADEME éditions (2010) 334 pp
6. Calvet, 2000J.-C. CalvetInvestigating soil and atmospheric plant water stress using physiological and micrometeorological data setsAgric. For. Meteorol., 103 (3) (2000), pp. 229–247
7. Calvet and Soussana, 2001J.-C. Calvet, J.-F. SoussanaModelling CO2-enrichment effects using an interactive vegetation SVAT schemeAgric. For. Meteorol., 108 (2) (2001), pp. 129–152
8. Calvet et al., 2004J.-C. Calvet, V. Rivalland, C. Picon-Cochard, J.-M. GuehlModelling forest transpiration and CO2 fluxes – response to soil moisture stressAgric. For. Meteorol., 124 (3–4) (2004), pp. 143–156 http://dx.doi.org/10.1016/j.agrformet.2004.01.007
9. Calvet et al., 2007J.-C. Calvet, N. Fritz, F. Froissard, D. Suquia, A. Petitpa, B. PiguetIn situ soil moisture observations for the CAL/VAL of SMOS: the SMOSMANIA networkProceedings of the International Geoscience and Remote Sensing Symposium, IGARSS, Barcelona (2007) http://dx.doi.org/10.1109/IGARSS.2007.4423019
10. Calvet et al., 2008J.-C. Calvet, A.-L. Gibelin, J.-L. Roujean, E. Martin, P. Le Moigne, H. Douville, J. NoilhanPast and future scenarios of the effect of carbon dioxide on plant growth and transpiration for three vegetation types of southwestern FranceAtmos. Chem. Phys., 8 (2008), pp. 397–406 http://dx.doi.org/10.5194/acp-8-397-2008
11. Calvet et al., 2012J.-C. Calvet, S. Lafont, E. Cloppet, F. Souverain, V. Badeau, C. Le BasUse of agricultural statistics to verify the interannual variability in land surface models: a case study over France with ISBA-A-gsGeosci. Model Dev., 5 (2012), pp. 37–54 http://dx.doi.org/10.5194/gmd-5-37-2012
12. Canal et al., 2014N. Canal, J.-C. Calvet, B. Decharme, D. Carrer, S. Lafont, G. PigeonEvaluation of root water uptake in the ISBA-A-gs land surface model using agricultural yield statistics over FranceHydrol. Earth Syst. Sci., 18 (2014), pp. 4979–4999 http://dx.doi.org/10.5194/hess-18-4979-2014
13. Carrer et al., 2012D. Carrer, S. Lafont, J.-L. Roujean, J.-C. Calvet, C. Meurey, P. Le Moigne, I. TrigoIncoming solar and infrared radiation derived from METEOSAT: impact on the modelled land water and energy budget over FranceJ. Hydrometeorol., 13 (2012), pp. 504–520 http://dx.doi.org/10.1175/JHM-D-11-059.1
14. Carrer et al., 2013D. Carrer, J.-L. Roujean, S. Lafont, J.-C. Calvet, A. Boone, B. Decharme, C. Delire, J.-P. Gastellu-EtchegorryA canopy radiative transfer scheme with explicit FAPAR for the interactive vegetation model ISBA-A-gs: impact on carbon fluxesJ. Geophys. Res. Biogeosci., 118 (2013), pp. 888–903 http://dx.doi.org/10.1002/jgrg.20070
15. Déqué and Somot, 2010M. Déqué, S. SomotWeighted frequency distributions expressing modelling uncertainties in the ENSEMBLES regional climate experimentsClim. Res., 44 (2–3) (2010), pp. 195–209 http://dx.doi.org/10.3354/cr00866
16. Durand et al., 1993Y. Durand, E. Brun, L. Merindol, G. Guyomarc'h, B. Lesaffre, E. MartinA meteorological estimation of relevant parameters for snow modelsAnn. Glaciol., 18 (1993), pp. 65–71
17. Feng and Hu, 2004S. Feng, Q. HuChanges in agro-meteorological indicators in the contiguous United States: 1951–2000Theor. Appl. Climatol., 78 (2004), pp. 247–264 http://dx.doi.org/10.1007/s00704-004-0061-8
18. Gabrielle et al., 2002B. Gabrielle, R. Roche, P. Angas, C. Cantero-Martinez, L. Cosentino, M. Mantineo, M.

Langensiepen, C. Hénault, P. Laville, B. Nicoullaud, G. GosseA priori parametrisation of the CERES soil-crop models and tests against several European data setsAgronomie, 22 (2002), pp. 25–38 http://dx.doi.org/10.1051/agro:2002003

19. Garrigues et al., 2015S. Garrigues, A. Olioso, J.-C. Calvet, E. Martin, S. Lafont, S. Moulin, A. Chanzy, O. Marloie, S. Buis, V. Desfonds, N. Bertrand, D. RenardEvaluation of land surface model simulations of evapotranspiration over a 12-year crop succession: impact of soil hydraulic and vegetation propertiesHydrol. Earth Syst. Sci., 19 (2015), pp. 3109–3131 http://dx.doi.org/10.5194/hess-19-3109-2015

20. Gibelin et al., 2006A.-L. Gibelin, J.-C. Calvet, J.-L. Roujean, L. Jarlan, S.O. LosAbility of the land surface model ISBA-A-gs to simulate leaf area index at the global scale: comparison with satellites productsJ. Geophys. Res., 111 (D18) (2006), p. D18102 http://dx.doi.org/10.1029/2005JD006691

21. Gibelin et al., 2008A.-L. Gibelin, J.-C. Calvet, N. ViovyModelling energy and CO2 fluxes with an interactive vegetation land surface model – evaluation at high and middle latitudesAgric. For. Meteorol., 148 (2008), pp. 1611–1628

22. Gouache et al., 2012D. Gouache, X. Le Bris, M. Bogard, O. Deudon, C. Pagé, P. GateEvaluating agronomic adaptation options to increasing heat stress under climate change during wheat grain filling in FranceEur. J. Agron., 39 (2012), pp. 62–70 http://dx.doi.org/10.1016/j.eja.2012.01.009

23. Gutiérrez et al., 2009E. Gutiérrez, D. Gutiérrez, R. Morcuende, A.L. Verdejo, S. Kostadinova, R. Martinez-Carrasco, P. PérezChanges in leaf morphology and compositions with future increases in CO2 and temperature revisited: wheat in field chambersJ. Plant Growth Regul., 28 (2009), pp. 349–357 http://dx.doi.org/10.1007/s00344-009-9102-y

24. Habets et al., 2013F. Habets, J. Boé, M. Déqué, A. Ducharne, S. Gascoin, A. Hachour, E. Martin, C. Pagé, E. Sauquet, L. Terray, D. Thiéry, L. Oudin, P. ViennotImpact of climate change on the hydrogeology of two basins in northern FranceClim. Change, 121 (4) (2013), pp. 771–785 http://dx.doi.org/10.1007/s10584-013-0934-x

25. Harding et al., 2015A.E. Harding, M. Rivington, M.J. Mineter, S.F.B. TettAgro-meteorological indices and climate model uncertainty over the UKClim. Change, 128 (2015), pp. 113–126 http://dx.doi.org/10.1007/s10584-014-1296-8

26. IFN, 2011IFNInventaire forestier national, une nouvelle partition écologique et forestière du territoire métropolitain: les sylvoécorégions. L'IF. 26(2011) 8 pp. Available on: <http://inventaire-forestier.ign.fr/spip/IMG/pdf/IF_SER_web.pdf>

27. IPCC, 2000IPCC SRESN. Nakićenović, R. Swart (Eds.), Special Report on Emissions Scenarios: A Special Report of Working Group III of the Intergovernmental Panel on Climate Change, Cambridge University Press, UK (2000)

28. Jacob et al., 2014D. Jacob, J. Petersen, B. Eggert, A. Alias, O.B. Christensen, L.M. Bouwer, A. Braun, A. Colette, M. Déqué, G. Georgievski, E. Georgopoulou, A. Gobiet, L. Laurent Menut, G. Nikulin, A. Haensler, N. Hempelmann, C. Jones, K. Keuler, S. Kovats, N. Kröner, S. Kotlarski, A. Kriegsmann, E. Martin, E. van Meijgaard, C. Moseley, S. Pfeifer, S. Preuschmann, C. Radermacher, K. Radtke, D. Rechid, M. Rounsevell, P. Samuelsson, S. Somot, J.-F. Soussana, C. Teichmann, R. Valentini, R. Vautard, B. Weber, P. YiouEURO-CORDEX: new high-resolution climate change projections for European impact researchReg. Environ. Change, 14 (2014), pp. 563–578 http://dx.doi.org/10.1007/s10113-013-0499-2

29. Lafont et al., 2012S. Lafont, Y. Zhao, J.-C. Calvet, P. Peylin, P. Ciais, F. Maignan, M. WeissModelling LAI, surface water and carbon fluxes at high-resolution over France: comparison of ISBA-A-gs and ORCHIDEEBiogeosciences, 9 (2012), pp. 439–456 http://dx.doi.org/10.5194/bg-9-439-2012

30. Leakey et al., 2009A.D.B. Leakey, E.A. Ainsworth, C.J. Bernacchi, A. Rogers, S.P. Long, D.R. OrtElevated CO2 effects on plant carbon, nitrogen, and water relations: six important lessons from FACEJ. Exp. Bot., 60 (2009), pp. 2859–2876 http://dx.doi.org/10.1093/jxb/erp096

31. Loustau et al., 2005D. Loustau, A. Bosc, A. Colin, J. Ogée, H. Davi, C. François, E. Dufrêne, M. Déqué, E. Cloppet, D. Arrouays, C. Le Bas, N. Saby, G. Pignard, N. Hamza, A. Granier, N. Bréda, P. Ciais, N. Viovy, F. DelageModeling climate change effects on the potential production of French plains forests at the sub-regional levelTree Physiol., 25 (2005), pp. 813–823

32. Masson et al., 2013V. Masson, P. Le Moigne, E. Martin, S. Faroux, A. Alias, R. Alkama, S. Belamari, A. Barbu, A. Boone, F. Bouyssel, P. Brousseau, E. Brun, J.-C. Calvet, D. Carrer, B. Decharme, C. Delire, S. Donier, K. Essaouini, A.-L. Gibelin, H. Giordani, F. Habets, M. Jidane, G. Kerdraon, E. Kourzeneva, M. Lafaysse, S. Lafont, C. Lebeaupin Brossier, A. Lemonsu, J.-F. Mahfouf, P. Marguinaud, M. Mokhtari, S. Morin, G. Pigeon, R. Salgado, Y. Seity, F. Taillefer, G. Tanguy, P. Tulet, B. Vincendon, V. Vionnet, A. VoldoireThe SURFEX v7.2 land and ocean surface platform for coupled or offline simulation of Earth surface variables and fluxesGeosci. Model Dev., 6 (2013), pp. 929–960 http://dx.doi.org/10.5194/gmd-6-929-2013

33. Pagé and Terray, 2010C. Pagé, L. TerrayNouvelles projections climatiques à échelle fine sur la France pour le 21ème siècle: les scénarii SCRATCH2010. Technical Report TR/CMGC/10/58Centre Européen de Recherche et de Formation Avancée en Calcul Scientifique (CERFACS), Toulouse, France (2010) Available at: <http://www.cerfacs.fr/~page/publications/report_cerfacs_regional_scenarii_scratch2010.pdf>

34. Pagé et al., 2010C. Pagé, L. Terray, J. Boédsclim: A Software Package to Downscale Climate Scenarios at Regional Scale Using a Weather-Typing Based Statistical Methodology. Technical Report TR/CMGC/09/21Centre Européen de Recherche et de Formation Avancée en Calcul Scientifique (CERFACS), Toulouse, France

(2010) Available at: <http://www.cerfacs.fr/~page/dsclim/dsclim_doc-latest.pdf>

35. Quintana-Segui et al., 2008P. Quintana-Segui, P. Lemoigne, Y. Durand, E. Martin, F. Habets, M. Baillon, C. Canellas, L. Franchisteguy, S. MorelAnalysis of near surface atmospheric variables: validation of the SAFRAN analysis over FranceJ. Appl. Meteorol. Clim., 47 (2008), pp. 92–107

36. Song et al., 2014X. Song, B. Cui, G. Yang, H. FengComparison of winter wheat growth with multi-temporal remote sensing imagery35th International Symposium on Remote Sensing of Environment. IOP Conf. Series: Earth and Environmental Science, vol. 17 (2014) http://dx.doi.org/10.1088/1755-1315/17/1/012044

37. Soussana et al., 2010J.-F. Soussana, A.I. Graux, F.N. TubielloImproving the use of modelling for projections of climate change impacts on crops and pasturesJ. Exp. Bot., 61 (2010), pp. 2217–2228 http://dx.doi.org/10.1093/jxb/erq100

38. Szczypta et al., 2014C. Szczypta, J.-C. Calvet, F. Maignan, W. Dorigo, F. Baret, P. CiaisSuitability of modelled and remotely sensed essential climate variables for monitoring Euro-Mediterranean droughtsGeosci. Model Dev., 7 (2014), pp. 931–946 http://dx.doi.org/10.5194/gmd-7-931-2014

39. Trnka et al., 2015M. Trnka, P. Hlavinka, M.A. SemenovAdaptation options for wheat in Europe will be limited by increased adverse weather events under climate changeJ. R. Soc. Interface, 12 (2015), p. 20150721 http://dx.doi.org/10.1098/rsif.2015.0721

40. Tubiello et al., 2007F.N. Tubiello, J.-F. Soussana, S.M. Howden, W. EasterlingCrop and pasture response to climate changeProc. Natl. Acad. Sci. U.S.A., 104 (2007), pp. 19686–19690 http://dx.doi.org/10.1073/pnas.0701728104

41. Yin, 2002X. YinResponses of leaf nitrogen concentration and specific leaf area to atmospheric CO2 enrichment: a retrospective synthesis across 62 speciesGlob. Change Biol., 8 (7) (2002), pp. 631–642

CHAPTER 10

MODELING THE EFFECT OF A HEAT WAVE ON MAIZE PRODUCTION

Uran Chung, Sika Gbegbelegbe, Bekele Shiferaw, Richard Robertson, Jin I. Yun, Kindie Tesfaye, Gerrit Hoogenboom, and Kai Sonder

Keywords: Geo-spatial crop modeling, DSSAT 4.5, HPC cluster, Global maize price, Food security

INTRODUCTION

The extreme weather event that occurred in the USA in 2012 had a significant impact on maize production. This was confirmed by the United States Department of Agriculture (USDA), which announced a reduction of USA maize production by 13% in 2012 compared to the reported 2011 production (USDA, 2013a). Many reports have tried to understand this reduction in maize production in the USA due to this weather extreme using a range of analysis techniques. However, earlier in the spring of 2012, the USDA and other studies predicted that maize production would increase by 20% in 2012 (FAO, 2012a). The heat wave, which started in May and June 2012, severely affected the USA Corn Belt in July and resulted in a reduction of maize yield. In addition, the drought that followed the heat wave in October 2012 impacted residual soil moisture for the following crop, affected maize yields in 2012 and could also impact USA maize production in 2013.

Extreme events such as heat waves, droughts, tornadoes, tsunami, and hurricanes have affected the USA for a long time. Moreover, the effects of such disasters have been well documented. Between 1980 and 2003, the USA experienced 58 weather-related disasters in which overall losses reached $1 billion dollars at the time of the event (Anderson and Bell, 2011 and Lott and Ross, 2005). The USA and National Climatic Data Center (NCDC) have collected a variety of information detailing these events, categorized by type of event.

The USA heat wave in 2012 could have a large impact on food security in other parts of the world as the USA accounts for 40% of global maize production. The USA is the leading maize exporter in the world and between 2005 and 2010 accounted for 50% of global maize exports (FAOSTAT, 2013). Moreover, concerns over the effect of extreme events on global food security are high with the projections implying that climate change may increase the frequency, duration, and intensity of heat waves (Meehl and Tebaldi, 2004).

Understanding how heat waves affect crop production is key to preparing communities for heat waves and estimating the potential impact of heat waves due to climate change. Heat waves are usually defined as prolonged periods of extreme heat, although no consistent definition exists regarding the temperature threshold, temperature metric, and number of days used to define heat waves. For example, most studies on heat waves have used thresholds of mean temperature (Hajat et al., 2006, Lobell et al., 2013 and Elliott et al., 2013), apparent temperature (Smoyer, 1998), or combinations of thresholds of apparent and minimum temperatures (Robinson, 2001 and Weisskopf et al., 2002). Use of inconsistent heat wave definitions leads to different time periods being classified as heat waves, hindering comparison and synthesis of results across studies. Further, heat waves differ in their intensity (degree of heat) and duration, although most studies use measures of intensity and duration to define a heat wave.

A heat wave is more important among other extreme events because it is usually followed by drought. For example, the USA Corn Belt suffered from a severe drought that began after high temperatures hit between May and August 2012. Therefore, the 2012 USA maize crop was adversely affected by a severe drought across much of the Corn Belt, resulting in reduced total production and lowered average yield. Despite the

Citation: Uran Chung, Sika Gbegbelegbe, Bekele Shiferaw, Richard Robertson, Jin I. Yun, Kindie Tesfaye, Gerrit Hoogenboom, Kai Sonder, Modeling the effect of a heat wave on maize production, Weather and Climate Extremes, Volumes 5–6, October 2014, Pages 67-77, ISSN 2212-0947, http://dx.doi.org/10.1016/j.wace.2014.07.002.

drought, the overall quality of the final 2012 maize production was good (USDA, 2013a).

Ciais et al. (2005) evaluated the heat wave impact on crop production in Europe in 2003 using remote sensing techniques. Many experiments at the chamber- and field-level have been carried out to assess crop productivity and mitigation. Nuttall et al. (2013)studied wheat growth under heat wave conditions (heat shock) and increasing CO_2levels in the Australian dryland environment. Process-based crop modeling is superior to other methods because it simulates the biophysical responses of crops using local crop management practices (i.e., irrigation and fertilizers). However, there are a few studies that have used process-based crop simulation models for impact assessments of the 2012 heat wave in the USA. Lobell et al. (2013) used a statistical model based on accumulated Extreme Degree Days (EDD) and APSIM-Maize. Elliott et al. (2013) used the CSM-CERES-Maize model of DSSAT to assess the 2012 heat wave of the USA at a county level.

The goal of this study is to determine the biophysical impact of weather extremes on maize production in the USA and assess the secondary effects of such weather extreme on world maize prices, production, consumption and trade.

MATERIALS AND METHODS

Spatial Biophysical Framework

Climate extreme weather data selection

According to the NCDC, severe heat waves that have led to billion-dollar disasters occurred in the USA ten times between 1980 and 2012 (Lott and Ross, 2005). Based on the study conducted by Lott and Ross (2005) we selected 8 years, (1980, 1986, 1988, 1993, 1998, 1999, 2001, and 2002) and then added two additional years (2006 and 2012) with more recent extreme heat waves. We obtained the weather data from the NCDC for these 10 years across the USA.

Identification and Characterization of heat Waves

June and July are critical months for maize production in the USA. This is the period when there is a change from the vegetative to reproductive stage and grain filling is initiated. Data on monthly precipitation, and maximum and minimum temperatures between June and August for 10 years during which heat waves occurred were collected. Among the ten years selected, three years (2001, 2002, 2006, and 2012) were outside of the baseline window which reflects the average monthly climate data that has been recorded between 1950 and 2000. Since recorded monthly rainfall and temperatures between June and August for 1980, 1986, 1988, 1993, 1998, and 1999 were not too different to those of 2001, 2002, 2006 and 2012 (data not shown), the data were averaged across the 10 years. The average values of rainfall and temperature for the years involving heat waves were then compared with those of the baseline climate (1950–2000).

Map Generation for Baseline and future Weather Extreme

The period for the baseline climate was from 1950 to 2000. The baseline climate grids were obtained from the WorldClim Global Climate Data Center (WorldClim Global Climate Data Center, 2013) and additional details from Hijmans et al. (2005). For each month between June and August, we first interpolated and replaced the values to the baseline climate with the arithmetic average maximum and minimum temperature and precipitation for 10 years during which heat waves occurred. Hence, we generated one climate scenario that combines the climate variables reflecting heat waves between June and August and baseline climate variables for the other months: 'Baseline EW'. The climate variables were mapped at 5 arc-minutes spatial resolution (10 km at equator). We did not change other variables such as solar radiation and the number of rainy days since data collection of irradiation and the number of rainy days was difficult and there were few methods of interpolation.

Climate change was also considered in this study. Nelson et al. (2010) estimated the impact of various Global Climate Models on future precipitation and temperature; the two key variables affecting crop yields. Based on their estimation, the combination of the CSIRO-Mk3.0 climate model for 2050 and the B1 emission scenario, from the Special Report on Emissions Scenario (SRES) (Nakicenovic et al., 2000), leads to the smallest changes in mean precipitation and temperature compared to the baseline climate. Similarly, the combination of the MIROC 3.2 climate model with the A1B emission scenario around 2050 leads to the largest changes in mean precipitation and temperature compared to the baseline climate. Hence, the range of crop yields under climate change should be encompassed by the yields generated under the CSIRO-B1 and MIROC-A1 climate scenarios.

The CSIRO-Mk3.0 and MIROC 3.2 climate models were used in this study and the climate grids were generated by the Consultative Group on International Agricultural Research (CGIAR) Research Program on Climate Change, Agriculture and Food Security (CCAFS) climate data archive (CCAFS, 2013). More details on these models are provided by Gordon et al. (2010) for CSIRO-Mk3.0 and Shiogama et al., 2010 for MIROC 3.2. These models were combined with the A1B and B1 emission scenarios. Between June and August, we again replaced the values from future climates with the mean maximum and minimum temperatures and precipitation from 10 years of heat wave occurrence. Hence, four future climate scenarios that incorporated the occurrence of heat waves were developed at the same spatial resolution of the baseline: 'CSI B1 EW', 'MIR B1 EW', 'CSI A1 EW' and 'MIR A1 EW'. 'CSI B1 EW' relates to the combination of CSIRO-Mk3.0 climate model, the B1 emission scenario and the extreme weather event; namely the combination of the heat wave, while 'MIR A1 EW' relates to the combination of the MIROC 3.2 climate model, the A1B emission scenario and the extreme weather event.

Crop Model

A process-based crop model (DSSAT CSM-CERES-Maize v4.5) (Hoogenboom et al., 2010; Jones et al., 2003) was used to estimate the impact of the extreme event on USA maize production. CSM-CERES-Maize is a popular crop model that is able to simulate crop growth with minimum weather datasets, for example maximum and minimum daily temperature, daily solar radiation, daily total precipitation, and the number of rainy days. Additional variables on soil and crop management practices, including crop variety used, are needed in DSSAT. A hybrid maize variety called Garst 8808, whose genetic parameters have already been estimated for DSSAT and used in another study (Nelson et al., 2010), was also used in this study. Additional information on soil profile and other crop management practices used in the biophysical simulations are described by Nelson et al. (2010) and Gbegbelegbe et al. (submitted for publication).

Run High Performance Computer (HPC) cluster

Maize production across the USA was simulated at the 5 arc-minutes resolution scale. The simulations were performed in the High Performance Computing (HPC) clusters. Large volumes of spatial data is needed to run CSM-CERES-Maize in the HPC clusters and detailed explanations of these processes can be found in Nelson et al. (2010) andGbegbelegbe et al. (submitted for publication).

Association between heat wave and maize production

Simulated maize yields across the USA Corn Belt states were extracted which were affected by heat for the following climate scenarios: baseline climate (Baseline), 'Baseline EW', 'CSI B1 EW', 'MIR B1 EW', 'CSI A1 EW', and 'MIR A1 EW'. The results were then sorted across the USA Corn Belt states. The USA Corn Belt states which were affected by the extreme event comprise 14 states from North Dakota to Texas. The simulated biophysical results were compared to the impact of the extreme event with the observed USDA 2012 yield report which was released on February 21, 2013 (USDA, 2013a). The reported production and yield of 2012 are aggregated across rainfed and irrigated areas. Regardless of rainfed and irrigated cultivation, the values in the USDA report were the sum of yields in each state. Hence, recorded rates on irrigated and rainfed areas in the USA (USDA, 2007) were used to estimate recorded irrigated and rainfed production and yield separately.

RESULTS

Biophysical Results

Characteristics of Heat Waves

Maximum temperatures from the baseline climate were compared with those related to the 'Baseline EW' climate scenario. The deviations related to the maximum temperature were about 1.0 °C across all three months (June–August) (Table 1). However, when the comparison involves the highest maximum temperatures between the baseline and 'Baseline EW' climate scenarios, the deviations across June to August reached over 9.0 °C in some cases (Table 1). Moreover, the deviations were highest for the months of June and July. In the case of minimum temperatures, the deviations across all three months were also about 1.6 °C. On a month-to-month basis, the deviations were highest for July. The deviations for monthly total precipitation were more than 40 mm for the months of June and July: there was less precipitation in June and July under the 'Baseline EW' climate scenario compared to the baseline climate (Table 1). Usually, heat waves are accompanied by drought. Lobell et al. (2013) found that the water supply/demand ratio was reduced in August 2012 for the region in the USA affected by the heat wave. Precipitation had been decreasing since May and the drought peaked in September 2012 (Elliott et al., 2013). Similarly, in our study, precipitation for all three months under the 'Baseline EW' scenario across the Corn Belt (Table 2) was reduced by half.

Table 1: Deviations of monthly maximum and minimum temperature and precipitation from June to August in 'Baseline EW'. Deviations are subtracted from the baseline climate

States	Deviations											
	Maximum temperature (°C)						Minimum temperature (°C)			Precipitation (mm)		
	Average			Highest			Average			Monthly total		
	June	July	August	June	July	August	June	July	August	June	July	August
Iowa (IA)	0.3	1.0	0.8	9.2	9.5	8.6	1.4	2.9	−0.6	−41	−83	−43
Illinois (IL)	−0.5	0.7	0.4	9.2	9.3	7.8	0.1	2.8	−0.3	−46	−78	−7
Nebraska (NE)	1.0	1.6	0.8	12.0	8.9	8.7	2.5	2.6	−0.5	−49	−62	−41
Minnesota (MN)	0.6	1.4	1.1	8.5	8.4	7.7	2.5	4.1	0.6	3	−22	−53
Indiana (IN)	0.2	1.0	0.8	11.6	9.9	7.2	0.2	3.1	−0.2	−69	−62	9

South Dakota (SD)	0.3	0.5	0.5	10.5	8.9	9.6	2.2	3.4	−0.1	−36	−35	−25
Kansas (KS)	1.0	1.1	1.4	12.2	8.9	8.9	1.6	2.6	−0.7	−47	−70	−16
Ohio (OH)	0.7	1.4	1.3	10.9	9.3	6.9	1.1	3.3	0.6	−47	−42	−21
Wisconsin (WI)	0.6	1.5	1.3	9.3	10.3	7.7	2.6	4.1	1.1	−40	−30	−47
Missouri (MO)	0.9	1.2	1.6	11.0	9.3	8.7	0.9	3.1	0.2	−59	−68	−24
Michigan (MI)	1.6	1.7	1.7	10.7	9.6	8.3	2.1	3.2	1.2	−16	−9	−13
Texas (TX)	0.5	1.5	2.0	7.8	5.8	6.1	1.4	1.6	2.0	−7	2	−1
North Dakota (ND)	0.3	0.5	0.3	9.0	8.8	8.4	1.5	3.4	−0.3	−17	−31	−24
Kentucky (KY)	1.3	1.8	1.8	11.4	9.5	5.4	−0.4	2.8	0.3	−68	21	−23
Average	0.6	1.2	1.1	10.2	9.0	7.9	1.4	3.1	0.2	−38	−41	−23
Average June to August	1.0			9.0			1.6			−34		

Table 2: Major maize producing states (Corn Belt) in the USA between 2005 and 2010.

States	Average yield (kg/ha)	Average production (1000 t)	State share (%)
Iowa (IA)	10,754	56,565	18.5
Illinois (IL)	10,377	50,597	16.5
Nebraska (NE)	10,178	35,385	11.5
Minnesota (MN)	10,429	30,301	9.9
Indiana (IN)	9980	22,970	7.5
South Dakota (SD)	7929	13,550	4.4
Kansas (KS)	8410	12,678	4.1
Ohio (OH)	9666	12,605	4.1
Wisconsin (WI)	9185	11,081	3.6
Missouri (MO)	8484	9955	3.2
Michigan (MI)	8892	7569	2.5
Texas (TX)	8191	6318	2.1
North Dakota (ND)	7647	5607	1.8
Kentucky (KY)	8703	4139	1.4
Others[a]	8747	27,228	8.9
United States	9625	306,546	100

a 34 other states outside the Corn Belt states (Alaska and Hawaii were excluded).

A comparison of temperature and precipitation values was also made between the baseline and future climates (Fig. 1). The average maximum temperatures of June to August in the 'CSI A1 EW' climate scenario were similar to those of June–August in the 'Baseline EW'. On the other hand, the average maximum temperature was substantially higher under the 'MIR A1 EW' climate scenario compared to the baseline climate for June. The trend of average maximum temperature of 'MIR A1 EW' was quite different compared with that of two scenarios (Fig. 1a).

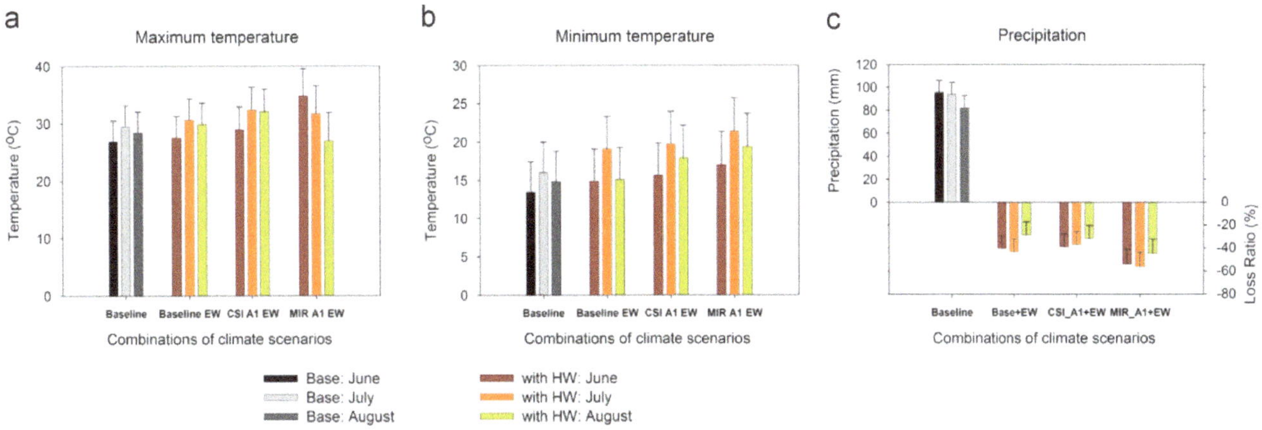

Figure 1: Climate scenarios with extreme heat wave from June to August in the Corn Belt (maximum temperature (a), minimum temperature (b), precipitation and its loss ratio based on baseline (c)). The baseline illustrates the normal climate and "Baseline EW" means the normal climate applied with extreme heat wave for 10 years and 'CSI A1 EW' and 'MIR A1 EW' express the future climate scenarios with extreme heat events under CSIRO-Mk3.0 and MIROC 3.2 of 2050 (2041–2060), respectively. The emission group of these future scenarios is A1B. HW=heat wave.

For minimum temperature, average monthly values were higher under the 'MIR A1 EW' climate scenario compared to the baseline climate between June and August (Fig. 1b). For precipitation, monthly averages were substantially lower (40%) under the three scenarios compared to the baseline climate (Fig. 1c), in particular the loss ratio of precipitation in July reached 60% under the 'MIR A1 EW' compared to the baseline climate. Hence, the risk of water stress should be higher under the future climate scenarios compared to the baseline scenario for rainfed maize.

Association Between Heat Waves and Maize Production

Table 3 shows the recorded grain yields for 26 years, especially around the Corn Belt. Between 1980 and 2000, maize production was much smaller in 1988 and 1993 than in 1980, 1986, 1998, and 1999 even though there was no large change in harvested area. Production in the Corn Belt was much smaller in 1983 compared to other years even though there were no weather extremes recorded in 1983: the country was rather going through an economic recession. After 2000 (Table 3), maize production has been increased by improving crop managements (i.e., irrigation, fertilizer etc.); there was also the reduction of maize production in the years when heat waves occurred even though the harvested area did not change largely. Maize production in 2012 was much smaller than in 2001, 2002 and 2006 even if there were no large changes in planted and harvested area. The simulated average grain yield across the USA was 12,009 kg/ha under the baseline climate (Table 4). This value of yield was not much different from the recorded average maize yield for the USA, which stood at around 9239 kg/ha of the recorded average grain yield in 2011 and 9625 kg/ha for the five year average (2005–2010) (Table 4). In addition, the recorded grain yield in 2012, when the heat wave occurred, and the simulated 'Baseline EW' grain yield were not too different. Overall, it can be anticipated that the recorded yields from USDA report in 2013 (USDA, 2013a) validate the geo-spatial crop model on USA maize production: the simulated grain yields of the baseline and 'Baseline EW' in the HPC cluster are consistent with recorded yields.

Table 3: The recorded production and (area) from 1980 to 2012

Year	Iowa	Illinois	Nebraska	Minnesota	Indiana	S. Dakota	Kansas	Ohio	Wisconsin	Missouri	Michigan	Texas	N. Dakota	Kentucky
1980*	37.2 (5.4)	27.1 (4.6)	15.3 (2.9)	15.5 (2.5)	15.3 (2.5)	3.1 (0.9)	3.0 (0.5)	11.2 (1.6)	8.8 (1.4)	2.8 (0.8)	6.3 (1.1)	3.0 (0.5)	0.4 (0.1)	2.6 (0.6)
1981	44.7 (5.6)	36.9 (4.6)	20.1 (2.7)	18.9 (2.7)	16.6 (2.4)	4.6 (1.0)	3.8 (0.5)	9.1 (1.5)	9.6 (1.4)	5.4 (0.8)	6.9 (1.2)	3.2 (0.4)	1.1 (0.2)	3.8 (0.6)
1982	40.4 (5.3)	38.7 (4.6)	19.6 (2.8)	18.7 (2.6)	20.7 (2.6)	4.9 (1.1)	3.6 (0.5)	12.1 (1.6)	9.2 (1.4)	5.2 (0.8)	7.8 (1.1)	3.0 (0.5)	0.9 (0.2)	4.0 (0.6)
1983	18.9 (3.5)	15.9 (3.2)	11.9 (2.0)	9.3 (1.8)	8.7 (1.9)	2.7 (0.8)	2.2 (0.4)	5.7 (1.1)	5.7 (0.9)	1.9 (0.6)	4.2 (0.7)	2.7 (0.4)	0.7 (0.2)	1.2 (0.4)

1984	3.7 (5.2)	31.7 (4.4)	20.5 (2.8)	17.5 (2.6)	17.9 (2.4)	4.7 (1.1)	3.0 (0.4)	11.7 (1.6)	8.8 (1.3)	3.9 (0.8)	5.6 (1.1)	3.7 (0.6)	1.1 (0.3)	3.7 (0.6)
1985	43.4 (5.5)	39.0 (4.6)	24.2 (3.0)	18.4 (2.5)	19.2 (2.5)	6.4 (1.2)	3.6 (0.4)	13.0 (1.6)	9.1 (1.4)	6.9 (1.0)	7.3 (1.1)	4.0 (0.6)	1.0 (0.2)	4.0 (0.6)
1986*	41.3 (5.0)	35.7 (4.3)	22.8 (3.0)	18.0 (2.6)	17.7 (2.4)	5.9 (1.3)	4.6 (0.6)	12.1 (1.6)	9.3 (1.6)	7.1 (1.0)	6.5 (1.1)	3.8 (0.6)	1.3 (0.4)	3.6 (0.7)
1987	33.2 (4.2)	30.5 (3.7)	20.6 (2.6)	16.1 (2.2)	16.0 (1.9)	5.8 (1.3)	3.6 (0.5)	9.2 (1.3)	8.4 (1.4)	6.2 (0.9)	4.7 (0.9)	3.4 (0.5)	1.2 (0.3)	3.0 (0.5)
1988*	22.8 (4.6)	17.8 (4.0)	20.8 (2.8)	8.8 (2.3)	10.5 (2.1)	3.4 (1.3)	3.7 (0.5)	6.5 (1.3)	3.3 (1.4)	3.9 (0.9)	2.8 (0.8)	3.3 (0.6)	0.6 (0.3)	−2.0 (0.5)
1989	36.7 (5.1)	33.6 (4.4)	21.5 (3.0)	17.8 (2.5)	17.6 (2.2)	4.8 (1.4)	3.9 (0.6)	8.7 (1.3)	7.9 (1.5)	5.6 (1.0)	5.7 (0.9)	3.8 (0.7)	0.9 (0.4)	3.5 (0.5)
1990	39.7 (5.2)	33.5 (4.3)	23.7 (3.1)	19.4 (2.7)	1.8 (2.3)	5.9 (1.4)	4.8 (0.6)	10.6 (1.5)	9.0 (1.5)	5.2 (0.8)	6.0 (1.0)	3.3 (0.7)	0.9 (0.3)	3.0 (0.5)
1991	36.3 (5.1)	29.9 (4.5)	25.2 (3.3)	18.3 (2.7)	13.0 (2.3)	6.1 (1.5)	5.2 (0.7)	8.3 (1.5)	9.7 (1.5)	5.4 (0.9)	6.4 (1.1)	4.2 (0.7)	1.3 (0.4)	2.8 (0.6)
1992	48.4 (5.3)	41.8 (4.5)	27.1 (3.4)	18.8 (2.9)	22.3 (2.5)	7.0 (1.5)	6.6 (0.7)	12.9 (1.5)	7.8 (1.6)	8.2 (1.0)	6.1 (1.1)	5.1 (0.7)	0.9 (0.4)	4.4 (0.6)
1993*	22.4 (4.9)	33.0 (4.2)	19.9 (3.2)	8.2 (2.5)	18.1 (2.2)	4.1 (1.4)	5.5 (0.8)	9.2 (1.4)	5.5 (1.4)	4.2 (0.9)	6.0 (1.0)	5.4 (0.8)	0.4 (0.3)	3.2 (0.6)
1994	49.0 (5.3)	45.4 (5)	29.3 (3.5)	23.3 (2.8)	21.8 (2.5)	9.3 (1.5)	7.7 (0.9)	12.4 (1.5)	11.1 (1.5)	7.0 (1.0)	6.6 (1.0)	6.1 (0.9)	1.4 (0.3)	4.0 (0.5)
1995	35.6 (4.7)	28.7 (4.1)	21.7 (3.2)	18.6 (2.7)	15.2 (2.2)	4.9 (1.1)	6.2 (0.9)	9.5 (1.3)	8.8 (1.5)	3.8 (0.7)	6.3 (1.0)	5.5 (0.8)	1.0 (0.3)	3.1 (0.5)
1996	43.6 (5.1)	37.3 (4.5)	30.1 (3.4)	22.1 (3.0)	17.0 (2.3)	9.4 (1.6)	9.1 (1.0)	7.8 (1.2)	8.5 (1.6)	9.0 (1.1)	5.5 (1.1)	5.1 (0.8)	1.7 (0.4)	3.8 (0.5)
1997	41.7 (4.9)	36.2 (4.5)	28.8 (3.6)	21.6 (2.8)	17.8 (2.4)	8.3 (1.5)	9.4 (1.1)	12.1 (1.5)	10.2 (1.6)	7.6 (1.1)	6.5 (1.0)	6.1 (0.8)	1.5 (0.3)	3.0 (0.5)
1998*	44.9 (5.1)	37.4 (4.3)	31.5 (3.6)	26.2 (3.0)	19.3 (2.3)	10.9 (1.6)	10.6 (1.2)	12.0 (1.4)	10.3 (1.5)	7.2 (1.1)	5.8 (0.9)	4.7 (1.0)	2.2 (0.4)	3.4 (0.5)
1999*	44.7 (4.9)	37.9 (4.4)	29.3 (3.5)	25.1 (2.9)	19.0 (2.3)	9.3 (1.5)	10.7 (1.3)	10.2 (1.4)	10.4 (1.5)	6.3 (1.1)	6.4 (0.9)	5.8 (0.8)	1.9 (0.3)	3.1 (0.5)
2000	43.9 (5.0)	42.4 (4.5)	25.8 (3.4)	24.5 (2.9)	20.6 (2.3)	10.8 (1.7)	10.5 (1.4)	12.3 (1.4)	9.2 (1.4)	10.1 (1.2)	6.1 (0.9)	6.0 (0.8)	2.6 (0.4)	4.1 (0.5)
2001*	42.3 (4.7)	41.9 (4.5)	28.9 (3.3)	20.5 (2.8)	22.5 (2.3)	9.4 (1.5)	9.8 (1.4)	11.1 (1.4)	8.4 (1.4)	8.8 (1.1)	5.1 (0.9)	4.3 (0.6)	2.1 (0.4)	4.0 (0.5)

2002*	49.9 (5.0)	38.0 (4.5)	23.9 (3.4)	26.7 (2.9)	16.0 (2.2)	7.7 (1.8)	7.4 (1.3)	6.4 (1.3)	9.9 (1.5)	7.2 (1.1)	5.9 (0.9)	5.2 (0.8)	2.9 (0.5)	2.7 (0.5)
2003	47.5 (5.0)	46.0 (4.5)	28.6 (3.3)	24.7 (2.9)	20.0 (2.3)	10.9 (1.8)	7.6 (1.2)	12.2 (1.3)	9.3 (1.5)	7.7 (1.2)	6.6 (0.9)	4.9 (0.7)	3.3 (0.6)	3.8 (0.5)
2004	57.0 (5.1)	53.0 (4.8)	33.5 (3.3)	28.5 (3.0)	23.6 (2.3)	13.7 (1.9)	11.0 (1.3)	12.5 (1.4)	9.0 (1.5)	11.9 (1.2)	6.5 (0.9)	5.9 (0.7)	3.1 (0.7)	4.4 (0.5)
2005	54.9 (5.2)	43.4 (4.9)	32.3 (3.4)	30.3 (3.0)	22.6 (2.4)	11.9 (1.8)	11.8 (1.5)	11.8 (1.4)	10.9 (1.5)	8.4 (1.3)	7.3 (0.9)	5.4 (0.8)	3.9 (0.6)	4.0 (0.5)
20068	52.1 (5.1)	46.2 (4.6)	29.9 (3.3)	28.0 (3.0)	21.5 (2.2)	7.9 (1.8)	8.8 (1.4)	12.0 (1.3)	10.2 (1.5)	9.2 (1.1)	7.3 (0.9)	4.5 (0.7)	3.9 (0.7)	3.9 (0.5)
2007	60.4 (5.7)	58.0 (5.3)	37.4 (3.8)	29.1 (3.4)	24.9 (2.6)	13.8 (2.0)	12.9 (1.6)	13.8 (1.6)	11.2 (1.6)	11.6 (1.4)	7.3 (1.1)	7.4 (0.9)	6.9 (1.0)	4.4 (0.6)
2008	55.6 (5.4)	54.1 (4.9)	35.4 (3.6)	30.0 (3.1)	22.2 (2.3)	14.9 (1.9)	12.4 (1.6)	10.7 (1.3)	10.0 (1.5)	9.7 (1.1)	7.5 (1.0)	6.4 (0.9)	7.2 (1.0)	3.9 (0.5)
2009	61.5 (5.5)	52.2 (4.9)	40.0 (3.7)	31.6 (3.1)	23.7 (2.3)	17.9 (2.0)	15.2 (1.7)	13.9 (1.4)	11.4 (1.6)	11.3 (1.2)	7.9 (1.0)	6.5 (1.0)	5.1 (0.8)	4.8 (0.5)
2010	54.7 (5.4)	49.4 (5.1)	37.3 (3.7)	32.8 (3.1)	22.8 (2.4)	14.5 (1.8)	14.8 (2.0)	13.5 (1.4)	12.8 (1.6)	9.4 (1.3)	8.0 (1.0)	7.7 (0.9)	6.3 (0.8)	3.9 (0.5)
2011	59.9 (5.7)	49.4 (5.1)	39.0 (4.0)	30.5 (3.3)	21.3 (2.4)	16.6 (2.1)	11.4 (2.0)	12.9 (1.4)	13.2 (1.7)	8.9 (1.3)	8.5 (1.0)	3.5 (0.8)	5.5 (0.9)	4.6 (0.6)
2012*	47.7 (5.7)	32.7 (5.2)	32.8 (4.0)	34.9 (3.5)	15.2 (2.5)	13.6 (2.5)	9.6 (1.9)	11.4 (1.6)	10.1 (1.8)	6.3 (1.5)	8.1 (1.1)	5.1 (0.7)	10.7 (1.5)	2.6 (0.7)

The columns with asterisk (*, i.e. 1980, 1986, 1988, 1998, 1999, 2001, 2002, 2006, and 2012) are years that extreme weather event occurred. Unit of production is million tons and unit of harvested area is million hectares. Source: USDA National Agricultural Statistics Service.

Table 4: Recorded and simulated grain yield and loss ratio for maize in the USA

	Recorded				Simulated					
					With Extreme weather event					
	2009	2010	2011	2012	Baseline	Base-line EW (1950–2000)	B1 (2050)		A1B (2050)	
							CSIRO	MIROC	CSIRO	MIROC
Grain yield (kg/ha)	10,369	9590	9239	7745	12,009	8545	7659	7643	7491	5081
Loss ratio (%)				−20		−29	−36	−36	−38	−58

Source: USDA National Agricultural Statistics Service (USDA, 2013a).

The extreme weather event under the baseline climate would decrease maize yield by 29% (8545 kg/ha) compared to the baseline climate (Table 4). If the climate extreme occurs under future climate, average grain yields under the CSIRO-Mk3.0 and MIROC 3.2 climate models would decrease by 36% under the B1 emission scenario compared to the baseline climate model. Under the A1B emission scenario, average grain yields under the CSIRO-Mk3.0 and MIROC 3.2 models would decrease by 38% and 58%, respectively, compared to the baseline climate (Table 3). Moreover, average grain yields under the 'CSI A1 EW' and 'CSI B1 EW' future climate scenarios are similar with yield reductions of 36% and 38%, respectively; however, there was a substantial difference between average grain yields under the 'MIR A1 EW' (58%) and 'MIR B1 EW' (36%) (Table 4).

Fig. 2 illustrates the simulated grain yield of irrigation with and without an extreme heat wave across all affected regions in the USA. The spatial irrigated maize yields in the Corn Belt states were between 5029 kg/ha and 12,573 kg/ha under the baseline climate (Fig. 2a), and between 5026 kg/ha and 12,566 kg/ha under the 'Baseline EW' scenario (Fig. 2b). When the climate extreme occurs under climate change, grain yield for irrigated maize ranged between 5089 kg/ha and 12,722 kg/ha under the 'CSI A1 EW', and between 5000 kg/ha and 12,499 kg/ha under the 'MIR A1 EW' scenario (Fig. 2c and d). In the case of rainfed maize, the simulated grain yields were between 3776 kg/ha and 9439 kg/ha for the baseline, and between 3171 kg/ha and 7928 kg/ha under the 'Baseline EW', respectively (figures not shown). Also, under the 'CSI A1 EW' and 'MIR A1 EW', the simulated grain yields were between 3132 kg/ha and 7463 kg/ha, and between 3006 kg/ha and 6013 kg/ha, respectively (figures not shown). Therefore, rainfed maize appears to be more sensitive to high temperature and drought. Overall, the spatial crop modeling framework performed well, as the simulated average maize yield under the baseline climate was similar to the reported average grain yield by the USDA (Table 4).

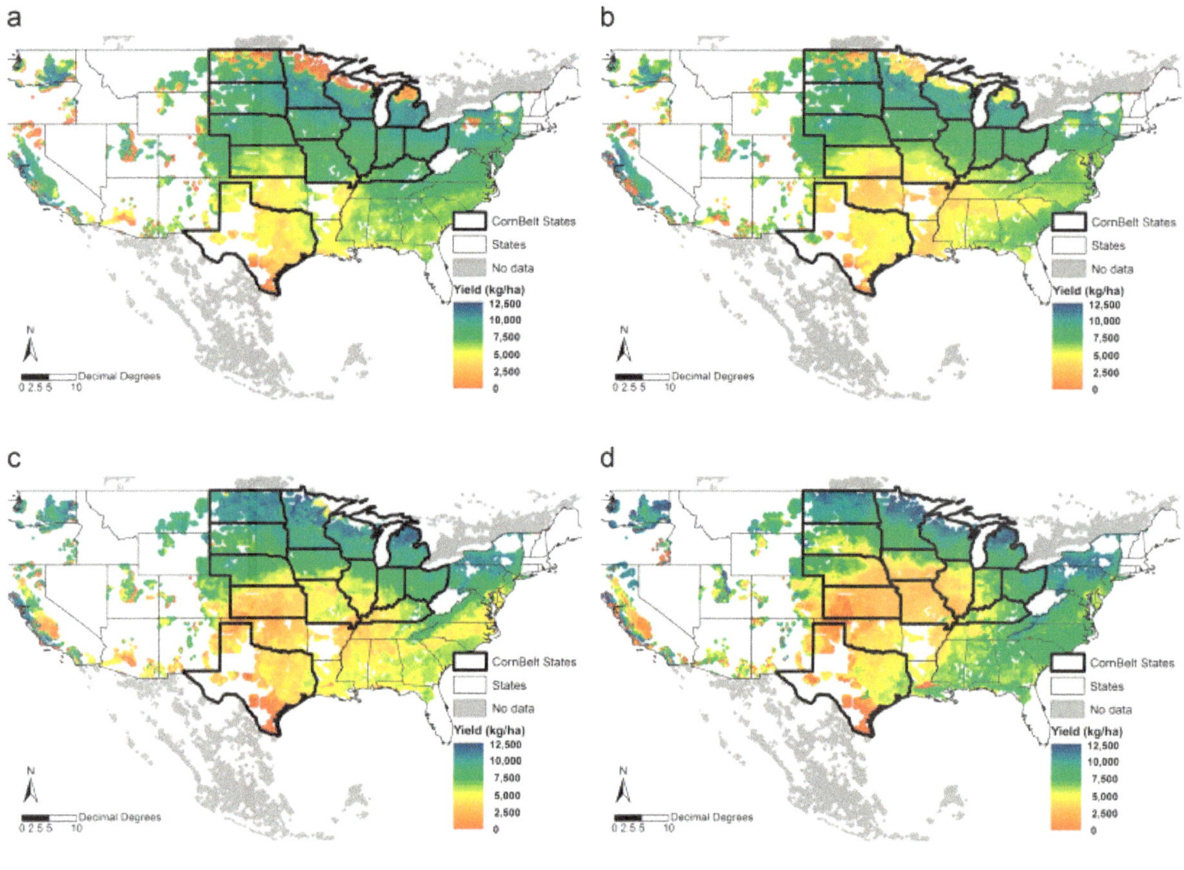

Figure 2: The maps express the spatial changes of grain yields under each condition and irrigated cropping system. Top-left is for the baseline climate (a), top-right for 'Baseline EW' (b), bottom-left for 'CSI A1 EW' under extreme heat (c), and bottom-right for 'MIR A1 EW' under extreme heat (d).

Socio-economic Results: Trend Analysis to Assess Impact of Weather extreme in 2012

Linkages between world Maize Prices and the Weather Extreme in the USA

At the beginning of 2012, historically low maize inventories were recorded in the USA, the largest maize producer and exporter in the world. However, a projected increase in global maize production, driven mainly by increased maize plantings in the USA and a record maize harvest in Brazil, kept world maize prices relatively stable during the first half of 2012 (Fig. 3). Other factors behind the projected increase in global maize production included stable maize production in China, the largest maize producer in the world after the USA, and projected increased maize plantings in the EU (FAO, 2012b).

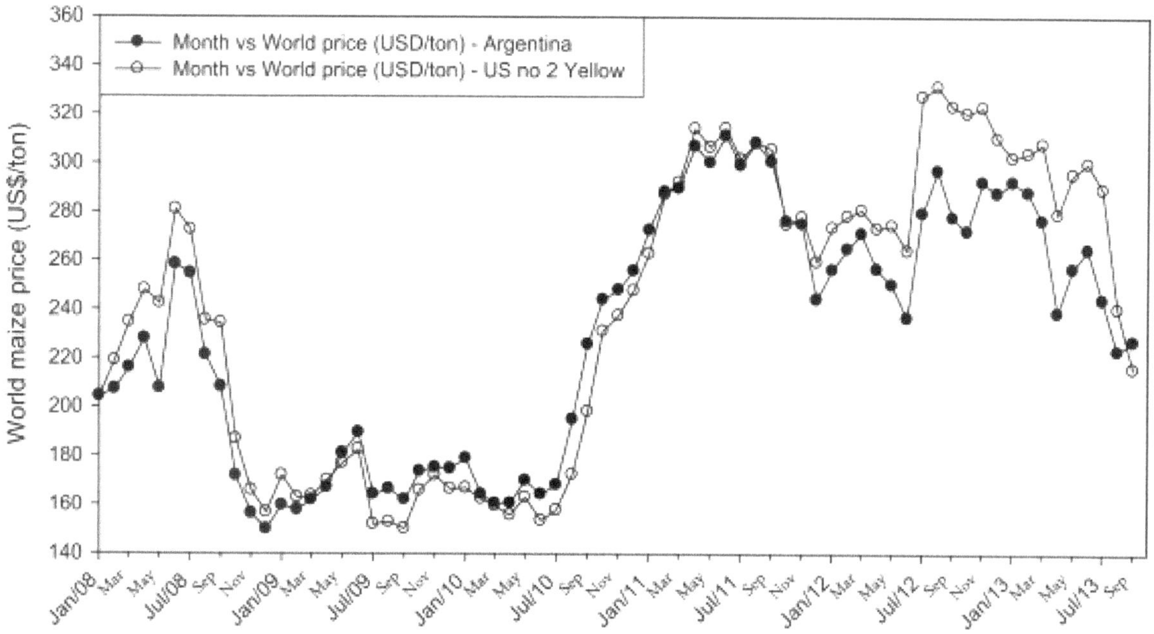

Figure 3: Trends in world maize prices.

In early 2012, the USDA projected that USA maize production would increase substantially in 2012/13 compared to 2011/12 (WASDE, 2012). More specifically, planted maize area would increase by 4%; maize yields would increase by 13%; the combination of these two factors would lead to maize production increasing by 20% or 53 million tons during the 2012/13 season (USDA, 2012a). Such projections usually support governments' efforts in assessing the food security status in their countries and identify import needs. They also help investors along the food value chain make better investment decisions, including the negotiation of fairer prices.

By July 2012, extreme and persistent drought had been recorded in June 2012 across the central and eastern Corn Belt. In addition, extreme heat from late June to early July was recorded across the central plains of the USA (USDA, 2012b). The combination of extreme heat and dryness lingered through July (USDA, 2012c). In August 2012, the USDA re-adjusted their projections, which involved a decrease of 16% in maize yields in 2012/13 compared to 2011/12. Hence, despite an increase of 5% in planted maize area, total maize production was projected to be 12% lower in 2012/13 compared to 2011/12 (USDA, 2012d).

Between June and August 2012, as knowledge on the effect of the weather extreme on the USA maize crop spread, the two indicators of the world maize price, namely the price of the USA no. 2 yellow maize and that of the Argentina yellow maize (Up River), increased by 25% each (Fig. 3). In August 2012, the USA no. 2 yellow maize, reached an all-time high of USD $331 per ton; it was 18% higher than its peak of USD $281 per ton which was recorded during the food price crisis in 2008. With a value of USD $297 per ton, the price of the Argentina yellow maize was 15% higher than its peak of USD $258 per ton during the food price crisis in 2008.

The rising maize prices were also fueled by reduced maize production in the EU due to a summer drought and low maize reserves worldwide. By November 2012, world maize reserves for the 2011/12 season were expected to be at their lowest since 2006/2007 (FAO, 2012c).

Effects of rising maize prices on maize consumption, trade and stocks in the USA

The sharp increase in world maize prices does not seem to have had substantial negative effects on maize consumption and retail food prices within the USA. USA maize production is estimated to have decreased by 13% or 40 million tons in 2012/13 vs. 2011/12 (WASDE, 2014). On the other hand, domestic maize consumption has barely changed: it is estimated to have decreased by 5% or 15 million tons (WASDE, 2014).

A strong USA dollar in 2012–2013 (USDA, 2012) made exports less attractive compared to the domestic markets; in addition, USA policies that have been boosting the biofuel industry in recent years (Piesse and Thirtle, 2009 and Yano et al., 2010) implied a strong domestic maize demand. Hence, domestic maize consumption reached 263.6 million tons (Fig. 4). On the other hand, USA maize exports decreased by 53% to 18.6 million tons (Fig. 4). Given that the total amount of maize utilized by the USA in 2013 amounted to 282.1 million tons (domestic consumption and exports) against an estimated total production of 273.8 million tons, the country had to supplement its production by importing 4 million tons of maize and reducing its maize stocks by 4.3 million tons (Fig. 4).

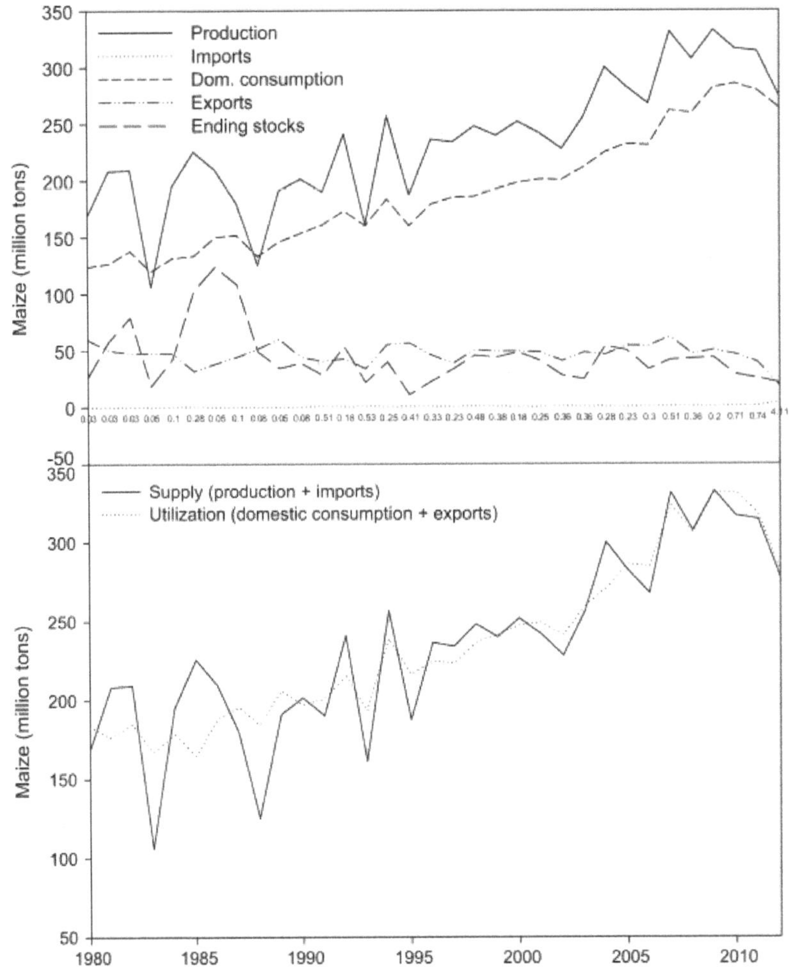

Figure 4: Impact of weather extremes on US maize production, consumption and trade. Source: USDA World Agricultural Supply and Demand Estimates – from 1981 to 2014.

The weather extreme of 2012 is similar to the one which occurred in 1993, in terms of its effects on USA maize production, consumption, trade and stocks. The 1993 weather extreme led to the highest year-to-year relative decrease in maize production since 1980. More specifically, USA maize production fell by 33% in 1993, compared to 1992; by contrast, USA maize consumption that same year fell only by 7.6% over the same period. USA maize production amounted to 160.9 million tons in 1993; the same amount of maize was domestically consumed within the USA, that same year (Fig. 4). In addition, the country decreased its maize exports by 20% to a low of 33.7 million tons (Fig. 4). In order to meet all its utilization requirements in 1993, the country had to reduce its stocks by 32.1 million tons and import 533 thousand tons of maize (Fig. 4). The key difference between the 2012 and 1993 event relates to the heavier reliance of the USA on maize stocks to meet its maize utilization in 1993. This is not surprising as USA maize ending stocks in 1992 stood at 53.7 million tons and were twice as high as those of 2011 (Fig. 4).

By contrast, the weather extreme of 1988 occurred when the USA had large maize stocks which stood at 108 million tons and hence were 4.3 times higher than ending maize stocks in 2011 (Fig. 4). The weather extreme of 1988 reduced maize production by 30% compared to 1987. In this case, the country slightly reduced their maize consumption but increased their maize exports and even reduced maize imports. That year, the USA heavily drew down its stocks to supplement its maize production so as to meet its total maize utilization (Fig. 4).

Effect of the Weather Extreme on World Maize Production, Consumption and Trade

The substantial reduction in USA maize exports implied a reduction in the supply of maize in international markets (FAO, 2012c). Recent estimates imply that world maize export decreased by 24% (23 million tons) between 2011/12 and 2012/13 (USDA, 2013b). Overall, world maize production decreased by 24 million tons whereas maize consumption decreased by 11.5 million tons. Hence, the world had to rely on world maize reserves to meet its consumption requirements: world maize stocks decreased by 7% between the beginning and the end of the 2012/13 season (USDA, 2013b).

World maize prices remained high after their peak in August 2012. Eleven months after the price spike, the USA no. 2 yellow maize was only 11% lower compared to August 2012; the Argentina yellow maize was only 9% lower (Fig. 3). This suggests that poor countries that depend on imports to meet their maize consumption requirements had to contend with high maize prices for much of 2012/13. Some of these countries include vulnerable countries in the Caribbean, northern Africa and western Asia, where the ratio of imports to maize consumption has remained above 50% for the last years (Table 5). Other countries have low import dependency ratios;

however, they are vulnerable because they would be too poor to afford highly-priced maize. This would be true for some countries in Sub-Saharan Africa (SSA), which import white maize; the latter is usually consumed as food and is sold at a premium over yellow maize. A detailed analysis involving a global multi-market and multi-commodity model would be necessary to estimate the impact of the weather extreme on food security across vulnerable and import-dependent countries.

Table 5: Trend in import to consumption ratio for developing regions

Region	2005	2006	2007	2008	2009
Sub-Saharan Africa (SSA)					
Eastern Africa	5.3	10.3	5.4	7.5	15.3
Central Africa	12.8	11.1	8.4	9.0	11.5
Southern Africa	4.7	14.0	15.7	7.0	5.8
Western Africa	3.4	3.4	2.7	2.5	2.6
CWANA					
Central Asia	1.2	1.8	4.2	6.1	4.9
Northern Africa	61.7	57.4	61.8	55.3	58.0
Western Asia	54.2	57.2	63.9	58.2	61.3
East and South East Asia					
Eastern Asia	18.7	18.7	17.4	16.8	16.1
South-eastern Asia	11.2	19.2	13.6	11.5	12.0
South Asia	10.8	10.5	13.8	12.1	17.3
LAC region					
Caribbean	77.0	81.5	78.2	78.9	77.5
Central America	26.5	30.1	31.2	30.9	28.6
South America	11.2	13.7	13.7	11.3	13.6

CWANA=Central and West Asia and North Africa; LAC=Latin America and Caribbean.

Socio-economic Results: Assessing Impact of the Weather Extreme in 2050

Some of the key drivers that would determine the effect of the weather extreme, were it to occur in 2050, include the cumulative effect of climate change on maize production; population and income growth worldwide; and the ability of maize farming systems to adapt to climate change through the adoption of improved maize technologies. The weather extreme could still have devastating effects if it were to occur in 2050. Food security worldwide could be enhanced between the 2000s and 2050s, due to favorable economic growth. However, the gains in enhancing food security might be lost to the adverse weather extremes brought by climate change. Hence, one more weather extreme, that would substantially reduce maize production in the USA, would worsen food security in countries already weakened by climate change.

However, some of the developing countries that would be vulnerable to the weather extreme in 2012 might not experience any loss if the weather extreme occurred in the 2050s under climate change. In such countries, the improvement in national food security over the years could outweigh the negative effect of climate change. In addition, world maize trade might change by 2050 with less dependence on the USA as an exporter. Under circumstances where the USA ends up accounting for 20% or less of world maize exports by 2050, a weather extreme that would affect maize production in the USA might have more subdued effects on world maize prices and hence on food security across vulnerable countries. The adoption of improved maize technologies coupled with adequate policies that facilitate technology adoption could also change the structure of world maize trade and hence dampen the negative effects of the weather extreme. For example, high-yielding maize germplasm with resistance to the abiotic and biotic stresses brought by climate change (Shiferaw et al., 2011) might help sustain maize production in some regions and increase production in others.

The impact of future weather extremes would also depend on the terms of trade that would exist in 2050. Some countries, namely the ones in Latin America and the Caribbean, could join forces to adjust their production and consumption patterns so as to reduce their maize import dependency by 2050. Under such scenario, a weather extreme that would substantially reduce USA maize production could have relatively insignificant effects on global food security.

DISCUSSION

Characteristics of extreme weather events in the Corn Belt of the USA

Crop yield is a result of the interactions in the soil-plant-atmosphere continuum of crop models. Weather data is the driving factor in crop modeling and it is important to use accurate weather data inputs in models. Although extreme events are common features of weather, they usually have negative consequences ranging from tolerable to disastrous conditions depending on the magnitude and duration of the extreme events. Since 1980, in the USA alone, heat wave events have occurred more than ten times. In 2012, maize-growing farmers in the USA experienced the highest temperatures on record since 2000. In addition, the heat wave was followed by severe drought. According to the latest USA Drought Monitor report, more than two-thirds of the contiguous United States was under drought conditions, the highest level since record-keeping began in January 2000; indeed soil moisture was reduced by half compared to normal years (Lobell et al., 2013).

Our results illustrate the considerable utility of the process-based spatial bio-economic framework, as a tool for assessing the impacts of extreme heat waves on maize production. Our approach for heat wave generation also underlines the usefulness of monthly weather data, even though the methodology is a simple arithmetic method. However, we need further studies on solar radiation and the number of rainy days that we did not change in the study and more high resolution climate maps focused on country level since the spatial resolution is a key decision in agricultural systems modeling (Zhao et al., 2013). Additionally, we should clearly address four basic properties: validation, uncertainty, credibility, and clarify to generate climate maps with extreme events such as spatial (i.e., km or degree on the gridded maps) and temporal resolutions (i.e., daily or monthly weather data) (Knutti et al., 2008 and Elliott et al., 2013).

Biophysical Impact of Extreme Weather Events in the Corn Belt of the USA

Depending on the cultivar and the cultivated area of maize, rainfall requirements can vary much. However, maize generally requires at least 90 to 300 mm of average precipitation between June and August in rainfed areas (Shaw, 1988, Dowswell et al., 1996 and Hartkamp et al., 2000). The range of normal temperature for good growth and development of maize is between 26 °C and 32 °C (Hartkamp et al., 2000). Also, if soil moisture is adequate, maize can grow well even under high temperatures, varying between 32 °C and 38 °C. However, precipitation when the heat wave occurred was just half of that reported for the baseline climate in Fig. 2. The high temperatures and reduced precipitation for three months decreased actual 2012 grain yield to an average 7745 kg/ha—20% less than the average yield over 2009–11 (10,369 kg/ha, 9590 kg/ha and 9239 kg/ha in 2009, 2010 and 2011, respectively, giving an average of 9733 kg/ha), a period when there was no weather extreme (Table 4). The simulated grain yield of 'Baseline EW' was 8545 kg/ha and the simulated grain yield of the baseline climate was 12,009 kg/ha, a reduction of 29%. The 9% difference between recorded and simulated yield losses results from the baseline overestimation typical of most models, given that the baseline represents optimal conditions, and is actually less than the differences projected by several other leading models (Table 4). Thus, the simulation faithfully represents the effects of a weather extreme such as high temperatures from June to August—when ears are developing, a crucial period for maize growth—and can be used to predict grain yield losses from a heat wave. Our study involved the use of daily weather data generated based on average monthly data. Indeed, not using recorded daily weather data might lead to an underestimation of the biophysical impact of weather extremes. However, our results are consistent with those of Elliott et al. (2013) who used recorded daily weather data and found that the 2012 weather extreme reduced USA maize yields by 25% relative to trend.

The input data used to reflect weather extremes were the same regardless of the climate models used. In reality, future weather extremes might involve greater heat and/or larger rainfall reduction. Hence, the results on future weather extremes can be considered as lower bounds on the impact of such extremes on maize yields in the USA.

Recently, studies on heat stress and drought tolerance have been being robustly investigated. We also need additional research on the effect of weather extremes on food security in the developing world if the weather extremes occur in the developing world in the near future. Another area of future research consists of simulating the biophysical effect of weather extremes using future farm technologies adapted to future climates.

Socio-economic Analysis of the Weather Extreme

Since the USA is the major maize producer and exporter worldwide, shocks that affect its maize supply are likely to impact international maize markets and by extension, maize systems and food security in other regions of the world. The weather extreme which affected USA maize production in 2012 likely influenced the 25% increase in world maize prices between June and August 2012. However, other factors might have also been at play, including speculation.

The weather extreme also led to a reduction in global maize trade and consumption in 2012/13 compared to 2011/12. Countries most likely to have experienced adverse effects from the weather extreme, including reduced food consumption and hence worsened food insecurity, include poor countries with high maize import dependency ratios. Other vulnerable countries have lower import dependency ratios but would be too poor to afford highly-priced maize.

The trend analysis used in this study highlights the linkages between the weather extreme and maize consumption and trade worldwide. However, it is not enough to pinpoint the exact effect of the weather extreme. Some careful analysis involving geo-spatially disaggregated economic modeling would be needed to isolate the potential impact of the weather extreme on food security across the developing world.

The study also highlighted some of the factors that could mitigate or fuel the negative effect of the weather extreme, were it to occur 40 years from now. Some of these factors include socio-economic growth over the years and the adoption of improved maize technologies. Adaptation to weather extremes would include policies aimed at supporting countries to cope with the aftermath of the weather extremes and policies that would enhance the adaptive capacity of countries. Social protection and safety net programs to protect the vulnerable would fall under coping-related policies. They include cash transfers to vulnerable households where markets work; where markets do not work, food aid might be the best option. Other programs targeting the vulnerable include those on school feeding and food-for-work. Medium- to long-term interventions would include the replenishment of food stocks in the world's breadbaskets: these food stocks need to be high enough to counter the negative effect of weather extremes on food production. If the USA ending maize stocks of 2011 were similar to those of 1992, the USA would have been able to use these stocks to cover the production loss brought by the 2012 weather extreme.

CONCLUDING REMARKS

The objectives of this study were to estimate the biophysical impact of the 2012 weather extreme in the USA on maize production in the country and assess its related effects on food security across the developing world. If the climate extreme occurs under the baseline climate, our estimates suggest a 29% reduction in maize yields across the USA. The socio-economic analysis suggests that the extreme climate of 2012 that occurred in the USA is likely to increase food insecurity among poor communities where maize provides a substantial portion of daily caloric intake and where households cannot easily adjust their food consumption patterns in the face of increased maize scarcity. Our results indicate that food insecurity would be more severe in SSA. Nelson et al. (2010)assessed the economic impact under four climate scenarios while we used only two scenarios for this assessment.

However, although this challenge is a good example due to the reasonable results achieved under the HPC cluster, there still remain many challenges to improve the gridded weather data and crop models on the HPC cluster. Chung et al. (2011) suggested the method that can predict the flowering date of a cherry tree by inputting gridded (i.e., spatial) real-time weather data with the normal years data (i.e., historical weather data). Extreme weather events, such as high temperatures or severe drought will occur many times in the future and it is difficult to forecast in real-time. However, we anticipate that the real-time forecasting technology will be improved and it would be easier to directly run the crop model on the HPC cluster. Also, the real-time result from HPC cluster can be useful in adaptive planning, strategies, and economic decisions in the mid- to long-term perspective.

The socio-economic analysis demonstrated that the weather extreme in the USA likely affected maize prices, consumption and trade worldwide. It also highlighted the potential pathways through which the weather extreme could affect food security across the developing world. However, additional research involving a process-based economic model that can capture the terms of trade of the key staple crops across the globe would be needed to isolate the socio-economic impact of the weather extreme.

ACKNOWLEDGMENTS

This study was supported by the CGIAR Research Program on Climate Change, Agriculture and Food Security (CCAFS:A4007).

REFERENCES

1. Anderson, G.B., Bell, M.L., 2011. Heat waves in the United States: mortality risk during heat waves and effect modification by heat wave characteristics in 43 U.S. communities. Environ. Health Perspect. 119 (2), 210–218.
2. CCAFS., 2013. GCM Downscaled Data Portal. Available at: ⟨http://www.ccafs-cli mate.org⟩.
3. Chung, U., Mack, L., Yun, J.I., Kim, S.-H., 2011. Predicting the timing of cherry blossoms in Washington, DC and mid-Atlantic states in response to climate change. PLoS ONE 6 (11), e27439. http://dx.doi.org/10.1371/ journal. pone.0027439.
4. Ciais, P., Reichstein, M., Viovy, N., Granier, A., Ogee, J., Allard, V., Aubinet, M., Buchmann, N., Bernhofer, C., Carrara, A., Chevallier, F., De Noblet, N., Friend, A.D.,Friedlingstein, P., Grunwald, T., Heinesch, B., Keronen, P., Knohl, A., Krinner, G., Loustau, D., Manca, G., Matteucci, G., Miglietta, F., Ourcival, J.M., Papale, D., Pilegaard, K., Rambal, S., Seufert, G., Soussana, J.F., Sanz, M.J., Schulze, E.D., Vesala, T., Valentini, R., 2005. Europe-wide reduction in primary productivitycaused by the heat and drought in 2003. Nature 437, 529–533.
5. Dowswell, C.R., Paliwal, R.L., Cantrell, R.P., 1996. Maize in the Third World. WestviewPress, p. 268.
6. Elliott, J., Glotter, M., Best, N., Boote, K., Jones, J., Rosenzweig, C., Smith, L.A., Foster, I., 2013. Predicting agricultural impacts of large-scale drought: 2012 and the casefor better modeling. The Center for Robust Decision Making on Climate andEnergy Policy No. 13-01, http://dx.doi.org/10.6084/m9.figshare.501263.
7. FAO, 2012a. Food Outlook. Food and Agriculture Organization, Rome. FAO, 2012b. Food Outlook: Global Market Analysis – May 2012.
8. Food and Agriculture Organization, Rome. FAO, 2012c. Food Outlook: Global Market Analysis – November 2012. Food and Agriculture Organization, Rome.
9. FAOSTAT, 2013. Statistical Databases on Global Food Production and Trade. Foodand Agriculture Organization, Rome. Gbegbelegbe, S., Cammarano, D., Asseng, S., Robertson, R., Chung, U., Adam, M., Abdalla, O., Payne, T., Reynolds, M., Shiferaw, B., Nelson, G., (submitted). Baseline simulations for global wheat production: capturing genetic diversity of cultivars. (submitted for publication).
10. Gordon, H., O'Farrell, S., Collier, M., Dix, M., Rotstayn, L., Kowalczyk, E., Hirst, T., Watterson, I., 2010. The CSIRO Mk3. 5 Climate Model. Centre for Australian Weather and Climate Research, CSIRO and the Bureau of Meteorology.
11. Hajat, S., Armstrong, B., Baccini, M., Biggeri, A., Bisanti, L., Russo, A., Paldy, A., Menne, B., Kosatsky, T., 2006. Impact of high temperatures on mortality: is there an added heat wave effect? Epidemiology 17, 632–638.
12. Hartkamp, A.D., White, J.W., Rodriguez, A., Banziger, M., Srinivasan, G., Granados, G., Crossa, J., 2000. Maize production environments revisited: a GIS-based approach. CIMMYT, Mexico, D.F (33 p.).
13. Hijmans, R.J., Cameron, S.E., Parra, J.L., Jones, P.G., Jarvis, A., 2005. Very high resolution interpolated climate surfaces for global land areas. Int. J. Climatol. 25, 1965–1978.
14. Hoogenboom, G., Jones, J.W., Wilkens, P.W., Porter, C.H., Hunt, L.A., Boote, K., Singh, U., Uryasev, O., Lizaso, J.I., Gijsman, A.J., 2010. Decision Support System for Agrotechnology Transfer (DSSAT) Version 4.5.0.43 [CD-ROM]. University of Hawaii, Honolulu, Hawaii.
15. Jones, J.W., G. Hoogenboom, C.H. Porter, K.J. Boote, W.D. Batchelor, L.A. Hunt, P.W. Wilkens, U. Singh, A.J. Gijsman, and J.T. Ritchie. 2003. DSSAT Cropping System Model. European Journal of Agronomy 18:235-265.
16. Knutti, R., Allen, M.R., Friedlingstein, P., Gregory, J.M., Hegerl, G.C., Meehl, G.A., Meinshausen, M., Murphy, J.M., Plattner, G.-K., Raper, S.C.B., Stocker, T.F., Stott,P.A., Teng, H., Wigley, T.M.L., 2008. A review of uncertainties in globaltemperature projections over the twenty-first century. J. Clim. 21, 2651–2663.
17. Lobell, D.B., Hammer, G.L., McLean, G., Messina, C., Roberts, M.J., Schlenker, W.,2013. The critical role of extreme heat for maize production in the United States. Nat. Clim. Change 3, 1–5.

18. Lott, N., Ross, T., 2006. Tracking and evaluating U.S. billion dollar weather disasters 1980–2005. The 86th AMS annul meeting: Environmental Risk and Impacts on Society: Successes and Challenges., Atlanta, Georgia.

19. Meehl, G.A., Tebaldi, C., 2004. More intense, more frequent, and longer lasting heat waves in the 21st century. Science 305, 994–997.

20. Nakicenovic, N., Alcamo, J., Davis, G., Vries, B.D., Fenhann, J., Gaffin, S., Gregory, K., Grubler, A., 2000. Special report on emissions scenarios : a special report of Working Group III of the Intergovernmental Panel on Climate Change. Cambridge University Press, Cambridge.

21. Nelson, G.C., Rosegrant, M.W., Palazzo, A., Gray, I., Ingersoll, C., Robertson, R., Tokgoz, S., Zhu, T., Nelson, n.d. Gerald C., Rosegrant, Mark W., Palazzo, Amanda, Gray, Ian, Ingersoll, Christina, Robertson, Richard, Tokgoz, Simla, Zhu, Tingju,

22. Sulser, Timothy B., Ringler, Claudia, Msangi, Siwa, You, Liangzhi, 2010. Food Security, Farming, and Climate Change to 2050: Scenarios, Results, Policy Options, http://dx.doi.org/10.2499/9780896291867.

23. Nuttall, J., Brady, S., Brand, J., Leary, G.O., Fitzgerald, G.J., 2013. Heat waves and wheat growth under a future climate. Aust. Soc. Agron. The 16th Australian Agronomy Conference: Climate Change. Available at: /http://www.regional. org.au/au/asa/2012/climate-change/8085_nuttalljg.htmS.

24. Piesse, J., Thirtle, C., 2009. Three bubbles and a panic: an explanatory review of recent food commodity price events. Food Policy 34 (2), 119–129.

25. Robinson, P.J., 2001. On the definition of a heat wave. J. Appl. Meteorol.y 40, 762–775.

26. Shaw, R.H., 1988. Climate requirement. In: Sprague, G.H., Dudley, J.W. (Eds.), Corn and Corn Improveent. CAB Direct. Shiferaw, B., Prasanna, B., Hellin, J., Banziger, M., 2011. Feeding a hungry world: past successes and future challenges to global food security in maize. Food Secur. 3, 307–327.

27. Shiogama, H., Emori, S., Takahashi, K., Nagashima, T., Ogura, T., Nozawa, T., Takemura, T., 2010. Emission scenario dependency of precipitation on global warming in the MIROC3.2 model. J. Clim. 23, 2404–2417.

28. Smoyer, K.E., 1998. A comparative analysis of heat waves and associated mortality in St. Louis, Missouri –1980 and 1995. Int. J. Biometeorol. 42, 44–50.

29. USDA, 2007. United States Summary and State Data. United States Department of Agriculture.

30. USDA, 2012. U.S. Drought 2012. Farm and Food Impacts. United States Department of Agriculture, Washington DC.

31. USDA, 2012a. Projected World and USA Supply and Use of Agricultural Commodities — Database – May 2012. United States Department of Agriculture.

32. USDA, 2012b. World Agricultural Supply and Demand Estimates – July 2012. United States Department of Agriculture, Washington DC.

33. USDA, 2012c. World Agricultural Supply and Demand Estimates – August 2012. United States Department of Agriculture, Washington DC.

34. USDA, 2012d. Projected World and USA Supply and Use of Agricultural Commodities – Database – August 2012. United States Department of Agriculture.

35. USDA, 2013a. Crop Production 2012 Summary. United States Department of Agriculture.

36. USDA, 2013b. Projected World and USA Supply and Use of Agricultural Commodities– Database – September 2013.

37. WASDE, 2012. May WASDE Report 2012/13 Outlook & Market Impacts. WASDE, 2014. March 2014 WASDE Report.

38. Weisskopf, M.G., Anderson, H.A., Foldy, S., Hanrahan, L.P., Blair, K., Török, T.J.,Rumm, P.D., 2002. Heat wave morbidity and mortality, Milwaukee, Wis, 1999 vs1995: an improved response? Am. J. Publ. Health 92, 830–833.

39. WorldClim Global Climate Data Center, 2013. Available at: ⟨http://www.worldclim.org⟩. Yano Y., Blandford D., Surry Y., 2010. Do Current U.S. Ethanol Policies Make Sense? Policy Issue.

40. Zhao, G., Bryan, B.A., King, D., Luo, Z., Wang, E., Bende-Michl, U., Song, X., Yu, Q., 2013. Large-scale, high-resolution agricultural systems modeling using a hybrid approach combining grid computing and parallel processing. Environ. Model. Softw. 41, 231–238.

CHAPTER 11 IMPACT OF CLIMATE CHANGE ON ESTUARIES

Peter E. Robins, Martin W. Skov, Matt J. Lewis, Luis Giménez, Alan G. Davies, Shelagh K. Malham, Simon P. Neill, James E. McDonald, Timothy A. Whitton, Suzanna E. Jackson, Colin F. Jago

Keywords: Climate change, Estuaries, Estuarine habitats, Water quality, Biodiversity, UK

INTRODUCTION

Estuaries are process centres for coastal hydrological, biogeochemical and biological cycles and their biological productivity rivals those of tropical rainforests and coral reefs (Cai, 2011). The high productivity combines with the provisioning of important ecosystem services, such as flood protection and providing recreational space, to maintain a Good Environmental Status (GEnS) so that they are socio-economically valuable and sustainable (Barbier et al., 2011 and Elliott et al., 2015). Estuaries are also biologically important; many in the UK are biodiversity rich and most host a series of habitats that are important nursery areas for commercial fisheries species (Beck et al., 2001), or feeding and breeding areas for terrestrial populations, such as birds (Sharps et al., 2015). Many estuarine habitats have statutory protection and all have roles in prolific estuarine foodwebs and the cycling of organic material (Kaiser et al., 2005). The ecological functions of providing habitat and nursery areas are attributed to the discontinuity in productivity, sediment characteristics, and turbidity, when compared with the surrounding coasts (Power et al., 2000). Estuaries are known to trap, filter and recycle suspended particulate matter (Ittekkot, 1988), as well as components that are potentially harmful to human health, such as nutrients and pathogenic viruses. In the UK, estuarine processes are directly affected by intrinsic anthropogenic activities, such as aquaculture and shellfish fisheries, and indirectly by activities such as catchment agriculture, forestation/deforestation, urbanisation and coastal management decisions (Attrill, 1998).

Estuaries are commonly defined as "*a semi-enclosed coastal body of water which has free connection to the open sea, extending into the river as far as the limit of tidal influence and within which seawater is measurably diluted with freshwater derived from land drainage*" (Dyer, 1997). This definition has shortcomings: tidal influence often exceeds the salt water intrusion in the river, and the seaward region of fresh water influence (ROFI) may extend offshore, such as in the Severn Estuary. Thus, estuaries can be further classified based on salinity structure (well mixed through to stratified), tidal range (microtidal to hypertidal) and geomorphology (Appendix 2). A review of 155 UK 'estuaries' classified sites by their geomorphology and tidal type (Fig. 1; Appendix 1;Davidson et al. (1991)). Mesotidal estuaries (2 m < tidal range < 4 m) are typical in the eastern UK, whereas macrotidal (4 m < tidal range < 6 m) or hypertidal (tidal range > 6 m) estuaries occur in the west, and are generally connected to catchments that respond quickly to rainfall events (Brown et al., 1991).

The Intergovernmental Panel on Climate Change (IPCC) is 'virtually certain' that global temperatures will increase during the 21st century, with climate models projecting UK shelf seas to warm by between 1.5 °C and 4 °C by 2100 (Jenkins et al., 2009). In turn, global mean sea-levels are projected to rise by between 44 and 74 cm, depending on our future carbon dioxide emissions (IPCC, 2013), with UK mean sea-level closely matching the global trend (Woodworth et al., 2009). Climate models also predict changes to precipitation pattern in the UK, with more inter-regional and inter-seasonal variability, albeit with high model uncertainty and less confidence than sea-level projections (IPCC, 2013); but with high impact, such as the damaging storm clusters experienced during the winter of 2013/2014 along the western coasts of the UK (Wadey et al., 2014). It follows that the long-term climate-driven impacts that UK estuaries face include a deterioration in water quality and alterations to habitat composition, which could ultimately affect their ecological integrity (Kennish, 2002).

Citation: Peter E. Robins, Martin W. Skov, Matt J. Lewis, Luis Giménez, Alan G. Davies, Shelagh K. Malham, Simon P. Neill, James E. McDonald, Timothy A. Whitton, Suzanna E. Jackson, Colin F. Jago, Impact of climate change on estuaries, http://dx.doi.org/10.1016/j.ecss.2015.12.016.

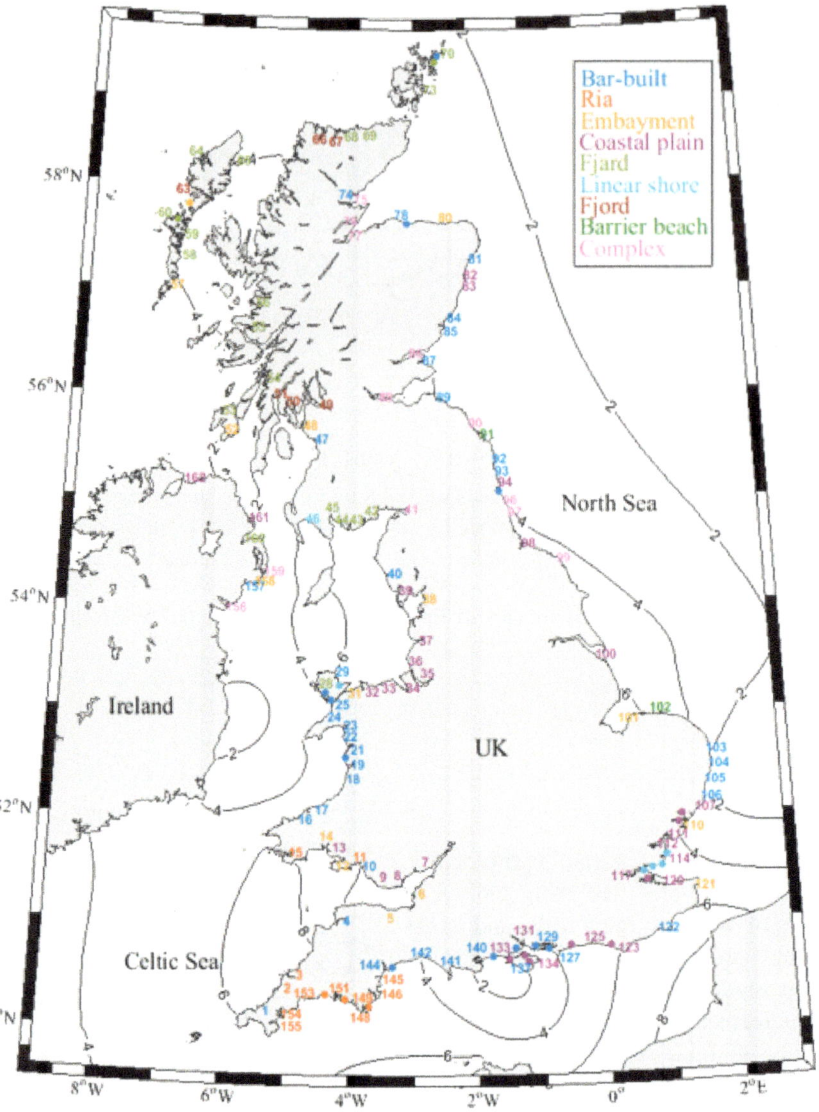

Figure 1: Location and geomorphological classification of 162 UK estuaries, as defined by Davidson et al. (1991), who did not include estuaries in Northern Ireland. Tidal type can be inferred from tidal range (m) contours (calculated from a 3D ocean model, simulated and validated by Lewis et al., 2014), classified by Davies (1964) as: microtidal (tidal range < 2 m), mesotidal (2 m–4 m), macrotidal (4 m–6 m), and hypertidal (>6 m).

In this paper, we review current understanding of how anthropogenic climate change will impact UK estuaries and their complex and important environments. Trends suggest that, over the 21st century, estuarine ecosystems will be mostly impacted by habitat loss, as a consequence of climate change and urban expansion. Other projected high priority problems include excessive nutrient and sewage inputs to estuaries, leading to eutrophication, hypoxia and anoxia (Kennish, 2002). Further damaging problems will arise from over-fishing, chemical contaminant fluxes in urbanised regions, freshwater diversions, the introduction of invasive species, and coastal subsidence and erosion. Although it is difficult to quantify, these impacts on estuarine ecosystems are at least in part linked to climate change.

We amalgamate current knowledge from peer-reviewed literature on climate change trends and projections for the UK, and discuss how climate change has and will impact upon both physical and ecological processes in UK estuaries. This review paper is structured as follows: In Section 2, we summarise past trends and future projections of UK coastal climate change, relevant to physical estuarine processes; Sections 3 and 4review observed trends and projected impacts of climate change to the physical and ecological processes in UK estuaries, respectively; finally, discussions and conclusions are summarised in Sections 5 and 6, respectively.

CLIMATE CHANGES TO PHYSICAL FORCING

We focus on sea surface temperature (SST), sea-level rise and atmospheric shifts as the main climate drivers that influence UK estuarine environments and ecosystems, and their interactions. Table 1 summarises observed trends in the drivers over the past century and projected changes for the 21st century, and estimates the level of confidence associated with each projection. Fig. 2 illustrates the key expected spatial and seasonal variability across the UK, according to the UK Climate Change Projections Report (UKCP09; Lowe et al., 2009).

Table 1: Observed trends and future projections of climate change to UK estuaries. The level of confidence (low, medium, or high) has been ascribed, based on the amount of evidence and the level of agreement of the evidence.

Process	Observed trend for the UK	*CONFIDENCE* and information sources	Projected 21st century change for the UK	*CONFIDENCE* and information sources
Sea surface temperature	+0.7 °C (1971–2010)	*HIGH* IPCC (2013) Jenkins et al. (2009)	+1.5 °C to +4 °C	*HIGH* Lowe et al. (2009)
Sea-level rise	+1 to +3 mm yr^{-1} during last century	*HIGH* Cazenave and Nerem (2004) Church et al. (2004) Woodworth and Blackman (2004) Church and White (2006) Menendez and Woodworth (2010) IPCC (2013)	+0.44 m to +0.74 m (Steric and eustatic processes) +1.9 m (Ice sheet melt)	*HIGH* Worth et al. (2006) Lowe et al. (2009) Woodworth et al. (2009) Lewis et al. (2011) IPCC (2013) *LOW* Jevrejeva et al., 2014 Knight et al., 2015
Storm surge	Pole-ward shift of storm tracks. Increased intensity/ decreased frequency extra-tropical cyclones. High variability in storm surges Clustering of extreme sea-levels	*MEDIUM* McCabe et al. (2001) *HIGH* Paciorek et al. (2002) Geng and Sugi (2003) Woodworth and Blackman (2004) Woodworth et al. (2009) Menendez and Woodworth (2010) *MEDIUM* Allan et al. (2009) *LOW* Wadey et al. (2014).	Centennial changes in extreme water levels are marginally significant (i.e., are of the same order as the natural climatological variability).	*LOW* Bijl et al. (1999) Alexandersson et al. (2000) Wakelin et al. (2003) Worth et al. (2006) Debernard and Røed (2008) Lowe et al. (2009) Howard et al. (2010) Lewis et al. (2011)
River flows	1969–2008: Increased flow during autumn and winter, decreased flow during spring, no trends during summer Rainfall (river flows) occurring in clustered events	*HIGH* Keef et al. (2009) Hannaford and Buys (2012) Hannaford (2015) Watts et al. (2015) *LOW* Jenkins et al. (2008) Burt and Ferranti (2012) Jones et al. (2013) Watts et al. (2015)	Increase in winter mean flow by up to 25% Decrease in summer mean flow by 40–80%	*MEDIUM* Fowler and Kilsby (2007) Jenkins et al. (2009) Prudhomme and Davies (2009a; 2009b) Fowler and Wilby (2010) Christierson et al. (2012) Prudhomme et al. (2012) Smith et al. (2013) Kay et al. (2013a,b)

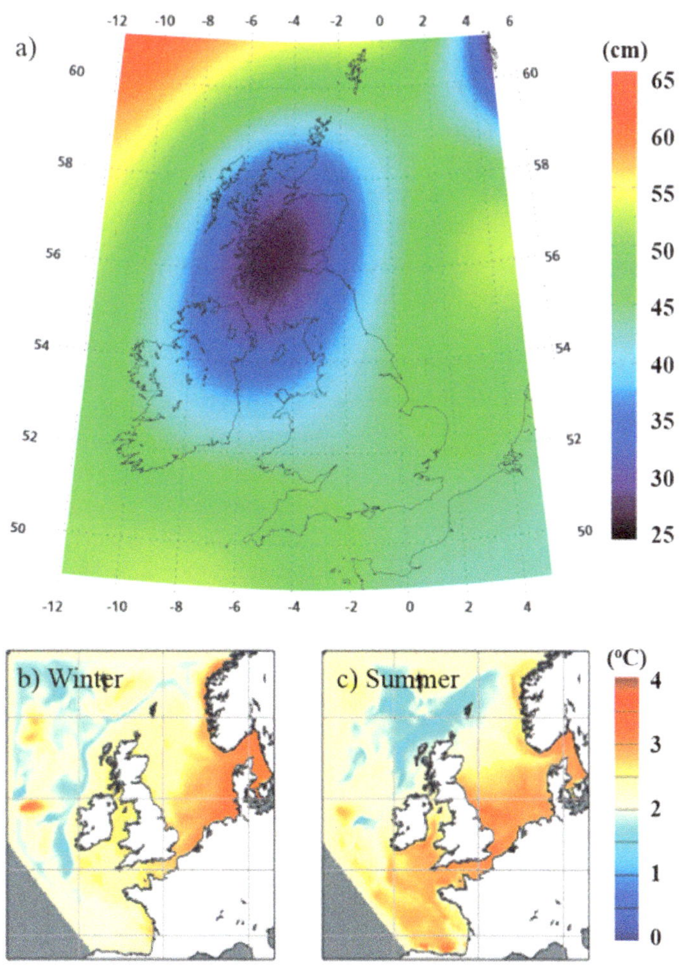

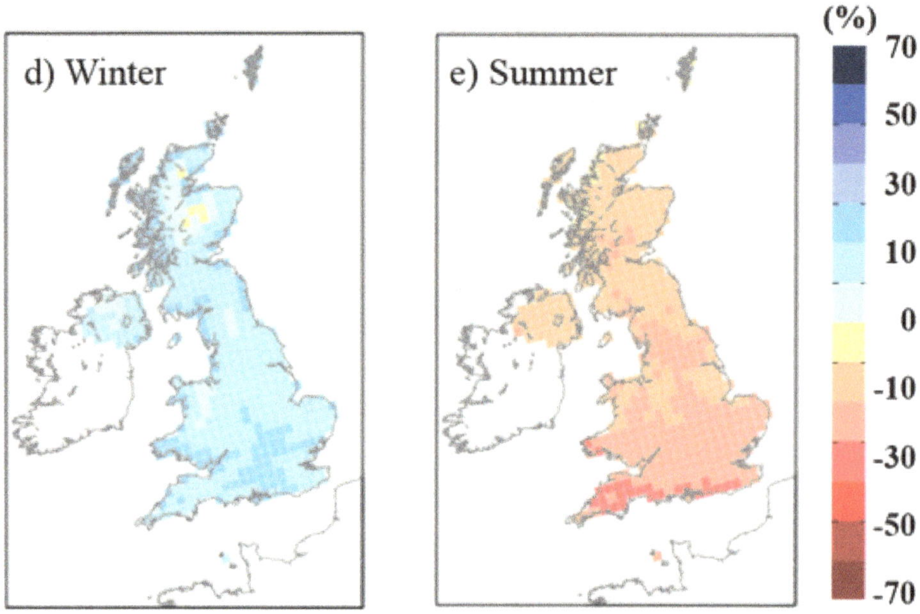

Figure 2: A selection of UKCP09 projected changes to UK climate drivers over the 21st century, showing spatial and seasonal variability (all figures published by Lowe et al. (2009)). (a) Shows relative sea-level change (cm) projections by 2100 (combined land movements and absolute sea-level change estimates from the central simulation (out of 11 ensemble members) for the IPCC medium emissions scenario). The middle panels show simulated seasonal mean trends in SST (°C), from (1961–1990) to (2070–2098), for (b) winter and (c) summer. Bottom panels show projected changes in mean precipitation (%) during (d) winter and (e) summer, at the 50% probability levels, for the 2080s under the IPCC medium emissions scenario.

Temperature

The warming of the global climate system is unequivocal (Jenkins et al., 2009). Data reveals that SST has warmed since the 1970s, at a rate of 0.11 °C per decade, and there is some evidence that SST has warmed since the 1870s (IPCC, 2013). Despite considerable year-to-year and even decade-to-decade variability, Jenkins et al. (2009)have shown that there is a clear UK coastal SST increase of about 0.7 °C over the past three decades, based on several long-term and short-term data sets.

Climate models project global mean SST to increase by 0.6–2 °C by 2100, depending on emission scenarios (low, medium, or high) and taking into account model uncertainty and seasonal variability (IPCC, 2013). Over the UK shelf edge regions and northern North Sea, the UKCP09 report projected temperature increases of 1.5–2.5 °C to occur this century, with larger increases of 2.5–4 °C in the Celtic, Irish and southern North Sea – a trend of 0.3 °C per decade (Lowe et al., 2009), e.g., Fig. 2b and c. However, a potential shutdown of the North Atlantic current (Gulf Stream) would result in much lower winter SST in the UK – but consensus among current climate model projections is that this is considered a low-probability, high-impact event (Wood et al., 2003). There has been a 1–3 °C rise in river water temperatures over UK and Europe during the past 100 years; since river water temperatures closely positively correlate with air temperatures (Whitehead et al., 2009).

Sea-level Rise

Global mean sea level has risen at a rate of 1–2 mm yr^{-1} since 1900 (Church et al., 2004 and IPCC, 2013), with an apparent change in rate to 3 mm yr^{-1} during the past 30 years, determined from satellite altimetry measurements (Cazenave and Nerem, 2004 and Church and White, 2006) and analysis of tide gauges worldwide (Woodworth and Blackman, 2004 and Menendez and Woodworth, 2010). Climate models, assuming different emission scenarios, projected with "medium confidence" that global mean sea levels will increase by 44–74 cm by 2100 (IPCC, 2013); an increase on the 18–59 cm previously projected by IPCC (2007). Alternative estimates based on vast ice sheet melt suggest, with low confidence, that sea level could rise by 1.9 m by 2100 (Jevrejeva et al., 2014), although recent studies suggest this trend is increasingly more likely (Knight et al., 2015).

Sea-level rise is not spatially uniform, with variability due to steric and eustatic processes, as well as vertical land movements such as isostatic adjustment (e.g.,Shennan and Horton, 2002). It is this relative mean sea-level which is most important for coastal communities (Lewis et al., 2011). In the UK, observed sea level trends are broadly consistent with the global average (Woodworth et al., 2009). Although, the effects of vertical land movements after the last ice age will augment this trend in the far southwest UK (by 1.2 mm yr^{-1}) and reduce it in parts of Scotland (by +2 mm yr^{-1}) (Shennan and Horton, 2002). Therefore, estuaries in southeast England are most at risk from the combined effects of land sinking and sea levels rising (Davidson et al., 1991), e.g., Fig. 2a. What is clear from recent analysis is that relative mean sea level around the UK is rising, albeit at a spatial varying rate, which could have significant consequences on estuarine systems (e.g., Robins et al., 2014), and shall be considered in Sections3 and 4.

Storm Surges

Analysis of atmospheric observations, and application of this data to models, show that the second half of the twentieth century experienced a pole-ward shift in the majority of storm tracks across the Northern Hemisphere (McCabe et al., 2001). Further, there was an increase in intensity, but a decrease in frequency, of extra-tropical cyclones (Paciorek et al., 2002 and Geng and Sugi, 2003). However, storm surge incidents in the UK were un-related with changes to extra-tropical cyclones: extreme water levels were driven by a rise in mean sea-level, rather than an increase in storminess (Woodworth et al., 2009 and Menendez and Woodworth, 2010). Therefore, it is uncertain whether the UK storm surge frequency has changed beyond natural levels of variability (Allan et al., 2009) – yet, some limited evidence is now emerging of clustering of extreme sea-levels (Wadey et al., 2014). For instance, there was an exceptional number of extreme high waters during the 2013/14 winter in the UK (Wadey et al., 2014).

Mean sea-level rise will change the dynamics of storm surge generation and propagation (Mcinnes et al., 2003) and estuarine flood risk and inundation (Senior et al., 2002). Models indicate future centennial changes in extreme water levels, which pose the greatest flood risk to estuaries, will be moderate and of the same order as the natural climatological variability (Worth et al., 2006, Debernard and Røed, 2008 and Lewis et al., 2011). The uncertainty within atmospheric models is high due to the parameterization of sub grid-scale processes, such as cloud formations and inter-decadal variability (Howard et al., 2010 and Lowe et al., 2009), and quantification of decadal variability of storm surge climates (Alexandersson et al., 2000 and Wakelin et al., 2003); therefore, future changes to the storm surge climate in the UK appear to be unresolved at present.

Rainfall and River Flow

River flows represent the integrated response of all hydrometeorological processes acting upon a catchment, and are therefore a key indicator of potential impacts of climate change to UK estuaries (Hannaford, 2015). Atmospheric warming has been shown byWatts et al. (2015) to intensify the hydrological cycle; although there is a complex, non-linear process leading from increasing temperatures, through changes in precipitation to river flow response (Hannaford, 2015). Past trends show that, while annual average rainfall in the UK is unchanged since the 1960›s, winter rainfall has intensified and increasingly occurred in clustered events (Burt and Ferranti, 2012, Jones et al., 2013 and Watts et al., 2015). For the same period, river-flows have increased in winter also, especially in upland western areas; autumn flows rose and spring flows decreased slightly, while there was no consistent change in the summer (Hannaford and Buys, 2012, Hannaford, 2015 and Watts et al., 2015). For past extreme events, data from extensive UK river-gauge networks show that the relationship between precipitation and river flow is spatially variable and dependent on catchment characteristics (Keef et al., 2009).

Many studies have investigated potential changes to future UK river flow rates (e.g., Kay et al., 2009, Prudhomme and Davies, 2009a and Prudhomme and Davies, 2009b), including impacts to groundwater (e.g., Jackson et al., 2011), and more recent studies have analysed the latest UKCP09 future river flow data series (1 km-resolution and catchment-bias-corrected; Prudhomme et al., 2012). The scientific consensus on projected future UK precipitation patterns is the "wetter winters and drier summers" signal (Arnell, 2003 and Christierson et al., 2012), with some spatial variability in the UK to this pattern. Indeed, the UKCP09 Report explains that little change is expected in the median precipitation amount by 2100, but significant changes in the trends of winter and summer precipitation are expected (Jenkins et al., 2009) (Fig. 2d and e). The report generally found a slight increase in the winter mean precipitation and a decrease in summer mean precipitation for the western UK (Jenkins et al., 2009). River catchment models indicate river flows are projected to reduce in summer (by 40–80%) and increase in winter (by up to 25%), again particularly in mountainous regions of western UK (Fowler and Wilby, 2010, Christierson et al., 2012 and Prudhomme et al., 2012). These projections were biased through inability to capture monthly precipitation climates (Prudhomme and Davies, 2009a, Prudhomme and Davies, 2009b and Smith et al., 2013). Furthermore, when future precipitation patterns are projected through to future river flow rates there is considerable uncertainty in GCM projections (Prudhomme and Davies, 2009a), as well as downscaling and catchment model uncertainties (Kay et al., 2009 and Prudhomme and Davies, 2009b).

CLIMATE IMPACTS ON THE PHYSICAL ESTUARINE ENVIRONMENT

Flooding and Inundation

Increased estuarine flood risk and inundation occurrence are caused primarily by a combination of increased sea-level (sea-level rise, storm surge and changes to the wave climate) and changes to incidences of high river flow events (Lewis et al., 2013 and Quinn et al., 2014). Estuarine flooding can also be the volumetric combination of storm surge and extreme river flow that can drastically increase flood hazard (e.g.,Maskell et al., 2013). Further, there is evidence for a positive relation in some UK estuaries between storm surge and peak river flows (Samuels and Burt, 2002, Senior et al., 2002 and Svensson and Jones, 2002), suggesting that surges should to be taken into account for flood risk estimation (Keef et al., 2009 and Zheng et al., 2013). Both sea-level and river flow climates in the UK appear to be changing due to climate change, potentially resulting in increased flood risk to UK estuaries (King, 2004). Potential negative impacts of estuarine flooding to ecosystem goods and services, in the absence of mitigation, include: damage to infrastructure and population displacement (Hanson et al., 2011 and Nicholls et al., 2011), increased erosion and loss of land to the sea (Zhang et al., 2004), pollution and increased human health risk, and significant loss of wetland habitats and ecosystems (Day et al., 1995 and Nicholls et al., 1999).

Modelling approaches are important for predicting flood risk, assessing impact, and mitigation decision-making – where risk is generally defined as the probability of the occurrence multiplied by the consequences. It is important to simulate the nonlinear coupling of surge-wave-river events, since static (sea-level-only) events will underestimate the cost significantly (Prime et al., 2015). Using published climate change scenarios, a UK panel (King, 2004) found that a combination of sea-level rise and increased storminess will allow storm surges to propagate much further inland, so that UK estuaries will be subject to both higher water levels and more energetic wave attack. However, Worth et al. (2006) simulated no significant increased storm surge on the east coast of the UK over the next century. Although the likelihood of extreme sea-level rise (up to 1.9 m over the 21st century) due to rapid ice sheet melt is apparently low (Jevrejeva et al., 2014), the consequence would be an increase to the projected risk; for the case of the UK›s largest estuary – the Severn, the projected flood risk has been reported by Quinn et al. (2013) to increase by 30%.

A key consideration for estuarine flood risk and inundation occurrence, under accelerating rates of sea-level rise is the potential for geomorphological features and flood depths to adjust, through sediment transport (i.e., the sediment flux into or out of an estuary), estuarine morphology (erosion and accretion), and sediment compaction (Dickson et al., 2007 and Shennan et al., 2012). Sediment infilling will cause flood risk to increase incrementally over long (decadal) timescales, in which case mitigation measures can be foreseen and employed in advance to reduce risk. However, based on analysis of historical data, Pye and Blott (2006) showed that morphological changes on the Suffolk coast (around Blyth, Ore-Alde-Butley, and Deben Estuaries) over the past 50 years have actually been lower than historically. Instead, rapid morphological changes (and potential increased flood risk) will result from extreme surge/wave/river events (Horrillo-Caraballo et al., 2013). Interestingly, Dissanayake et al. (2015) explain that the initial storm in a cluster of storms has most morphological impact on the coastline. In addition to these natural trends, shoreline protection has greatly reduced the input of new sediment sources, resulting in much contemporary coastal erosion (Nicholls et al., 2013). The capacity for estuaries to respond to sea-level rise will depend on landward constraints to transgression, available accommodation space and available sediment supply (marine and fluvial) (Nicholls et al., 2013). Rossington et al. (2007) predicted the maximum rate of sea-level rise that UK estuaries can experience without losing all intertidal areas. Their results suggest that some estuaries are more vulnerable to sea-level rise (e.g., Southampton Water) and others are more resilient (e.g., the Ribble), depending on estuary area, sediment supply, sediment transport potential and human interference. Some studies of intertidal and saltmarsh areas suggest low resilience to sea-level rise (e.g., Orford and Pethick, 2006, French, 2006 and Marani et al., 2007).

Projected increases in UK average temperatures will increase evapotranspiration throughout many UK catchments, which may alter vegetation types and reduce soil moisture – all of which could reduce runoff into estuaries and potentially dampen flood risk slightly (Boorman, 2003). As an example, modelled catchment runoff from IPCC climate change scenarios in the Ouse-Humber Estuary showed increased evapotranspiration (5% in winter, 15% in summer), although total runoff generally increased by up to 30% in winter due to projected increases in rainfall (Boorman, 2003).

From our review, it is clear that uncertainty is high in predicting future changes to physical drivers in UK estuaries, due to the spatial variability associated with projected changes to rainfall patterns. Furthermore, considering GCM model uncertainty, downscaling uncertainties, hydrological model uncertainties and land use changes, as well as variability of the climate system (e.g., Kay et al., 2009, Prudhomme and Davies, 2009a,Prudhomme and Davies, 2009b and Christierson et al., 2012), prediction of future estuary changes requires further research to determine confidence in future river flow rates. Furthermore, surge–rainfall association requires further investigation (Zheng et al., 2014) and their implications to biology, water quality and sediment transport processes (Lowe et al., 2009 and Robins et al., 2014).

Hydrodynamics and Mixing

In the UK, the majority of the main estuaries are relatively shallow, and are classified as either mesotidal, macrotidal or hypertidal, in terms of tidal forcing (Davidson et al., 1991 and Prandle, 2009). The strong tidal forcing produces a vertically well-mixed water column, though weak vertical stratification is sometimes evident, e.g., approaching slack high water in the Conwy (Howlett et al., 2015), or more commonly during the ebb tide depending on the river flow (Pye and Blott, 2014). Typically, the tide becomes asymmetric upon entering the estuary due to shallow water frictional effects – resulting in a shorter, stronger flooding phase and a longer, weaker ebbing phase (Dronkers, 1986 and Moore et al., 2009). Influences of waves on hydrodynamics and mixing are usually small within the estuary due to wave dissipation over the UK shelf, particularly over offshore sand banks (Carini et al., 2015), and prevented from penetrating into bar-built estuaries because of the constricted entrance. Of high importance for hydrodynamics and mixing are variabilities in river flows – particularly due to the nature and frequency of extreme flow events (Struyf et al., 2004). Rivers in the UK tend to rise very quickly after rainfall, especially in the north and west, but flowrates reduce more slowly; as such, UK estuaries are often in a state of non-equilibrium in terms of stratification and salt balance, and nutrient/material fluxes; processes which are critical for ecosystem goods and services (Prandle, 2009).

Consequently, projected alterations to sea-levels and river flows due to climate change could have significant implications for estuarine functioning (Whitehead et al., 2009). Modelling in the Conwy by Robins et al. (2014) shows interaction of sea-level rise with maximal projected changes in river flow over the 21st century will potentially increase the saline intrusion length. Sea-level rise combined with long periods of low flow conditions (drought) simulated the saline intrusion and turbidity maximum to be pushed further up-estuary from today's limit and retaining river-borne substances in the estuary for extended periods. On the other hand, increased river flows simulated the saline intrusion to be pushed down-estuary, and potentially flush nutrients and pollutants offshore (Robins et al., 2014). These findings relate to similar studies in the US by Yang et al. (2015). The extent of the saline pumping effect is dependent on the strength and frequency of river flow events, which are projected to be more variable in the UK with climate change. Correspondingly, Prandle and Lane (2015) investigated 96 UK estuaries and concluded that a sea-level rise of 1 m will increase the saline intrusion length by >7% in deep estuaries and by >25% in estuaries shallower than 10 m. Further, a change in river flow of 25% (either increase or decrease) will have significant effects on both vertical mixing and salinity intrusion. Accordingly, estuarine recovery after rainfall events restores a steady state salt distribution (MacCready, 1998 and Hetland and Geyer, 2004). However, few observations of estuarine tidal-pumping/recovery have been documented in the UK (e.g., Simpson et al., 2001) and there is generally a lack of knowledge about how estuarine hydrodynamics react to extreme climatic events for UK case studies.

Estuarine hydrodynamics in the majority of UK systems is further complicated by the presence of estuarine fronts – small-scale regions which exhibit significant changes in hydrographic variables (O›Donnell, 1993). Estuarine fronts have significant roles in estuarine circulation, and affect sediment grain size distribution (Neill, 2009) and larval dispersal (Robins et al., 2012). Since estuarine fronts are driven by horizontal salinity gradients (controlled by river flow and tidal range), it is anticipated that the role of fronts in estuarine processes will become increasingly important within the context of 21st century climate change, when extremes in river flow, coupled with increased estuarine tidal range due to sea-level rise, are likely.

Sediment Transport and Morphology

The development of UK estuaries has typically resulted from the latest post-glacial rise in sea level (of the order hundreds of metres), which inundated the UK coastline and drowned river mouths (Dyer, 1997). However, Tessier et al. (2012) have demonstrated that the majority of morphological development across northwest European estuaries has occurred following centennial-scale periods of major climate change (enhanced storminess), in addition to the (lesser) continual effect of relative sea-level rise, given an available sediment supply. In the present climate, the morphological development of such systems, other than impacts caused by human activities, has been determined by the changing relative importance of the river, wave and tidal influences (Prandle, 2009), with tidal flow asymmetry being considered as the main factor influencing net sediment transport and hence being responsible for the stability of the system (Hoitink et al., 2003). Both the magnitude and duration of the flood and ebb tides affect the morphological outcome and feedback effects can keep an estuary in a state of dynamic equilibrium, oscillating between morphological states (Pethick, 1994), subject to the influence of sea-level rise. According to Lanzoni and Seminara (1998) and more recently Pye and Blott (2014), deep or bar-built estuaries tend to generate ebb-current dominance, whereas in shallow or wide-mouthed estuaries, flood-current dominance occurs, which will affect sediment transport pathways. But classification of flood/ebb flow asymmetry (e.g.,Friedrichs and Aubrey, 1988) does not necessarily give an accurate indication of the net sediment transport.

State-of-the-art, process-based, sediment transport models are believed to have the correct general behaviour over the wide range of wave-current conditions encountered on site. However, the morphological models in which these processes are embedded are capable of predicting only the statistics defining channel geometry and meandering on time scales of decades or longer. The seabed is inherently unstable and morphological modelling outcomes are capable of reconstructing only the distribution of channels and sand flats that characterise many shallow tidally-dominated estuaries. Small bathymetric irregularities introduced at the start of a numerical simulation evolve into larger scale bed features. McCann et al. (2011) demonstrated how a tidal channel/sand flat network appears spontaneously and exhibits 'self-organisation' linked to the formation of ebb-dominant channels and flood-dominant shoals. In McCann's simulations sediment was exported seaward in the deeper channels, while net transport was shown to be landward (i.e., up-estuary) on the tidal flats.

Robins and Davies (2010) attempted to systematise this general behaviour, classifying estuary regions as 'flood-dominant' if the asymmetry in the tide causes a net, up-estuary, sediment accumulation (because the stronger flood flow may be above the threshold of sediment motion for longer than the ebb flow). They referred to the opposite behaviour as 'ebb-dominant'. They demonstrated that an estuary is likely to become more flood-dominant as the tidal wave travels landward, while providing clear evidence of the role of ebb channels in promoting ebb dominance in net sediment transport. Their modelling suggested that, for idealised estuary shapes characterising mesotidal systems found along the UK coast, net flood-dominance in sediment transport occurs in regions of shallower water (e.g., tidal flats and salt marshes in the upper estuary) where $a/h > 1.2$ (a = local tidal amplitude, h = local, cross-sectionally averaged, water depth), while net ebb-dominance occurs in regions of deeper water (Robins and Davies, 2010). However, the magnitude of net ebb-dominance was less pronounced than the net flood-dominance. Sea-level rise caused the threshold value of a/h to be reduced, with the estuary shifting from ebb-dominance towards flood-dominance, with less transport overall. Reducing the sediment grain size increased both the gross and net sediment transport but did not significantly affect the transport pattern. Importantly, the inclusion of an ebb channel promoted ebb-dominance in the channel. Applying these outcomes to a realistic case (the Dyfi Estuary), Robins and Davies (2010) concluded that, due to a projected sea-level rise of 1 m in the next 100 years, greater flood-dominance (i.e., net sediment accumulation into the estuary) but less sediment transport overall would occur. In a similar vein, Pye and Blott (2014) postulated generally that sediment accumulation would continue until the creation of a central channel and high flats, which would switch the system from flood-to ebb-dominance. Pye and Blott (2014) also stated that the effects of extreme river events and storm surges on the long-term sediment estuarine budget remain unclear and require further research.

The mobilisation of historic, consolidated, heavy metal contaminants in semi-industrialised UK estuaries (e.g., from mudflats and saltmarshes in the Humber and Mersey) poses a serious risk to water quality and human health (Lee and Cundy, 2001). On the other hand, a potential positive outcome is that this re-mobilisation, which is often a result of extreme stormy weather, accelerates sediment exchange with the open sea (e.g., the exchange of mercury-rich fine sediments from the Mersey Estuary into Liverpool Bay (Harland et al., 2000)).

Fluxes of Nutrients

Biogeochemical cycles vary over both time and space. There is a requirement in the UK to further understand such variability and nutrient coupling and processing in different catchment types to aid development of mitigation actions under climate change. Temporal changes include diurnal cycling and event based changes between high and low flow conditions, seasonal, decadal and longer. Spatially, estuarine biogeochemical cycles form a continuum from freshwater to marine, transporting and processing many forms of macronutrients with discontinuities in the retention and storage of macronutrients (Statham, 2012).

Estuaries, therefore, play a crucial role in biogeochemical cycling of nutrients, which in turn influence the growth and resilience of plants and animals (Nedwell et al., 1999; Malham et al., 2014).

The impact of nutrient loading in UK estuaries depends on the physical and geological setting (Jickells et al., 2000) and the physical and chemical conditions prevalent at any particular time. Importantly, projections for the UK suggest that winter river flows will increase by between 4% and 9% by 2020, with summer decrease of roughly 11% (1–32% range), based on IPCC climate change projections (Whitehead et al., 2009). Nutrient projections suggest a possible rise of macronutrient concentrations (such as nitrogen and phosphorus) under climate change conditions, with consequent impacts of augmented eutrophication, hypoxia and harmful algal blooms (Tappin, 2002 and Statham, 2012). Projected 21st century sea-level rise, combined with reduced river flows (projected during summer in the UK), could heighten the likelihood of estuarine nutrient trapping, in between river flushing events (Robins et al., 2014). Conversely, nutrient retention will be less likely during winter if projected increases in river flow frequency occur, which could counteract the effects of sea-level rise to some extent (Struyf et al., 2004). Nutrification in summer is more prone to cause eutrophication than in winter: only summer production is nutrient limited; winter production is light limited. Thus, the dry summer effect has the greatest potential for augmenting eutrophication.

However, current understanding, let alone prediction, of biogeochemical processes in estuaries is highly complex and requires further study (Jickells et al., 2000, Andrews et al., 2006, Najjar et al., 2010 and Statham, 2012). Overall, more is known about fluxes of inorganic than organic nutrients (Jarvie et al., 2012). Further research is needed on bacterial cycling of organic nutrients (Asmala et al., 2014), nutrient reactivity (Valdemarsen et al., 2014), and on nutrient speciation and interaction with fresh and salt water under different pH and dissolved organic matter conditions (Morgan et al., 2012). Catchment-estuary model coupling methods are improving (Uncles, 2003 and Huang et al., 2013), and there is currently a UK drive in data generation (e.g., NERC Macronutrients Cycles Programme), which will prove valuable for model validation.

CLIMATE IMPACTS ON ESTUARINE ECOSYSTEMS

Human Health: Pollutants and Pathogenic Microorganisms

By 2100, an increase in average temperatures, together with an increase in winter rainfall and longer drought periods will ultimately have impacts upon public health risk to UK estuaries, associated with water-borne and food-borne diseases caused by microbial pathogens (McMichael et al., 2006). There are over 100 recognised types of microbial pathogen found in contaminated water used for recreation and as potable water (Rose et al., 2001), with the potential for food-borne transmission via seafood (Marques et al., 2010). Projected increased precipitation and flooding may result in untreated sewage discharge via combined sewer overflows (CSOs) and increased agricultural runoff (Perciasepe, 1998). In addition, toxic pollutants, ammonia, organic solids and oxygen-demanding substances are also discharged into the water system (Friedman-Huffman and Rose, 1999 and Stachel et al., 2004) and may increase concurrently with high precipitation events. There are precedents for precipitation events preceding outbreaks of water-borne disease (Curriero et al., 2001 and Rose, 2005); for example, outbreaks of the protozoan pathogen *Cryptosporidium* have been observed following extreme precipitation events, which reduces the efficiency of water treatment processes resulting in increased human health risk (Rose et al., 2001). In addition, faecal indicator organism concentrations have been found to be higher during a winter of high precipitation, when compared with the rest of that year (Lipp et al., 2001), and high precipitation also results in an increase in the viral load of estuarine environments, particularly after CSOs, where a significant increase in Adenovirus and Norovirus concentrations have been observed in receiving waters (Rodríguez et al., 2012). It is well established that increased rainfall may influence the concentrations of toxic metals, organic chemicals, algal toxins, and human pathogen contaminants in seafood (Marques et al., 2010). Conversely, drought conditions will result in decreased river flow and lower concentrations of microbial contaminants (Rose et al., 2001). Sea-level rise may lead to the loss of coastal wetlands, which filter nutrients, microbial agents, and chemical agents (McMichael et al., 2006). However, the implications of sea-level rise on the fate of microbial, nutrient and chemical agents and their interactions with the wider ecological dynamics of the system are poorly understood, and could be addressed by integrated catchment-river-estuary models (Rose et al., 2001).

The cholera model perhaps represents the best-studied mechanism for understanding the role of climate on infectious disease transmission (Lipp et al., 2002). *Vibrio cholerae*(the bacterial species responsible for cholera) and other *Vibrio* spp. are endemic in marine and coastal environments and demonstrate strong associations with warmer sea surface temperatures that drive their seasonality and ecological distribution (Rose et al., 2001). Furthermore, favourable environmental conditions (dependent on nutrients and temperature) promote the formation of algal blooms that act as a reservoir for *Vibrio* spp. and can cause reversion from a quiescent to infectious state (Rose et al., 2001). Future temperature shifts, therefore, have the potential to expand the prevalence of *V. cholera* at both temporal and geographical scales (Lipp et al., 2002). However, cholera outbreaks have disappeared from much of the developed world (including the UK), and the endemicity of cholera in the tropics and subtropics exemplifies the favourable environmental conditions that drive their ecological distribution. In contrast to cholera, which is autochthonous in estuarine systems and favours dry weather and warmer temperatures, the prevalence of other bacterial and viral pathogens of allochthonous origin are governed by different conditions. For example, faecal indicator organisms demonstrate enhanced survival in estuarine environments at lower temperatures and their abundance has a strong correlation with high precipitation events (Lipp et al., 2001 and Malham et al., 2014). Further research is required to understand the projected effects of future climate scenarios on estuarine pathogen load and subsequent human health risk.

Estuarine Habitats

There is little evidence-based research on the effects of climate change drivers on estuarine habitats, except for the effects of temperature and sea-level rise (Mieszkowska et al., 2013 and Fagherazzi et al., 2012). Here, we highlight current knowledge for the most researched habitats, the majority of which are intertidal.

Many UK salt marshes were lost to agricultural expansion and coastal development in the 18th and 19th centuries (Doody, 2008). The total area of 40,522 ha in England and Wales was still declining, by <100 ha y^{-1} at the last survey (1989–2009; EA, 2011), mainly due to losses in the southeast associated with isostatic adjustment. Other geographical areas had marsh expansion over the same period (EA, 2011) and it is not known whether these patterns of change are associated with emergent climate change. In principle, climate change might be both 'good' and 'bad' for salt marshes. A moderate rise in the annual average temperature can boost vegetation productivity and favour marsh expansion, although a greater frequency of dry summer spells could depress plant growth through desiccation and evaporation-driven increases in sediment salinity (Gedan and Bertness, 2010). Increased frequency of high rainfall events is likely to boost riverine sediments supply to marshes, which is a key stimulant of marsh vertical growth (Fagherazzi et al., 2013). Conversely, increased storminess could accelerate erosion of marshes located in wide estuaries, where the fetch, and thus potential wave energy, is higher. Evidence suggest marshes are well-capable of keeping pace with sea-level rise if given the space to transgress inland, and if exceptional land subduction and depressed sediment supplies are absent (Fagherazzi et al., 2012). Coastal squeeze remains the greatest threat to salt marshes in terms of sea-level rise, the elimination of which is linked to policy decisions, such as shoreline management planning. Marsh vertical and horizontal growth is strongly dependent on sediment supply, which, if diminished, can switch marshes from accreting to eroding and preclude their keeping pace with sea-level rise (Bouma et al., 2014). Section 2 showed sediment supply is governed by hydrodynamics, determining for instance the up-estuary transport of marine sediments, as well as the down-estuary transport of catchment and riverine particles. Tendencies of increased fluvial transport of sediments bode well for the capacity of marshes to keep pace with sea-level rise. Emergent evidence for the past 70 years show that many marshes on the west coast of England and Wales have undergone dynamic changes in areal extent, but the collective area has expanded, particularly since the 1960s (Martin Skov, *pers. comm.*). These changes are not linked to isostatic adjustment, which is commonly invoked as a cause for national variation in marsh erosion rates (EA, 2011). Future projections of changes to estuarine function might do well to carefully consider the causes for such regional differences in ecosystem change, and the implications to estuarine management.

Rocky shores in the UK are predominantly located on open coasts, although they do occur in some estuaries. The following findings are mainly from coastal shores, but many are likely to be transferable to estuarine settings. The area cover of intertidal rocky shores is likely to decline with sea-level rise: as many shorelines increase in steepness with elevation, the distance between the low and high tidal mark, and thus the intertidal area, will decrease with sea-level rise (Jackson and McIlvenny, 2011). Temperature rise is already having some effects. Throughout the UK, a number of rocky shore species have been gradually moving north in response to rise in temperature, punctuated by cool years in which the advance has ceased, or reversed (Mieszkowska et al., 2013). The tendency is for species that are intolerant to warming to retreat northwards to remain in cooler conditions, while Mediterranean/Lusitanian species are coming up from the south, facilitated by an increasingly benign climate (Mieszkowska et al., 2013). Similar changes are occurring in other estuarine habitats, such as *Sabellaria* reefs (Frost et al., 2004), as well as in terrestrial and marine systems (Hiddink and Hofstede, 2008). Changes to community composition brings with it the potential for change in biodiversity and ecological functioning (organic production, degradation, habitat provisioning), as altered composition and biodiversity changes the interactions between species and their ecological roles (Balvanera et al., 2006). Rocky shore habitats are competed for by algae and grazing molluscs, and the relative dominance of these groups is determined by environmental conditions: increased wave exposure and higher temperatures favour animals (Hawkins et al., 2008). Thus, predictions are that algae will become less prevalent in a warming and more hydrologically energetic climate, which could diminish overall biological production of estuarine rocky shores (Hawkins et al., 2009).

Seagrass beds in Europe have dramatically declined over the past century, by 50–80% (Short and Wyllie-Echeverria, 1996). In the UK, there is no research-based evidence on how the area cover of seagrass beds will respond to climate change (Mieszkowska et al., 2013), although they are considered to be vulnerable to a plethora of changes associated with climate change (Jones et al., 2011). Current patterns of change are variable: there are indications of areal increases in Scotland and Wales (Smith et al., 2002 and Howsen, 2009), but retreat in the Isle of Wight (Mieszkowska et al., 2013). Seagrasses elsewhere have suffered from climate-related changes in temperature, storminess and prevailing winds (Bjork et al., 2008). Mediterranean seagrass beds are predicted to decline in response to temperature stress (Marba and Duarte, 2010). In the UK, forecasted temperatures are well within the tolerance levels for *Zostera noltii* (37 °C, Massa et al., 2009); however, *Zostera marina*, the other UK species, stops growing at >20 °C and starts dying at > 25 °C (Greve et al., 2003), and dieback during summer heat waves have been observed as far north as Denmark (Greve et al., 2003 and Winters et al., 2011). It is not unreasonable to expect similar patterns in the UK (Mieszkowska et al., 2013).

Sedimentary habitats account for 90–95% of estuarine intertidal habitats and are powerhouse production areas of invertebrates that feed higher trophic levels, including birds (McLusky and Elliott, 2004). Sediment habitats are already undergoing geomorphological change from climatic drivers including increasing wave height, surges and sea-level rise (Lowe et al., 2009); they are likely to respond to changes in river flow, although there are no detailed studies to support this prediction in the UK. A projected more energetic wave climate is likely to steepen benthic profiles (Mieszkowska et al., 2013) and diminish the abundance and diversity of sediment living animals through coarsening of sediments (Pethick and Crooks, 2000, Fujii and Raffaelli, 2008, Yates et al., 2010 and Fujii,

2012). Mudflats are thought to be particularly sensitive to climate change, through an assortment of drivers, including shifts in sediment supply and sediment coarsening (Gubbay and Earll, 2010 and Jones et al., 2011), and the risk of coastal squeeze as enhanced wave and tidal energy require mudflats to move up-estuary to less energetic areas (Pethick and Crooks, 2000). Modelling indicates an upstream shift in the position of the most productive intertidal flats of the Humber estuary could reduce estuarine production by 23%, because the area for flats is reduced by the narrowing of the estuary (Fujii and Raffaelli, 2008 and Fujii, 2012). Accurate predictions are difficult to make given the influences of factors such as inter-estuary variation in geomorphology, which regulates hydrological forcing and sediment supply, spatial variation in catchment processes, which regulate sediment supply and water chemistry, and the interactions of biological and physical variables on benthic organisms (Fujii, 2012). For instance, a decline in productivity with sediment coarsening might be countered by increased temperature and milder winters that boost benthic production, (Fujii, 2012) and/or by upper reaches of estuaries becoming more saline and accommodating of invertebrate diversity and biomass (Ysebaert et al., 2003).

The honeycomb worm *Sabellaria* sp. makes bulky biogenic reefs in intertidal and subtidal rocky areas. Reefs are colonised by an array of invertebrates and are a priority habitat for protection in the UK. *Sabellaria* is a 'warm-water' genus and the reefs have been extending northwards in the UK in recent decades, as a response to less severe winter temperatures (Frost et al., 2004); a pattern which is likely to continue into the future. Temperature rise is unlikely to markedly affect the distribution of UK mussel beds, as predicted conditions will well within their temperature tolerance; however, mussel condition might well be affected by eutrophication (Mieszkowska et al., 2013).

The potential impact of climate change on estuarine habitats is somewhat determined by their physical nature (biogenic, non-biogenic, hard, soft) and position (intertidal, subtidal, location in estuary). The persistence of biogenic habitats, such as mussel beds, is governed by the physiological tolerances of the habitat-forming species. Non-biogenic systems can continue to exist despite climate change, although some might become biologically unavailable. Thus, rocky shores will persist as a habitat, but are projected to decrease in area due to coastal squeeze (Jackson and McIlvenny, 2011). Intertidal habitats are naturally more influenced by short-term changes in weather than sub-tidal habitats. Ultimately, changes in community composition induced by climate change might affect the resilience of systems to other sources of environmental change, such as eutrophication and coastal development, although the effects of multi-stressor impacts on UK estuaries is limited. Major research gaps also exist on how biological feedbacks that affect physical conditions in estuaries will be influenced by sea-level rise. For instance, saltmarsh colonisation alters the forcing and distribution of hydrological energy, which facilitates further expansion of marshes and ultimately influences the tidal prism and sediment distribution. Whether or not such biological feedback mechanisms will be enhanced or reduced by change in climatic conditions is currently not known.

Larvae in Estuaries

Environmental cues synchronize the reproductive cycle of many marine invertebrates and fish (Kingsford et al., 2002). A broad diversity of taxa have the capacity to detect variations in current direction, magnetism, water pressure, water chemistry (e.g., amino acids, salinity), sound and vibration (e.g., waves breaking or fish assemblages), and light gradients (Forward et al., 2001). Some organisms can detect multiple cues (e.g., decapods and fishes) and are likely to have integrated sensory responses (Kingsford et al., 2002). The type of response may be endogenous and regulated by an internal clock (e.g., tidal stream transport), or a direct response to the environment (e.g., diel transport) (Queiroga and Blanton, 2005). Further, the phasing and extent of vertical migration both change throughout ontogeny (Queiroga and Blanton, 2005). These cyclic cues may affect a number of reproductive parameters, including sex determination, gametogenesis, spawning, transport, settlement, retention, connectivity, and metamorphosis (Lawrence and Soame, 2004).

For estuarine species, primary environmental cues are tidal currents and photoperiod. European shore crab larvae in estuaries within the North Sea use selective tidal stream transport to swim to the surface during the flood phase of the tide, which aids in-shore dispersal towards estuarine feeding grounds (Moksnes et al., 2014). Conversely, decapod crustacean larvae may swim to the surface during the ebb phase of the tide and migrate offshore (Queiroga and Blanton, 2005). In other regions, shore crab display nocturnal (diel) vertical migration to avoid predation (Moksnes et al., 2014). Estuarine species live in a highly dynamic environment where recruitment variability is a key determinant of population dynamics. Environmental requirements for successful recruitment may differ between co-occurring species, and therefore species may be advantaged or disadvantaged under climate change (Allen et al., 2008 and Jenkins et al., 2015). Climate change impacts on UK estuaries are likely to alter the environmental cues, which in turn will affect their reproductive parameters. For example, the phase relationship between temperature of the water column and photoperiod directly controls the metabolism and lifecycles of aquatic organisms, with most biological processes operating faster at higher temperatures, leading to shorter pelagic larval durations and time to metamorphosis (Whitehead et al., 2009 and González-Ortegón and Giménez, 2014).

Variability in salinity has been documented to both increase and decrease time to metamorphosis, depending on species (Forward et al., 2001). The estuarine copepod*Eurytemora affinis* (which exhibits a combination of tidal, salinity, and temperature cues) will change its migration in response to the location of the estuarine salinity maximum (Hough and Naylor, 1991). Indeed, evidence from Chaalali et al. (2013) shows climate-driven up-estuary migration of the species in the Gironde Estuary, France. The presence of aquatic vegetation and increased humic acids (a decomposition product of terrestrial plant material; Fox, 1981), both accelerate the time to metamorphosis of brachyuran crabs (Forward et al., 2001). Alternatively, cues that delay metamorphosis include hypoxia and chemical odours from potential predators (Forward et al., 2001). Thus, species that cue reproduction based on these signals are likely to be particularly vulnerable in the future (Lawrence and Soame, 2004). Consequently, climate change may cause local extirpations of populations in the extreme, or more likely impact fecundity, spawning success, and recruitment significantly.

However, change will ultimately depend on the relative speed of adaptation to climate change (Allen et al., 2008) and the degree of mixing between populations across the Metapopulation (Lawrence and Soame, 2004). Some larvae of estuarine species appear to be resilient to environmental variability and may be able to tolerate changes related with climate change. For instance, an important species, inhabiting brackish waters in the UK is *Palaemon varians*; this is a shrimp found in saltmarshes, coastal ponds and lagoons. Adults of *P. varians* can potentially extend the upper thermal range through acclimation (Ravaux et al., 2012). Larvae of this species appear to be exported and develop in estuarine-coastal areas (Fincham, 1979). Larvae show a wide range of salinity tolerance (5–42: Antonopoulou and Emson, 1988). In addition, larvae appear to tolerate high temperatures as well as food limitation. Metabolic rates increase with temperature (range: 5–25 °C), but total energy losses, calculated from oxygen consumption are lower than at low temperature. In this shrimp, energy reserves at the time of hatching appear to confer larvae a high capacity to tolerate food limitation (Oliphant and Thatje, 2014). Oliphant and Thatje (2013) and Oliphant et al. (2014) showed that there is sufficient variability among broods in per offspring investment to produce variability in development and starvation tolerance; if such variability is heritable, it may confer this species a sufficient capacity to adapt to variable conditions of food, temperature, and salinity.

Besides the hypothesised effect of environmental factors as cues, such factors operate as stressors, in particular in the early life history stages of marine invertebrates. Meteorological forcing, such as extreme rainfall events, promote salinity decreases in estuaries. Thus, low salinities can reduce growth rates especially during the larval stages, if these are weak osmoregulators or osmoconformer (e.g., decapod crustaceans:Torres et al., 2011). Low salinity, experienced by parents or at the egg/embryo stage, can have knock-on effects along the life cycle of estuarine organisms (Giménez, 2006,Jensen et al., 2014 and Chaparro et al., 2014). For instance, low salinity at the embryonic stage can affect the capacity of larvae to osmoregulate (Charmantier et al., 2002). Exposure of larvae to low salinity also affects juvenile size and growth rate (Giménez et al., 2004 and Rey et al., 2015).

It is likely that multiple stressors will operate on estuarine organisms in a synergistic way (Dolbeth et al., 2011 and Przeslawski et al., 2015). As a consequence of warming, temperature may modify the sensitivity to other environmental stressors, but also thermal sensitivity should be modulated by exposure to additional stressors (Sokolova and Portner, 2007). For instance, under low salinity, larvae may be more vulnerable to pollutants (González-Ortegón et al., 2013) or food limitation (González-Ortegón and Giménez, 2014). On the other hand, increased temperatures may increase the tolerance of low salinities (González-Ortegón and Giménez, 2014). Overall, an assessment of the impact of changes in estuaries will require an understanding of the correlated changes in environmental variables such as salinity and temperature.

Biota

Species living in estuaries experience a naturally highly variable physical environment that is often close to their physiological tolerances (Attrill and Rundle, 2002). Despite this, projected climate changes may have significant impacts on estuarine species and their ecosystems. Moreover, invasive species to estuaries have increased during the last 20 years, due to global temperature warming (Austin et al., 2010), leading to some dramatic effects on native assemblages (Thompson et al., 2002). Problems associated with invasive species, especially pathogens, are projected to increase over the next few decades (e.g., McDowell et al., 2014).

Projected sea-level rise and increased winter river discharge for the UK are thought to be the most important impacts of climate change on species, principally through associated habitat changes (Kimmerer, 2002 and IPCC, 2014). Habitat loss from sea-level rise, such as coastal squeeze in the absence of managed retreat, is likely to have important implications for estuarine food webs (Fujii, 2007, Fujii and Raffaelli, 2008 and Fujii, 2012). Although only from an individual study site in the Humber estuary, Fujii (2007)demonstrates that a ‘squeezed' location can support fewer species and biomass of keystone benthic macroinvertebrates per unit area due to the narrower, steeper and less muddy beach profile, when compared with adjacent undefended shorelines. Yamanaka et al. (2013) have supported these findings with a study based on three estuaries with meiofauna included with macrofauna in the study. Evidence does appear to support that coastal squeeze will decrease benthos biomass; however Yamanaka et al. (2013) found that local conditions were important in describing macrobenthic biomass in intermediate-to-steep sloping estuary shores. This means that predicting the effect of coastal squeeze on macrobenthos without knowledge of complex local processes and conditions, and their interaction with the physical drivers affected by coastal squeeze, is likely complex if to be made reliably. Such observations do not incorporate how other feedbacks, such as changes in primary production, will interact with these trends.

Estuarine foodwebs may also be impacted by habitat modification from increased river discharge, changes in river-borne materials, and variations in the position and intensity of estuarine turbidity maxima and salinity stratification (Kirby, 2010 and Uncles et al., 2014). Changes in the salinity experienced by fauna in estuaries is likely to have an impact on species diversity and distribution: a projected increase in salinity range may reduce species diversity, as found in the Thames Estuary (Attrill, 2002), and move the focus of biomass maximum higher up estuaries (Fujii, 2007). Additionally, the occurrence of hypoxic events and the vulnerability of species in estuaries to these events is likely to increase (Rabalais et al., 2009 and Vaquer-Sunyer and Duarte, 2011). Key components such as benthic filter and deposit feeders have been observed to be, and are predicted to be, negatively affected through these changes (Fujii and Raffaelli, 2008). These functional groups are very important for nutrient cycling and supporting higher trophic levels such as predatory fish and birds (Austin and Rehfisch, 2003), and so can have far reaching implications for British estuarine ecology if affected beyond natural variation. Turbidity has been suggested to be important for survival and behaviour of mid-water organism such as juvenile fish through food supply and vulnerability to visual predators, which may present further implications to changes in river discharge and sediment suspension (Power et al., 2000).

In addition to these habitat modifications, climate changes to temperature may also indirectly affect populations and food webs (Walther et al., 2002). For example, projected milder UK winters may negatively impact infaunal bivalve reproductive output and recruitment directly and through affecting predator phenology (Philippart et al., 2003 and Beukema and Dekker, 2014). Predatory shore crabs *Carcinus maenas* and brown shrimp *Crangon crangon* recruit earlier after mild winters, which can result in a greater overlap with their post-larval bivalve prey and reduces bivalve and flatfish recruitment success, when compared with periods with proceeding colder winters (Power et al., 2000 and Beukema and Dekker, 2014). This again may reduce the biomass of benthic macroinvertebrate species that support higher trophic levels and the estuarine ecosystem. However, as acknowledged already, the complexities of local systems need considering before broad climate change predictions are applied to individual estuaries.

DISCUSSION

Global climate models, linked to impact models which predict biophysical and socio-economic impacts, propose that stabilizing global carbon dioxide levels at 550 ppm by 2150 could significantly reduce most stressors on the water cycle, compared with unmitigated emissions scenarios; for example, reducing flooding frequency by some 80–90% along the most vulnerable coastlines, which include estuaries in southeast England (Arnell et al., 2002). Even though these emissions targets will be difficult to achieve, the UK are attempting to show leadership, and are seeking international commitment to reduce carbon dioxide and other greenhouse gas emissions through the United Nations Framework Convention on Climate Change (UNFCCC). The UK government has committed to reducing the country's emissions by 60% from 1990 levels by 2050 – achieved by reducing the amount of energy we consume and by substantially increasing our use of renewable energy resources (King, 2004).

Even with these initiatives, in the short term (the 21st century) some consequences of climate change appear to be inevitable for the UK under any future emissions scenario (Arnell et al., 2002), such as sea-level rise. If we define the risk of climate-driven impact as the probability of the occurrence multiplied by the magnitude of the consequence, it is obvious that the level of risk is different for each affected party and for each region. Thus, the magnitude of potential climate-driven impacts on UK estuaries, described in this paper, and hence the level of risk, require further quantification in order to inform management strategies. Regarding flood risk and flood mitigation, Elliott et al. (2014)point out that the UK has a strategy to try to protect urban and industrial coasts but deem protection of farmland not to be economically justifiable. Therefore, rural areas require local adaptation strategies that rely on soft engineering and improved community awareness.

More effective coastal management in the UK requires reducing the potentially large uncertainties, associated with climate change projections based on limited simulation of climatic variability and complex impact predictions, so that levels of risk can be assigned accurately (Arnell et al., 2002). For example, unresolved temporal and spatial variability within climate models needs to be parameterised in the boundary forcing conditions of local-scale impact models (e.g., Lewis et al., 2011). In terms of predicting climate-driven impact, many researchers are using species distribution modelling to examine how ranges may shift under climate change. However, this approach has been criticised because most models have not included non-climate stressors, such as species competition and predation, habitat loss and land-use changes, eutrophication and invasive species, that could themselves influence distributions (McDowell et al., 2014). Additionally, state-of-the-art, process-based, morphological models are, in general, capable of predicting only the statistical behaviour defining channel geometry and meandering on time scales of decades or longer. Therefore, we must understand the relative importance of climatic and non-climatic variables in controlling present-day distributions to predict the impacts of climate change, and also understand the associated uncertainties so that we can measure the potential risk and improve management strategies.

CONCLUSIONS

In this review, we summarise published literature of historic observations and current anthropogenic climate change projections for the 21st century, and the anticipated processes and impacts to UK estuarine systems, including physical, biological, and ecological changes. Fig. 3 summarises the likely main, but not all, pathways from climatic drivers to processes to impacts on UK estuaries. There is also no indication from Fig. 3 of the scale of impact; for example, the consequences of flooding may be more significant than shifts in frontal positions, although this is difficult to quantify.

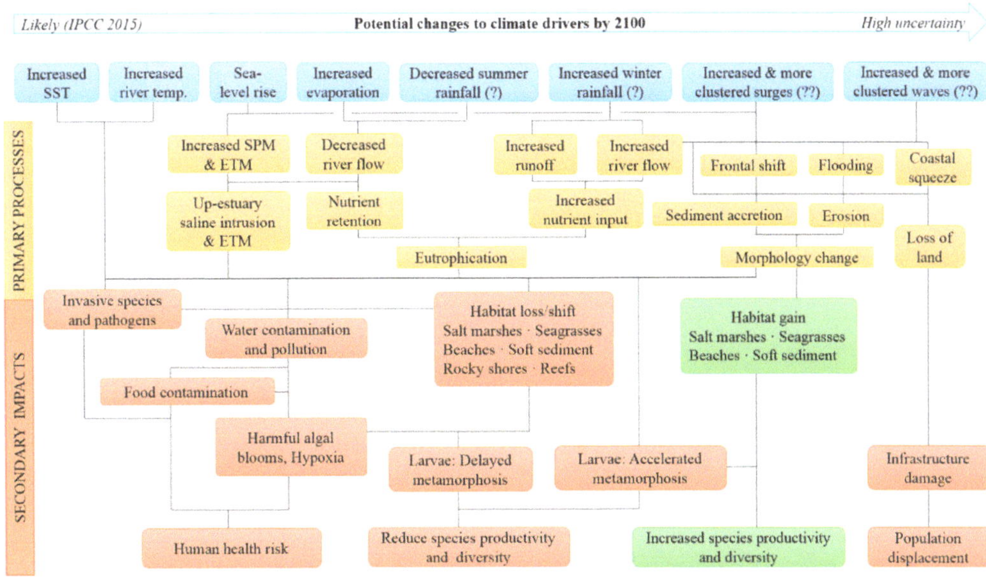

Figure 3: Flow diagram showing the main climatic drivers, and primary and secondary impacts of climate change to UK estuaries. Pathways (black lines) are directed *from* the bottom/side *to* the top of a 'driver/impact box'. Predominantly positive impacts are coloured green. (For interpretation of the references to colour in this figure legend, the reader is referred to the web version of this article.)

In terms of the climatic drivers shown in Fig. 3, sea surface temperatures around the UK have increased by 0.7 °C over the past 30 years, and are projected with high confidence to increase further by between 1.5 °C and 4 °C, by 2100, depending on our carbon emissions. Sea-levels have risen by 1–3 mm yr^{-1} during the 20th century, and are projected with high confidence to rise further by between 0.44 m and 0.74 m, by 2100, again depending on our carbon emissions but also with significant spatial variability. Precipitation and storminess have shown high variability over the past few decades, with a general pole-ward shift and increased intensity/decreased frequency of events. Although UK river flows have generally increased during autumn and winter over the past 50 years, flow rates have decreased during spring with no detectable trends during summer. Future UK projections, albeit with low confidence and high variability, are for more clustering of events, with increased winter precipitation, river flow, surge and wave intensity, and longer dry and calm periods in summer.

Potential changes to physical, *primary*, processes (see Fig. 3) include increased flooding and coastal squeeze, caused by increased sea-level, changing surge and wave climates, and changing river flow events. The capacity for estuaries to respond to flooding will depend in the short term on available 'floodable' space, and in the long term on morphodynamics relative to sea-level rise (e.g., sediment supply and accretion, and coastal realignment). Sea-level rise will likely cause a shift towards net sediment accretion in UK estuaries, but with reduced transport overall. Turbulent mixing – critical for water quality and coastal ecology – is primarily controlled by river flow variability. Hence, potential alterations to river flows will increase the temporal variability and length scales of estuarine fronts, stratification and mixing. For example, longer dry periods in summer, in combination with sea-level rise will push the salinity maximum further up-estuary and enhance estuarine nutrification, with negative impacts of eutrophication, harmful algal blooms and hypoxia. On the other hand, clustering of storm events will alter the temporal mixing variability in winter.

Some of the *secondary impacts* on the estuarine ecosystem (see Fig. 3) may ultimately result in a risk to human health, associated with ecological disruption and heightened water-borne and food-borne microbial pathogens concentrations, mainly due to projected increased temperatures and changing river flows. Changes in these climate drivers are also likely to influence the concentrations of toxic metals, organic chemicals, algal toxins, and human pathogen contaminants in seafood. The primary impact of coastal squeeze is the greatest threat to already declining UK salt marsh habitats, which in turn may reduce their species diversity and food webs. Some intertidal habitats and biogenic reefs – often priority habitats for protection in the UK – have gradually moved north in response to rising temperatures, bringing changes in biodiversity and ecological functioning – for example, algae in rocky shores will likely become less prevalent. Projected increases in mild winters may negatively impact infaunal bivalve reproduction. Invasive species to estuaries have increased, due to temperature warming, affecting native assemblages and introducing pathogens. However, temperature rise is unlikely to affect species that live within their temperature tolerances, such as mussels. Sedimentary habitats, particularly mudflat-living species, are sensitive to an assortment of climate change drivers and primary impacts including wave attack and sediment coarsening. Changing climate drivers are likely to alter environmental cues which synchronize the reproductive cycle of many marine invertebrates and fish. Estuarine species that cue reproduction based on salinity signals are likely to be particularly vulnerable to climate change. Eventually, climate change may have implications on the conservation status of UK estuaries.

We recognise there is variety and diversity of estuarine ecosystems in Britain, and variety in drivers of change that require more detailed quantification/prediction. We therefore recommend that future research should focus on improving our understandings of the primary and secondary impacts to UK estuaries, on a case-by-case basis, caused by potential changes to regional climatic drivers; for example, through improved modelling and model coupling, accompanied by long-term monitoring. Current UK research funding, such as the NERC Macronutrient Cycles Programme, aims to contribute to this via integrated climate-catchment-river-estuary process modelling combined with intensive sampling across several UK systems. Some emerging outcomes of this programme are that climate-driven, coupled, soil-catchment-river models are required to accurately predict future river flows and nutrient fluxes. Further, uncertainties associated with commonly-used daily-averaged river flows to infer impact downstream in the estuary appear to be significant, emphasising the need for catchment modelling.

ACKNOWLEDGEMENTS

Author contributions: Abstract and Introduction (all); Section 2 (P. Robins and M. Lewis); Section 3 (P. Robins, M. Lewis, S. Neill, A. Davies, S. Malham); Section 4 (P. Robins, J. MacDonald, M. Skov, L. Gimenez, T. Whitton). This review paper was commissioned by the Environment Agency of behalf of the Natural Environment Research Councils (NERC) Living with Environmental Change Programme (LWEC). P. Robins, A. Davies, S. Malham, T. Whitton, S. Neill and C. Jago wish to acknowledge the support of the SEACAMS (Sustainable Expansion of the Applied Coastal and Marine Sectors; www.seacams.ac.uk, Grant Number 80366) research project at Bangor University, the Welsh Government, the Higher Education Funding Council for Wales, the Welsh European Funding Office, and the European Regional Development Fund Convergence Programme. P. Robins, S. Malham, J. MacDonald, and C. Jago wish to acknowledge the NERC-funded Macronutrients Cycles Programme (Turf2surf project: www.turf2surf.org). S. Neill, P. Robins, M. Lewis and M. Skov wish to acknowledge the support the Sêr Cymru National Research Network for Low Carbon, Energy and the Environment (NRN-LCEE). We would also like to thank the reviews for their constructive comments on a previous version of this paper.

REFERENCES

1. Alexandersson et al., 2000 H. Alexandersson, H. Tuomenvirta, T. Schmith, K. Iden Trends of storms in NW Europe derived from an updated pressure data set Clim. Res., 14 (2000), pp. 71–73
2. Allen et al., 2008 R.M. Allen, Y.M. Buckley, D.J. Marshall Offspring size plasticity in response to intraspecific competition: an adaptive maternal effect across life-history stages Am. Nat., 171 (2008), pp. 225–237
3. Allan et al., 2009 R. Allan, S. Tett, L. Alexander Fluctuations in autumn–winter severe storms over the British Isles: 1920 to present Int. J. Climatol., 29 (2009), pp. 357–371
4. Andrews et al., 2006 J.E. Andrews, D. Burgess, R.R. Cave, E.G. Coombes, T.D. Jickells, D.J. Parkes, R.K. Turner Biogeochemical value of managed realignment, Humber estuary, UK Sci. Total Environ., 371 (2006), pp. 19–30
5. Antonopoulou and Emson, 1988 E. Antonopoulou, R. Emson The combined effects of temperature and salinity on survival, moulting and metamorphosis of the larval stages of three species of palaemonid prawns J.S. Ryland, P.A. Tyler (Eds.), Reproduction, Genetics and Distributions of Marine Organisms: Proc. 23rd Eur. Mar. Biol. Symp., Swansea, 5–9 September 1988 (1988)
6. Arnell et al., 2002 N.W. Arnell, M.G.R. Cannell, M. Hulme, R.S. Kovats, J.F.B. Mitchell, R.J. Nicholls, M.L. Parry, M.T.J. Livermore, A. White The consequences of CO2 stabilisation for the impacts of climate change Clim. Change, 53 (2002), pp. 413–446
7. Arnell, 2003 N.W. Arnell Relative effects of multi-decadal climatic variability and changes in the mean and variability of climate due to global warming: future streamflows in Britain J. Hydrol., 270 (2003), pp. 195–213
8. Attrill, 1998 M.J. Attrill A Rehabilitated Estuarine Ecosystem: the Environment and Ecology of the Thames Estuary Kluwer Academic Publisheres, London (1998)
9. Attrill, 2002 M.J. Attrill A testable linear model for diversity trends in estuaries J. Anim. Ecol., 71 (2002), pp. 262–269 Attrill and Rundle, 2002 M.J. Attrill, S.D. Rundle Ecotone or ecocline: ecological boundaries in estuaries Estuar. Coast. Shelf Sci., 55 (2002), pp. 929–936
10. Austin and Rehfisch, 2003 G.E. Austin, M.M. Rehfisch The likely impact of sea level rise on waders (Charadrii) wintering on estuaries J. Nat. Conserv., 11 (2003), pp. 43–58
11. Austin et al., 2010 J. Austin, L. Zhang, R.N. Jones, P. Durack, W. Dawes, P. Hairsine Climate change impact on water and salt balances: an assessment of the impact of climate change on catchment salt and water balances in the Murray-Darling Basin, Australia Clim. change, 100 (2010), pp. 607–631
12. Asmala et al., 2014 E. Asmala, D.G. Bowers, R. Autio, H. Kaartokallio, D.N. Thomas Qualitative changes of riverine dissolved organic matter at low salinities due to flocculation J. Geophys. Res. Biogeosci., 119 (2014), pp. 1919–1933
13. Balvanera et al., 2006 P. Balvanera, A.B. Pfisterer, N. Buchmann, J.S. He, T. Nakashizuka, D. Raffaelli, B. Schmid Quantifying the evidence for biodiversity effects on ecosystem functioning and services Ecol. Lett., 9 (2006), pp. 1146–1156
14. Barbier et al., 2011 E.B. Barbier, S.D. Hacker, C. Kennedy, E.W. Koch, A.C. Stier, B.R. Silliman The value of estuarine and coastal ecosystem services Ecol. Monogr., 81 (2011), pp. 169–193
15. Beck et al., 2001 M.W. Beck, K.L. Heck, K.W. Able, A.D. Childers, D.B. Eggleston, B.M. Gillanders, B. Halpern, C.G. Hays, K. Hoshino, T.J. Minello, R.J. Orth, P.F. Sheridan, M.P. Weinstein The identification, conservation, and management of estuarine and marine nurseries for fish and invertebrates Bioscience, 51 (2001), pp. 633–641
16. Beukema and Dekker, 2014 J. Beukema, R. Dekker Variability in predator abundance links winter temperatures and bivalve recruitment: correlative evidence from long-term data in a tidal flat Mar. Ecol. Prog. Ser., 513 (2014), pp. 1–15

17. Bijl et al., 1999 W. Bijl, R. Flather, J.G. de Ronde, T. Schmith Changing storminess? An analysis of long-term sea level data sets Clim. Res., 11 (1999), pp. 161–172
18. Bjork et al., 2008 M. Bjork, F. Short, E. McLeod, S. Beer Managing Seagrasses for Resilience to Climate Change IUCN, Gland, Switzerland (2008)
19. Boorman, 2003 L.A. Boorman Saltmarsh Review. An Overview of Coastal Saltmarshes, Their Dynamic and Sensitivity Characteristics for Conservation and Management JNCC Report No. 334, pp. 114. ISSN 0963-8091 (2003)
20. Bouma et al., 2014
21. T.J. Bouma, J. van Belzen, T. Balke, Z. Zhu, L. Airoldi, A.J. Blight, A.J. Davies, C. Galvan, S.J. Hawkins, S.P.G. Hoggart, J.L. Lara, I.J. Losada, M. Maza, B. Ondiviela, M.W. Skov, E.M. Strain, R.C. Thompson, S. Yang, B. Zanuttigh, L. Zhang, P.M.J. Herman Identifying knowledge gaps hampering application of intertidal habitats in coastal protection: opportunities & steps to take Coast. Eng., 87 (2014), pp. 147–157
22. Brown et al., 1991 J. Brown, W.R. Turrell, J.H. Simpson Aerial surveys of axial convergent fronts in UK estuaries and the implications for pollution Mar. Pollut. Bull., 22 (1991), pp. 397–400
23. Burt and Ferranti, 2012 T.P. Burt, E.J.S. Ferranti Changing patterns of heavy rainfall in upland areas: a case study from northern England Int. J. Climatol., 32 (2012), pp. 518–532
24. Cai, 2011 W.J. Cai Estuarine and coastal ocean carbon paradox: CO2 sinks or sites of terrestrial carbon incineration? Annu. Rev. Mar. Sci., 3 (2011), pp. 123–145
25. Carini et al., 2015 1 R.J. Carini, C.C. Chickadel, A.T. Jessup, J. Thomson Estimating wave energy dissipation in the surf zone using thermal infrared imagery J. Geophys. Res. Oceans (2015 May 1)
26. Cazenave and Nerem, 2004Cazenave, R.S. NeremPresent-day sea level change: observations and causes Rev. Geophys., 42 (2004) http://dx.doi.org/10.1029/2003RG000139Chaalali et al., 2013
27. Chaalali, X. Chevillot, G. Beaugrand, V. David, C. Luczak, P. Boët, A. Sottolichio, B. Sautour Changes in the distribution of copepods in the Gironde estuary: a warming and marinisation consequence? Estuar. Coast. Shelf Sci., 134 (2013), pp. 150–161
28. Chaparro et al., 2014 O.R. Chaparro, C.J. Segura, S.J.A. Osores, J.A. Pechenik, L.M. Pardo, V.M. Cubillos Consequences of maternal isolation from salinity stress for brooded embryos and future juveniles in the estuarine direct-developing gastropod Crepipatella dilatata Mar. Biol., 161 (2014), pp. 619–629
29. Charmantier et al., 2002 G. Charmantier, L. Giménez, M. Charmantier-Daures, Anger Ontogeny of osmoregulation, physiological plastic-ity and larval export strategy in the grapsid crab Chas-magnathus granulata (Crustacea Decapoda) Mar. Ecol. Prog. Ser., 229 (2002), pp. 185–194
30. Christierson et al., 2012 B.V. Christierson, J.-P. Vidal, S.D. Wade Using UKCP09 probablistic climate information for UK water resource planning J. Hydrol., 424–425 (2012), pp. 48–67
31. Church et al., 2004 J.A. Church, N.J. White, R. Coleman, K. Lambeck, J.X. Mitrovica Estimates of the regional distribution of sea level rise over the 1950-2000 period J. Clim., 17 (2004), pp. 2609–2625
32. Church and White, 2006 J.A. Church, N.J. White A 20th century acceleration in global sea level rise Geophys. Res. Lett., 33 (2006) http://dx.doi.org/10.1029/2005GL024826
33. Curriero et al., 2001 F.C. Curriero, J.A. Patz, J.B. Rose, S. Lele The association between extreme precipitation and waterborne disease outbreaks in the United States, 1948-1994 Am. J. Public Health, 91 (2001), pp. 1194–1199
34. Davidson et al., 1991 N.C. Davidson, D.A. Laffoley, J.P. Doody, L.S. Way, J. Gordon, R. Key, M.W. Pienkowski, R. Mitchell, K. Duff Nature Conservation and Estuaries in Great Britain Nature Conservancy Council, Peterborough (1991)
35. Davies, 1964 Davies A morphogenic approach to world shorelines Z. Geomorphol., 8 (1964), pp. 27–42
36. Day et al., 1995 J.W. Day, D. Pont, P.F. Hensel, C. Ibañez Impacts of sea-level rise on deltas in the Gulf of Mexico and the Mediterranean: the importance of pulsing events to sustainability Estuaries, 18 (1995), pp. 636–647
37. Debernard and Røed, 2008 J. Debernard, L.P. Røed Future wind, wave and storm surge climate in the Northern Seas: a revisit Tellus, 60A (2008), pp. 427–438
38. Dickson et al., 2007 M.E. Dickson, M.J. Walkden, J.W. Hall Systemic impacts of climate change on an eroding coastal region over the twenty-first century Clim. Change, 84 (2007), pp. 141–166
39. Dissanayake et al., 2015 P. Dissanayake, J. Brown, H. Karunarathna Impacts of storm chronology on the morphological changes of the Formby beach and dune system, UK Nat. Hazards Earth Syst. Sci. Discuss., 3 (2015), pp. 2565–2597
40. Dolbeth et al., 2011 M. Dolbeth, P.G. Cardoso, T.F. Grilo, M.D. Bordalo, D. Raffaelli, M.A. Pardal Long-term changes in the production by estuarine macrobenthos affected by multiple stressors Estuar. Coast. Shelf Sci., 92 (2011), pp. 10–18
41. Doody, 2008 P. Doody Saltmarsh Conservation, Management and Restoration Springer, London (2008)
42. Dronkers, 1986 J. Dronkers Tidal asymmetry and estuarine morphology Neth. J. Sea Res., 20 (1986), pp. 117–131
43. Dyer, 1997 K.R. Dyer Estuaries: a Physical Introduction (second ed.)John Wiley and Sons/Wiley & Sons, Chichester (1997) ISBN: 0-471-9741-4. 195 pp.
44. EA, 2011 EA The Extent of Saltmarsh in England and Wales: 2006–2009 , 978-1-84911-233-8Environment Agency, Bristol (2011)
45. Elliott et al., 2014 M. Elliott, N.D. Cutts, A. Trono A typology of marine and estuarine hazards and risks as vectors of change: a review for vulnerable coasts and

their management Ocean Coast. Manag., 93 (2014), pp. 88–99

46. Elliott et al., 2015 M. Elliott, Á. Borja, A. McQuatters-Gollop, K. Mazik, S. Birchenough, J.H. Andersen, S. Painting, M. Peck Force majeure: will climate change affect our ability to attain good environmental status for marine biodiversity? Mar. Pollut. Bull., 95 (2015), pp. 7–27
47. Fagherazzi et al., 2012 S. Fagherazzi, M.L. Kirwan, S.M. Mudd, G.R. Guntenspergen, S. Temmerman, A. D'Alpaos, J. Koppel, J.M. Rybczyk, E. Reyes, C. Craft, J. Clough Numerical models of salt marsh evolution: ecological, geomorphic, and climatic factors Rev. Geophys., 50 (2012) http://dx.doi.org/10.1029/2011RG000359
48. Fagherazzi et al., 2013 S. Fagherazzi, P.L. Wiberg, S. Temmerman, E. Struyf, Y. Zhao, P.E. Raymond Fluxes of water, sediments, and biogeochemical compounds in salt marshes Ecol. Process, 2 (2013), p. 3
49. Forward et al., 2001 R.B. Forward, R.A. Tankersley, D. Rittschof Cues for metamorphosis of Brachyuran crabs: an overview Am. Zool., 41 (2001), pp. 1108–1122
50. Fincham, 1979 A.A. Fincham Larval development of British prawns and shrimps (Crustacea: Decapoda:Natantia). 2. Palaemonetes (Palaemonetes) varians (Leach, 1814) and morphological variation Bull. Br. Mus. (Nat. Hist.) Zool., 35 (1979), pp. 163–182
51. Fowler and Kilsby, 2007 H.J. Fowler, C.G. Kilsby Using regional climate model data to simulate historical and future river flows in northwest England Clim. Change, 80 (2007), pp. 337–367
52. Fowler and Wilby, 2010 H.J. Fowler, R.L. Wilby Detecting changes in seasonal precipitation extremes using regional climate model projections: implications for managing fluvial flood risk Water Resour. Res., 46 (2010) http://dx.doi.org/10.1029/2008WR007636
53. Fox, 1981 L.R. Fox Defence and dynamics in plant-herbivore systems Am. Zool., 21 (1981), pp. 853–886
54. French, 2006 J. French Tidal marsh sedimentation and resilience to environmental change: exploratory modelling of tidal, sea-level and sediment supply forcing in predominantly allochthonous systems Mar. Geol., 235 (2006), pp. 119–136
55. Friedman-Huffman and Rose, 1999 D.E. Friedman-Huffman, J.B. Rose The Continuing Threat of Waterborne Pathogens Providing Safe Drinking Water in Small Systems: Technology, Operations, and Economics (1999), p. 11
56. Friedrichs and Aubrey, 1988 C.T. Friedrichs, D.G. Aubrey Non-linear tidal distortion in shallow well-mixed estuaries: a synthesis Estuar. Coast. Shelf Sci., 27 (1988), pp. 521–545
57. Frost et al., 2004 M.T. Frost, R. Leaper, N. Mieszkowska, P. Moschella, J. Murua, C. Smyth, S.J. Hawkins Recovery of a Biodiversity Action Plan Species in Northwest England: Possible Role of Climate Change, Artificial Habitat and Water Quality Amelioration Sabellaria alveolata English Nature (2004)
58. Fujii, 2007 T. Fujii Spatial patterns of benthic macrofauna in relation to environmental variables in an intertidal habitat in the Humber estuary, UK: developing a tool for estuarine shoreline management Estuar. Coast. Shelf Sci., 75 (2007), pp. 101–119
59. Fujii, 2012 T. Fujii Climate change, sea-level rise and implications for coastal and estuarine shoreline management with particular reference to the ecology of intertidal benthic macrofauna in NW europe Biology (Basel), 1 (2012), pp. 597–616
60. Fujii and Raffaelli, 2008 T. Fujii, D. Raffaelli Sea-level rise, expected environmental changes, and responses of intertidal benthic macrofauna in the Humber estuary, UK Mar. Ecol. Prog. Ser., 371 (2008), pp. 23–35
61. Gedan and Bertness, 2010 K.B. Gedan, M.D. Bertness How will warming affect the salt marsh foundation species Spartina patens and its ecological role? Oecologia, 164 (2010), pp. 479–487
62. Geng and Sugi, 2003 Q. Geng, M. Sugi Possible change of extratropical cyclone activity due to enhanced greenhouse gases and sulfate aerosols-study with a high-resolution AGCM J. Clim., 16 (2003), pp. 2262–2274
63. Giménez, 2006 L. Giménez Phenotypic links in complex life cycles: conclusions from studies with decapod crustaceans Int. Comp. Biol., 46 (2006), pp. 615–622
64. Giménez et al., 2004 L. Giménez, K. Anger, G. Torres Linking life history traits in successive phases of a complex life cycle: effects of larval biomass on early juvenile development in an estuarine crab Chasmagnathus granulata Oikos, 104 (2004), pp. 570–580
65. González-Ortegón et al., 2013 E. González-Ortegón, J. Blasco, L. Le Vay, L. Giménez A multiple stressor approach to study the toxicity and sub-lethal effects of pharmaceutical compounds on the larval development of a marine invertebrate J. Hazard. Mater., 263 (2013), pp. 233–238
66. González-Ortegón and Giménez, 2014 E. González-Ortegón, L. Giménez Environmentally mediated phenotypic links and performance in larvae of a marine invertebrate Mar. Ecol. Prog. Ser., 502 (2014), pp. 185–195
67. Greve et al., 2003 T.M. Greve, J. Borum, O. Pedersen Meristematic oxygen variability in eelgrass (Zostera marina) Limnol. Oceanogr., 48 (2003), pp. 210–216
68. Gubbay and Earll, 2010 S. Gubbay, R. Earll Assessing the Vulnerability of Marine Biodiversity in Wales to the Impacts of Climate Change: Workshop Report CCW Contract Science Report No: 934, 21pp CCW, Bangor (2010)
69. Hannaford, 2015 J. Hannaford Climate-driven changes in UK river flows A review of the evidence Prog. Phys. Geogr., 39 (2015), pp. 29–48
70. Hannaford and Buys, 2012 J. Hannaford, G. Buys Seasonal trends in UK river flow regimes and extremes in the UK J. Hydrol., 475 (2012), pp. 158–174
71. Harland et al., 2000 B.J. Harland, D. Taylor, A. Wither The distribution of mercury and other trace metals in the sediments of the Mersey Estuary over 25 years 1974–1998 Sci. Total Environ., 253 (2000), pp. 45–62

72. Hawkins et al., 2008 S.J. Hawkins, P.J. Moore, M.T. Burrows, E. Poloczanska, N. Mieszkowska, R.J.H. Herbert, S.R. Jenkins, R.C. Thompson, M.J. Genner, A.J. Southward Complex interactions in a rapidly changing world: responses of rocky shore communities to recent climate change Clim. Res., 37 (2008), pp. 123–133

73. Hawkins et al., 2009 S.J. Hawkins, H.E. Sugden, N. Mieszkowska, P.J. Moore, E. Poloczanska, R. Leaper, R.J.H. Herbert, M.J. Genner, P.S. Moschella, R.C. Thompson, S.R. Jenkins, A.J. Southward, M.T. Burrows Consequences of climate-driven biodiversity changes for ecosystem functioning of North European rocky shores Mar. Ecol. Prog. Ser., 396 (2009), pp. 245–259

74. Hanson et al., 2011 S. Hanson, R. Nicholls, N. Ranger, S. Hallegatte, J. Corfee-Morlot, C. Herweijer, J. Chateau A global ranking of port cities with high exposure to climate extremes Clim. Change, 104 (2011), pp. 89–111

75. Hiddink and Hofstede, 2008 J.G. Hiddink, R.T. Hofstede Climate induced increases in species richness of marine fishes Glob. Change Biol., 14 (2008), pp. 453–460

76. Hetland and Geyer, 2004 R.D. Hetland, W.R. Geyer An idealized study of the structure of long, partially mixed estuaries J. Phys. Oceanogr., 34 (2004), pp. 2677–2691

77. Hoitink et al., 2003 A.J.F. Hoitink, P. Hoekstra, D.S. Van Maren Flow asymmetry associated with astronomical tides: implications for the residual transport of sediment J. Geophys. Res. Oceans, 108 (2003), p. C10

78. Horrillo-Caraballo et al., 2013 J.M. Horrillo-Caraballo, D.E. Reeve, D. Simmonds, A. Fox, R. Thompson, S. Hoggart, S.S.H. Kwan, D. Greaves Application of a source-pathway-receptor-consequence (SPRC) methodology to the Teign Estuary, UK J. Coast. Res. (2013)

79. Hough and Naylor, 1991 A.R. Hough, E. Naylor Field studies on retention of the planktonic copepod Eurytemora affinis in a mixed estuary Mar. Ecol. Prog. Ser., 76 (1991), pp. 115–122

80. Howlett et al., 2015 E.R. Howlett, D.G. Bowers, J. Malarkey, C.F. Jago Stratification in the presence of an axial convergent front: causes and implications Estuar. Coast. Shelf Sci., 161 (2015), pp. 1–10

81. Howard et al., 2010 T. Howard, J. Lowe, K. Horsburgh Interpreting century-scale changes in North Sea storm surge climate derived from coupled model simulations J. Clim., 23 (2010), pp. 6234–6247

82. Howsen, 2009 C. Howsen Intertidal SAC Monitoring, Camarthen Bay and Estuaries SAC CCW Marine Monitoring Report no 79 Countryside Council for Wales, Bangor (2009)

83. Huang et al., 2013 G. Huang, J. Zhou, B. Lin, Q. Chen, R. Falconer Distributed numerical hydrological and hydrodynamic modelling for large river catchment Proceedings of the 35th IAHR World Congress (2013), pp. 2420–2431

84. IPCC, 2007 IPCC Synthesis Report: Contribution of Working Groups I, II and III to the Fourth Assessment Report of the Intergovernmental Panel on Climate Change IPCC, Geneva (2007) 104pp

85. IPCC, 2013 IPCC Summary for policymakers T.F. Stocker, D. Qin, G.-K. Plattner, M. Tignor, S.K. Allen, J. Boschung, A. Nauels, Y. Xia, V. Bex, P.M. Midgley (Eds.), Climate Change 2013: the Physical Science Basis. Contribution of Working Group I to the Fifth Assessment Report of the Intergovernmental Panel on Climate Change, Cambridge University Press, United Kingdom and New York, NY, USA (2013)

86. IPCC, 2014 IPCC C.B. Field, V.R. Barros, D.J. Dokken, K.J. Mach, M.D. Mastrandrea, T.E. Bilir, M. Chatterjee, K.L. Ebi, Y.O. Estrada, R.C. Genova, B. Girma, E.S. Kissel, A.N. Levy, S. MacCracken, P.R. Mastrandrea, L.L. White (Eds.), Climate Change 2014: Impacts, Adaptation, and Vulnerability. Part A: Global and Sectoral Aspects. Contribution of Working Group II to the Fifth Assessment Report of the Intergovernmental Panel on Climate Change, Cambridge University Press, Cambridge, United Kingdom and New York, NY, USA (2014) 1132 pp.

87. Ittekkot, 1988 V. Ittekkot Global trends in the nature of organic matter in river suspensions Nature, 332 (1988), pp. 436–438

88. Jackson and McIlvenny, 2011 A.C. Jackson, J. McIlvenny Coastal squeeze on rocky shores in northern Scotland and some possible ecological impacts J. Exp. Mar. Biol. Ecol., 400 (2011), pp. 314–321

89. Jackson et al., 2011 C.R. Jackson, R. Meister, C. Prudhomme Modelling the effects of climate change and its uncertainty on UK Chalk groundwater resources from an ensemble of global climate model projections J. Hydrol., 399 (2011), pp. 12–28

90. Jarvie et al., 2012 H.P. Jarvie, T.D. Jickells, R.A. Skeffington, P.J.A. Withers Climate change and coupling of macronutrient cycles along the atmospheric, terrestrial, freshwater and estuarine continuum Sci. Total Environ., 434 (2012), pp. 252–258

91. Jenkins et al., 2008 G.J. Jenkins, M.C. Perry, M.J.O. Prior The Climate of the United Kingdom and Recent Trends Met Office Hadley Centre, Exeter, UK (2008)

92. Jenkins et al., 2009 G. Jenkins, J. Murphy, D. Sexton, J. Lowe UK Climate Projections: Briefing Report Met Office Hadley Centre, Exeter (2009) http://ukclimateprojections.defra.gov.uk/

93. Jenkins et al., 2015 G.P. Jenkins, D. Spooner, S. Conron, J.R. Morrongiello Differing importance of salinity stratification and freshwater flow for the recruitment of apex species of estuarine fish Mar. Ecol. Prog. Ser., 523 (2015), pp. 125–144

94. Jensen et al., 2014 N. Jensen, R.M. Allen, D.J. Marshall Adaptive maternal and paternal effects: gamete plasticity in response to parental stress Funct. Ecol., 28 (2014), pp. 724–733

95. Jevrejeva et al., 2014 S. Jevrejeva, A. Grinsted, J.C. Moore Upper limit for sea level projections by 2100 Environ. Res. Lett., 9 (2014), p. 104008

96. Jickells et al., 2000 T. Jickells, J. Andrews, G. Samways, R. Sanders, S. Malcolm, D. Sivyer, R. Parker, D. Nedwell, M. Trimmer, J. Ridgway Nutrient fluxes through the

Humber estuary-Past, present and future AMBIO A J. Hum. Environ., 29 (2000), pp. 130–135

97. Jones et al., 2011 D. Jones, V. Bain, S. Dawson, T. Watt Assessing the Vulnerability of Marine Biodiversity in Wales to the Impacts of Climate Change. Workshop Report CCW contract Science Report no 934 Countryside Council for Wales, Bangor (2011)

98. Jones et al., 2013 M.R. Jones, H.J. Fowler, C.G. Kilsby, S. Blenkinsop An assessment of changes in seasonal and annual extreme rainfall in the UK between 1961 and 2009 Int. J. Climatol., 33 (2013), pp. 1178–1194

99. Kaiser et al., 2005 M.J. Kaiser, M.J. Attrill, S. Jennings, D.N. Thomas, D.K.A. Barnes, A.S. Brierley, J.G. Hiddink, H. Kaartokallio, N.V.C. Polunin, D.G. Raffaelli Marine Ecology, Processes, Systems and Impacts Oxford University Press, Oxford (2005)

100. Kay et al., 2009 A.L. Kay, H.N. Davies, V.A. Bell, R.G. Jones Comparison of uncertainty sources for climate change impacts: flood frequency in England Clim. Change, 92 (2009), pp. 41–63

101. Kay et al., 2013a A.L. Kay, V.A. Bell, E.M. Blyth, S.M. Crooks Davies HN andReynard NS A hydrological perspective on eva-poration: historical trends and future projections in Brit-ain J. Water Clim Change, 4 (3) (2013), pp. 193–208

102. Kay et al., 2013b A.L. Kay, S.M. Crooks, H.N. Davies Prudhomme C andReynard NS Probabilistic impacts of climatechange on flood frequency using response surfaces. I: England and Wales Reg. Environ. Change, 14 (3) (2013), pp. 1215–1227

103. Keef et al., 2009 C. Keef, C. Svensson, J.A. Tawn Spatial dependence in extreme river flows and precipitation for Great Britain J. Hydrol., 378 (2009), pp. 240–252

104. Kennish, 2002 M.J. Kennish Environmental threats and environmental future of estuaries Environ. Conserv., 29 (2002), pp. 78–107

105. Kimmerer, 2002 W.J. Kimmerer Effects of freshwater flow on abundance of estuarine organisms: physical effects or trophic linkages? Mar. Ecol. Prog. Ser., 243 (2002), pp. 39–55

106. King, 2004 D.A. King Climate change science: adapt, mitigate, or ignore? Science, 303 (2004), pp. 176–177

107. Kingsford et al., 2002 M.J. Kingsford, J.M. Leis, A. Shanks, K.C. Lindeman, S.G. Morgan, J. Pineda Sensory environments, larval abilities and local self-recruitment Bull. Mar. Sci., 70 (2002), pp. 309–340

108. Kirby, 2010 R. Kirby Distribution, transport and exchanges of fine sediment, with tidal power implications: Severn Estuary, UK Mar. Pollut. Bull., 61 (2010), pp. 21–36

109. Knight et al., 2015 P. Knight, T. Prime, J.M. Brown, K. Morrissey, A.J. Plater Application of flood risk modelling in a web-based geospatial decision support tool for coastal adaptation to climate change Nat. Hazards Earth Syst. Sci. Discuss., 3 (2015), pp. 1615–1642

110. Lanzoni and Seminara, 1998 S. Lanzoni, G. Seminara On tide propagation in convergent estuaries J. Geophys. Res., 103 (1998), pp. 30793–30812

111. Lawrence and Soame, 2004 A.J. Lawrence, J.M. Soame The effects of climate change on the reproduction of coastal invertebrates Ibis, 146 (2004), pp. 29–39

112. Lee and Cundy, 2001 S.V. Lee, A.B. Cundy Heavy metal contamination and mixing processes in sediments from the Humber Estuary, Eastern England Estuar. Coast. Shelf Sci., 53 (2001), pp. 619–636

113. Lewis et al., 2011 M.J. Lewis, K. Horsburgh, P. Bates, R. Smith Quantifying the uncertainty in future coastal flood risk estimates for the UK J. Coast. Res., 27 (2011), pp. 870–881

114. Lewis et al., 2013 M.J. Lewis, P. Bates, K. Horsburgh, J. Neal, G. Schumann A storm surge inundation model of the northern Bay of Bengal using publicly available data Q. J. R. Meteorol. Soc., 139 (2013), pp. 358–369

115. Lewis et al., 2014 M.J. Lewis, S.P. Neill, M.R. Hashemi, M. Reza Realistic wave conditions and their influence on quantifying the tidal stream energy resource Appl. Energy, 136 (2014), pp. 495–508

116. Lipp et al., 2001 E.K. Lipp, R. Kurz, R. Vincent, C. Rodriguez-Palacios, S.R. Farrah, J.B. Rose The effects of seasonal variability and weather on microbial fecal pollution and enteric pathogens in a subtropical estuary Estuaries, 24 (2001), pp. 266–276

117. Lipp et al., 2002 E.K. Lipp, A. Huq, R. Colwell Effects of global climate on infectious disease: the cholera model Clin. Microbiol. Rev., 15 (2002), pp. 757–770

118. Lowe et al., 2009 J.A. Lowe, T.P. Howard, A. Padaens, J. Tinker, J. Holt, S. Wakelin, G. Milne, J. Leake, J. Wolf, K. Horsburgh UK Climate Change Projections Science Report: Marine and Coastal Projections Met Office, Hadley Centre, Exeter (2009)

119. MacCready, 1998 P. MacCready Estuarine adjustment to changes in river flow and tidal mixing J. Phys. Oceanogr., 29 (1998), pp. 708–726

120. Malham et al., 2014 S.K. Malham, P. Rajko-Nenow, E. Howlett, K.E. Tuson, T.L. Perkins, D.W. Pallett, H. Wang, C.F. Jago, D.L. Jones, J.E. McDonald The interaction of human microbial pathogens, particulate material and nutrients in estuarine environments and their impacts on recreational and shellfish waters Environ. Sci. Process. Impacts, 16 (2014), p. 2145

121. Marani et al., 2007 M. Marani, A. D'Alpaos, S. Lanzoni, L. Carniello, A. Rinaldo Biologically-controlled multiple equilibria of tidal landforms and the fate of the Venice lagoon Geophys. Res. Lett., 34 (2007), p. L1140

122. Marba and Duarte, 2010 N. Marba, C.M. Duarte Mediterranean warming triggers seagrass (Posidonia oceanica) shoot mortality Glob. Change Biol., 16 (2010), pp. 2366–2375

123. Marques et al., 2010 A. Marques, M.L. Nunes, S.K. Moore, M.S. Strom Climate change and seafood safety: human health implications Food Res. Int., 43 (2010), pp. 1766–1779

124. Maskell et al., 2013 J. Maskell, K. Horsburgh, M. Lewis, P. Bates Investigating river-surge interaction in idealised estuaries J. Coast. Res., 30 (2013), pp. 248–259

125. Massa et al., 2009 S.I. Massa, S. Arnaud-Haond, G.A. Pearson, E.A. Serrao Temperature tolerance and survival of intertidal populations of the seagrass Zostera noltii (Hornemann) in Southern Europe (Ria Formosa, Portugal) Hydrobiologia, 619 (2009), pp. 195–201
126. McCabe et al., 2001 G.J. McCabe, M.P. Clark, M.C. Serreze Trends in Northern Hemisphere surface cyclone frequency and intensity J. Clim., 14 (2001), pp. 2763–2768
127. McCann et al., 2011 D.L. McCann, A.G. Davies, J.D. Bennell Bed roughness feedback in TELEMAC-2D and SISYPHE Proceedings of the XVIII Telemac and Mascaret User Club (2011), pp. 99–104
128. McDowell et al., 2014 W.G. McDowell, A.J. Benson, J.E. Byers Climate controls the distribution of a widespread invasive species: implications for future range expansion Freshw. Biol., 59 (2014), pp. 847–857
129. Mcinnes et al., 2003 K.L. Mcinnes, K.J.E. Walsh, G.D. Hubbert, T. Beer Impact of sea-level rise and storm surges on a coastal community Nat. Hazards, 30 (2003), pp. 187–207
130. McLusky and Elliott, 2004 D.S. McLusky, M. Elliott The Estuarine Ecosystem; Ecology, Threats and Management (third ed.)OUP, Oxford (2004) pp. 216
131. A.J. McMichael, R.E. Woodruff, S. Hales Climate change and human health: present and future risks Lancet, 367 (2006), pp. 859–869
132. Menendez and Woodworth, 2010 M. Menendez, P.L. Woodworth Changes in extreme high water levels based on a quasi-global tide-gauge dataset J. Geophys. Res., 115 (2010), p. C10011
133. Mieszkowska et al., 2013 N. Mieszkowska, L. Firth, M. Bentley Impacts of climate change on intertidal habitats Mar. Clim. Change Impacts Partnersh. Sci. Rev., 2013 (2013), pp. 180–190
134. Moksnes et al., 2014 P.-O. Moksnes, H.C.K. Tryman, R. Hordoir, P.R. Jonsson Larval behavior and dispersal mechanisms in shore crab larvae (Carcinus maenas): local adaptations to different tidal environments? Limnol. Oceanogr., 59 (2014), pp. 588–602
135. Moore et al., 2009 R.D. Moore, J. Wolf, A.J. Souza, S.S. Flint Morphological evolution of the Dee Estuary, Eastern Irish Sea, UK: a tidal asymmetry approach Geomorphology, 103 (2009), pp. 588–596
136. Morgan et al., 2012 B. Morgan, A.W. Rate, E.D. Burton Water chemistry and nutrient release during the resuspension of FeS-rich sediments in a eutrophic estuarine system Sci. Total Environ., 432 (2012), pp. 47–56
137. Najjar et al., 2010 R.G. Najjar, C.R. Pyke, M.B. Adams, D. Breitburg, C. Hershner, M. Kemp, R. Howarth, M.R. Mulholland, M. Paolisso, D. Secor, K. Sellner, D. Wardrop, R. Wood Potential climate-change impacts on the Chesapeake Bay Estuar. Coast. Shelf Sci., 86 (2010), pp. 1–20
138. Neill, 2009 S.P. Neill A numerical study of lateral grain size sorting by an estuarine front Estuar. Coast. Shelf Sci., 81 (2009), pp. 345–352
139. Nedwell et al., 1999 D.B. Nedwell, T.D. Jickells, M. Trimmer, R. Sanders Nutrients in Estuaries, Advances in Ecological Research, 08 (1999), pp. 60191–60199 http://dx.doi.org/10.1016/S0065-2504
140. Nicholls et al., 1999 R.J. Nicholls, F.M.J. Hoozemans, M. Marchand Increasing flood risk and wetland losses due to global sea-level rise: regional and global analyses Glob. Environ. Change Human Policy Dimens., 9 (1999), pp. S69–S87
141. Nicholls et al., 2011 R.J. Nicholls, N. Marinova, J.A. Lowe, S. Brown, P. Vellinga, D. de Gusmão, J. Hinkel Sea-level rise and its possible impacts given a 'beyond 4 degrees C world' in the twenty-first century Philos. Trans. R. Soc. A, 369 (2011), pp. 161–181
142. Nicholls et al., 2013 R.J. Nicholls, I.H. Townend, A.P. Bradbury, D. Ramsbottom, S.A. Day Planning for long-term coastal change: experiences from England and Wales Ocean Eng., 71 (2013), pp. 3–16
143. O'Donnell, 1993 J. O'Donnell Surface fronts in estuaries: a review Estuaries, 16 (1993), pp. 12–39
144. Oliphant and Thatje, 2014 A. Oliphant, S. Thatje Energetic adaptations to larval export within the brackish-water palaemonine shrimp, Palaemonetes varians Mar. Ecol. Prog. Ser., 505 (2014), pp. 177–191
145. Oliphant and Thatje, 2013 A. Oliphant, S. Thatje Per offspring investment implications for crustacean larval development: evolutionary insights into endotrophy and abbreviated development Mar. Ecol. Prog. Ser., 493 (2013), pp. 207–217
146. Oliphant et al., 2014 A. Oliphant, M. Ichino, S. Thatje The influence of per offspring investment (POI) and starvation on larval developmental plasticity within the palaemonid shrimp, Palaemonetes varians Mar. Biol., 161 (2014), pp. 2069–2077
147. Orford and Pethick, 2006 J.D. Orford, J.S. Pethick Challenging assumptions of coastal habitat formation in the 21st century Earth Surf. Process. Landf., 31 (2006), pp. 1625–1642
148. Paciorek et al., 2002 C.J. Paciorek, J.S. Risbey, V. Ventura, R.D. Rosen Multiple indices of Northern Hemisphere cyclone activity, winters 1949-99 J. Clim., 15 (2002), pp. 1573–1590
149. Perciasepe, 1998 R. Perciasepe EPA: ocean protection challenges – 'Daunting but manageable' Sea Technol., 39 (1998), pp. 25–29
150. Pethick, 1994 J.S. Pethick Estuaries and wetlands: function and form R.A. Falconer, P. Goodwin (Eds.), Wetland Management, Thomas Telford, London (1994), pp. 75–87
151. Pethick and Crooks, 2000 J.S. Pethick, S. Crooks Development of a coastal vulnerability index: a geomorphological perspective Environ. Conserv., 27 (2000), pp. 359–367
152. Philippart et al., 2003 C.J.M. Philippart, H.M. van Aken, J.J. Beukema, O.G. Bos, G.C. Cadée, R. Dekker Climate-related changes in recruitment of the bivalve Macoma balthica Limnol. Oceanogr., 48 (2003), pp. 2171–2185

153. Power et al., 2000 M. Power, M.J. Attrill, R.M. Thomas Environmental factors and interactions affecting the temporal abundance of juvenile flatfish in the Thames Estuary J. Sea Res., 43 (2000), pp. 135–149

154. Prandle, 2009 D. Prandle Estuaries: Dynamics, Mixing, Sedimentation, and Morphology University Press, Cambridge (2009)

155. Prandle and Lane, 2015 D. Prandle, A. Lane Sensitivity of estuaries to sea level rise: vulnerability indices Estuar. Coast. Shelf Sci., 160 (2015), pp. 60–68

156. Prime et al., 2015 T. Prime, J.M. Brown, A.J. Plater Physical and economic impacts of sea-level rise and low probability flooding events on coastal communities PLoS One, 10 (2015), p. e0117030

157. Prudhomme and Davies, 2009a C. Prudhomme, H. Davies Assessing uncertainties in climate change impact analyses on the river flow regimes in the UK. Part 1: baseline climate Clim. Change, 93 (2009), pp. 177–195

158. Prudhomme and Davies, 2009b C. Prudhomme, H. Davies Assessing uncertainties in climate change impact analyses on the river flow regimes in the UK. Part 2: future climate Clim. Change, 93 (2009), pp. 197–222 Prudhomme et al., 2012

159. C. Prudhomme, S. Dadson, D. Morris, J. Williamson, G. Goodsell, S. Crooks, L. Boelee, H. Davies, G. Buys, T. Lafon, G. Watts Future flows climate: an ensemble of 1-km climate change projections for hydrological application in Great Britain Earth Syst. Sci. Data, 4 (2012), pp. 143–148

160. Przeslawski et al., 2015 R. Przeslawski, M. Byrne, C. Mellin A review and meta-analysis of the effects of multiple abiotic stressors on marine embryos and larvae Glob. Change Biol., 21 (2015), pp. 2122–2140

161. Pye and Blott, 2006 K. Pye, S.J. Blott Coastal processes and morphological change in the Dunwich-Sizewell area, Suffolk, UK J. Coast. Res., 22 (2006), pp. 453–473

162. Pye and Blott, 2014 K. Pye, S.J. Blott The geomorphology of UK estuaries: the role of geological controls, antecedent conditions and human activities Estuar. Coast. Shelf Sci., 150 (2014), pp. 196–214

163. Quinn et al., 2013 N.D. Quinn, P.D. Bates, M. Siddall The contribution to future flood risk in the Severn Estuary from extreme sea level rise due to ice sheet mass loss J. Geophys. Res., 118 (2013), pp. 5887–5898

164. Quinn et al., 2014 N.D. Quinn, M. Lewis, M. Wadey, I. Haigh Assessing the variability in extreme storm-tide time-series for coastal flood risk assessment J. Geophys. Res., 119 (2014), pp. 4983–4998

165. Queiroga and Blanton, 2005 H. Queiroga, J. Blanton Interactions between behaviour and physical forcing in the control of horizontal transport of decapod crustacean Larvae Adv. Mar. Biol., 47 (2005), pp. 107–214

166. Rabalais et al., 2009 N.N. Rabalais, R.E. Turner, R.J. Diaz, D. Justic Global change and eutrophication of coastal waters ICES J. Mar. Sci., 66 (2009), pp. 1528–1537

167. Ravaux et al., 2012 J. Ravaux, N. Léger, N. Rabet, M. Morini, M. Zbinden, S. Thatje, B. Shillito Adaptation to thermally variable environments: capacity for acclimation of thermal limit and heat shock response in the shrimp Palaemonetes varians J. Comp. Physiol. B, 182 (2012), pp. 899–907

168. Rey et al., 2015 F. Rey, G.M.S. Neto, R. Rosa, H. Queiroga, R. Calado Laboratory trials reveal that exposure to extreme raining events prior to metamorphosis affect the post-settlement performance of an estuarine crab Estuar. Coast. Shelf Sci., 154 (2015), pp. 179–183

169. Robins and Davies, 2010 P.E. Robins, A.G. Davies Morphological controls in sandy estuaries: the influence of tidal flats and bathymetry on sediment transport Ocean Dyn., 60 (2010), pp. 503–517

170. Robins et al., 2012 P.E. Robins, S.P. Neill, L. Giménez A numerical study of marine larval dispersal in the presence of an axial convergent front Estuar. Coast. Shelf Sci., 100 (2012), pp. 172–185

171. Robins et al., 2014 P.E. Robins, M.J. Lewis, J.H. Simpson, E.R. Howlett, S.K. Malham Future variability of solute transport in a macrotidal estuary Estuar. Coast. Shelf Sci., 151 (2014), pp. 88–99

172. Rodríguez et al., 2012 R.A. Rodríguez, P.M. Gundy, G.K. Rijal, C.P. Gerba The impact of combined sewage overflows on the viral contamination of receiving waters Food Environ. Virol., 4 (2012), pp. 34–40

173. Rose et al., 2001 J.B. Rose, P.R. Epstein, E.K. Lipp, B.H. Sherman, S.M. Bernard, J.A. Patz Climate variability and change in the United States: potential impacts on water- and foodborne diseases caused by microbiologic agents Environ. Health Perspect., 109 (2001), pp. 211–221

174. Rose, 2005 J.B. Rose Identification and characterization of biological risks for water Management of Intentional and Accidental Water Pollution, Springer, Netherlands (2005) 2006. 85–101

175. Rossington et al., 2007 S.K. Rossington, R.J. Nicholls, M.A.F. Knaapen, Z.B. Wang Morphological behaviour of UK estuaries under conditions of accelerating sea-level rise River, Coastal and Estuarine Morphodynamics (2007) Enchede, Netherlands

176. Samuels and Burt, 2002 P.G. Samuels, N. Burt A new joint probability appraisal of flood risk Proc. Int. Civ. Eng. Water Mar. Eng., 154 (2002), pp. 109–115

177. Senior et al., 2002 C.A. Senior, R.G. Jones, J.A. Lowe, C.F. Durman, D. Hudson Predictions of extreme precipitation and sea-level rise under climate change Philos. Trans. R. Soc., 360 (2002), pp. 1301–1311

178. Sharps et al., 2015 E. Sharps, J. Smart, M.W. Skov, A. Garbutt, J.G. Hiddink Light grazing of saltmarshes is a direct and indirect cause of nest failure in common redshank Tringa totanus IBIS, 157 (2015), pp. 239–249

179. Shennan and Horton, 2002 I. Shennan, B. Horton Holocene land and sea level changes in Great Britain J. Quat. Sci., 17 (2002), pp. 511–526

180. Shennan et al., 2012 I. Shennan, G. Milne, S. Bradley Late Holocene vertical land motion and relative sea-level changes: lessons from the British Isles J. Quat. Sci., 27 (2012), pp. 64–70

181. Short and Wyllie-Echeverria, 1996 F.T. Short, S. Wyllie-Echeverria Natural and human-induced disturbance of seagrasses Environ. Conserv., 23 (1996), pp. 17–27

182. Simpson et al., 2001 J.H. Simpson, R. Vennell, A.J. Souza The salt fluxes in a tidally energetic estuary Estuar. Coast. Shelf Sci., 52 (2001), pp. 131–142

183. Smith et al., 2002 P.M. Smith, R.O.D. Dixon, M.P. Cochrane Plant Life of Edinburgh and the Lothians Edinburg University Press, Edinburgh (2002)

184. Smith et al., 2013 A. Smith, P. Bates, J. Freer, F. Wetterhall Future flood projection: investigating the application of climate models across the UK Hydrol. Process, 28 (2013), pp. 2810–2823

185. Sokolova and Portner, 2007 I.M. Sokolova, H.O. Portner The role of thermal environment in stress adaptation and tolerance: integration of multiple stressors J. Therm. Biol., 32 (2007), p. 117

186. Stachel et al., 2004 B. Stachel, R. Götz, T. Herrmann, F. Krüger, W. Knoth, O. Päpke, U. Rauhut, H. Reincke, R. Schwartz, E. Steeg, S. Uhlig The Elbe flood in August 2002-occurrence of polychlorinated dibenzo-p-dioxins, polychlorinated dibenzofurans (PCDD/F) and dioxin-like PCB in suspended particulate matter (SPM), sediment and fish Water Sci. Technol., 50 (2004), pp. 309–316

187. Statham, 2012 P.J. Statham Nutrients in estuaries – an overview and the potential impacts of climate change Sci. Total Environ., 434 (2012), pp. 213–227

188. Struyf et al., 2004 E. Struyf, S.V. Damme, P. Meire Possible effects of climate change on estuarine nutrient fluxes: a case study in the highly nitrified Schelde estuary (Belgium, The Netherlands) Estuar. Coast. Shelf Sci., 52 (2004), pp. 131–142

189. Svensson and Jones, 2002 C. Svensson, D.A. Jones Dependence between extreme sea surge, river flow and precipitation in eastern Britain Int. J. Climatol., 22 (2002), pp. 1149–1168

190. Tappin, 2002 A.D. Tappin An examination of the fluxes of nitrogen and phosphorus in temperate and tropical estuaries: current estimates and uncertainties Estuar. Coast. Shelf Sci., 55 (2002), pp. 885–901

191. Tessier et al., 2012 B. Tessier, I. Billeaud, P. Sorrel, N. Delsinne, P. Lesueur Infilling stratigraphy of macrotidal tide-dominated estuaries. Controlling mechanisms: sea-level fluctuations, bedrock morphology, sediment supply and climate changes (the examples of the Seine estuary and the Mont-Saint-Michel Bay, English Channel, NW France) Sediment. Geol., 279 (2012), pp. 62–73

192. Thompson et al., 2002 R.C. Thompson, T.P. Crowe, S.J. Hawkins Rocky intertidal communities: past environmental changes, present status and predictions for the next 25 years Environ. Conserv., 29 (2002), pp. 168–191

193. Torres et al., 2011 G. Torres, L. Giménez, K. Anger Growth, tolerance to low salinity, and osmoregulation in decapod crustacean larvae Aquat. Biol., 12 (2011), pp. 249–260

194. Uncles, 2003 R.J. Uncles From catchment to coastal zone: examples of the application of models to some long-term problems Sci. Total Environ., 314 (2003), pp. 567–588

195. Uncles et al., 2014 R.J. Uncles, J.A. Stephens, C. Harris Freshwater, tidal and wave influences on a small estuary Estuar. Coast. Shelf Sci., 150 (2014), pp. 252–261

196. Valdemarsen et al., 2014 1 T. Valdemarsen, C.O. Quintana, E. Kristensen, M.R. Flindt Recovery of organic-enriched sediments through microbial degradation: implications for eutrophic estuaries Mar. Ecol. Prog. Ser., 503 (2014 Jan 1), pp. 41–58

197. Vaquer-Sunyer and Duarte, 2011 R. Vaquer-Sunyer, C.M. Duarte Temperature effects on oxygen thresholds for hypoxia in marine benthic organisms Glob. Change Biol., 17 (2011), pp. 1788–1797

198. Wadey et al., 2014 M.P. Wadey, I.D. Haigh, J.M. Brown A century of sea level data and the UK's 2013/14 storm surges: an assessment of extremes and clustering using the Newlyn tide gauge record Ocean Sci. Discuss., 11 (2014), pp. 1995–2028

199. Wakelin et al., 2003 S.L. Wakelin, P.L. Woodworth, R.A. Flather, J.A. Williams Sea level dependence on the NAO over the NW European Continental Shelf Geophys. Res. Lett., 30 (2003) http://dx.doi.org/10.1029/2003GL017041

200. Walther et al., 2002 28 G.R. Walther, E. Post, P. Convey, A. Menzel, C. Parmesan, T.J. Beebee, J.M. Fromentin, O. Hoegh-Guldberg, F. Bairlein Ecological responses to recent climate change Nature, 416 (6879) (2002 Mar 28), pp. 389–395

201. Watts et al., 2015 G. Watts, R. Battarbee, J.P. Bloomfield, J. Crossman, A. Daccache, I. Durance, J.A. Elliott, G. Garner, J. Hannaford, D. Hannah, T. Hess, C. Jackson, A. Kay, M. Kernan, J. Knox, J. Mackay, D. Monteith, S. Ormerod, J. Rance, M. Stuart, A. Wade, S. Wade, K. Weatherhead, P. Whitehead, R. Wilby Climate change and water in the UK – past changes and future prospects Prog. Phys. Geogr., 39 (2015), pp. 6–28

202. Whitehead et al., 2009 P.G. Whitehead, R.L. Wilby, R.W. Battarbee, M. Kernan, A.J. Wade A review of the potential impacts of climate change on surface water quality Hydrol. Sci. J., 54 (2009), pp. 101–123

203. Winters et al., 2011 G. Winters, P. Nelle, B. Fricke, G. Rauch, T.B. Reusch Effects of a simulated heat wave on photophysiology and gene expression of high- and low-latitude populations of Zostera marina Mar. Ecol. Prog. Ser., 435 (2011), pp. 83–95

204. Wood et al., 2003 R.A. Wood, M. Vellinga, R. Thorpe Global warming and thermohaline circulation stability Philos. Trans. R. Soc. Lond. A Math. Phys. Eng. Sci., 361 (2003), pp. 1961–1975

205. Woodworth and Blackman, 2004 P.L. Woodworth, D.L. Blackman Evidence for systematic changes in extreme high water since the mid-1970s J. Clim., 17 (2004), pp. 1190–1197

206. Woodworth et al., 2009 P.L. Woodworth, F.N. Teferle, R.M. Bingley, I. Shennan, S.D. Williams Trends in UK

mean sea level revisited Geophys. J. Int., 176 (2009), pp. 19–30

207. Worth et al., 2006 K. Worth, R. Weisse, H. von Storch Climate change and North Sea storm surge extremes: an ensemble study of storm surge extremes expected in a changed climate projected by four different regional climate models Ocean Dyn., 56 (2006), pp. 3–15

208. Yamanaka et al., 2013 T. Yamanaka, D. Raffaelli, P.C. White Non-linear interactions determine the impact of sea-level rise on estuarine benthic biodiversity and ecosystem processes PLoS One, 8 (2013), p. e68160

209. Yang et al., 2015 Z.Q. Yang, T.P. Wang, N. Voisin, A. Copping Estuarine response to river flow and sea-level rise under future climate change and human development Estuar. Coast. Shelf Sci., 156 (2015), pp. 19–30

210. Yates et al., 2010 C.J. Yates, J. Elith, A.M. Latimer, D. Le Maitre, G.F. Midgley, F.M. Schurr, A.G. West Projecting climate change impacts on species distributions in megadiverse South African Cape and Southwest Australian Floristic Regions: opportunities and challenges Austral Ecol., 35 (2010), pp. 374–391

211. Ysebaert et al., 2003 T. Ysebaert, P.M.J. Herman, P. Meire, J. Craeymeersch, H. Verbeek, C.H.R. Heip Large-scale spatial patterns in estuaries: estuarine macrobenthic communities in the Scheldt estuary, NW Europe Estuar. Coast. Shelf Sci., 57 (2003), pp. 335–355

212. Zhang et al., 2004 K.Q. Zhang, B.C. Douglas, S.P. Leatherman Global warming and coastal erosion Clim. Change, 64 (2004), pp. 41–58

213. Zheng et al., 2013 F. Zheng, S. Westra, S.A. Sisson Quantifying the dependence between extreme rainfall and storm surge in the coastal zone J. Hydrol., 505 (2013), pp. 172–187

214. Zheng et al., 2014 F. Zheng, S. Westra, M. Leonard, S.A. Sisson Modeling dependence between extreme rainfall and storm surge to estimate coastal flooding risk Water Resour. Res., 50 (2014), pp. 2050–2071

CHAPTER
12

CHANGES IN PRECIPITATION EXTREMES PROJECTED BY A 20-KM MESH GLOBAL ATMOSPHERIC MODEL

Akio Kitoh, Hirokazu Endo

Keywords: Heavy precipitation, Climate change, High-resolution, GCM

INTRODUCTION

Both too much water and too little water are of great concern for human life because the contrasts in precipitation between wet and dry regions and between wet and dry seasons are projected to increase in a coming future world (Intergovernmental Panel on Climate Change (IPCC), 2013). Warmer climate should theoretically lead to more precipitation extremes due to increasing atmospheric water vapor content (Allen and Ingram, 2002 and Allan and Soden, 2008). The intensity of precipitation extremes, however, depends on not only water vapor content but also on atmospheric environmental changes (O›Gorman and Schneider, 2009). Based on global climate model (GCM) simulations of the Coupled Model Intercomparison Project Phase 3 (CMIP3, Meehl et al., 2007), it is assessed that the frequency of heavy precipitation or the proportion of total rainfall from heavy rainfalls will likely increase in the 21st century over many areas of the globe (IPCC, 2012). The same assessment was made with the CMIP5 (Taylor et al., 2012) models under Representative Concentration Pathways (RCPs) such that extreme precipitation events over most of the mid-latitude land masses and over wet tropical regions will very likely become more intense and more frequent by the end of this century, as global mean surface temperature increases (IPCC, 2013).

The CMIP5 models have better performance than the previous CMIP3 models in simulating precipitation extremes in the present climate (Sillmann et al., 2013a). A part of this improvement may come from increasing horizontal resolution, about 280 km in CMIP3 versus about 200 km in CMIP5. Spatial distribution of precipitation indices is reasonably reproduced but differences are still found in precipitation intensity where the magnitude of precipitation extremes is underestimated by climate models (Sillmann et al., 2013a and Mehran et al., 2014).

In a future warming world, change rate of heavy precipitation amounts will generally increase more than that of annual mean precipitation (Tebaldi et al., 2006, Sun et al., 2007 and Sillmann et al., 2013b). The increasing rate of annual mean precipitation at the end of the 21st century projected by CMIP5 models in RCP8.5 scenario is 9% (median value), while that in simple daily intensity index (SDII) defined as annual total precipitation divided by the number of wet days is 12% and that in annual maximum 5-day precipitation total (R5d) is 20% (Sillmann et al., 2013b). This increasing rate depends on the scenario, where the change ratio of R5d is 6% in RCP2.6 and 10% in RCP4.5, respectively. Similarly, 20-year or 50-year return values of daily precipitation are projected to increase and 20-year or 50-year return periods for the present precipitation events will reduce in the future almost everywhere except for subtropical dry regions (Kharin et al., 2013 and Toreti et al., 2013). The 20-year return values increase about 6% per a degree change in annual mean surface temperature, but with considerable inter-model variability (Kharin et al., 2013). Large variability is also known among models in simulated increase of precipitation extremes, particularly in summer when organized convection matters (Toreti et al., 2013).

Both thermodynamical and dynamical factors are responsible for regional precipitation changes, with the former dominating the latter for extreme precipitation changes (Emori and Brown, 2005). Increase in maximum atmospheric water vapor due to temperature rise is a principal factor for increasing

Citation: Akio Kitoh, Hirokazu Endo, Changes in precipitation extremes projected by a 20-km mesh global atmospheric model, http://dx.doi.org/10.1016/j.wace.2015.09.001.

the intensity of individual precipitation events (Kunkel et al., 2013). There are additional factors such as changes in updrafts, which controls changes in moisture flux convergence, and soil moisture over land. Representation of precipitation, particularly its intensity, depends on the horizontal resolution of models. Coarse horizontal resolution of GCMs tends to prevent simulating realistic extreme events (Sun et al., 2006 and Min et al., 2011). High-resolution modeling or statistical downscaling is then necessary to project weather and climate extremes and their future changes under global warming.

Dynamical downscaling is able to simulate more realistic, though not perfect, extreme events. Regional climate models (RCMs) would add values on more realistic topography and lower boundary conditions and better representation of dynamical processes (e.g.,Takayabu et al., 2015). Therefore RCMs are widely used to project future changes in extremes. For example, Fowler et al. (2007) used six RCMs integrations under the SRES A2 scenario to downscale extreme precipitation changes over Europe at the end of the 21st century. Horizontal resolution of these RCMs is about 50 km. They found that all RCMs project increases in magnitude of extreme precipitation from 1-day to 10-day duration for most of Europe. It is also found that the magnitude of change is influenced by the driving GCMs but projected scatter is moderated by the RCM probably due to better representation of extreme events in RCM. RCMs also give different spatial pattern of precipitation extremes than that obtained by parent GCMs.

Another way for downscaling is to use an atmospheric general circulation model (AGCM) with horizontally high-resolution (Kitoh et al., 2009 and Kitoh et al., 2015). Higher resolution model is needed to better reproduce precipitation climatology such as the East Asia summer rain band (Kitoh and Kusunoki, 2008), over India (Rajendran et al., 2013), South America (Kitoh et al., 2011), Central America and the Caribbean (Nakaegawa et al., 2014). Kamiguchi et al. (2006) analyzed future changes of precipitation indices between the present climate and the future climate at the end of the 21st century under the IPCC Special Report on Emission Scenario (SRES) A1B scenario by two 10-year simulations with the global 20-km mesh AGCM developed at the Meteorological Research Institute (MRI) version 3.1 (MRI-AGCM3.1, Mizuta et al., 2006). They found significantly increased heavy precipitation in South Asia, the Amazon and West Africa. Over the Amazon, an increase in dry spell is found during the dry season in the future, which is related to the Walker circulation changes associated with the El Niño-like SST changes in the future.

The MRI-AGCM has been updated recently (MRI-AGCM3.2, Mizuta et al., 2012), which improved heavy precipitation climatology around the tropical western Pacific and the global distribution of tropical cyclones (Murakami et al., 2012). In this paper, after evaluating the present-day precipitation extremes of 20-km mesh MRI-AGCM3.2, we analyze projected future changes in precipitation extremes at the end of the 21st century. The scenario is based on RCP8.5, and four different spatial patterns in sea surface temperature (SST) changes are used as boundary conditions. This type of simulations, which uses the observed present-day interannually varying SST plus ensemble mean future SST changes obtained by CMIP-class models, can minimize the effects of climate model bias (Kitoh et al., 2015). In particular, the 20-km mesh MRI-AGCM has superiority in reproducing precipitation extremes in its present-day experiments; the results obtained in the future climate projections can be used widely for impact studies from future changes of extremes.

MODEL AND EXPERIMENT

MRI AGCM

We used the MRI-AGCM3.2 (Mizuta et al., 2012), which is the updated version from the MRI-AGCM3.1 (Mizuta et al., 2006). The model is based on a hydrostatic primitive equation system using a spectral transform method of spherical harmonics. The 20-km mesh version uses a triangular truncation at wave number 959 (TL959) in the horizontal, which has 1920×960 grid points. There are 64 layers in the vertical with a top at 0.01 hPa. A two-time-level semi-implicit semi-Lagrangian scheme is employed for time integration. For cumulus parameterization scheme, a new mass-flux type scheme (Yoshimura et al., 2015) is used. This Yoshimura scheme is based on Tiedtke (1989) scheme with a detailed entraining and detraining plume within a single grid cell, but also allows multiple convective updrafts with different heights to exist as with the Arakawa-Schubert (1974)type scheme. Prognostic variables include cloud water and cloud amount. The radiation code considers absorptions by greenhouse gases in the long wave scheme. The direct effect of aerosol (sulfate, black carbon, organic carbon, mineral dust, and sea salt) is included, but indirect effects are not considered in this experiment.

Experiment Method

We have conducted the Atmospheric Model Intercomparison Project (AMIP)-type simulations for the present climate (1979–2003) and the future climate at the end of the 21st century (2075–2099). The present climate simulation used the observed interannually varying monthly mean SST and sea-ice concentration during 1979–2003 based on the HadISST1.1 data (Rayner et al., 2003). For the future climate, the boundary SST data were prepared by superposing the future change in the multi-model ensemble of SST projected by CMIP5 multi-model dataset to the present-day observed SST. In short, the future SST is the sum of (i) the trend in the multi-model ensemble (MME) of SST projected by CMIP5 multi-model dataset, (ii) future change in MME of SST for 2075–2099 and (iii) de-trended observed SST for the period 1979–2003. See Mizuta et al., 2008 and Mizuta et al., 2014 for the details. Fig. 1a shows the annual mean SST changes of the 28 CMIP5 models. It is noted that there is a large contrast in SST warming between the Northern Hemisphere and the Southern Hemisphere, with the former warming more than the latter. In the tropics, relatively larger warming is noted in the central and eastern equatorial Pacific. Over the Indian Ocean, a warming in the western Indian Ocean is larger than that in the eastern Indian Ocean.

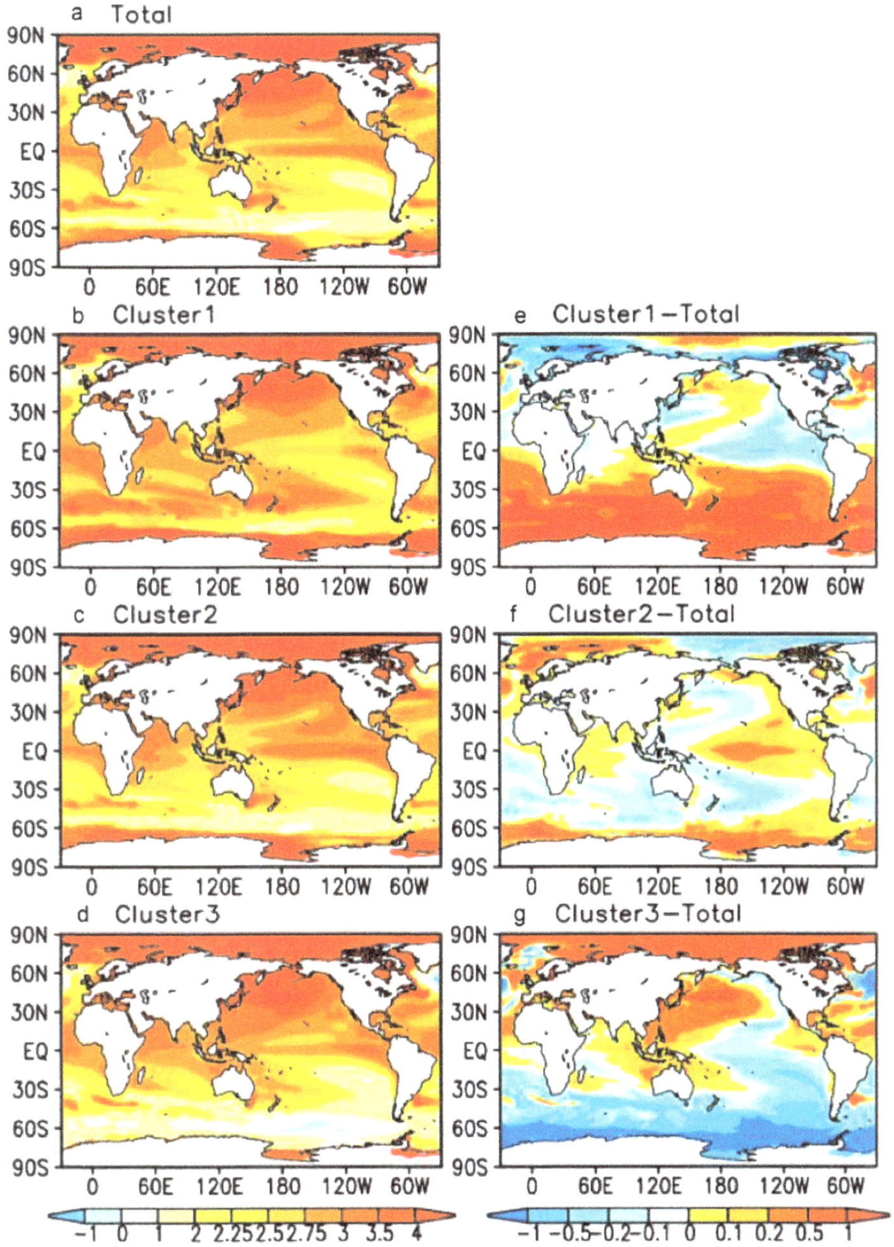

Figure 1: Annual-mean sea surface temperature changes (K) between the present (1979–2003 average, historical experiment) and the end of the 21st century (2075–2099 average, RCP 8.5 experiment), for (a) the composite of total 28 models, and (b, c, d) the composites of the three clusters of the models. (e, f, g) Differences for each cluster from the total mean.

In order to assess the uncertainty in projections, we made three other simulations with different SST spatial patterns in the future. Three SST patterns are obtained by a cluster analysis of 28 CMIP5 RCP8.5 experiments (Mizuta et al., 2014), and are shown in Fig. 1. In a clustering procedure, an equal weight is given in all 28 models. Values are normalized so that the tropical (30°S to 30°N) mean SST changes from the present (1979–2003 of historical experiment) to the end of the 21st century (2075–2099 of RCP8.5 experiment) become the same as that in the total 28 models mean. Then clustering is performed using normalized tropical SST changes because the effect of tropical SST should be larger than that of the mid- to high-latitude oceans. The clustering is finished when the final three clusters are obtained. Finally, 8, 14, and 6 models are classified as Cluster 1, Cluster 2, and Cluster 3, respectively. Cluster 1 is characterized by a nearly uniform warming in the both hemispheres, while cluster 3 is dominated by a larger warming in the Northern Hemisphere than in the Southern Hemisphere. Cluster 2 shows a larger warming over the central equatorial Pacific (so-called El Niño-like pattern). The CMIP5 models tend to show greater precipitation increases over larger SST warming ("warmer-get-wetter" pattern). These characteristics of the three clusters are discussed in Mizuta et al. (2014).

For the present climate experiment, we have performed two simulations with different initial conditions. For the future, four member simulations with different SST patterns were performed. Initial conditions are arbitrary chosen from a previous long-term climate simulation because memory of atmospheric initial conditions is lost within a month or so. A few months spin-up simulation is performed before starting the main integration period. In this study, we use the averages of two members for the present (50 years total), and the averages of four members for the future (100 years total). Sensitivity to different SST forcing on regional precipitation changes is also discussed. The annual mean SST changes between the present (1979–2003 mean) and the future (2075–2099 mean) averaged in the tropics (30°S to 30°N) is set to 2.74 °C, which is the average of 28 CMIP5 models of the RCP8.5 experiments (Mizuta et al., 2014). Simulated global annual mean surface air temperature changes by the MRI-AGCM3.2 between the present and the future is 3.49 °C.

Observed Data

For verification of the simulated precipitation, we used two observed dataset. One is the 0.25-degree resolution TRMM-3B42V7 (Huffman et al., 2007) for the period 1998–2013. The other is the GPCP-1DD v.1.2 of the Global Precipitation Climatology Project (GPCP) with a 1-degree by 1-degree grid (Huffman and Bolvin, 2009). This data set also covers 1998–2013. The data period is different between the observed data and the model simulation, but does not affect our study as we investigate climatological statistics.

Data Processing

We used five precipitation indices, out of which four extreme precipitation indices are the ones suggested by the Expert Team on Climate Change Detection and Indices (ETCCDI;Klein Tank et al., 2009), as shown in Table 1: average precipitation (Pav), simple precipitation daily intensity index (SDII), annual maximum 5-day precipitation total (R5d), annual maximum 1-day precipitation total (R1d), and annual maximum consecutive dry days (CDD).

Table 1: Definition of the precipitation extreme indices.

Index	Unit	Definition
Pav	mm/day	Annual mean precipitation
SDII	mm/day	Simple daily intensity index: Annual total precipitation divided by the number of wet days (daily precipitation≥1 mm)
R5d	mm	Annual maximum 5-day precipitation total
R1d	mm	Annual maximum 1-day precipitation total
CDD	day	Annual maximum number of consecutive dry days (daily precipitation<1 mm)

All the precipitation indices for the observations and models are calculated on their original grid. In addition, precipitation indices calculated from model data spatially averaged into 1-degree by 1-degree boxes are also calculated to check their resolution dependency.

Regional analysis is performed over the twenty-two land sub-regions used in the IPCC Fourth Assessment Report (Christensen et al., 2007), as shown in Fig. 2. For calculation of the regional change ratio for the indices, the indices on their original grid are averaged for different regions for the present (1979–2003) and the future (2075–2099), and then the changes in the regionally averaged indices are divided by the corresponding regionally averaged indices in the present.

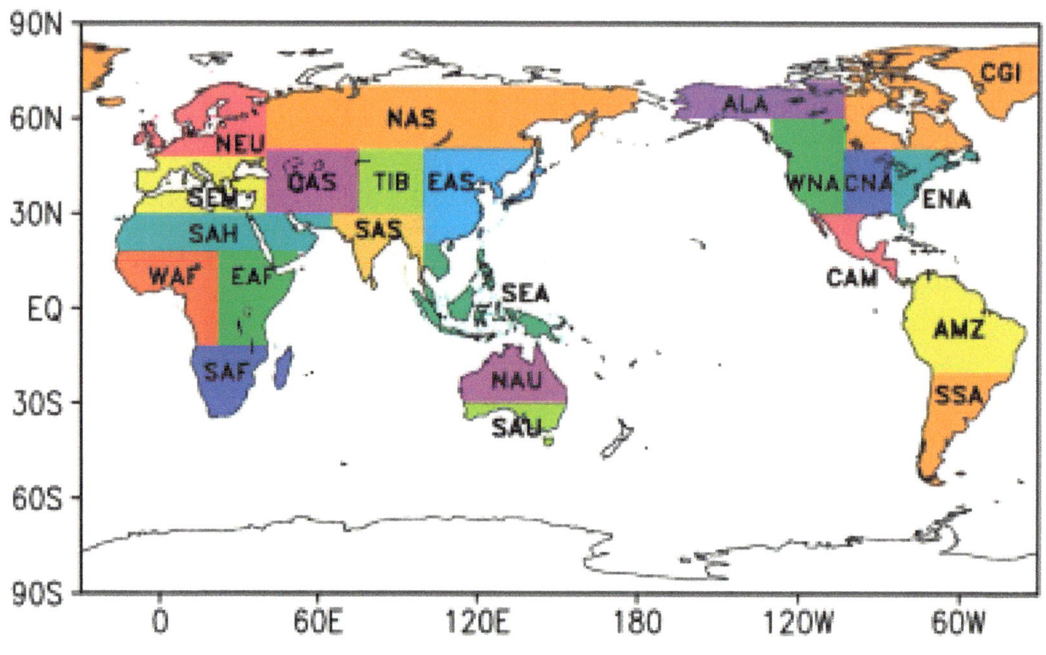

Figure 2: Regional domain used in this study.

EVALUATION OF PRESENT-DAY CLIMATE

MRI-AGCM3.2 shows a good skill in simulating dynamical and hydro-meteorological fields such as monsoon precipitation. The model also reproduces various characteristics of tropical cyclones such as their intensity and global distribution (Mizuta et al. 2012).

Here we evaluate five extreme precipitation indices, Pav, SDII, R5d, R1d and CDD, whose definition is shown in Table 1. Fig. 3 shows global distributions of these five indices based on daily precipitation data both for observations and the 20-km mesh MRI-AGCM3.2. Here extreme precipitation indices based on TRMM-3B42 are shown as 16-year averages for the period 1998–2013. We note that quantitatively there is large discrepancy in values between the two observed dataset, namely TRMM-3B42 and GPCP-1DD, where GPCP-1DD shows considerably smaller amount in precipitation extremes, as will be shown later.

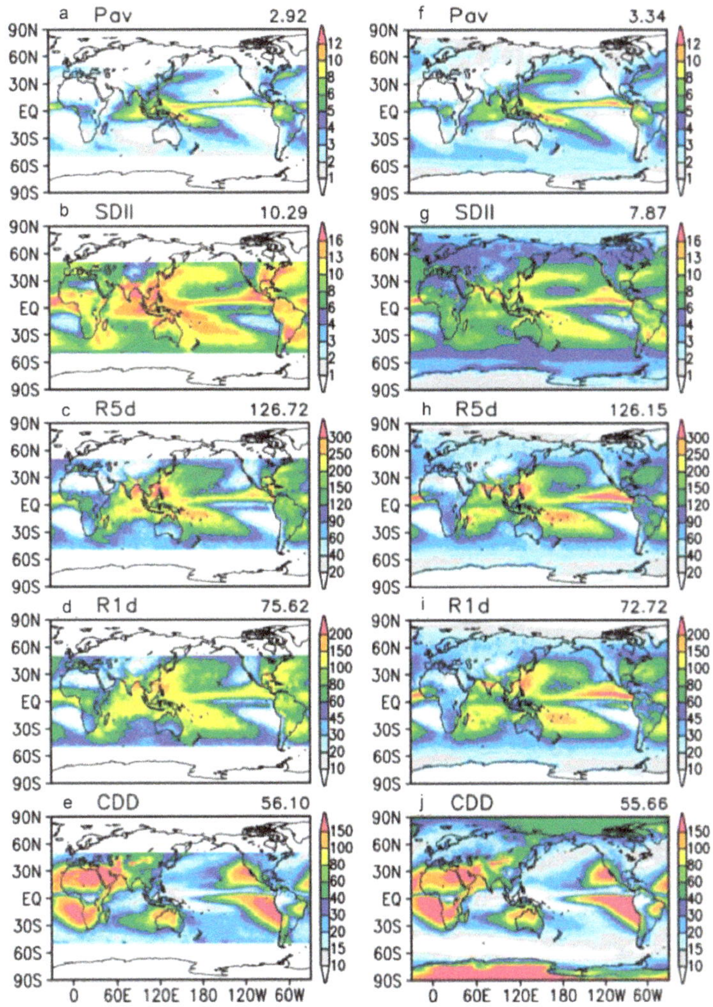

Figure 3: (left) Observed and (right) simulated annual precipitation indices. (a, f) Pav, (b, g) SDII, (c, h) R5d, (d, i) R1d, (e, j) CDD. Observations are based on TRMM-3B42 for the period 1998–2013. See Table 1 for units. Area averages for 50°S to 50°N are shown in upper right corner of each figure.

The model has reproduced the annual mean precipitation (Pav) very well in its geographical pattern (Fig. 3a and f). Simulated Pav averaged for 50 °S to 50 °N overestimated the TRMM observations by 10%. The MRI-AGCM3.2 shows maxima in annual precipitation over the oceans, such as the western tropical Pacific, Intertropical Convergence Zone (ITCZ), South Pacific Convergence Zone (SPCZ), the Indian Ocean, and mid-latitudes to the south of Japan and to the southeast of North America. Over land the model reproduces heavy rainfall area over Amazon. The 20-km mesh high-resolution enables a narrow precipitation area along the southern periphery of Himalaya. As discussed in Mizuta et al. (2012), monsoonal seasonal migration of precipitation zones between winter and summer is also well captured.

There is quantitatively large discrepancy in simple daily precipitation intensity (SDII) between the model and simulations (Fig. 3b and g). The MRI-AGCM3.2 significantly (more than 20% in the 50 °S to 50 °N mean) underestimated the TRMM observations. As the annual precipitation is about 10% larger in the model, this means the number of wet days (daily precipitation≥1 mm) being overestimated in the model, i.e., too many wet days with moderate-to-light precipitation (Mizuta et al. 2012). This too many drizzle-day syndrome (models produce precipitation too frequently and too lightly) is found in many GCMs (Dai, 2006 and Stephens et al., 2010). This may be related to models' horizontal resolution, because precipitation is calculated for a considerably larger domain than actual precipitating clouds.

Simulated annual maximum 5-day precipitation total (R5d) and annual maximum 1-day precipitation total (R1d), on the other hand, matches well the observations (Fig. 3 c vs. h and Fig. 3d vs. 3i). However, the model overestimated both R5d and R1d over the eastern tropical Pacific. TRMM observations show the maxima over the Philippines Sea, the South China Sea, and the Bay of Bengal. The model tends to underestimate R5d and R1d over the maritime continent and at the equator.

The current model (MRI-AGCM3.2) is much better than the previous version of the model (MRI-AGCM3.1). For example, the R5d value in SAS was 104.3 mm in MRI-AGCM3.1 (Kamiguchi et al., 2006) while it is 173.6 mm in our results. As TRMM and GPCP show 184.3 mm and 157.8 mm, respectively, our model gives much closer agreement with the observations.

Fig. 3e and j shows the CDD, maximum number of consecutive dry days (daily precipitation<1 mm). The model generally reproduces CDD well.

Fig. 4 shows the regional averages of observed and simulated annual precipitation indices. Twenty-two regional domains used in this study are shown in Fig. 2. Here we show two observations based on TRMM-3B42 and GPCP-1DD. Note that TRMM data does not cover high-latitudes. In order to check the resolution difference between the two observed datasets, simulated data re-gridded into 1-degree by 1-degree (red crosses) are also plotted together with original 20-km mesh data (black crosses). Spatial averaging into 1-degree by 1-degree causes the reduction in SDII and R1d. However, it does not affect much R5d, and very little in Pav. Differences between TRMM-3B42 and GPCP-1DD are particularly large in R1d, and larger than the differences between the original and re-gridded simulated values. This implies that the differences between TRMM-3B42 and GPCP-1DD are not solely responsible for their resolution.

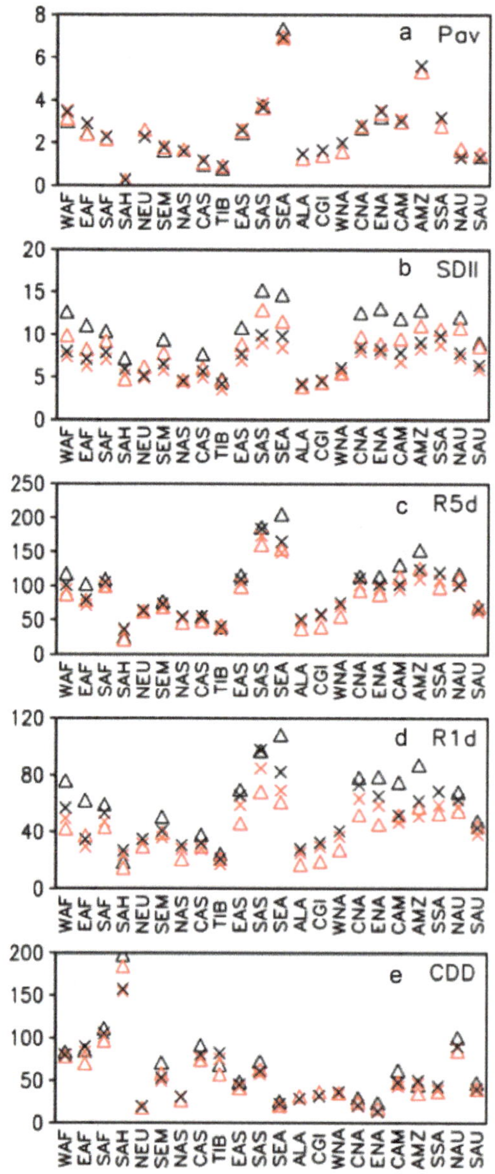

Figure 4: Regional averages of observed and simulated annual precipitation indices in (a) Pav, (b) SDII, (c) R5d, (d) R1d, (e) CDD. Observations are based on TRMM-3B42 (black triangles) and GPCP-1DD (red triangles) for the period 1998–2013. Simulated data re-gridded into 1-degree by 1-degree (red crosses) are also plotted together with original 20-km mesh data (black crosses). (For interpretation of the references to color in this figure legend, the reader is referred to the web version of this article.)

Comparison of regionally averaged Pav reveals that the MRI-AGCM3.2 reproduces regional annual mean precipitation amount very well (Fig. 4a). The model tends to overestimate Pav over dry regions (SAH, CAS, ALA, CGI, WNA). In other tropical and mid-latitude wet regions, the model reproduces the observed values within 20%.

Maxima in simulated Pav are found in SEA and AMZ, followed by SAS, as in the observations. Sillmann et al. (2013a) noted that CMIP5 models underestimated the total annual precipitation in SEA and overestimated in AMZ compared with HadEX2 dataset (Donat et al., 2013). Although we are using different observations, these biases are not seen in MRI-AGCM3.2 simulations.

Fig. 4b shows the regional means of SDII. The SDII estimated by GPCC-1DD are smaller than that calculated by TRMM-3B42 in all regions where data is available. Even with this large uncertainty in the observations, simulated SDII is underestimated in almost all regions. Exceptions are found in cold regions such as ALA, CGI and WNA. Largest SDII is observed in SAS, followed by SEA, while total precipitation (Pav) is largest in SEA, not in SAS. The MRI-AGCM3.2 simulated these characteristics.

Fig. 4c and d shows the regional means of R5d and R1d. The model reproduces regional differences in these two extreme precipitation indices, and simulated values are found in between the two observations in most areas. Maxima in R5d and R1d are located in SEA and SAS both in the observations and the model. There are ambiguities between TRMM and GPCP, namely, the R1d maximum is in SEA followed by SAS in TRMM but in opposite order in GPCP. The MRI-AGCM3.2 simulates the R1d maximum in SAS followed by SEA.

Regional distribution of CDD is also well simulated by the MRI-AGCM3.2 (Fig. 4e). Small CDD values in SEA are noted both in the observations and model simulation. A relatively long CDD appears in SAS, where mean value as well as excessive precipitation is large.

Overall five precipitation extremes indices simulated in MRI-AGCM3.2 in its present-day climate experiment are in good agreement with those observed. In the next section, we investigate future changes in these indices at the end of the 21st century in RCP8.5 scenario.

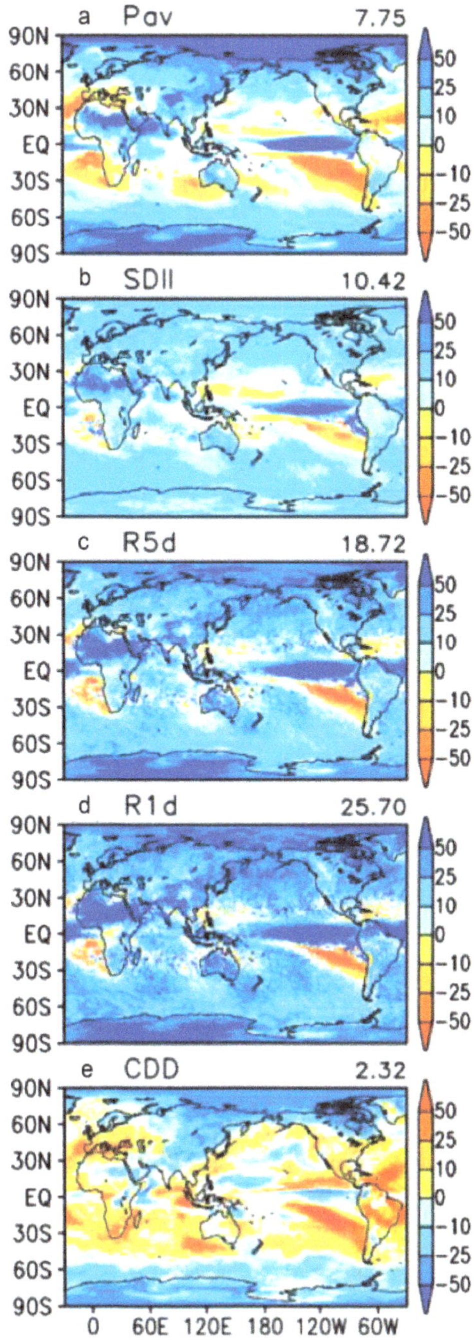

Figure 5: Future change ratio (%) in (a) Pav, (b) SDII, (c) R5d, (d) R1d, (e) CDD. Global averages are shown in upper right corner of each figure. Shadings denote where all four future members have the same sign of the change. A spatial nine-point smooth filter is applied to reduce small-scale noise. Global average values are shown in upper right corner of each figure.

FUTURE CHANGES IN PRECIPITATION EXTREMES

Fig. 5 shows the projected future changes in five precipitation indices at the end of the 21st century, as indicated by percent changes from the present values. The present corresponds to the average of two members for 1979–2003. The future is the average of four members for 2075–2099. Shadings denote where all four future members have the same sign of change compared to the averages of present-day simulations.

Global-scale spatial pattern of annual mean precipitation changes (Fig. 5a) is very similar to that of the CMIP5 multi-model mean assessed in IPCC (2013) and other references. Annual mean precipitation is projected to increase in the high latitudes, the equatorial Pacific Ocean and many of mid-latitude wet regions, while decrease in many mid-latitude and subtropical dry regions. This similarity, at least in its large-scale features, can be understood by the fact that our experiment strategy uses the CMIP5 multi-model mean SST changes and precipitation changes are strongly affected by SST anomalies (a warmer-get-wetter pattern; Xie et al., 2010; Chadwick et al., 2013).

Projected future changes in SDII (Fig. 5b) generally show a similar geographical pattern to that in Pav, but with smaller area in negative changes. The Mediterranean and the region to the southwest of Australia are among such areas. Area with negative future changes becomes much smaller in more extreme precipitation indices of R5d and R1d (Fig. 5c and d). Globally averaged change ratio become larger in R5d (18.72%) and more in R1d (25.70%) than that in SDII (10.42%) and Pav (7.75%). Normalized by the global annual mean surface air temperature change, they correspond to 2.22%/°C (Pav), 2.99%/°C (SDII), 5.36%/°C (R5d) and 7.36%/°C (R1d), respectively. The projected 18.72% increase in R5d for the period 2075–2099 in RCP8.5 scenario is within CMIP5 models range (Sillmann et al., 2013b). Almost all land grid points show positive changes in R5d and R1d with some exceptions near some coastal areas.

Fig. 5e shows the future changes in CDD. Annual CDD is projected to decrease in the high latitudes, eastern part of the Eurasian continent, and some oceanic areas such as the eastern equatorial Pacific. Areas with increasing CDD are much broader than where Pav decreases. This means that there are regions where annual mean precipitation is projected to increase, and at the same time the number of consecutive dry days becomes longer as already noted (IPCC, 2013).

Among the four SST clusters experiments, there are some differences in the magnitude of changes in extreme precipitation indices, although spatial patterns of precipitation changes are similar each other. It is found that the cluster 3 experiment gives the largest R1d, R5d and SDII, while the cluster 1 experiment gives the least. For example, the global mean R1d ranges from 23.95% (cluster 1) to 26.86% (cluster 3). The cluster 3 SST experiment resulted in heaviest precipitation in the central and eastern equatorial Pacific. It should be noted that uncertainty in future SST change pattern results in some quantitative differences.

Fig. 6 shows the percent changes in the five precipitation indices but for June–July–August (JJA) and December–January–February (DJF) seasons. Changes in Pav are again generally similar to CMIP5 results. Seasonal shift of projected changes is seen. Negative Pav anomalies cover whole Europe in JJA, while they are found in the Mediterranean region and North Africa in DJF. In South Africa, negative precipitation changes are projected in JJA season. In wintertime northern America, negative Pav anomalies are projected in southwestern US and Mexico. Over Amazon, precipitation will increase in summer (DJF), but decrease in winter (JJA).

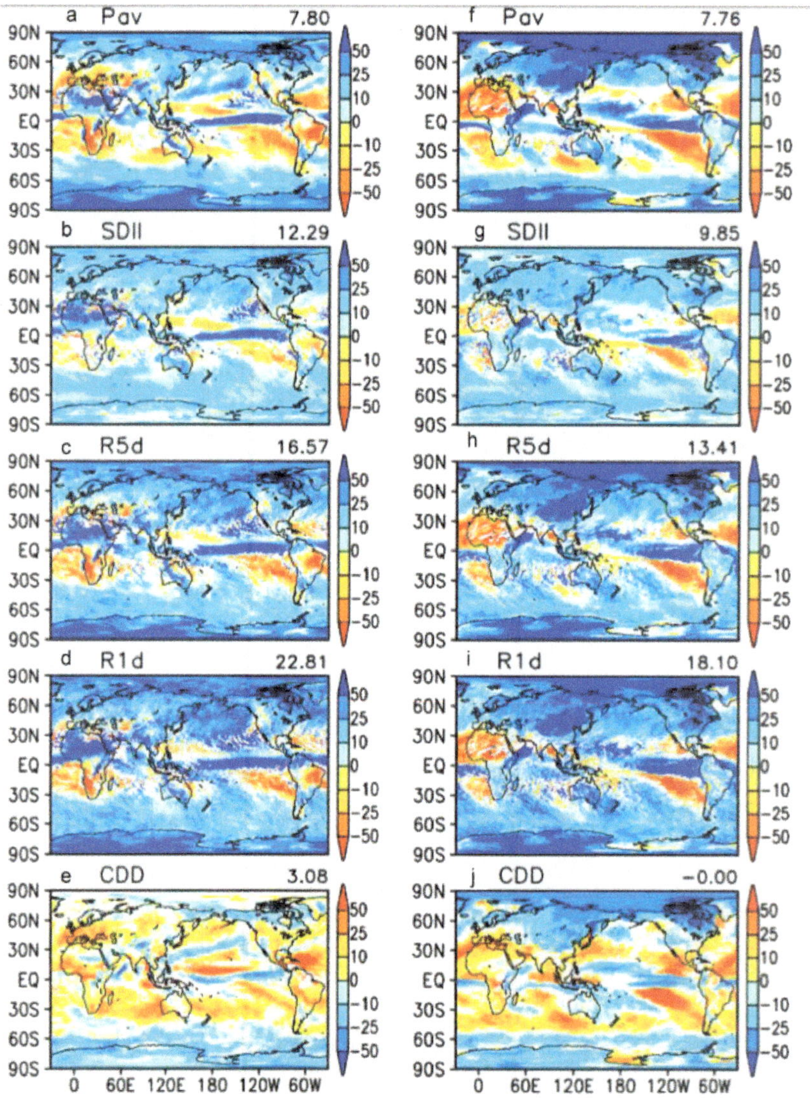

Figure 6: As in Fig. 5 but for (a–e) June–July–August, and (f–j) December–January–February.

The above contrast between summer and winter can also be seen in SDII, R5d and R1d in the low-latitudes where extremely large precipitation is projected to increase in the summer hemisphere land regions, and decrease in the winter hemisphere. The changes with opposite sign to CMIP5 projections are seen in the western tropical Pacific where SDII, R5d and R1d are projected to decrease in summer (JJA). Importance of coupled ocean–atmosphere process in the western tropical Pacific is demonstrated for seasonal prediction because atmospheric feedback on SST is critical in this region (Wang et al., 2005). Therefore, disagreement projected precipitation changes between the CMIP5 models and our experiment can be due to our experiment setting where we used fixed SST as lower boundary conditions. Large percentage changes appear in some dry regions such as Sahara due to very little precipitation in the present climate. CDD is projected to increase in northern summer in Europe, north Asia and North America.

Next we investigate future changes in regional average annual precipitation indices (Fig. 7). As we have made four members simulations for the future climate projections with different SST patterns, all four results are shown to check uncertainty of future regional precipitation extremes projections.

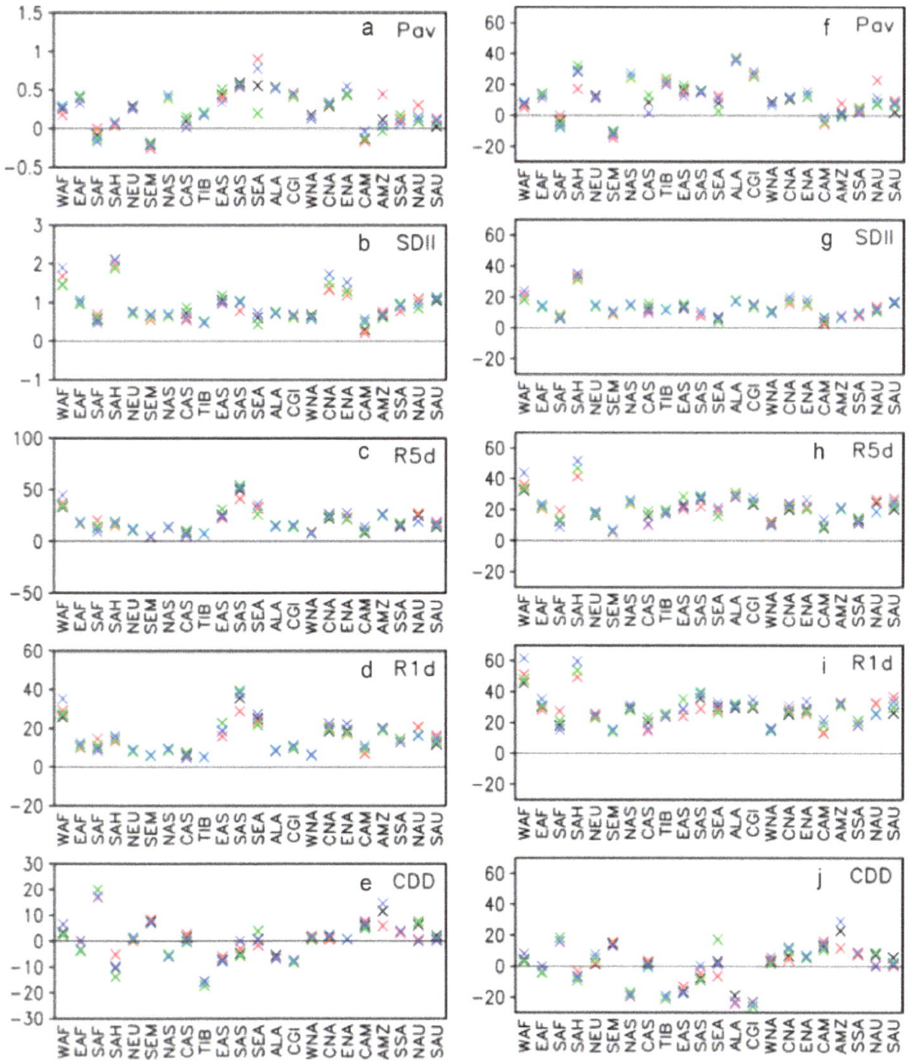

Figure 7: Future changes of regionally averaged precipitation indices (a–e) and their change ratio (%) (f–j). Projections using the future SST pattern of the 28 CMIP5 multi-model mean, cluster 1, cluster 2, and cluster 3 are represented in black, red, green, and blue, respectively. (For interpretation of the references to color in this figure legend, the reader is referred to the web version of this article.)

Fig. 7a shows the regionally averaged annual mean precipitation changes between the end of the 21st century and present, while Fig. 7b shows the percent change ratio. Annual mean precipitation (Pav) is projected to increase in almost all land domains, but it is projected to decrease in SAF, SEM and CAM. Regions with most consistent changes in Pav are SAS, ALA and CGI where standard deviation among four realizations is less than 10% of the ensemble mean changes. Large scatter among four realizations is found in AMZ, CAS, SAF, NAU and SEA. In some of the above domains, different SST patterns used have resulted in large precipitation changes, possibly related to large-scale circulation changes in the tropical Pacific. In AMZ, even sign of Pav change is different among realizations where the experiment with the cluster 2 SST shows decreased Pav while other members show increased Pav. The cluster 2 SST has a larger SST warming in the central and eastern tropical Pacific. This strong El Niño-like SST pattern may have resulted in precipitation decrease, which is consistent with the fact that in interannual time scale Amazon tends to have less precipitation in El Niño years. The SAF is also affected much by the SST patterns used. The cluster 1 SST has a large warming around South Africa than other clusters. Experiments with other SST clusters resulted in a Pav decrease in SAF, while warmer SST in the cluster 1 SST resulted in near zero precipitation changes.

Large positive Pav changes in magnitude are projected in SEA and SAS. Scatter among four experiments is large in SEA, where the experiment with the cluster 2 SST shows least positive Pav increase. Large positive ratio changes are shown in cold regions in high latitudes (ALA, CGI, NAS) and high mountain area (TIB).

Heavy precipitation indices (SDII, R5d, R1d) increase in all regional domains, even where mean precipitation decrease (SAF, SEM, CAM). SAS is the domain of the largest extreme precipitation (R5d and R1d) increase, followed by WAF, SEA. ENA, CNA and AMZ. SEM is the domain of the least increase in R5d.

It is noted that the largest SDII increase rate in SAH is due to a large increase rate in Pav while an almost no change in the number of wet days (daily precipitation amount≥1 mm), which is distinct in the summer season (not shown). WAF and SAH are the domains of largest ratio increases in R5d and R1d. Scatter among experiments in R1d and R5d is relatively small in SEA. This is in contrast to Pav, in which SEA shows large uncertainty among experiments.

Large increases in CDD are projected in SAF, AMZ, SEM and CAM. A projected increase in dry days over Amazon is consistent with Kamiguchi et al. (2006) who used the MRI-AGCM3.1 with SRES A1B multi-model mean SST anomalies. In our study, though, there are large scatter in AMZ among experiments associated with north-south contrast of future SST changes. Projected increase in CDD over Amazon is longest in cluster 3 where the Northern Hemisphere SST warms more than the Southern Hemisphere SST. More northward shift of ITCZ over the tropical Atlantic Ocean in cluster 3 may have resulted in a longer spell of dry period in Amazon.

On the other hand, large decreases in CDD are projected in TIB, SAH, CGI, EAS, ALA and NAS. In EAS, a decrease in CDD mainly comes from wetter condition in winter (Fig. 6).

As already noted, general tendency of future precipitation extremes projected by MRI-AGCM3.2 is consistent with CMIP3 and CMIP5 multi-model ensembles. However, inter-model variability is large in some regional domains. For example, model agreement in R5d changes is less in SAH and CAM in CMIP5 model, in which more than a quarter of models projects changes in opposite signs (Sillmann et al., 2013b). Uncertainty becomes larger in CDD changes (Sillmann et al., 2013b), being consistent with our results. Our result is generally consistent with the former findings with the MRI-AGCM3.1 (Kamiguchi et al., 2006) although quantitative differences exist due to model improvement and different scenarios used.

CONCLUDING REMARKS

In this paper, we used the four-member ensemble simulation with the global 20-km mesh AGCM at the end of the 21st century (2075–2099) under the RCP8.5 scenario. The present-day climate simulation showed good correspondence in magnitude of precipitation extremes indices such as R5d as compared with the observed estimates. A high spatial resolution of the model enabled the reproducibility of various weather systems (Oouchi et al., 2006 and Murakami and Sugi, 2010), but updated physics such as cumulus parameterization improved it quantitatively (Mizuta et al., 2012 and Murakami et al., 2012). Projected changes are shown in Murakami et al. (2012).

Spatial patterns of projected future changes in precipitation extremes are generally similar to that of annual mean precipitation, but area with decreasing becomes much smaller in more extreme precipitation indices of R5d and R1d. Globally averaged change ratio, normalized by the global annual mean surface air temperature change, is found to become larger in R5d (5.36%/°C) and more in R1d (7.36%/°C) than that in SDII (2.99/°C) and Pav (2.22%/°C).

For the future projection we used four different SST change patterns derived from CMIP5 model projections (Mizuta et al., 2014). It is found that some regions are strongly affected by SST patterns. For the Pav changes, largest variability is found in Southeast Asia for its magnitude, and in Amazon for its ratio change. Amazon is the only regional domain where some member experiment yielded different signs in changes. For R5d, South Asia is the domain with largest increase in magnitude. Large scatter among experiments is found in CDD. In this paper, we did not perform detailed analysis of atmospheric circulation changes associated with different SST change patterns. This definitely should be a future work.

In this experimental setting, we only covered some uncertainty coming from future SST change patterns. The different model physics is another source for uncertainty in future climate projections. Endo et al. (2012) had performed such an attempt and showed that large uncertainty in future precipitation projections in South Asia and Southeast Asia come from differences in model physics such as cumulus convection schemes.

We have made four ensemble experiments for future projections, which is not enough to cover uncertainty range and to get probabilistic regional climate information. About dozens or 100 realizations may be needed to achieve such climate change information. The 20-km mesh GCM is too heavy to perform many realizations. Instead we use the 60-km mesh version of the same model for ensemble experiments with many members. Although the 60-km mesh model itself has some limitation to quantitatively reproduce tropical cyclone intensity, some precipitation extremes are reasonably simulated (Endo et al., 2012). Therefore, a use of both 20- km and 60- km models would deliver meaningful information.

In our experimental setting, SST is prescribed as lower boundary conditions in the present and future simulations. The absence of air–sea interaction can cause distorted precipitation field and associated atmospheric circulations over certain regions such as the western tropical Pacific (Wang et al., 2005). This also may have resulted in too stronger tropical cyclones and associated precipitation and wind extremes. An introduction of air–sea introduction may remedy this problem and is under consideration.

ACKNOWLEDGMENTS

This work was conducted under the Development of Basic Technology for Risk Information on Climate Change of the SOUSEI Program of the Ministry of Education, Culture, Sports, Science, and Technology of Japan.

REFERENCES

1. Allan, R.F., Soden, B.J., 2008. Atmospheric warming and the amplification of precipitation extremes. Science 321, 1481–1484.
2. Allen, M.R., Ingram, W.J., 2002. Constraints on future changes in climate and the hydrologic cycle. Nature 419, 224–232.
3. Arakawa, A., Schubert, W.H., 1974. Interaction of a cumulus cloud ensemble with the large-scale environment. Part I. J. Atmos. Sci 31, 674–701.
4. Chadwick, R., Boutle, I., Martin, G., 2013. Spatial patterns of precipitation change in CMIP5: why the rich don't get richer in the tropics. J. Clim. 26, 3803–3822.
5. Christensen, J.H., Hewitson, B., Busuioc, A., Chen, A., Gao, X., Held, I., Jones, R., Kolli, R.K., Kwon, W.-T., Laprise, R., Magana Rueda, V., Mearns, L., Menendez, C.G.,Raisanen, J., Rinke, A., Sarr, A., Whetton, P., 2007. Regional climate projections.In: Solomon, S., Qin, D., Manning, M., Chen, Z., Marquis, M., Averyt, K.B., Tignor, M., Miller, H.L. (Eds.), Climate Change 2007: The Physical Science Basis. Contributionof Working Group I to the Fourth Assessment Report of the Intergovernmental Panel on Climate Change. Cambridge University Press, Cambridge,
6. United Kingdom, New York, NY, USA.
7. Dai, A., 2006. Precipitation characteristics in eighteen coupled climate models. J. Clim. 19, 4605–4630.
8. Donat, M.G., et al., 2013. Updated analyses of temperature and precipitation extreme indices since the beginning of the twentieth century: The HadEX2 dataset.J. Geophys. Res. Atmos. 118, 2098–2118. http://dx.doi.org/10.1002/
9. jgrd.50150.
10. Emori, S., Brown, S.J., 2005. Dynamic and thermodynamic changes in mean andextreme precipitation under changed climate. Geophys. Res. Lett. 32, L17706.http://dx.doi.org/10.1029/2004GL022306.
11. Endo, H., Kitoh, A., Ose, T., Mizuta, R., Kusunoki, S., 2012. Future changes and uncertainties in Asian precipitation simulated by multi-physics and multi-sea surface temperature ensemble experiments with high-resolution Meteorological Research Institute atmospheric general circulation models (MRIAGCMs).
12. J. Geophys. Res. 117, D16118. http://dx.doi.org/10.1029/2012JD017874.Fowler, H.J., Ekstrom, M., Blenkinsop, S., Smith, A.P., 2007. Estimating change in extreme European precipitation using a multimodel ensemble. J. Geophys. Res. 112, D18104. http://dx.doi.org/10.1029/2007JD008619.
13. Huffman, G.J., D.T., Bolvin, 2009. GPCP One-degree Daily Precipitation Data Set Documentation. oftp://precip.gsfc.nasa.gov/pub/1dd-v1.1/1DD_v1.1_doc.pdf4.
14. Huffman, G.J., Adler, R.F., Bolvin, D.T., Gu, G., Nelkin,E.J., Bowman, K.P., Hong, Y., Stocker, E.F., Wolff, D.B., 2007. The TRMM multi-satellite precipitation analysis:Quasi-global, multi-year, combined-sensor precipitation estimates at finer scale. J. Hydrometeor. 8, 38–55.
15. IPCC, 2012. Managing the risks of extreme events and disasters to advance climate change adaptation.. In: Field, C.B., Barros, V., Stocker, T.F., Qin, D., Dokken, D.J., Ebi, K.L., Mastrandrea, M.D., Mach, K.J., Plattner, G.-K., Allen, S.K., Tignor, M., Midgley, P.M. (Eds.), A Special Report of Working Group I and II of the Intergovernmental
16. Panel on Climate Change. Cambridge University Press, Cambridge, UK, New York, NY, USA (582 pp.).
17. IPCC, 2013. Climate change 2013: the physical science basis. In: Stocker, T.F., Qin, D.,Plattner, G.-K., Tignor, M., Allen, S.K., Boschung, J., Nauels, A., Xia, Y., Bex, V.,Midgley, P.M. (Eds.), Contribution of Working Group I to the Fifth Assessment Report of the Intergovernmental Panel on Climate Change. Cambridge University Press, Cambridge, United Kingdom, New York, NY, USAhttp://dx.doi. org/10.1017/CBO9781107415324 (1535 pp.).
18. Kamiguchi, K., Kitoh, A., Uchiyama, T., Mizuta, R., Noda, A., 2006. Changes in precipitation-based extremes indices due to global warming projected by a global 20-km-mesh atmospheric model. SOLA 2, 64–67. http://dx.doi.org/10.2151/ sola.2006-017.
19. Kharin, V.V., Zwiers, F.W., Zhang, X., Wehner, M., 2013. Changes in temperature and precipitation extremes in the CMIP5 ensemble. Clim. Change 119, 345–357. http://dx.doi.org/10.1007/s10584-013-0705-8.
20. Kitoh, A., Kusunoki, S., 2008. East Asian summer monsoon simulation by a 20-km mesh AGCM. Clim. Dyn. 31, 389–401. http://dx.doi.org/10.1007/ s00382-007-0285-2.
21. Kitoh, A., Ose, T., Kurihara, K., Kusunoki, S., Sugi, M., KAKUSHIN Team-3 Modeling Group, 2009. Projection of changes in future weather extremes using superhigh-resolution global and regional atmospheric models in the KAKUSHIN Program: Results of preliminary experiments. Hydrol. Res. Lett. 3, 49–53. http: //dx.doi.org/10.3178/hrl.3.49.
22. Kitoh, A., Kusunoki, S., Nakaegawa, T., 2011. Climate change projections over South America in the late 21st century with the 20 and 60 km mesh Meteorological Research Institute atmospheric general circulation model (MRI-AGCM). J. Geophys. Res. 116. http://dx.doi.org/10.1029/2010JD014920, D06105.
23. Kitoh, A., Ose, T., Takayabu, I., 2015. Dynamical downscaling for climate projection with high-resolution MRI AGCM-RCM. J. Meteorol. Soc. Jpn. . http://dx.doi.org/ 10.2151/jmsj.2015-022
24. Klein Tank, A.M.G., Zwiers F.W., Zhang X., 2009. Guidelines on Analysis of Extremesin a Changing Climate in Support of Informed Decisions for Adaptation. WCDMP-No.72, WMO-TD No.1500, (56 pp).

25. Kunkel, K.E., Karl, T.R., Easterling, D.R., Redm ond, K., Young, J., Yin, X., Hennon, P., 2013. Probable maximum precipitation (PMP) and climate change. Geophys. Res. Lett. 40, 1402–1408. http://dx.doi.org/10.1002/grl.50334.

26. Meehl, G.A., Covey, C., Delworth, T., Latif, M., McAvaney, B., Mitchell, J.F.B., Stouffer, R.J., Taylor, K.E., 2007. The WCRP CMIP3 multi-model dataset: a new era in climate change research. Bull. Am. Meteorol. Soc. 88, 1383–1394.

27. Mehran, A., AghaKouchak, A., Philips, T.J., 2014. Evaluation of CMIP5 continental precipitation simulations relative to satellite-based gauge-adjusted observations. J. Geophys. Res. Atmos 119, 1695–1707. http://dx.doi.org/10.1002/

28. 2013JD021152.

29. Min, S.K., Zhang, X., Zwiers, F.W., Hegerl, G.C., 2011. Human contribution to moreintense precipitation extremes. Nature 470, 378–381.

30. Mizuta, R., Oouchi, K., Yoshimura, H., Noda, A., Katayama, K., Yukimoto, S., Hosaka, M., Kusunoki, S., Kawai, H., Nakagawa, M., 2006. 20-km-mesh global climatesimulations using JMA-GSM model – Mean climate states. J. Meteorol. Soc. Jpn. 84, 165–185. http://dx.doi.org/10.2151/jmsj.84.165.

31. Mizuta, R., Adachi, Y., Yukimoto, S., Kusunoki, S., 2008. Estimation of future distribution of sea surface temperature and sea ice using CMIP3 multi-model ensemble mean. Tech. Rep. Meteorol. Res. Inst. 56, 28.

32. Mizuta, R., Yoshimura, H., Murakami, H., Matsueda, M., Endo, H., Ose, T., Kamiguchi, K., Hosaka, M., Sugi, M., Yukimoto, S., Kusunoki, S., Kitoh, A., 2012. Climate simulations using MRI-AGCM3.2 with 20-km grid. J. Meteorol. Soc. Jpn. 90A, 233–258. http://dx.doi.org/10.2151/jmsj.2012-A12.

33. Mizuta, R., Arakawa, O., Ose, T., Kusunoki, S., Endo, H., Kitoh, A., 2014. Classification of CMIP5 future climate responses by the tropical sea surface temperature changes. SOLA 10, 167–171. http://dx.doi.org/10.2151/sola.2014-035.

34. Murakami, H., Sugi, M., 2010. Effect of model resolution on tropical cyclone climate projections. SOLA 6, 73–76. http://dx.doi.org/10.2151/sola.2010-019.

35. Murakami, H., Wang, Y., Yoshimura, H., Mizuta, R., Sugi, M., Shindo, E., Adachi, Y.,Yukimoto, S., Hosaka, M., Kusunoki, S., Ose, T., Kitoh, A., 2012. Future changes in tropical cyclone activity projected by the new high-resolution MRI-AGCM. J. Clim. 25, 3237–3260.

36. Nakaegawa, T., Kitoh, A., Murakami, H., Kusunoki, S., 2014. Annual maximum 5-day rainfall total and maximum number of consecutive dry days over Central America and the Caribbean in the late twenty-first century projected by an atmospheric general circulation model with three different horizontal resolutions. Theor. Appl. Climatol. 116, 155–168. http://dx.doi.org/10.1007/s00704-013-0934-9.

37. O'Gorman, P.A., Schneider, T., 2009. The physical basis for increases in precipitation extremes in simulations of 21st –century climate change. Proc. Nat. Acad. Sci. USA 106, 14773–14777.

38. Oouchi, K., Yoshimura, J., Yoshimura, H., Mizuta, R., Kusunoki, S., Noda, A., 2006. Tropical cyclone climatology in a global warming climate as simulated in a 20 km-mesh global atmospheric model: frequency and wind intensity analyses. J. Meteorol. Soc. Jpn. 84, 259–276. http://dx.doi.org/10.2151/JMSJ.84.259.

39. Rajendran, K., Sajani, S., Jayasankar, C.B., Kitoh, A., 2013. How dependent is climate change projection of Indian summer monsoon rainfall and extreme events on model resolution? Curr. Sci. 104, 1409–1418.

40. Rayner, N.A., Parker, D.E., Horton, E.B., Folland, C.K., Alexander, L.V., Rowell, D.P.,Kent, E.C., Kaplan, A., 2003. Global analyses of sea surface temperature, sea ice, and night marine air temperature since the late nineteenth century. J. Geophys. Res. 108, 4407. http://dx.doi.org/10.1029/2002JD002670.

41. Sillmann, J., Kharin, V.V., Zhang, X., Zwiers, F.W., Bronaugh, D., 2013a. Climate extremes indices in the CMIP5 multimodel ensemble: Part 1. Model evaluation in the present climate. J. Geophys. Res. Atmos. 118, 1716–1733. http://dx.doi.org/ 10.1002/jgrd.50203.

42. Sillmann, J., Kharin, V.V., Zwiers, F.W., Zhang, X., Bronaugh, D., 2013b. Climate extremes indices in the CMIP5 multimodel ensemble: Part 2. Future climate projections. J. Geophys. Res. Atmos. 118, 2473–2493. http://dx.doi.org/10.1002/ jgrd.50188.

43. Stephens, G.L., L'Ecuyer, T., Forbes, R., Gettlemen, A., Golaz, J.-C., Bodas-Salcedo, A., Suzuki, K., Gabriel, P., Haynes, J., 2010. Dreary state of precipitation in global models. J. Geophys. Res. 115, D24211. http://dx.doi.org/10.1029/2010JD014532.

44. Sun, Y., Solomon, S., Dai, A., Portmann, R.W., 2006. How often does it rain? J. Clim. 19, 916–934.

45. Sun, Y., Solomon, S., Dai, A., Portmann, R.W., 2007. How often will it rain? J. Clim. 20, 4801–4818.

46. Takayabu, I., Kanamaru, H., Dairaku, K., Benestad, R., von Storch, H., Christensen, J. H., 2015. Reconsidering the quality and utility of downscaling. J. Meteorol. Soc. Jpn. (in press)

47. Taylor, K.E., Stouffer, R.J., Mehl, G.A., 2012. An overview of CMIP5 and the experiment design. Bull. Am. Meteorol. Soc. 93, 485–498, doi:10.1175/BAMS-d-11-00094.1.

48. Tebaldi, C., Hayhoe, K., Arblaster, J., Meehl, G., 2006. Going to extremes: An intercomparison of model-simulated historical and future changes in extreme events. Clim. Chang. 79, 185–211. http://dx.doi.org/10.1007/ s10584-006-9051-4.

49. Tiedtke, M., 1989. A comprehensive mass flux scheme for cumulus parameterization in large-scale models. Mon. Weather Rev. 117, 1779–1800.

50. Toreti, A., Naveau, P., Zampieri, M., Schindler, A., Scoccimarro, E., Xoplaki, E., Dijkstra, H.A., Gualdi, S., Luterbacher, J., 2013. Projections of global changes in precipitation extremes from Coupled Model Intercomparison Project Phase 5 models. Geophys. Res. Lett. . http://dx.doi.org/10.1002/grl.50940

51. Wang, B., Ding, Q., Fu, X., Kang, I.-S., Jin, K., Shukla, J., Doblas-Reyes, F., 2005. Fundamental challenge in simulation and prediction of summer monsoon rainfall. Geophys. Res. Lett. 32, L15711. http://dx.doi.org/10.1029/2005GL022734.
52. Xie, S.-P., Deser, C., Vecchi, G.A., Ma, J., Teng, H., Wittenberg, A.T., 2010. Global warming pattern formation: sea surface temperature and rainfall. J. Clim. 23, 966–986.
53. Yoshimura, H., Mizuta, R., Murakami, H., 2015. A spectral cumulus parameterization scheme interpolating between two convective updrafts with Semi-Lagrangian calculation of transport by compensatory subsidence. Mon. Weather Rev. 143, 597–621, doi:10.1175/MWR-d-14-00068.1.

CHAPTER 13

MAPPING THE VULNERABILITY HOTSPOTS TO FLOODING DISASTERS

Shada Elalem, Indrani Pal

Keywords: Vulnerability, Flooding, Disaster, Hindu-Kush Himalaya (HKH)

INTRODUCTION

Hydro-meteorological disasters are significant concerns around the world. The most recent Intergovernmental Panel on Climate Change (IPCC) Assessment Report (Field et al., 2014) on impacts, adaptation, and vulnerability highlighted eight major climate risks the world is facing today—the economic losses due to inland flooding being ranked as the *third* most major climate risk only after coastal flooding and food insecurity. The IPCC Special Report on Managing the Risks of Extreme Events and Disasters to Advance Climate Change Adaptation (SREX) report (Seneviratne et al., 2012) recognized that projected increases in temperature and heavy precipitation imply regional-scale changes in flood frequency and intensity. Global flood projections based on multiple Coupled Model Intercomparison Project version 5 – Global Climate Model (CMIP5 GCM) simulations coupled with global hydrology and land surface models showed flood hazards increasing over about half of the globe, but with great variability at the catchment scale (Dankers et al., 2014). All these scientific studies provide the general consensus that the climatic changes are expected to fuel the alterations of extreme weather event patterns at the local scale. On the other hand, changes in socio-economic and demographic patterns are also expected to significantly modify resultant losses. IPCC-Assessment Report 5 (AR5) drew on an emerging body of social science literature to suggest how policymakers might take practical steps to help communities adapt. The report also clearly specifies the fact that: *"the biggest single barrier to improving societal resilience to the vagaries of climate is Poverty"* (Field et al., 2014 and Kintisch, 2014). As such, we can better offer potential strategies to adapt and maximize the local resilience to a changing climate, as well as reduce vulnerability of the regions to damages from disasters by better preparation, with integrated knowledge on the natural and societal systems.

The Hindu-Kush Himalaya (HKH), popularly known as world's *"third pole"* for containing the third largest fresh water reserve, is on the front lines of climate change, like many other mountainous regions (NAS, 2012). One fifth of world's population lives on the HKH water resources in Asia (ICIMOD, 2014), and a great fraction of them are exposed to devastating flooding impacts every year. Several of those flooding events translate to disasters, resulting in huge economic and human losses (mortality and morbidity) (Pathak et al., 2008). Although, characteristics of extreme hydro-climatic events (for example: magnitude, timing, and duration) are important precursors to analyze with a goal of advancement of the early warning systems, it is also important to recognize that the magnitude of disaster losses also relate largely to the social drivers such as population and property exposure, and institutional conditions defining differential sensitivity and capacities to environmental stress (Eakin and Luers, 2006). While disasters are threatening for both rich and poor, levels of impacts and recovery differ from one region to another. For example, richer nations are better equipped, in general, to adapt to the same amount of economic losses as compared to the poorer nations (Mendelsohn et al., 2006). Therefore, disaster risk differs to large extent with the type and magnitude of vulnerability, ranging from frequency, timing and magnitude of biophysical indicators to population and property exposure to other socio-economic and institutional conditions.

Citation: Shada Elalem, Indrani Pal, Mapping the vulnerability hotspots to flooding disasters, http://dx.doi.org/10.1016/j.wace.2014.12.001.

Hence, the magnitude of *relative* losses, be it economic value or mortality or morbidity numbers, are expected to differ across locations over the HKH region. Many of the HKH mountain communities survive on less than $2 a day (Sharma, 2014) and it has more than 40% of the world's poor (Karki et al., 2011), who, as a result, remains extremely vulnerable to growing climatic and environmental uncertainties. The immense diversity within this region needs to be recognized: diversity of climates, and microclimates, hydrology and ecology, and, above all, the socio-economic conditions, human cultures and activities for designing appropriate initiatives (NAS, 2012).

Recently, one of the major National Academy of Sciences reports (NAS, 2012) pointed out that this important region is in trouble, because our scientific knowledge is limited. That is due to the fact that the hydro-climatic and socio-economic data available for this region has been scarce. However, with more and more satellite remote sensing data and reported socio-economic data being available, we are getting better equipped to generate more knowledge on how the extreme events vary across this important region and what factors (natural and/or social) play the dominant role to make this region vulnerable to flooding impacts.

The overarching goal of this research is *to detect the diversity of HKH region with respect to vulnerability of affected locations to flooding disaster occurrence, as well as their resultant economic and human impacts according to historical record.* Comparing spatio-temporal information of natural precursors as well as the impacts identifies the areas of composite vulnerability. Although historical climate variability may not directly correlate with current vulnerability of a region to flood disasters, neither that may directly associate with future vulnerability, because many factors together define vulnerability, identifying a region that experienced flood disasters in the past and present and their concurrent climate anomaly is our aim here because they together can make the same region susceptible to disaster risk in the future as well, given all other factors being unchanged or worse from the past. This can be used as the basic point for regional case studies on vulnerability and multifaceted adaptive management strategies. In addition, this study provides a room to further understand the hydro-climatic dynamics that are important in the context of anticipated global climate change that is poorly understood for the HKH region. Overall, the vulnerability maps produced in this study increases our basic understanding of the *'what, where and when'* disaster impacts occur over the locations and countries across the Hindu-Kush Himalaya region.

STUDY REGION

The Hindu-Kush Himalaya (HKH) region covers around 3,441,719 km^2 of area around eight countries: Afghanistan, Bangladesh, Bhutan, China, India, Myanmar, Nepal, and Pakistan (Fig. 1(A)). Around 35,110 km^2 of the total area is covered with snow and ice (ICIMOD, 2014). This region is considered the third largest deposit of ice and snow in the world, after Antarctica and the Arctic (NAS, 2012). These glaciers affect the region's hydrology because the melt water from the glaciers feed the rivers and streams, especially in the summer (NAS, 2012). The region has an access to diversity of natural resources such as largest rivers, fertile valleys, extreme heights, and beautiful landscapes view that makes this area having a high density of population (Karki et al., 2011).

Figure 1: Location and demographic information of Hindu-Kush Himalaya region in South-east Asia. (A) Geographic location ranging from Afghanistan in the west to Myanmar in the east. (B) River basin boundary.

HKH region includes about 60% of Afghanistan boundary, a fraction of Bangladesh, the entirety of Bhutan, part of China (about 17% of country area), part of India (11 mountain states and about 14% of the total country area), about 47% of Myanmar country area, the entirety of Nepal, and about 51% of Pakistan country area (Table 1). The HKH region also contains some of most densely populated areas (NAS, 2012). The population is concentrated mostly in India, Pakistan, China and Afghanistan respectively—almost 34% of the total populations of HKH region are concentrated in India (~72.36 million,Table 1). Although about 48% of the geographic area of HKH region falls in China, most population density are reported in Nepal, India, Bangladesh, and Pakistan respectively. The population density is extremely important in this study because it helps to look at where people are vulnerable to disaster impacts.

Table 1: Total geographic area of each country falling within HKH region (in km^2), population of each country living within HKH region (millions), population density of each country within HKH region (number of people living per km^2 geographic area), and proportion of total geographic area of each country falling within HKH region (%)

HKH Country	Total geographic area of each country falling within HKH region (km^2)	Population of each country living within HKH region (million)	Population Density of each country within HKH region (number of people living per km^2 geographic area)	Portion of total geographic area of each country falling within HKH region (%)
Afghanistan	390,475	28.48	73	60
Bangladesh	13,189	1.33	100	9
Bhutan	38,394	0.71	15	100
China	1,647,725	29.48	17	17
India	482,920	72.36	150	14
Myanmar	317,640	11.01	34	47
Nepal	147,181	27.8	189	100
Pakistan	404,195	39.36	97	51

The greater Himalayan region is a water resource for 1.3billion people living in that region (ICIMOD, 2014). It has about ten major river basins including Amu Darya in Afghanistan, Tajikistan, Turkmenistan, and Uzbekistan; Brahmaputra in China, India, Bhutan and Bangladesh; Ganges in India, Nepal, China and Bangladesh; Indus in China, India and Pakistan; Irrawaddy in Myanmar; Mekong in China, Myanmar, Laos, Thailand, Cambodia and Vietnam; Salween in China, Myanmar and Thailand; Tarim in Kyrgyzstan and China; Yangtze in China; and Yellow in China (Fig. 1(B)). Glacier melt contributes to about 50–80% of the water resource in the western part of Himalaya (e.g. Satluj and Indus rivers) (Jianchu et al. 2009), in contrast, the generally smaller central and eastern Himalayan catchments receive more than 80% of their annual runoff from rainfall and less than 20% from snowmelt (except the large Tsangpo/Brahmaputra catchment which receives about 34% of its annual discharge from snowmelt) (Bookhagen and Burbank, 2010). The rainfall gradient has different distribution from west to east. Monsoon rainfall in the summer (June to September) is predominant in the central and eastern part of HKH where up to 70% of annual rainfall is contributed by the summer monsoon, and westerly weather system or winter rains contribute to most of the ice, snow and river runoff in the western Himalaya and Pamir regions. Therefore, the hydro-meteorological drivers to river flows differ a great deal from the west to the east, which is divided near the Satluj valley (Bookhagen and Burbank, 2010).

DESCRIPTION OF THE DATA

Disaster Data

We downloaded the historical disaster information from The International Disaster Database (EM-DAT, 2014) to identify the reported flooding disasters that occurred in the entirety of the HKH region in the last 6 decades (1951–2013). EM-DAT is one of the Centers for Research on the Epidemiology of Disaster (CRED)'s core data, which provides objective basis information about the disasters such as the mortality, morbidity and the related economic damages. The World Health Organization (WHO) and the Belgian Government created the EM-DAT database in 1988 to serve the humanitarian agencies at the national and international levels. It has more than 18,000 world disasters recorded since 1900 to present, which might not be complete due to communication constraints, especially in the earlier decades of 20th century. They compiled this information from different sources such as UN agencies, government organizations, insurance companies, research institutions and press agencies (EM-DAT, 2014). The publicly available data we use for this research is processed and summarized in Table S1in the Supplementary Information. This include the start and end dates of disasters, the country of occurrence, and location information (name of the city or place where the disasters occurred), the type and sub types of the disasters (for example, types of flooding events—general or flash floods), the reported numbers of mortality and morbidity (note: there were many unreported human losses especially in the earlier decades), the associated economic losses in concurrent US dollar values (note: there were many unreported economic losses especially in the earlier decades), and the corresponding disaster file number (kept in record for future updates). The longitude and latitude information for each of 155 disaster locations in Table S1, were not available in EM-DAT. Hence we collected these information from the "I TOUCH MAP" for each disaster location of interest (http://itouchmap.com/?r=b). Furthermore, there were some reported disasters in EM-DAT database with missing dates and/or unreasonable lengths recorded in days/months. After carefully studying the history of disasters at each of those locations, as well as examining the spatially nearest neighbors, we excluded some of those cases (but recorded them for future scrutiny) for the hydro-meteorological analysis.

Hydro-climate Data

Because of political/geopolitical/bureaucratic factors, in many of the HKH countries, hydro meteorological data are treated as sensitive data from national defense point of view. In the absence of the river flow data, temperature and precipitation are the most critical natural factors to consider to study extreme weather precursors to flooding over the HKH region. Extreme rainfall events in the summer season (also known as monsoon period) dominate over central and eastern part of HKH, while temperature generally plays a central role for melting snow and glacier ice over the western part. In addition, winter and springtime mid-latitude westerly weather systems also govern over the western HKH. Therefore, part of HKH region falls at the interface of the monsoon and westerly weather system dominated areas, which is susceptible to flooding caused by either or both types of natural systems—the most recent example of that being early monsoon flooding in Uttarakhand area of northwest India.

We collected daily precipitation and monthly temperature data from multiple sources. First, we used high-resolution (0.25-degree grids, mm/day) precipitation data—Asian Precipitation Highly-Resolved Observation Data Integration Towards Evaluation (APHRODITE), created by the Research Institute for Humanity and Nature (RIHN) and Meteorological Research Institute of Japan Meteorological Agency (MRI-JMA), which we downloaded from the International Research Institute for Climate and Society data library (IRI, 2014). This data was available from 1 Jan 1951 to 31 Dec 2007. To fill in the data gaps between Jan 2008 and Dec 2013, we used satellite precipitation data—Tropical Rainfall Measuring Mission (TRMM), which is created by National Aeronautics and Space Administration (NASA) and available from 1 Jan 1998 to 31 Dec 2013. We downloaded NASA-TRMM from the IRI Data Library as well. To match the gridded precipitation data with disaster location information (Table S1), we picked four nearest grid points around each disaster location (lat/long in Table S1) and calculated the average precipitation for each recorded disaster length ± 7 days. For concurrent temperature analysis, we used a monthly dataset with 0.5-degree grids from Jan 1951 to Dec 2013, which was created by National Oceanic and Atmospheric Administration – National Centers for Environmental Prediction-Climate Prediction Center – Global Historical Climatology Network (NOAA NCEP CPC GHCN), and also downloaded from the IRI-data library (IRI, 2014). In this case, we considered the overlapping grid points for each disaster location and timing recorded in Table S1.

Demographic Information and National Socio-Economic Data

We collected regional demographic information from the International Center for Integrated Mountain Development (ICIMOD, 2014). This includes total population and population density (person per km^2) of each country living within HKH region, proportion of HKH boundary (in km^2) shared within each country, proportion of each country area falling within HKH region (refer to Table 1). We also collected the national socio-economic data from two different sources: 2012 Gross Domestic Product (GDP) value from the World Bank (The World Bank, 2014) and the yearly GDP information from 1961 to 2012 from the United Nations database (UN data, 2014).

RESULTS AND DISCUSSIONS

Spatio-temporal Pattern of "reported" Frequency of Flooding Disasters

Fig. 2(A) shows spatial pattern of total number of flooding disasters reported in each location over the HKH region within the study interval (1951–2013). This figure displays what location is vulnerable to flooding disaster occurrence as per historical record. High spatial variability is identified across the HKH. Parts of northern Afghanistan and north-western Pakistan in the west seem to be showing high vulnerability to disaster occurrence, as well as north of India and Nepal, and some locations within China and Bangladesh in the east of HKH. Decadal patterns are illustrated in Fig. 2(B), which indicates that most disaster cases are reported in the recent decade (2001–2013) as compared to the earlier decades. Most notably, the 2013 Uttarakhand flooding event in India and 2010 Pakistan flooding event, those made world headlines, fall within this decade.

A

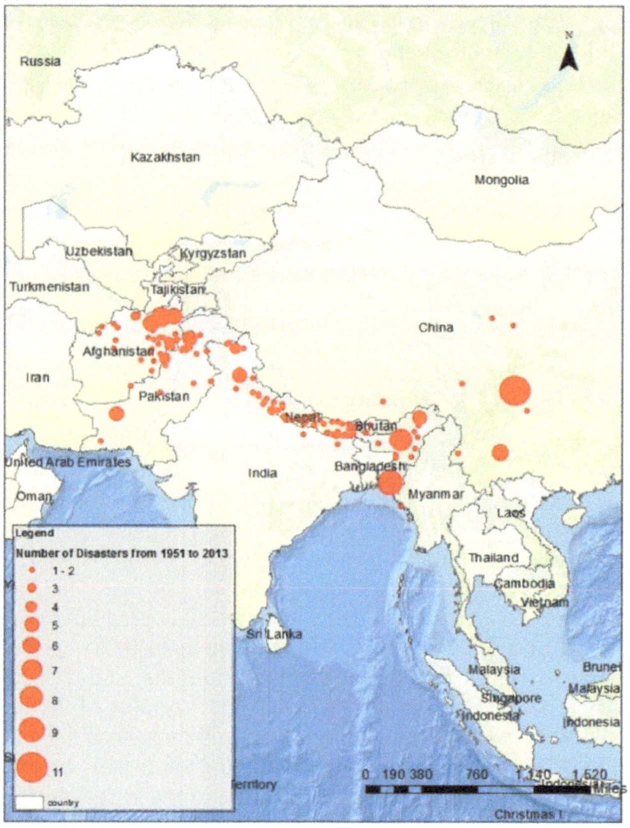

Figure 2: Spatio-temporal patterns of frequency of "reported" flooding disasters. (A) Total frequency of disasters over 1951–2013 showing high number of disasters recorded in Afghanistan and Pakistan in the west, India and Nepal in the center, and China, Bangladesh, Bhutan and Myanmar in the east. (B) Decade-wise spatial frequency patterns showing the highest number of disasters reported in the recent decade of 2001–2013.

The country specific yearly and decadal trends of frequency of disasters are in Fig. S1(A) and (B) in the supplementary information. These figures depict that there is a clear time trend in the reported frequency of flooding disasters over the entirety of the HKH region, of which most disasters recorded are for Afghanistan, China, India, Nepal, and Pakistan.

The positive time trends (annual and decadal) in frequency of reported disasters, as noted in Fig. 2(B) and Fig. S1, might be due to several reasons—one prominent and perhaps the most significant one for our case is that more number of events are reported in the recent decades due to better communication systems available. In addition, the agencies like CRED and the US Agency for International Development (USAID) began to record the disasters since the 1980s (Than, 2005). However, we do not at all rule out of the significant possibility of the impacts of socio-economic changes and climatic changes on the frequency of extreme weather events and resultant disasters (NAS, 2012), nor the institutional dysfunctions. We need to conduct further research in this area, using other publicly available databases (e.g. Dartmouth Flood Observatory, MunichRe, and others) to identify whether there are any consistent time trends in frequency of flooding disasters over HKH region those associate with climatic changes and socio-economic factors such as Governance failure, as it was the case for 2008 Koshi flood, and 2013 Uttarakhand flood causing major economic and human impacts.

Spatio-Temporal Pattern of "Reported" Economic Losses

In addition to disaster frequency, historical data also helps us to detect countries vulnerable to economic losses due to flooding disasters. We can do this in various different ways—one, by looking at "reported" total amount of economic losses per disaster frequency per country, which gives a snapshot view of "total economic impacts" per disaster reported. Secondly, magnitude of economic losses per disaster could be standardized with respect to a country's socio-economic status (e.g. gross domestic product/GDP). This can act as an important indicator of "relative impacts". This second indicator helps us to understand how a country "feels" about or able to deal with total economic losses due to a disaster occurrence as a function of the economic growth at the country level. Fig. 3 and Fig. 4 report these two sets of analyses outcomes respectively.

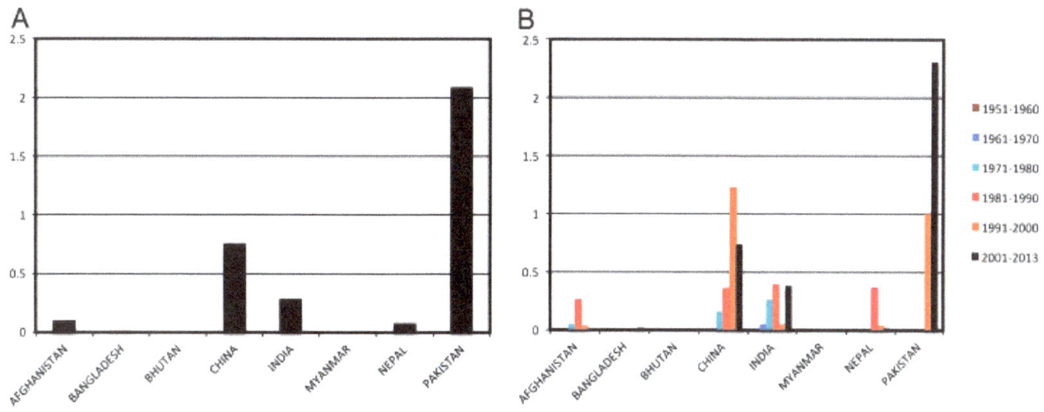

Figure 3: Total "reported" economic losses (in US dollar value) from flooding disasters. *Y*-axis indicates reported economic losses in billions of US dollars per recorded disaster occurrence within the time interval. (A) Country-wise total economic losses reported over 1951–2013 showing highest losses in Pakistan per disaster occurrence and followed by China and India. (B) Same as (A) but decade-wise patterns showing Pakistan had the highest amount of economic loss reposted in the recent most decade, which got higher from the earlier decade (1991–2000), China's reported losses decreased in the recent decade.

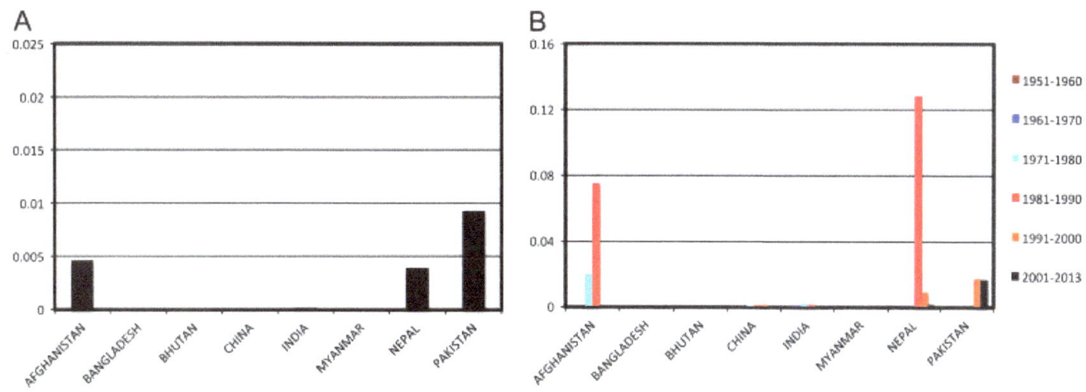

Figure 4: Total "reported" economic losses (in US dollar value) standardized with respect to 2012 GDP estimates (Fig. S3(A)). *Y*-axis indicates reported economic value in billions of US dollars per reported disaster occurrence within the time interval per country specific 2012 or concurrent decadal GDP values. (A) Country-wise "relative" economic losses indicating highest impacts in Pakistan and followed by Afghanistan and Nepal. (B) Decade-wise economic losses per disaster, standardized with respect to concurrent GDP values (shown in Fig. S3), showing Pakistan still having the highest economic impacts in the recent decade equaling to the previous decade (1991–2000), but for Nepal and Afghanistan, recorded economic losses were higher in 1981–1990 as compared to the recent decades.

According to the data displayed in Fig. 3(A), historically, Pakistan experienced the highest economic damages per disaster as a whole and followed by China and India. The most current decade reported the highest economic damage in Pakistan, as in Fig. 3(B), which includes the world's second worst flooding event in the decade—the 2010 event (IRIN, 2014). Furthermore, Fig. 4(A) indicates economic losses in US $ values standardized with respect to the 2012 gross domestic product (GDP) for each country. The most striking feature of this figure is that Pakistan still remains the most vulnerable country in terms of economic damages, per historical record, as compared to other HKH countries; however, this time we see Afghanistan and Nepal showing up higher economic damages as well. Because India and China are economically most advanced among the eight countries under study, the economic losses standardized with respect to their GDP value reduces the economic impacts drastically for them. Because Nepal and Afghanistan were much poorer in terms of their country economic growth in 1981–1990 (Fig. S3), economic impacts in those countries were much higher in that decade that is showing a decreasing pattern in the recent decade (Fig. 4(B)). Pakistan is showing the equal economic impacts in the recent 2 decades (1991–2013). Therefore relative economic impacts comparison shows an interesting way to look at how a country feels or has the capacity to deal with economic impacts of disasters.

Now, one can also argue that total economic losses recorded in the history could also be an indicator of how much an individual country is losing in total, and whether that amount has been changing over the time. Therefore, we also looked at the total dollar amount of losses without dividing them by the number of disasters and/or GDP (Fig. S2). We now notice that China is almost at par with Pakistan, a little bit higher in fact in terms of total money lost, because, for China the economic losses reported so far is the highest although they experienced much lower frequency of disasters within the HKH territory as compared to Pakistan or other countries over the study period (Fig. 2 and Fig. S1). However, historically reported disaster losses seemed to have decreased in China in the recent decade (2001–2013) as compared to previous decade but for Pakistan and India the opposites are being true (Fig. S2(A)). Fig. S2(B) shows total economic losses for each country and decade wise estimates standardized with respect to 2012 GDP values. Nepal is now showing total economic impacts at par with Pakistan but the recent decade was recorded to be highly devastating for Pakistan and that for Nepal has reduced over the decades.

These whole sets of analyses indicate that Pakistan's vulnerability is not only in terms of historically recorded high frequency of disasters, but also in terms of the economic losses. A country's economic status is a major decisive factor of how the impacts are being felt and/or dealt with. For example, Afghanistan and Nepal shows higher economic impacts once we standardize their economic damages with respect to the corresponding GDP values. Therefore, the scale of economic impacts is dependent both on a country's socio-economic status as well as the frequency of disasters, which is consistent with the findings reported by Mendelsohn (2006).

Spatio-temporal Pattern of "Reported" Human Losses (Morbidity and Mortality)

Historical data also helps us to understand which country is vulnerable to human losses due to flooding disasters. We define human impacts by morbidity and mortality numbers. Because of the high amount of missing data, we restrict our analysis for human losses within the last 2–3 decades. When we take a look at spatio-temporal trends of reported morbidity in the last two decades, Pakistan again takes the largest share of the pie in terms of the total people affected due to flooding disasters and that is followed by India and Bangladesh respectively (Fig. 5(A)).

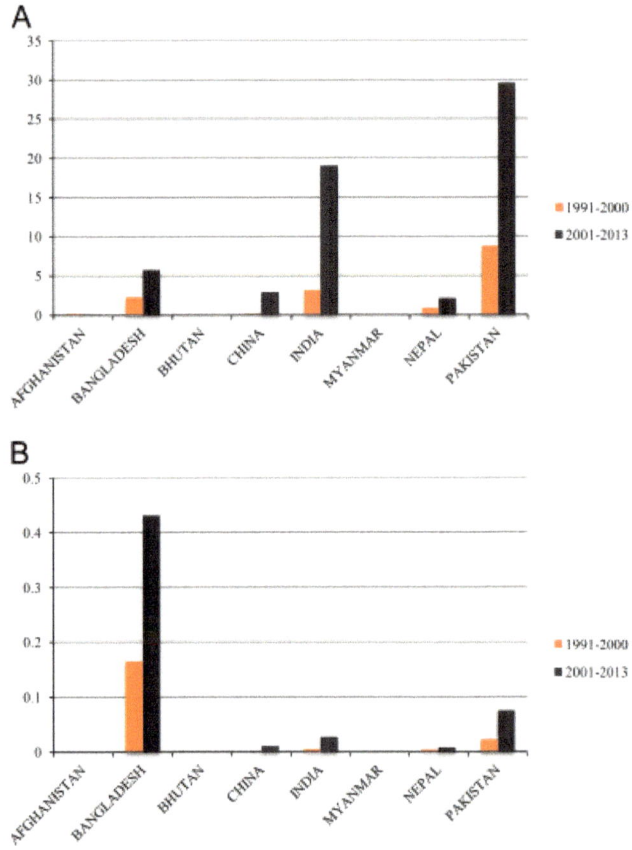

Figure 5: Reported human impacts due to flooding disasters (morbidity). *Y*-axis indicates reported total morbidity (in millions) (upper panel) and reported morbidity per year standardized with respect to total population currently living within HKH region (lower panel). (A) Total morbidity (affected population in millions) for each country in the last two decades showing the highest numbers recorded for Pakistan, and followed by India and Bangladesh (1991–2000 and 2001–2013). (B) Total morbidity per year standardized with respect to total population currently living within HKH region—indicating the highest fraction of people in Bangladesh among all other countries are most vulnerable to human impacts every year, which is followed by Pakistan.

It is well known that the human impacts are highly driven by the population exposed to flooding disasters (Federal Emergency Management Institute (FEMA, 2013)). Fig. 5(B) shows *relative human impacts*—the total morbidity per year standardized with respect to total population currently living within HKH region in each country boundary. This indicates the highest fraction of people in Bangladesh living within the HKH region has been recorded to be most vulnerable to morbidity per year per flooding disaster, which is followed by

Pakistan and other countries. Only the recent 2 decades (1991–2000 and 2001–2013) are shown in Fig. 5 because recorded morbidity numbers were missing in the earlier decades (Table S1).

Fig. 6 shows the decade specific (A) total reported mortality per country (in 1000s) over the HKH region and (B) total mortality per year standardized with respect to total population currently living within HKH region in each country border. Per historical record, the number of people killed was highest in India in the recent decade, which got much higher than the earlier decade, and that is followed by Pakistan, Nepal, Afghanistan and China respectively (Fig. 6(A)). To note here that the total number of people died for China decreased dramatically over the last three decades.

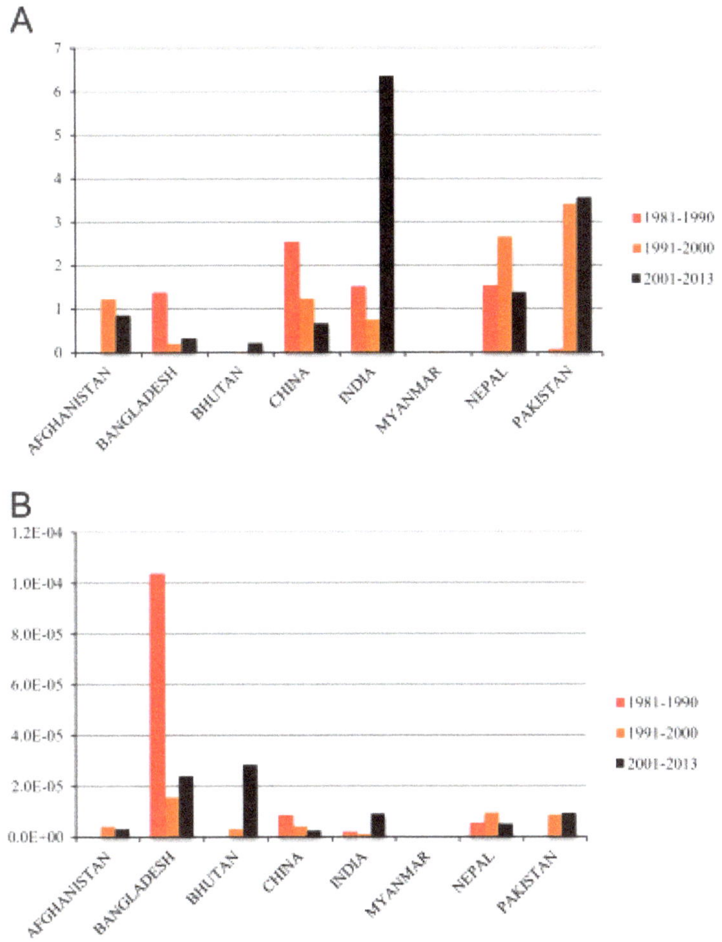

Figure 6: Reported human impacts due to flooding disasters (mortality). *Y*-axis indicates reported total mortality (in thousands) (upper panel) and reported mortality per year standardized with respect to total population currently living within HKH region (lower panel). (A) Total mortality (population died in thousands) for each country in the last three decades showing the highest number recorded for India in the recent decade that is followed by Pakistan and Nepal. (B) Total mortality per year standardized with respect to total population currently living within HKH region—indicating the highest fraction of people in Bhutan and Bangladesh living within HKH were affected in 2001–2013.

Fig. 6(B) displays *relative human impacts*—the fraction of people died every year with respect to the total population living within HKH region for each country. We notice that this number gets significantly higher for Bangladesh than other countries because population living in Bangladesh is much lesser than others but more people die there. Most notably, this number decreases for China and Bangladesh a great deal in the recent decade but it is still the second highest in Bangladesh. Now Bhutan is also showing the greatest vulnerability in terms of people died in the recent decade than the earlier decades, per historical record.

State of Concurrent Weather Systems

Timing or Seasonality

In addition to occurrence numbers and impacts, it is also important to look at *when* in a given year a disaster usually occurs at a given location. Historical data indicate that the flooding disasters have a tendency to occur mainly during the summer season (Fig. 7(A)). This is anticipated because most of the regions surrounding HKH get high rainfall during southwest monsoon. Because we include September in the fall season, when monsoon weather system is still active, many of the flooding disasters in the western and central region of HKH also occurred in the fall. However, most of the disasters in the extreme western part of HKH, and particularly over the northern Afghanistan, occurred in the spring. That is because westerly weather system is typically active in this region in the late winter and spring. This seasonality does not alter in the decade wise maps, shown inFig. 7(B).

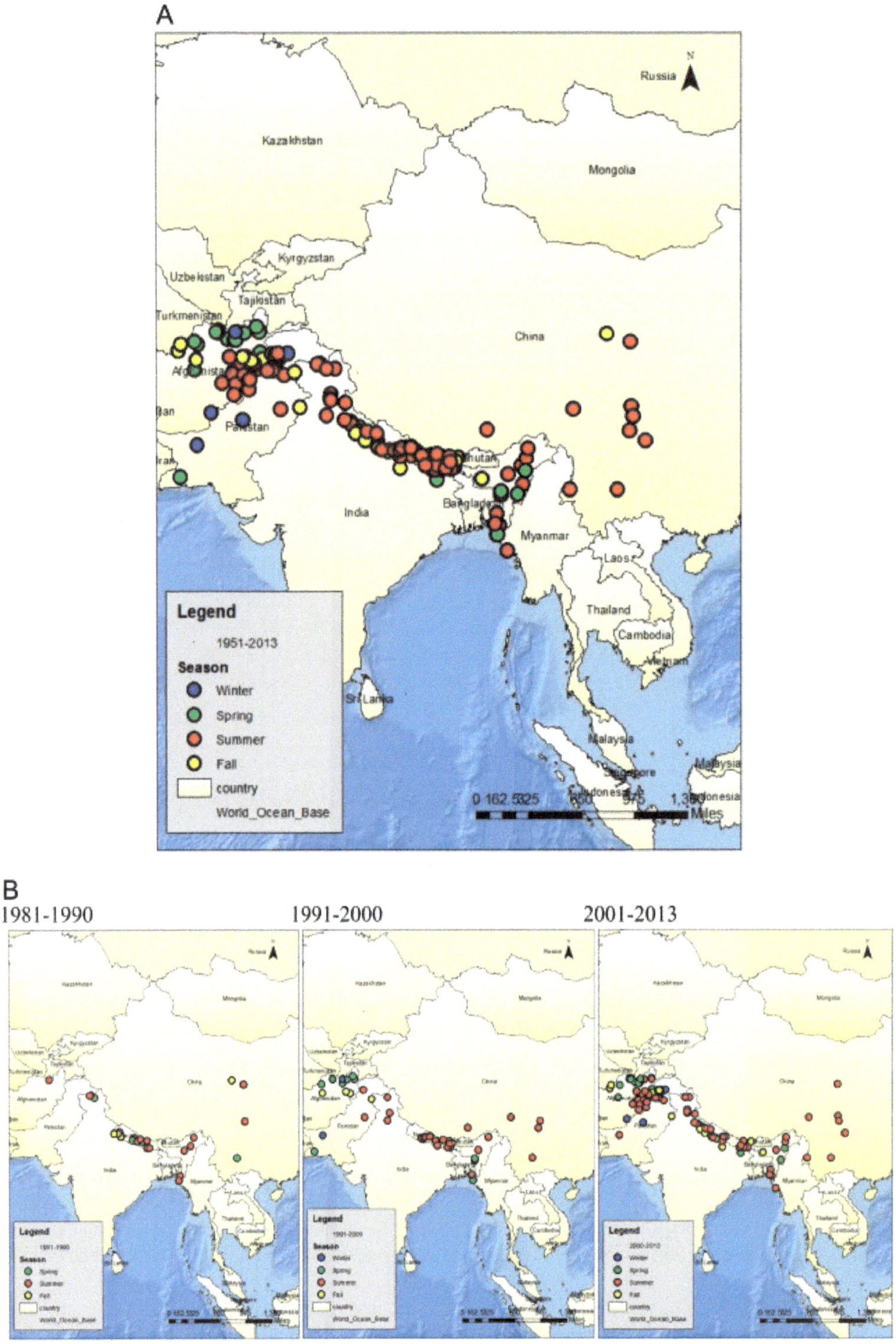

Figure 7: Timing of flooding disasters over the HKH region as per historical record; winter is Dec—Feb, spring is Mar—May, summer or monsoon is Jun—Aug, and autumn is Sep—Nov. (A) Average seasonality within 1951–2013 showing monsoon season dominance (red color), while spring (green color) and fall seasons (yellow color) being important for the north of Afghanistan. (B) Average decadal seasonality in the last 3 decades showing similar seasonal patterns as (A).

Seasonality of occurrences of flooding disasters specific to each country gives indication of the spatio-temporal heterogeneity in expected timing of disaster events. Fig. S4displays that the seasonality in Afghanistan is bi-modal—one is in spring and the other in the summer. For India, China, and Nepal, disasters occur mainly in the summer season (July and/or August), indicated by historical data. The summer season (July, August) is also important for Pakistan, but there have also been some cases reported in winter (February) too. In general, the entire HKH region is vulnerable to flooding events during the summer time (June to August).

Precipitation

Impact in a given location could also be a function of the length of a disaster (in days) as well as the magnitude of accumulated total precipitation in that interval for a given disaster. Accumulated precipitation magnitude per disaster differs across HKH region, as shown in Fig. 8(A). There is a clear spatial gradient in it, starting from the west to the east.

The highest amount of accumulated precipitation is noted in the eastern part of Himalaya falling within India and Bangladesh, which is also true in the decadal maps. If we incorporate the varying length of disasters, to study average precipitation intensity for a given disaster as indicated in Fig. 8(C), Fig. 8(A) changes to Fig. S5, which also indicates high heterogeneity of precipitation intensity (mm/day) across locations with higher intensity of precipitation recorded in northern Pakistan, the central and the eastern HKH.

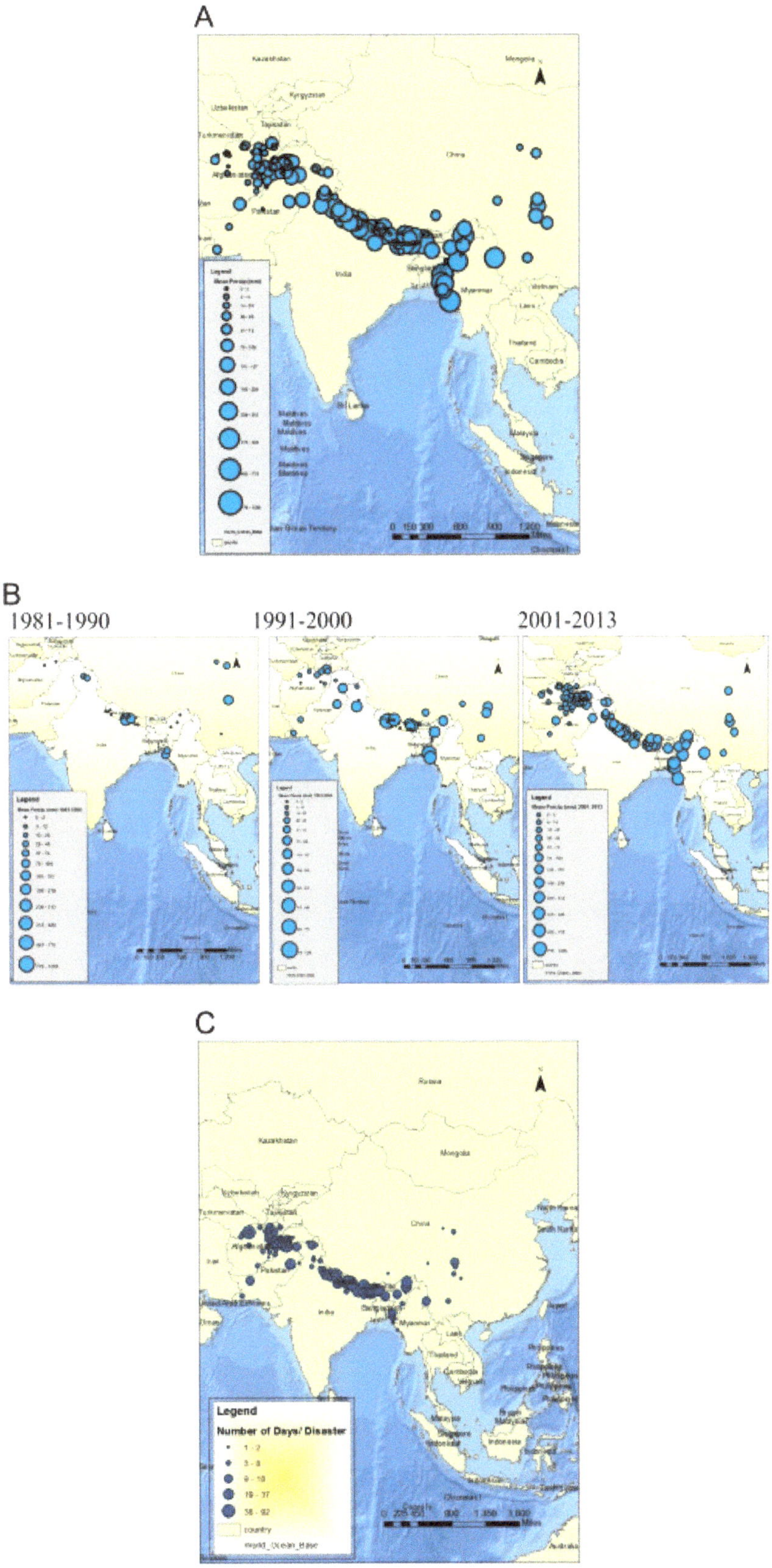

Figure 8: Concurrent daily precipitation characteristics. (A) Average magnitude of accumulated precipitation (mm) per disaster showing higher amount of precipitation is expected in the central and eastern HKH than in the western part. (B) Decadal average magnitude of accumulated precipitation per disaster. (C) Average length of disaster (in days) showing varying patterns across the HKH.

Temperature

It is also interesting to take a look at what the magnitude of monthly mean temperature status was during the disaster months. Fig. 9(A) and (B) shows that average temperature magnitude in a disaster month was generally recorded around 21 °C for most of the regions. This average number is low because we incorporate all the seasons and variability in mean monthly temperature, which was high in the western end than in the central or eastern regions (Fig. 9(C)). This is because flooding disasters peak in different seasons in the western part of HKH (spring/summer in Fig. 7(A) and Fig. S4) when temperatures can differ drastically. High temperature in the central and eastern part of HKH help bringing more monsoon moisture towards this region while in spring and winter, temperature in the western part take a dominant role to melt snow and glacier ice and contribute to flooding.

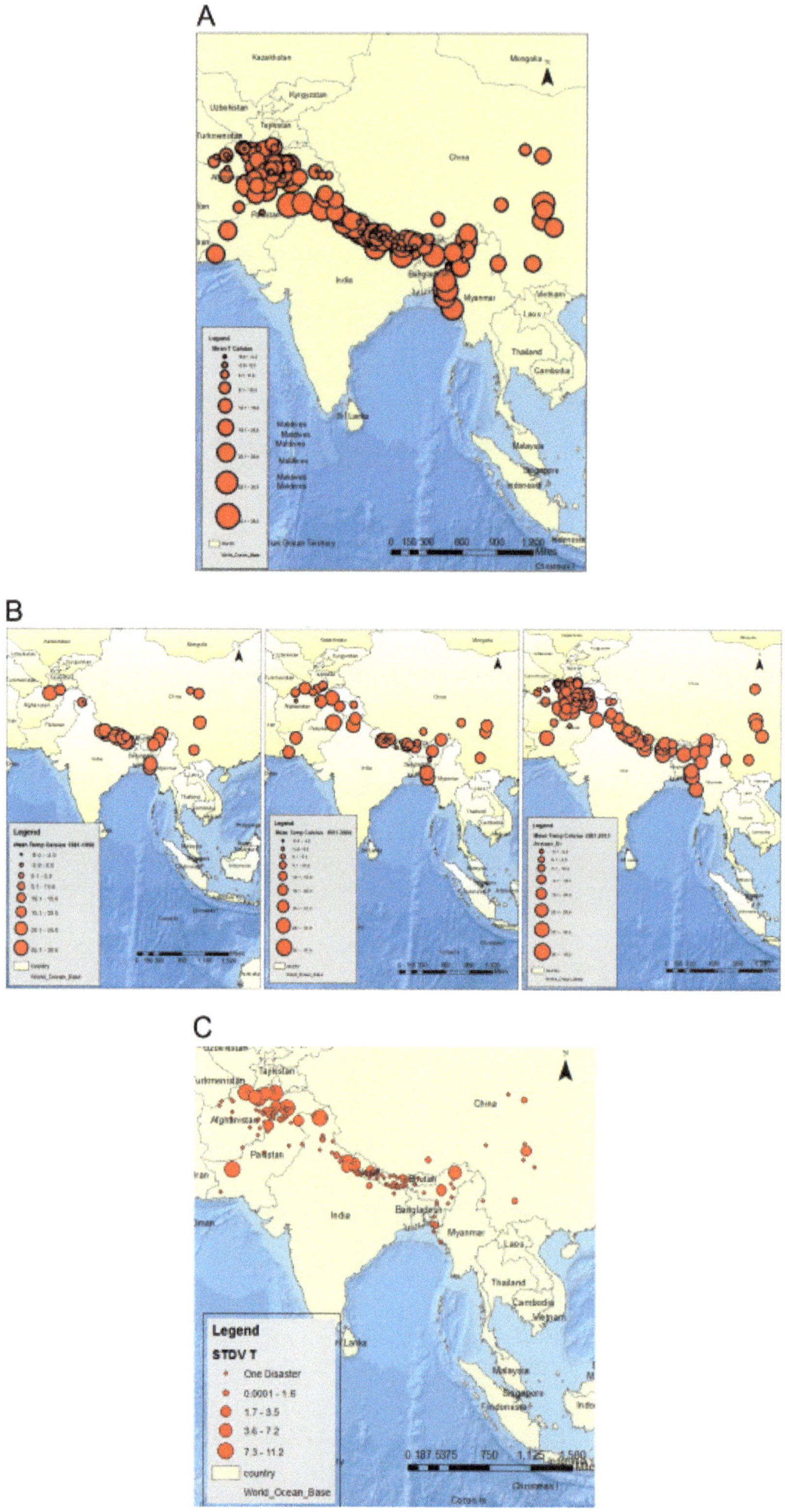

Figure 9: Concurrent monthly temperature characteristics. (A) Average magnitude of monthly mean temperature during disaster months. (B) Decadal average magnitude of monthly mean temperature during disaster months. (C) Standard deviation of monthly mean temperature during disaster months showing higher variation over the western part.

Fig. 8 and Fig. 9 are summarized in Fig. S6 where temperature and precipitation patterns are overlaid with each other to show composite and/or comparative impacts of meteorological variables. Fig. S6 indicates that high amount of accumulated precipitation is as important as the monthly temperature magnitude is during a flooding disaster for the eastern and central Himalaya, while for the western part, monthly temperature visibly takes a dominant role than the amount of precipitation. High temperature in the western part increases the glacier melt, especially during the beginning of May (Panday et al. 2011), which may enhance the frequency of disasters. Historical data also indicate that more than 50% of the total reported disasters in each country occurred when monthly temperatures exceeded the historical average (Fig. S7(A)), most of which occurred in the summer season (Fig. S7(B)).

SUMMARY

Mapping the vulnerability of flooding disasters in the Hindu-Kush Himalayas is of considerable interest to develop local adaptive management measures for the communities and building location specific resilience to disaster impacts. The range of topography in HKH makes climate and socio-economic data availability and their analysis at local scales a challenge. Here we take a bottom up approach where instead of considering the whole HKH region and analyzing their climate data, we mine a disaster database to identify the locations' those were reported to have experienced flooding disasters and understand patterns of their economic and human impacts. This way we identify and report the locations and/or regions, which are vulnerable to flooding disaster occurrence (*where*), timing in a year *when* those locations are vulnerable in general (or on average), type and magnitude of impacts (*what consequences*) they bring to corresponding localities and countries, and *what* climate factors might be playing a role (*to what they are vulnerable*).

High spatial variability in occurrence of flooding disasters was identified across the HKH. Parts of northern Afghanistan and north-western Pakistan in the western HKH showed high vulnerability to disaster occurrence, as well as north of India and Nepal, and some locations within China and Bangladesh in the east of HKH. Decadal patterns indicated that most disaster cases were reported in the recent decade (2001–2013) as compared to the earlier decades. Historically, Pakistan experienced the highest economic losses per disaster and followed by China and India. The most current decade reported the highest economic damage in Pakistan per disaster occurrence. However, when the economic costs of disasters are standardized with respect to a country's economy (2012 or corresponding decadal national GDP value), Afghanistan and Nepal showed up higher impacts in addition to Pakistan. Therefore, a country's economic status is a major decisive factor on how the impacts are being felt and/or dealt with.

Regarding human impacts, Pakistan again was largest in terms of total people affected due to flooding disasters in the history and that is followed by India and Bangladesh respectively. However, if population exposed to flooding disasters is considered, Bangladesh shows the highest vulnerability for the fraction of people living within the HKH region, which is followed by Pakistan and other countries. As regards to mortality, the number of people killed was highest in India in the recent decade, which got much higher than the earlier decade, and that is followed by Pakistan, Nepal, Afghanistan and China respectively. However, if one considers mortality numbers with respect to number of people living within the HKH region, this number gets significantly higher for Bangladesh than other countries because population living in Bangladesh territory of HKH is much lesser than other countries but more people die there due to impacts induced by different types of flooding (flash/riverine/cyclone).

Most of these flooding disasters occur in the summer during active South-west Monsoon that brings cloud bursts or incessant rains, and some occur in the spring, especially in northern Afghanistan, which experiences westerly weather system and/or snow and glacier melt and/or cloud bursts. Therefore, temperature seems to be most important climate factor in the western region while both precipitation and temperature are important for the rest of the HKH region. Temperature usually stays higher than the historic average in the disaster months in all over the HKH region, but the variability is higher in the west.

In summary, complex interactions between biophysical risk factors, socio-political, cultural and socio-economic conditions, and the governance/institutional issues (e.g. failure of Early Warning Systems (EWS)) play a dominant role to define and characterize the type and magnitude of vulnerability to flooding disaster occurrence and their impacts, and most importantly, there are differential capacities of regions to recuperate from past impacts and/or to cope with and adapt to future impacts, as also detailed in Eakin and Luers (2006). In addition, it is also desirable to determine which combinations of attributes best characterize the vulnerability of specific regions and develop suitable metrics that can be used in a variety of settings to reliably link systems attributes (e.g. biophysical and economic) to vulnerability outcomes. Therefore, this begs a multi-disciplinary, multi-dimensional, and multi-scale understanding of vulnerability in this diverse region of HKH that helps developing better maps for distribution of flood vulnerability, outcomes and potential impacts that can become a central tool for communicating the results to other academics, researchers, policy makers, and the community at large, in order for designing location specific management practices and developing adaptive capacity. We acknowledge uncertainty in the historical data records and therefore a verification of the time trends in losses will be conducted in future using other available data, as well as accounting for dynamic nature of vulnerability and representing other social indicators that may be determinant of vulnerability for particular areas. However, the fundamental importance of this study lies being the first of its kind for the Hindu-Kush Himalaya region and the synoptic view of presenting vulnerability of locations to flooding disasters, which is entirely non-existent in the literature or in any publicly available reports. Extensions of this work towards understanding the natural and other socio-economic factors and their combinations are in progress, with the goal of building a multi-scale prediction model for extreme precipitation and temperature and their resultant economic and human losses.

REFERENCES

1. Bookhagen, B., Burbank, D., 2010: Toward a Complete Himalayan Hydrological Budget: Spatiotemporal Distribution of Snowmelt and Rainfall and Their Impact on River Discharge, March, Page 23.
2. Dankers, R., Arnell, N., Clark, D., Falloon, P., Fekete, B., Gosling, S., Heinke, J., Kim, H., Masaki, Y., Satoh, Y., Stacke, T., Wada, Y., Wisser, D., 2014. First look at changes in flood hazard in the Inter-Sectoral Impact Model Intercomparison Project ensemble. PNAS 111 (9), 3257–3261.
3. Eakin, H., Luers, A.L., 2006. Assessing the vulnerability of social-environmental systems. Annu. Rev. Environ. Resour. 31, 365–394.
4. EM-DAT, 2014. The International Disaster Database, Center for Research on the Epidemiology of Disasters-CRED, Disaster List. ⟨http://www.emdat.be/disaster-list⟩. FEMA (Federal Emergency Management Institute), 2013. Hazard, Vulnerability, and Risk Analysis, Chapter 6, pages from 160 to 161.
5. Field, C., Barros, V., Mach, K., Mastrandrea, M., Aalst, M., Adger, N., Arent, D., Barnett, B., Betts, R., Bilir, E., Birkmann, J., Carmin, J., Chadee, D., Challinor, A., Chatterjee, M., Cramer, W., Davidson, D., Estrada, Y., Gattuso, J., Hijioka, Y., Guldberg, J., Huang, H., Insarov, G., Jones, R., Kovats, S., Romero, P., Nymand, J., Larsen, J.,
6. Losada, I., Marengo, J., McL ean, R., Mearns, L., Mechler, R., Morton, J., Niang, I., Oki, T., Mukarugwiza, J., Opondo, M., Poloczanska, E., Po □ rtner, H., Redsteer, M.,Reisinger, A., Revi, A., Schmidt, D., Shaw, R., Solecki, W., Stone, D., Stone, J., Strzepek, K., Suarez, A., Tschakert, P., Valentini, R., Vicuna, S., Villamizar, A., Vincent, K., Warren, R., White, L., Wilbanks, T., Wong, P., Yohe, G., 2014. Climate Change 2014: Impacts, Adaptation, and Vulnerability, Technical Summary, March, (IPCC), Cambridge University Press, Cambridge, UK, and New York, NY, USA, Pages from 16 to 17; from 25 to 27. ⟨http://ipcc-wg2.gov/AR5/images/ uploads/WGIIAR5-TS_FGDall.pdf⟩.
7. ICIMOD, 2014. Hindu kush Himalaya Region, Regional Information. ⟨http://www. icimzod.org/?q=1137⟩.
8. IRI Data Library, International Research Institute for Climate and Society, Earth institute| Columbia University, 2014. ⟨http://iri.columbia.edu⟩.
9. IRIN, 2014. DISASTERS: Pakistan Floods, One of the World's Worst, Humanitarian News and Analysis a Service of the UN Office for the Coordination of Humanitarian Affairs. ⟨http://www.irinnews.org/report/90142/disasters-pakistan-floods-one-of-theworld-s-worst⟩.
10. Jianchu, X., Arun, S., Mats, E., 2009. Climate Change and its Impacts on Glaciers andWater Resource Management in the Himalayan Region, Assessment of Snow,Glacier and Water Resources in Asia, International Hydrological Programme ofUNESCO and Hydrology and Water Resources Programme of WMO, pages from 45 to 50. ⟨http://w ww.mtnforum.org/sites/default/files/publication/files/5412.pdf⟩.
11. Karki, M., Sharma, S., Mahat, T., Aksha, S., Tuladhar, A., And 14 Case Study Authors, 2011. Hindu Kush Himalaya (HKH) Region, From Rio 1992 to 2012 and beyond: Sustainable Mountain Development, International Centre for Integrated Mountain Development (ICIMOD) Kathmandu, Nepal, September, pages from 4 to 5.⟨http://www.uncsd2012.org/content/documents/HKH_30_9_2011_0[1].pdf⟩.
12. Kintisch, E., 2014. In New Report, IPCC Gets More Specific About Warming Risks, Novim, April. ⟨http://www.novim.org/resources/novim-news/358-illinois-pension-problemhow-big-is-it-really-11⟩.
13. Mendelsohn, R., Dinar, A., Williams, L., 2006. The Distributional Impact of Climate Change on Rich and Poor Countries, Environmental and Development Economics 11:159–178, Cambridge University Press, pages from 173 to 176. ⟨http:// www.peconsulting.se/wp-content/uploads/2014/03/The-distributional-impact-of-climate-change-on-rich-and-poor-countries.pdf⟩.
14. NAS (National Academy of sciences), 2012. Himalayan Glaciers: Climate Change, Water Resources, and Water Security, pages from 1 to 3; from18 to 20; from 39 to 40; from 47 to 48, Page 83, ISBN 978-0-309-26098-5. ⟨http://www.nap.edu/catalog.php?record _id=13449⟩.
15. Panday, P., Frey, K., Ghimire, B., 2011. Detection of the timing and duration of snowmeltin the Hindu Kush-Himalaya using QuikSCAT, 2000–2008. Environ. Res. Lett. May(6), 024007. http://dx.doi.org/10.1088/1748-9326/6/2/024007 ⟨http://iopscience.iop.org/1748-9326/6/2/024007/fulltext/#erl379996bib25⟩.
16. Pathak, D., Gajurel, A.P., Mool, P.K., 2008. Climate Change Impacts on Hazards in theEastern Himalaya: Assessment of Climate Change Vulnerability of the Mountain Ecosystems of the Eastern Himalaya, Study Report, ICIMOD, Kathmandu.
17. Seneviratne, S.I., Nicholls, N., Easterling, D., Goodess, C.M., Kanae, S., Kossin, J., Luo, Y., Marengo, J., McInnes, K., Rahimi, M., Reichstein, M., Sorteberg, A., Vera, C., Zhang, X., 2012. Changes in climate extremes and their Impacts on the natural physical environment. In: Field, C.B., Barros, V., Stocker, T.F., Qin, D., Dokken, D.J., Ebi, K.L., Mastrandrea, M.D., Mach, K.J., Plattner, G.-K., Allen, S.K., Tignor, M.,Midgley, P.M. (Eds.), Managing the Risks of Extreme Events and Disasters to Advance Climate Change Adaptation. Cambridge University Press, Cambridge, UK, and New York, NY, USA (A Special Report of Working Groups I and II of theIntergovernmental Panel on Climate Change, (IPCC).
18. Sharma, S., 2014. No End To The Himalayan Blunder?, Climate Himalaya, Jan. ⟨http://chimalaya.org/2014/01/06/no-end-to-the-himalayan-blunder/#/⟩.
19. Than, K., 2005. Scientists: Natural Disasters Becoming More Common, Live science, October. ⟨http://www.livescience.com/414-scientists-natural-disasters-common. html⟩.
20. The World Bank, 2014. Countries and Economies, Data. ⟨http://data.wo rldbank.org/country⟩.
21. UN data, 2014. GDP (current US $), Source World Development Indicators | The World Bank. ⟨http://data.un.org/Data.aspx?q=gdp&d=WDI&f=Indicator_Code%3aNY.GDP.MKTP.CD#WDI⟩

CHAPTER 14

COMPARISON OF DROUGHT INDICES FOR APPRAISAL OF DROUGHT CHARACTERISTICS

Vinit K. Jain, Rajendra P. Pandey, Manoj K. Jain, Hi-Ryong Byun

Keywords: Drought, Effective drought index, Standardized precipitation index, Z-score, China *Z*-index, Rainfall departure, Time step of drought

INTRODUCTION

The preparedness and planning to cope with adverse impacts of a drought event depend on the information about its areal extent, severity and duration. This information can be obtained through drought monitoring and forecasting that is usually done using drought indices which provide quantitative information to decision makers about drought characteristics (Dogan et al., 2012). Over the years, a number of indices have been proposed for drought characterization (e.g. Palmer Drought Severity Index, PDSI;Palmer, 1965), Deciles (Gibbs and Maher, 1967), Crop Moisture Index (CMI; Palmer, 1968), Bhalme and Mooley Drought Index (BMDI; Bhalme and Mooley, 1980), Surface Water Supply Index (SWSI; Shafer and Dezman, 1982), Standardized Precipitation Index (SPI; Mckee et al., 1993), Effective Drought Index (EDI; Byun and Wilhite, 1999), Soil Moisture Deficit Index (SMDI; Narasimhan and Srinivasan, 2005), Reconnaissance Drought Index (RDI; Tsakiris et al., 2007). All of these indices are normally continuous functions of any one or more hydro-meteorological variables, viz. precipitation, temperature, soil water, streamflow, groundwater, potential evapotranspiration etc. (WMO, 1975).

Most of the available drought indices are developed for specific regions and have the limitation of use under different climatic conditions because of the inherent complexity of drought phenomena. For example, PDSI (Palmer, 1965), is extensively used in the United States, the RDDI (Gibbs and Maher, 1967) is operational in Australia, the China *Z*-Index (CZI) is used by the National Metrological Center in China (Wu et al., 2001) and SPI (Edwards and Mckee, 1997) is being used widely including India because of its adaptability to different time scales and climatic conditions. Several attempts have been made in the past to analyze the appropriateness in describing the droughts characteristics for a particular region using different indices (Keyantash and Dracup, 2002, Ntale and Gan, 2003, Morid et al., 2006, Barua et al., 2011 and Dogan et al., 2012) using multiple criterions of tractability, transparency, and dimensionality for comparison of various drought indices. Smakhtin and Hughes (2007) developed Spatial and Time Series Information modeling (SPATSIM) software for automated estimation of drought severity using five different drought indices. Pandey et al. (2008) used SPATSIM for a drought study of Orissa, India and found that EDI performed better than other DIs. Dogan et al. (2012) compared six meteorological drought indices and indicated that each drought index identified the drought characteristics differently. They observed the variation in severity values and duration of a drought event computed using different indices. Further, they concluded that EDI performed better for monthly rainfall changes in semi-arid Kenya closed basin, and Turkey. Also, Dogan et al. (2012) suggested that application of multiple time steps has significant role in assessment of regional drought characteristics. If one is interested in short term rainfall anomalies near real time drought monitoring, the shorter time step (i.e., 1-month, 3-month) should be used. However, if one needs to study the regional drought characteristics and to formulate area specific mitigation plan, he must use longer time steps, i.e. 6-, or 12- month. Many times the 1-month time step may lead to erroneous assessment of drought severity, because monthly rainfall deficiencies are common feature of arid regions. The present study is undertaken in the Ken

Citation: Vinit K. Jain, Rajendra P. Pandey, Manoj K. Jain , Hi-Ryong Byun, , Comparison of drought indices for appraisal of drought characteristics, http://dx.doi.org/10.1016/j.wace.2015.05.002.

River Basin, a drought prone region of central India, where the summer concentration of rainfall is very high, to examine the applicability of various drought indices for assessment of meteorological drought characteristics. The specific objectives of this study are (1) to compute severity of past drought events using various meteorological drought indices (2) to compute correlation between estimated drought severity at multiple time steps and (3) comparing the quantitative values of drought attributes obtained using various meteorological drought indices.

BRIEF DESCRIPTION OF THE STUDY AREA AND DATA USED

The study has been carried out for the Ken River Basin, located in drought prone region of central India. The study basin lies between north latitudes of 23° 07′ and 25° 30′ and east longitudes of 78° 30′ and 80° 40′. The river Ken originates at an altitude of 550.0 m above mean sea level (msl). The average elevation of the plain areas of the basin is about 328 m above msl. The study basin is situated in the semi-arid and dry sub-humid climatic region of India with a single rainy season (June-September) followed by dry winter, and then a very dry summer. The basin is fed by south-west monsoon which starts from middle of June and lasts till the end of September. The spatial distribution of rainfall in thc basin is highly variable. The average annual areal rainfall in the basin varies from 1250 mm to 800 mm with an estimated basin averaged annual rainfall of about 1165 mm. Approximately 91% of the annual rainfall is received during monsoon months. Continuous monthly rainfall data of 13 districts in and around the study basin for a period of 102 years from 1901 to 2002 is obtained and used in the present analysis. The mean monthly, seasonal and annual rainfall of all districts is presented in Table 1. Fig. 1 shows the location of the Ken River Basin in India along with the demarcation of districts jurisdiction for which rainfall data has been utilized.

Table 1: Mean monthly, annual and seasonal rainfall of various districts in and around the Ken River Basin.

District	Geographic Location		Mean monthly Rainfall of various districts (mm)												Annual	Seasonal (Jun–Sep)
	Longitude	Latitude	Jan	Feb	Mar	Apr	May	Jun	Jul	Aug	Sep	Oct	Nov	Dec		
Banda	80.31	25.48	22	11	11	4	5	89	321	390	173	26	12	9	1073	973
Chattarpur	79.59	24.92	22	13	10	4	5	106	352	396	182	26	16	9	1142	1036
Damoh	79.45	23.84	18	17	12	6	7	135	356	399	192	32	16	10	1200	1081
Fatehapur	80.82	25.93	20	11	10	4	5	87	295	348	164	30	9	8	990	893
Hamirpur	80.15	25.95	19	10	7	4	5	81	306	347	171	24	12	8	992	905
Jabalpur	79.95	23.17	19	23	19	7	12	162	354	381	193	40	17	12	1239	1090
Katni	80.40	23.84	22	23	17	7	12	151	350	399	193	34	13	13	1234	1093
Mahoba	79.87	25.29	21	11	8	4	5	90	335	369	177	23	16	9	1066	970
Narsimhapur	79.19	22.94	14	15	15	7	10	138	352	353	213	35	22	9	1183	1056
Panna	80.18	24.72	23	18	13	6	6	117	351	411	191	27	14	11	1188	1070
Raisen	77.78	23.33	11	9	7	4	10	115	408	372	232	22	20	9	1218	1127
Sagar	78.75	23.83	15	12	7	3	7	119	375	394	203	26	19	10	1191	1091
Satna	80.83	24.58	26	20	16	5	8	113	336	400	184	26	13	10	1159	1034
Ken River Basin as a whole			**20**	**15**	**11**	**5**	**6**	**116**	**352**	**395**	**190**	**28**	**16**	**10**	**1161**	**1052**

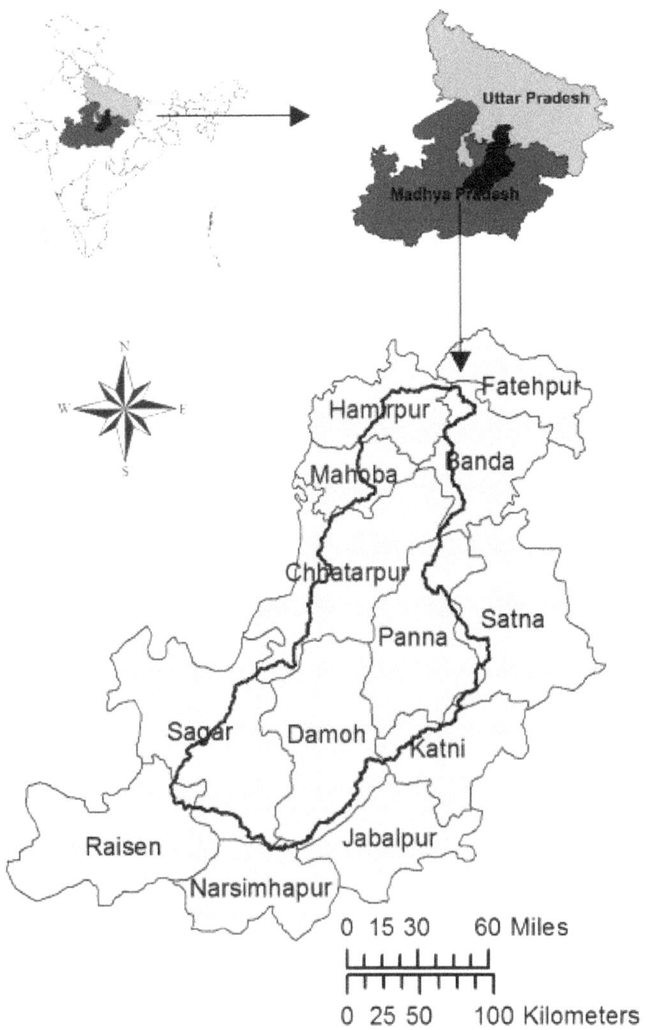

Figure 1: Location of the Ken River Basin (the darkest area in thick black line) and adjoining districts.

SELECTION OF DROUGHT INDICES

In the present study long term area weighted rainfall data for the period from 1901 to 2002 for 13 districts obtained from India Water Portal website has been used for evaluating meteorological drought indices. Six meteorological drought indices have been applied for computation of severity and duration of drought events. These indices are Rainfall Departure (RD), Z-Score, CZI, SPI, EDI and RDDI. A brief description of above indices is given below:

Rainfall Departure from the mean or median (RD)

Rainfall Departure (RD) is a good indicator of dry/wet conditions for a given time over specified areas. It is calculated by subtracting the long term average rainfall from monthly rainfall and dividing the difference by the long term average rainfall. The negative values represent the deficit and positive values represent the excess rainfall with respect to corresponding average. It is a straightforward measure of rainfall deviation from its long-term mean or median or a pre-determined specific percentage of mean based on the regional weather conditions. This method is generally applicable to define local weather condition for rainfall deficiencies of smaller duration and used in weather broadcasting. Rainfall Departure is easy to understand and simple to calculate. For example, Indian Meteorological Department (IMD) defines a year to be drought year if annual rainfall in that year is less than 75% of its long-term mean (Appa Rao, 1986). Droughts in South Africa are defined as periods with less than 70% of normal precipitation (Smakhtin and Hughes, 2004). RD is categorized into dry/wet categories based on the percent departure (deficit/excess) of rainfall over a period with respect to its corresponding long term average. This can be calculated on monthly or multi monthly time steps.

Standardized Precipitation Index (SPI)

The Standardized Precipitation Index (SPI) for any location is calculated, based on the long-term precipitation record for a desired period. The available long-term rainfall data is fitted to gamma probability distribution, which is then transformed to a normal distribution so that the mean SPI for the location and desired period is zero (Mckee et al., 1993). This transformed probability is the SPI value, which varies between +2.0 and −2.0, with extremes outside this range occurring at 5% of the time (Edwards and Mckee, 1997).

The SPI is perhaps the most popularly used drought index. Mckee et al. (1993)developed SPI to identify and monitor drought events using monthly rainfall data. It is intended to identify drought periods as well as the severity of droughts, at multiple time steps, such as at 1, 3, 6, 9, 12 or 24-months. However, the objective choice on the best time step may depend on the purpose of drought analysis. Due to its reliability, and ability to address drought at multiple time steps for a variety of climatic regions, SPI has been used extensively in various parts of the world (e.g. Guttman, 1999; Keyantash and Dracup, 2002; Ntale and Gan, 2003; Vicente-Serrano et al., 2004; Patel et al., 2007;Pandey et al., 2008; Mishra and Singh, 2009; Edossa et al., 2010; Roudier and Mahe, 2010; Stricevic et al., 2011; Zhai et al., 2010; etc.). However, continuous long term data of at least 30 years is required to compute SPI; it does not allow missing data. The complete procedure for estimation of SPI is available in Edwards and Mckee (1997).

China *Z*-Index (CZI)

China *Z*-Index (CZI) is extensively used by National Climate Centre (NCC) of China to monitor drought conditions throughout the country (Wu et al., 2001 and Dogan et al., 2012). CZI assumes that precipitation data follow the Pearson Type III distribution and is related to Wilson–Hilferty cube-root transformation (Wilson and Hilferty, 1931) from chi-square variable to the *Z*-scale (Kendall and Stuart, 1977). The value of CZI is calculated as

$$CZI = \frac{6}{C_{st}}\left(\frac{C_{st}}{2}ZScore_t + 1\right)^{1/3} - \frac{6}{C_{st}} + \frac{C_{st}}{6} \quad (1)$$

where, C_{st} is coefficient of skewness for 't' time step. 't' can be equal to 1-,2-,3- ….9-, 12-, 24 months etc. ZScore is the statistical Zscore and will be computed for the same time step 't'. However, NCC computes CZI only for 1-month time step. In the present study CZI will be computed for five time steps i.e. 1-, 3-, 6-, 9- and 12-month time step.

Many studies comparing the CZI with that of SPI and Z-score reported similar results (Wu et al., 2001 and Morid et al., 2006). Further, Wu et al. (2001) suggested that because of simplicity in calculating drought severity at monthly time step using CZI, it can be preferred over SPI, where rainfall data are often incomplete.

Statistical Z-Score (Z-Score)

This index is also as simple as RD and calculated by subtracting the long term mean from an individual rainfall value and then dividing the difference by the standard deviation. The Z-Score does not require adjusting the data by fitting the data to the Gamma or Pearson Type III distributions. Because of this, it is speculated that Z-Score might not represent the shorter time scales (Edwards and Mckee, 1997). Because of its simple calculation and effectiveness, Z-Score have been used in many drought studies (Akhtari et al., 2009,Komuscu, 1999, Morid et al., 2006, Patel et al., 2007, Tsakiris and Vangelis, 2004, Wu et al., 2001 and Dogan et al., 2012). Various researchers also acclaimed that it is as good as SPI and can be calculated on multiple time steps. It can also accommodate missing values in the data series like CZI.

Effective Drought Index (EDI)

The Effective Drought Index (EDI) was proposed by Byun and Wilhite (1999) to monitor the duration and severity of drought. Byun and Wilhite (1999) used a new concept of effective precipitation (EP). They defined effective precipitation as a function of current month's rainfall and weighted rainfall over a defined preceding period computed using a time dependent reduction function. They computed .EDI as a function of the amount of precipitation required to return to normal (PRN). Where, PRN is calculated from monthly effective precipitation and its deviation from the mean for each month. Thus, first step in calculation of EDI is to calculate EP. If P_i is rainfall '$m-1$' months before the current month and N is the duration of preceding period, then effective precipitation for the current month (EP_j) is given as

$$EP_j = \sum_{m=1}^{N}\left[\left(\sum_{i=1}^{m} P_i\right)/m\right] \tag{2}$$

For example if $N=3$ then $EP=P_1+(P_1+P_2)/2+(P_1+P_2+P_3)/3$,where P_1, P_2 and P_3 are rainfall values during current month(j), previous month and 2 months before respectively. Then average and standard deviation of EP values for each month are calculated and the time series of EP values is converted to deviations from the mean (DEP):

$$DEP_j = EP_j - \overline{EP_j} \tag{3}$$

The PRN_j values are then calculated as follows:

$$PRN_j = \frac{DEP_j}{\sum_{i=1}^{N}\left(\frac{1}{i}\right)} \tag{4}$$

The summation term is the sum of the reciprocals of all the months in the duration N (i.e. for N =3 months), this term will be equal to $\frac{1}{1}+\frac{1}{2}+\frac{1}{3}$. Finally EDI is calculated as follows:

$$EDI = PRN/\sigma_{PRN} \tag{5}$$

where, σ_{PRN} is the standard deviation of the corresponding month's PRN values.

Originally, EDI was developed to monitor drought condition on daily time step (Akhtari et al., 2009, Kalamaras et al., 2010, Kim and Byun, 2009, Kim et al., 2009, Morid et al., 2006, Roudier and Mahe, 2010 and Lee et al., 2012). Subsequently, it was extended for monthly drought monitoring (Smakhtin and Hughes, 2004, Morid et al., 2007, Pandey et al., 2008 and Deo and Byun, 2014). The EDI and SPI use similar classification of drought severity. The EDI is based on the concept of effective precipitation (EP) that is calculated by a time dependent reduction function on daily/monthly precipitation and requires specifying at least 30-years data for calculation of mean effective precipitation. For calculation of monthly EDI, time step is defined by the unit of time (originally daily, but monthly in this study), the rainfall for the current month is given more weight and the weights in decreasing order are given to each preceding month's rainfall values. Using the concept of reduction function, the onset and end date of drought can be defined clearly. The software to calculate the EDI at monthly and daily time step is freely available at http://atmos.pknu.ac.kr/~intra2/.

Rainfall Deciles Based Drought Index (RDDI)

The RDDI was originally suggested by Gibbs and Maher (1967) for investigating rainfall deficiency as per criteria set by the Australian Bureau of Meteorology. Deciles are calculated from actual rainfall series. First, rainfall values of each calendar month (or sum of rainfall values for a group of months for multiple time step) are ranked from lowest to highest and a cumulative frequency distribution is constructed. The distribution is then split into 10 deciles (10% slices). The first decile that has the top rainfall values indicates wettest months in the series, the last decile indicates driest months in the series. Deciles have been used in many drought indices evaluation studies (e.g., Keyantash and Dracup, 2002; Morid et al., 2006; Smakhtin and Hughes, 2007; Mpelasoka et al., 2008;Pandey et al., 2008; Barua et al., 2011; Dogan et al., 2012).The use of decile is advantageous due to simplicity in its computation. However, Keyantash and Dracup (2002) noted that this simplicity may leads to conceptual difficulties. For example, it is reasonable for a drought to terminate when observed rainfall is close to or above normal conditions. However, minor amounts of precipitation during non-monsoon periods, during which little or no precipitation is common (e.g., summer along the West Coast) could activate the first stopping rule, even though the absolute quantity of precipitation is trivial and does not terminate the water deficit. Therefore, climates with highly seasonal precipitation may not be well suited to rainfall deciles.

METHODOLOGY

The present investigation is carried out to compare the six drought indices for applicability in the Ken River Basin. The common procedural steps followed for comparison of drought indices are as under:

Calculation of Drought Indices

The drought severity is calculated using SPI, Z-Score, CZI, RD, RDDI for 1, 3, 6, 9, and 12-month time steps and monthly EDI. In this study, the application of five drought indices for five different time steps and one drought index (EDI) for self-defined single time step resulted in 26 drought index time series.

Comparison of DI values using Pearson correlation coefficients

The Pearson Product-Moment Correlation Coefficient, commonly termed as correlation coefficient (r), is a measure of co-linearity between two arrays and most widely used test statistics. Correlation Coefficient is computed pair wise between all 26 time series for the entire period of record creating a cross correlation matrix. It is calculated using Eq. (6).

$$r = \frac{1}{n}\sum_{i=1}^{n}\left(\frac{x_i - \bar{x}}{S_x}\right)\left(\frac{y_i - \bar{y}}{S_y}\right)$$

where, xi and yi represent the values of arrays with 'n ' number of elements being compared and $\bar{x}$ and $\bar{y}$ are the mean values of two arrays and Sx and Sy are the standard deviation of xi and yi respectively. The r measures the degree of similarity in variation about the means of two values.

Comparison of the Drought Characteristics Identified by each Drought Index

Various indices have been used to identify number of drought months (i.e. the number of months falling under moderate, severe and extreme dry categories for which value of drought index is −1 or less) and maximum duration of drought event (i.e. continuous period of such drought months) in the given district. Subsequently, the above drought attributes estimated using different indices for various districts have been compared.

Comparison of Drought Indices During Historical Drought Period

Drought severity values computed using various indices are plotted for the historical drought period and compared to assess the suitability of different indices in identifying the onset and termination of drought.

The procedure is applied for comparison of various drought indices for different districts in and around the Ken River Basin.

RESULTS AND DISCUSSION

Category of Drought Severity

The SPI is most widely used drought index in various parts of the world and is found to be suitable for Asia region too (Smakhtin and Hughes, 2004). EDI is relatively new drought index and have not been tested extensively for its applicability in Indian region, however some studies (Morid et al., 2006, Mishra and Singh, 2009 and Dogan et al., 2012) reported that EDI is more responsive to the emerging drought and perform better. RD is used for short term weather broadcasting and for annual and seasonal drought assessment in India. In addition to above, three other indices i.e. Z-Score, CZI, and RDDI are also used because of their simplicity in calculation. In India, most of the rainfall occurs during the monsoon season and there is high seasonal variability from month to month. Therefore, it is decided to compute these indices for five time steps i.e. 1-, 3-, 6-, 9-, 12-month time steps, because occurrence of monthly rainfall deficiency is common feature of climate and 1-month time step may not describe the drought situation appropriately. The drought index values computed using these indices help in understanding the level of severity of a drought event. The various DIs have different range of values for defining the severity of a drought event. Therefore, for inter comparison of various indices, droughts are generally categorized into moderate, severe or extreme category depending upon the estimated values of index. The range of values for each severity category of various indices is shown in Table 2.

Table 2: Dry and wet categories of various drought indices based on index value

Category	Range of drought index values					
	EDI	SPI	Z-Score	CZI	RD (%)	RDDI
Extremely dry	≤−2.0	≤−2.0	≤−2.0	≤−2.0	<−60	≤10
Severely dry	−1.99 to −1.5	−1.99 to −1.5	−1.99 to −1.5	−1.99 to −1.5	−60 to <−40	10–20
Moderately dry	−1.49 to −1.0	−1.49 to −1.0	−1.49 to −1.0	−1.49 to −1.0	−40 to <−30	20–30
Normal	−0.99 to 0.99	−0.99 to 0.99	−0.99 to 0.99	−0.99 to 0.99	−30 to 30	30–70
Moderately wet	1.0 to 1.49	1.0 to 1.49	1.0 to 1.49	1.0 to 1.49	>30 to 40	70–80
Very wet	1.5 to 1.99	1.5 to 1.99	1.5 to 1.99	1.5 to 1.99	>40 to 60	80–90
Extremely wet	≥2.0	≥2.0	≥2.0	≥2.0	>60	≥90

Comparison of Drought Indices using Drought Severity Values

The comparison of various drought indices has been made by comparing the computed values of drought indices. For this purpose the Pearson correlation coefficients between paired time series of the severity values estimated using all drought indices are computed. For example severity values estimated using EDI are paired with severity values estimated using SPI-1 and correlation coefficient (0.52) is computed for this pair to form one cell of the matrix. Likewise the severity values estimated using SPI-1 is paired with severity values estimated using other DIs for all time steps and so on. This created a cross correlation matrix of 26 rows and 26 columns (5 indices X 5time steps and EDI with single time step=26). The first drought severity value using various indices for 1-, 3-, 6-, 9- and 12-month time steps are obtained from the months of January-1901, March-1901, June-1901, September-1901 and December-1901, respectively. Therefore, the correlation between index values of different time steps is computed from the month of Jan-1902 to December-2002, in order to maintain the equal number of data points for all time steps. The similar correlation matrices are prepared for all districts and correlation matrix for district Sagar is shown in Table 3, as illustration.

Table 3: Sample correlation matrix of Pearson product-moment correlation coefficients computed between paired values of drought index time series for Sagar district

Row no.	Drought Index	EDI	SPI-1	SPI-3	↔		RDDI-6	RDDI-9	RDDI-12
	1	**2**	**3**	**4**	**25**		**26**	**27**	
1	EDI	1.00	0.45	0.64	...	...	0.72	0.81	0.78
2	SPI-1	0.45	1.00	0.51	...	...	0.34	0.27	0.23
3	SPI-3	0.64	0.51	1.00	...	...	0.58	0.45	0.40
4	SPI-6	0.81	0.37	0.63	...	...	0.91	0.68	0.60
5	SPI-9	0.92	0.30	0.50	...	...	0.67	0.91	0.79
6	SPI-12	0.87	0.26	0.43	...	...	0.61	0.78	0.91
↕	...	...	...	...	...	...	...	...	...
	...	...	...	...	...	...	...	...	...
19	RD-6	0.69	0.32	0.53	...	...	0.77	0.51	0.44
20	RD-9	0.89	0.30	0.47	...	...	0.60	0.84	0.70
21	RD-12	0.88	0.26	0.44	...	...	0.59	0.76	0.88
22	RDDI-1	0.36	0.88	0.41	...	...	0.30	0.23	0.19
23	RDDI-3	0.56	0.49	0.91	...	...	0.57	0.45	0.39
24	RDDI-6	0.72	0.34	0.58	...	...	1.00	0.69	0.61
25	RDDI-9	0.81	0.27	0.45	...	...	0.69	1.00	0.81
26	RDDI-12	0.78	0.23	0.40	...	...	0.61	0.81	1.00
27	Average correlation	**0.72**	**0.46**	**0.59**	...	...	**0.60**	**0.60**	**0.57**
28	Average correlation of same time steps*		**0.92**	**0.95**	...	...	**0.88**	**0.91**	**0.92**

The average in column SPI-3 is computed from the average correlation of SPI-3 with 3-monthly time step of DIs such as SPI-3, Z-Score-3, CZI-3, RD-3, RDDI-3 and average in column RDDI-6 is computed from the average correlation of RDDI-6 with 6-monthly time step of DIs such as SPI-6, Z-Score-6, CZI-6, RD-6 and RDDI-6.

From Table 3 it can be noted that the values of correlation coefficients of EDI with other indices (column 2) are more than that of the values of correlation coefficient with SPI-1 (column 3), except correlation of SPI-1 with other DIs of 1-month time step. For instance, correlation of EDI with SPI-3 is equal to 0.64, whereas, correlation of SPI-1 with SPI-3 is equal to 0.51. Likewise, values of correlation coefficient of EDI (each row of column 2) with other indices are also higher than the correlation of SPI-3 with other indices (column 4), except correlation of SPI-3 with other DIs of 3-month time step. The similar results are seen with other indices too. It indicates that correlation coefficient of EDI paired with other indices is comparatively higher (except for 1-month time step of other DIs). Further it is interesting to see that, the correlation of EDI with other indices increases with the increase in their time step up to 9-month. The correlation coefficient of EDI with other indices is more than 0.8 for the 9-month time step. The EDI shows highest correlation coefficient of 0.93 with Z-Score-9 (values not shown in Table 3). The correlation coefficients of EDI with 12-month time steps of other DIs varies from 0.78 to 0.88.. The correlation coefficient of EDI paired with other DIs for selected five time steps are shown in Fig. 2.

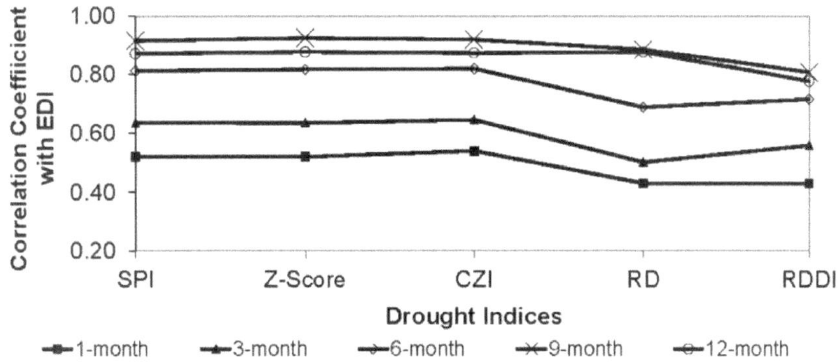

Figure 2: Correlation of EDI with various drought indices for five time steps.

The Fig. 2 clearly indicates that correlation of EDI with 9, and 12-month time steps of SPI, Z-Score, CZI, RD and RDDI is more than 0.7. Also, correlation of EDI with 1 and 3-month time steps of SPI, CZI, Z-Score, RD and RDDI is more than 0.5 and 0.6 respectively. This suggests that EDI is comparable with other DIs for detection of long and short term droughts, though it always uses more than 12 months duration for its computation (time dependent reduction function). However, correlation of EDI with RD and RDDI is comparatively less for lower time steps. This suggests that RD and RDDI are comparable with EDI for higher time steps only.

On the other hand, very high correlation coefficient is observed between the paired values of SPI-1 with Z-Score-1 (r=0.92), CZI-1(*r*=0.97), RD-1 (*r*=0.83) and RDDI-1 (*r*=0.88). Similarly, correlation coefficient of SPI-3 with Z-Score-3 (*r*=0.96), CZI-3 (*r*=0.99), RD-3 (*r*=0.88) and RDDI-3 (*r*=0.91) are also very good. Other indices also exhibit high values of correlation coefficients paired with the estimated values of index for same time step. It indicates that for same time step most of the indices are better correlated (*r*>0.8). However, the estimate of SPI-1 values does not show good correlation with that of 3-, 6-, 9- or 12-month time steps of different DIs. Similarly, SPI-3 does not show good correlation with 1-, 6- 9- or 12-month time steps of other DIs. Likewise, poor correlation has been observed between estimates of other DIs for dissimilar time steps. This revealed that the estimates obtained from various DIs may not be comparable for different time steps.

Further, row 27 in Table 3 presents the average correlation coefficients estimated between various DIs for different time steps. For instance, average of column 2 i.e. EDI (0.72) represents the average correlation of EDI, with different DIs for all time steps. Similarly, column 4 i.e. SPI-3 (0.59) represents the average correlation of SPI-3 with different DIs for all time steps and so on. This analysis indicated that the highest average correlation is obtained for EDI and this also confirms that EDI is comparatively better correlated with other DIs with time steps of 1-,3-,6-,9- and 12-month.

Similarly, the average value of correlation coefficient of a DI for a given time step with different DIs is shown in row 28 in the Table 3. For example, in the column 3 row 28 ofTable 3, the value 0.92 is the average of paired correlations of SPI-1 and SPI-1, SPI-1 and Z-Score-1, SPI-1 and CZI-1, SPI-1 and RD-1, and SPI-1 and RDDI-1 (i.e. for 1-month time step). Similarly, in column 25 row 28 of Table 3, the value 0.88 is obtained from the average correlation of RDDI-6 with 6-monthly time step of DIs i.e. SPI-6, Z-Score-6, CZI-6, RD-6 and RDDI-6. Likewise all other averages in row 28 of Table 3 are obtained from the correlations of same time steps. The high value of average correlation coefficient (in Row 28) indicated that most of the DIs are better correlated with the DIs of same time step. Further, it is seen that the value of average correlation is relatively higher in longer time steps.

The average values of correlation in 1-month time step column of each index (i.e. column 3, 8, 13, 18 and 23 of row 27 in Table 3) are further averaged to represent the 1-month time step average correlation of DIs. Similarly, average values of correlation coefficients of DIs for 3-, 6-, 9- and 12-month time steps have been computed. These computed values of correlation coefficient of DIs for different time step are plotted for different districts in Fig. 3. Further, the mean value of correlation coefficient for each time step is shown as continuous thick blue line in Fig. 3.

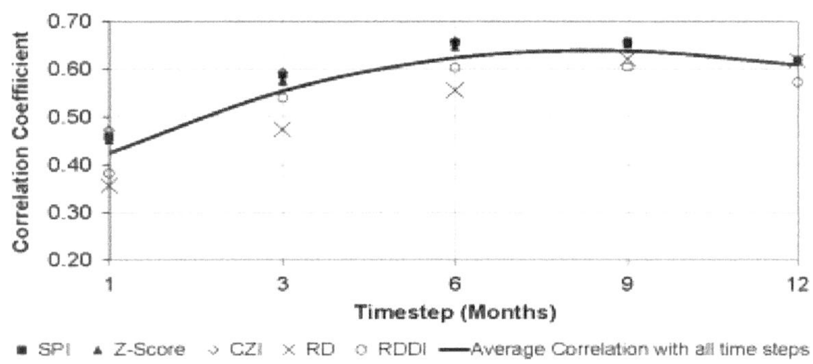

Figure 3: Average correlations of a drought index with a given time step with other indices for all time steps and average correlation of all indices for a given time step.

It is evident from the Fig. 3 that the mean value of correlation coefficients of DIs is maximum for 9-month (0.64) time steps followed by 6 months (0.62) and 12 months (0.61). From this analysis, it may be concluded that the 9-month time step may be more appropriate for comparison of DIs. It appears to be justified because in major parts of India, once the monsoon season is over (in September) significant rainfall is received only after 9 months i.e. onset of next monsoon.

In the analysis, RDDI showed relatively poor relationship with other DIs even for same time steps. The similar results are obtained for other districts too. This suggests that the application of RDDI is not suitable for any of the district in study basin. On the other hand, EDI is better correlated with other DIs for all time steps and best correlation of EDI with other DIs was found for 9-month time step (Fig. 2).

Appraisal of drought characteristics

As stated earlier that droughts are characterised with their duration and severity. Therefore, the drought months are identified usingvarious indices during period 1902–2002 and are presented in Table 4.

Table 4: Number of drought months identified with each drought index for all districts in and around the Ken River Basin.

Drought Index	Ban-da	Chattar-pur	Damoh	Fateh-pur	Hamir-pur	Jabal-pur	Kat-ni	Ma-hoba	Nar-sim-hapur	Pan-na	Raisen	Sagar	Sat-na
EDI	201	220	214	172	205	190	196	225	201	206	202	206	187
SPI-1	143	157	206	175	140	254	255	137	259	206	167	226	173
Z-Score-1	83	69	71	81	87	73	98	85	73	69	61	61	83
CZI-1	102	111	105	100	102	163	178	111	97	117	77	87	143
RD-1	598	598	593	617	593	548	556	588	606	585	649	626	571
RDDI-1	375	375	369	365	381	373	372	382	376	370	367	370	368
SPI-3	192	202	195	183	190	186	182	200	189	193	181	195	184
Z-Score-3	189	197	179	172	192	178	182	202	168	192	135	154	176
CZI-3	200	209	193	190	207	194	192	213	190	202	178	189	193
RD-3	432	431	422	421	413	366	390	413	438	409	469	427	0
RDDI-3	366	367	367	366	365	368	366	366	367	367	367	367	366
SPI-6	191	210	194	177	197	200	184	206	199	195	189	195	195
Z-Score-6	196	212	190	188	203	198	187	217	190	204	172	183	198
CZI-6	200	215	197	192	206	200	189	220	198	204	186	191	200
RD-6	285	308	294	282	293	248	250	301	307	292	350	303	254
RDDI-6	361	365	365	362	360	367	366	362	367	365	367	366	363
SPI-9	193	208	199	177	194	185	181	209	217	196	191	198	191
Z-Score-9	199	219	201	183	207	184	180	217	209	207	186	196	194
CZI-9	201	219	202	185	207	189	181	218	215	208	190	198	195
RD-9	209	220	194	215	209	152	147	212	204	207	261	199	167
RDDI-9	361	362	364	361	360	364	363	361	364	362	364	364	362
SPI-12	204	202	183	179	200	195	172	210	240	200	208	201	180
Z-Score-12	209	211	183	189	210	190	177	214	227	208	199	199	182
CZI-12	209	211	183	189	210	194	177	215	237	208	207	201	182
RD-12	163	167	146	164	156	89	100	169	158	164	225	158	115
RDDI-12	361	361	361	361	360	361	361	360	361	361	361	361	361

It is very interesting to note that number of identified drought months increased with the increase of time step for SPI, Z-Score and CZI, whereas, RD identified maximum number of drought months for 1-month time step and decreased considerably with the increase of time step. However, RDDI identified approximately equal number of drought months with each time step and estimated relatively more number of drought months as compared to other DIs. As discussed in preceding sections, most DIs are better correlated for 9-month time step. Also, it is found that the number of drought months identified using EDI and other DIs for 9-month time step (except RDDI) are nearly equal. The number of drought months obtained using different DIs for 9-month time step for various district in the study basin are plotted in Fig. 4.

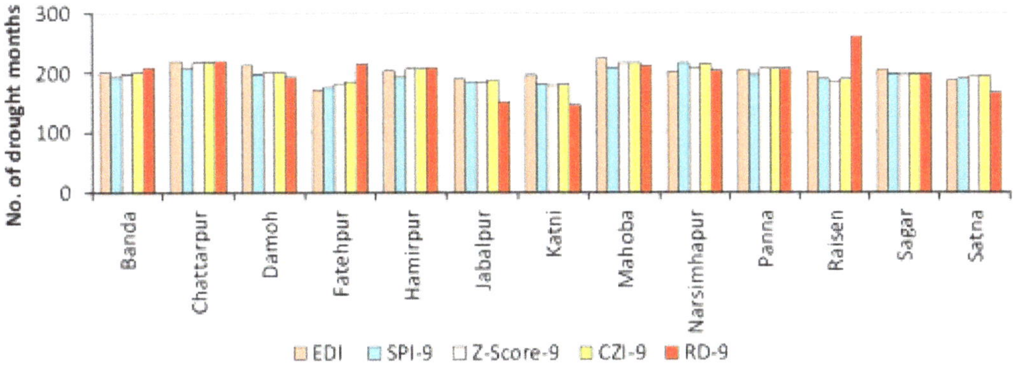

Figure 4: Comparison of number of drought months identified using EDI and 9-monthly time step of various indices for different districts.

It is clear from Fig. 4 that drought months identified using various DIs for 9-month time step for differentdistricts are within a close range except in case of Raisen district.It further supports the conclusion that 9-monthly time step provides reasonable estimates of drought severity and duration for comparative analysis of various indices in the Ken River Basin.

The identified drought months are segregated under different drought severity categories according to the severity value (mentioned in Table 2) of each index i.e. Moderate, Severe and Extreme for various districts in study basin. The plots of number of drought months identified under different severity classes using various indices for Sagar district are shown as percentage of period under consideration in Fig. 5(a–e).

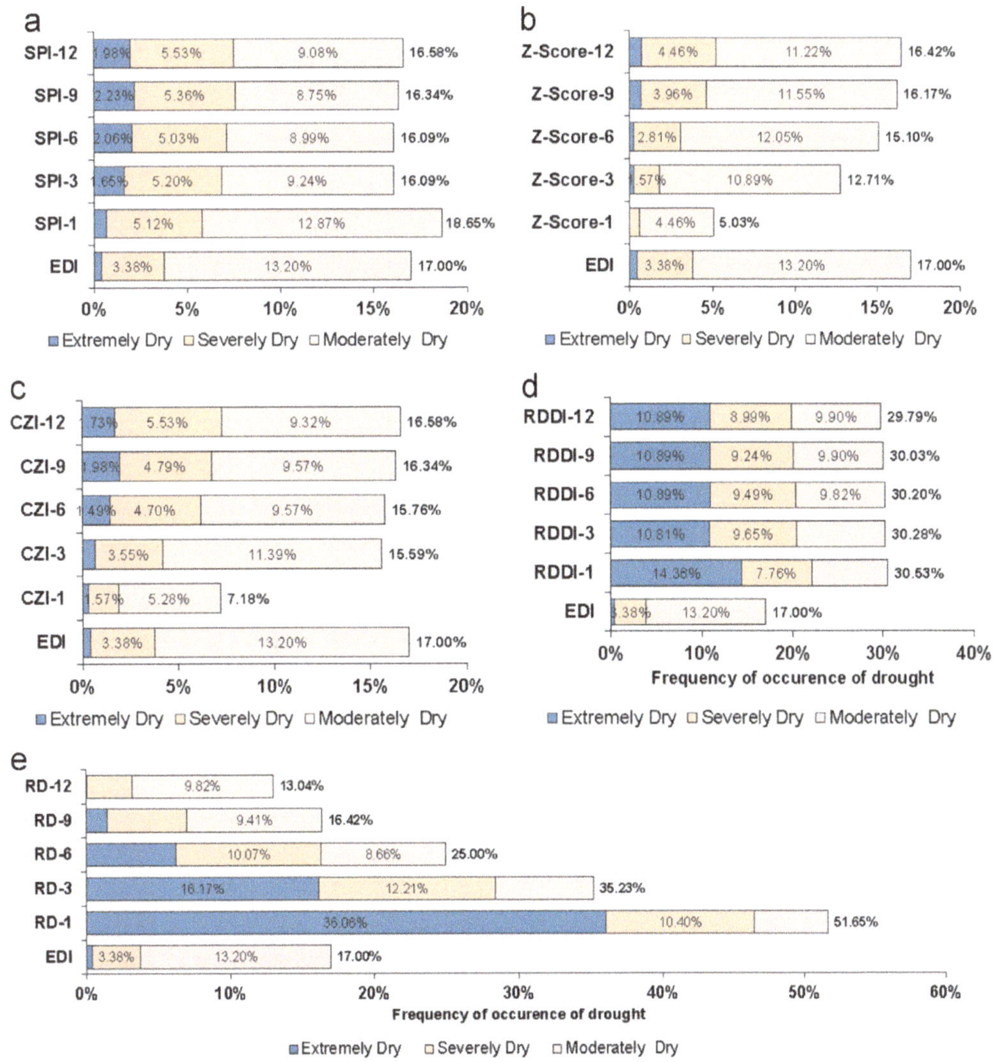

Figure 5: Comparison of percentage of drought severity categories identified using EDI and 1, 3, 6, 9 and 12-month time step of (a) SPI (b) Z-Score (c) CZI (d) RDDI and (e) RD for Sagar district.

It can be noted from Fig. 5(a–c) the drought months as percent of total number of months under consideration obtained using SPI, Z-Score and CZI for 6, 9, and 12-month time step are very close ranging from 15.01% to 16.58%. However, there are significant differences in number of drought months with moderate, severe and extreme severity categories even for higher time steps of DIs. Also, RDDI (Fig. 5d) identified approximately equal number of drought months under different severity categories for all time steps. On the other hand, RD identified more number of drought months at shorter time step and significantly less number of drought months under extreme category with increase in time step (Fig. 5e). The estimates of severity categories using RDDI and RD are significantly different from the estimates of drought months using SPI, CZI and Z-Score. This means that RD and RDDI are not suitable for the Ken River basin where summer concentration of precipitation is very high.

In addition to the number of total drought months, the maximum duration of a single drought event (continuous period of drought months) are also identified using various indices for all time steps. The maximum duration of identified drought event increases at higher time step. The analysis indicated that maximum duration of drought event varies for each district. Some of the districts with longer duration are more prone to drought as compared to other districts. Maximum duration of identified drought event using DIs for 9-month time step for various district are presented in Fig. 6.

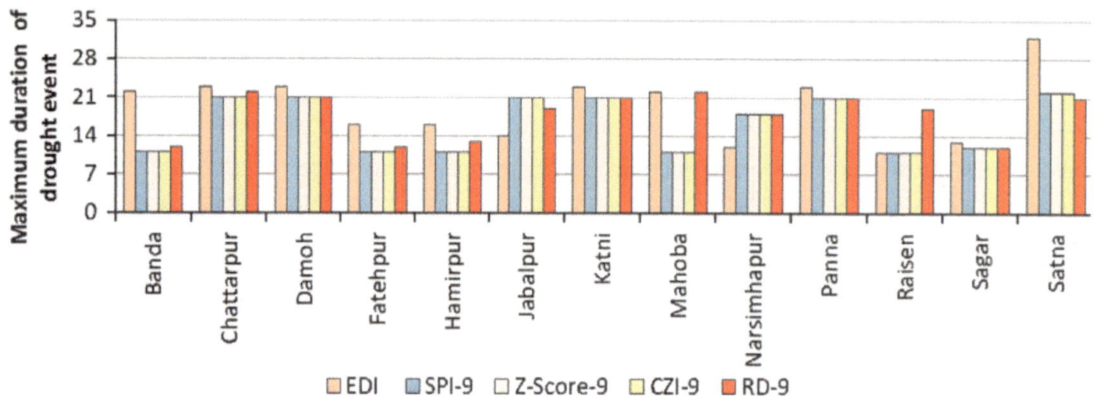

Figure 6: Comparison of the maximum duration of drought events identified using EDI and 9-month time step of various indices for different districts.

Fig. 6 indicates that Satna district has faced drought event of maximum duration i.e. 32 months estimated using EDI. Also, Panna, Katni, Chattarpur and Damoh districts have faced drought events for maximum duration of continuous 23 months.

Estimates of drought duration obtained using EDI appears to be slightly more as compared to other DIs for most of the districts. However, these estimates from other DIs are within a close range of drought duration.

The scrutiny of estimates revealed that the reason of longer drought duration estimate by EDI is because of the early detection of the drought onset as compared to other DIs. Also, the drought is not terminated by EDI, until rainfall during non-monsoon months is sufficient high to meet the water deficiencies required for termination of drought. It is due to the reason that EDI considers the reduction of water with time passing. For example, the rainfall 9 month before is summed in SPI with same weight of it one month before. However, EDI considers it not with same weight but with reduced weight because water is diminished by evaporation, run-off etc for 8 months. Therefore EDI detects drought earlier than other DIs.

Comparative Evaluation of Drought Indices in Real Case

Once a drought is triggered during the monsoon months, it is often continued till the arrival of next monsoon. It may also however terminate if significantly excess rainfall occurs during non-monsoon month of the year. It is simply because the rainfall deficiencies created during monsoon months are not usually met from rainfall during non-monsoon months. In order to establish the suitability of a particular DI and the time step in defining the onset, termination and severity quantification of drought event, the values of various indices are plotted for the drought years for various districts. The plots of comparison for EDI, SPI, CZI and Z-Score for 1-month and 9-monthtime steps for district Sagar are shown in Fig. 7(a) and (b) respectively. Also, analysis presented in preceding sections indicated that EDI provides relatively better assessment of drought for study area. Therefore, the performance of other DIs has been discussed in comparison to the values of EDI. Specific discussions are presented here for the occurrence of drought event during January-1991 to January-1994. The results are then validated based on the time series of monthly rainfall compared to monthly average rainfall for 1-month time step. Similarly, DIs with higher time steps are validated based on multi-monthly rainfall (obtained from sum of previous months rainfall depending upon the time step) compared to average of multi-monthly rainfall of the corresponding period (fixed for particular month). For example, for 6-month time step, the rainfall value of June month is the sum of five previous months' rainfall plus June month rainfall and the multi-monthly average is the average of all June months in the 6-monthly time series. Similarly for 9-month time step the rainfall value of December month is sum of April–December months' rainfall and the multi-monthly average is the average of all December months in the 9-monthly time series (Fig. 7b).

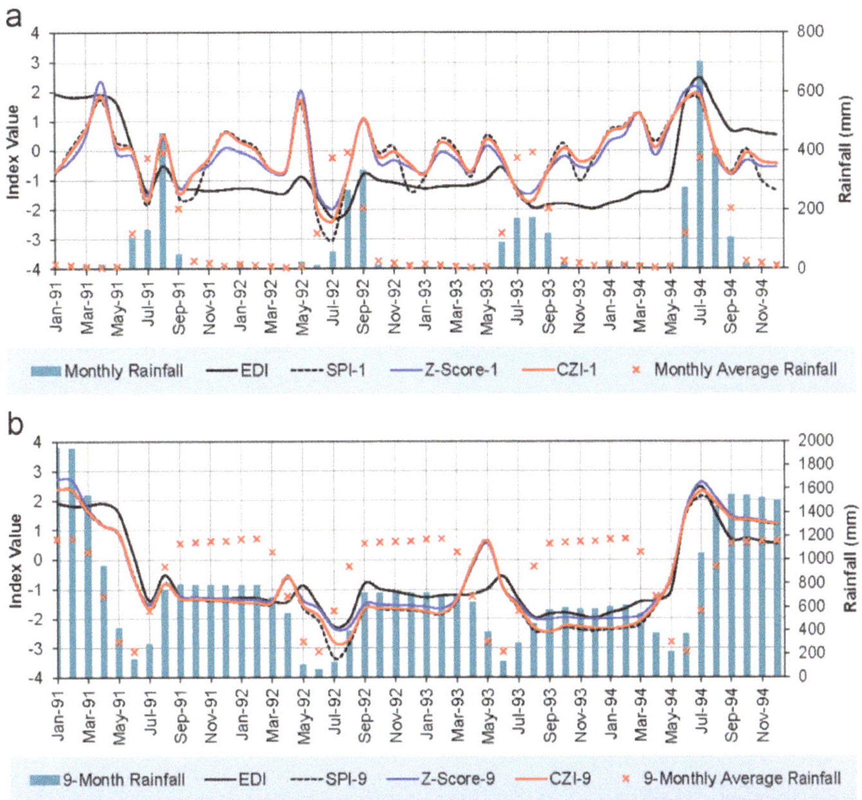

Figure 7: Comparison of EDI, SPI, Z-Score and CZI along with rainfall deficiencies for the corresponding periods of (a) one month time step and (b) 9-month time step during the historical drought events for Sagar district.

It can be seen from Fig. 7(a) that during July-1991 there had been significant rainfall deficit (−64%). The EDI, SPI, CZI and *Z*-Score represents drought situation appropriately. However, rainfall during the month of August-91 was slightly above the monthly average value (+16%). The SPI, CZI and *Z*-Score indicated this month as moderately wet (0.43–0.54), the termination of drought in August-91 and re-onset in September-91, while EDI identified it under normal range (−0.52). This is simply because the consideration of previous deficit accumulation is important and it is considered in case of EDI.

Subsequently, the period between September-1991 to April-1992 (except Dec-91) was again a deficit period and therefore, drought started in July-1991 is still continued in September-1991 and onward. Further the rainfall values for the month December-1991 is 12 mm against average of 10 mm. Since, this excess is insignificant and insufficient to terminate the drought; this situation is very well indicated by EDI which indicated this period as drought without terminating the on-going drought event. On the other hand, *Z*-Score-1, CZI-1both in October-1991 and SPI-1 in November-1991 indicated the termination of drought.

Further, during January-1992 to April-1992, no significant rainfall has been received during this period, however, SPI-1, *Z*-Score-1 and CZI-1 indicated this period as normal, whereas, EDI rightly indicated this period under drought condition. In May-1992, only small surplus of precipitation made DIs to represent this month under wet category except EDI, because it is in dry season and this surplus was insufficient to fulfill previously accumulated deficit. Rainfall during the period June-1992 (89% deficit), July-1992 (84% deficit) to August-1992 (32% deficit) was again continuously less than average and SPI (−2.28) identified it as extreme drought during Jun-92. However, the same event (June-1992) is described under severe category by EDI (−1.55) because of less significant rainfall during June month and with increased severity during July-1992 (−2.28) and August-1992 (−2.0). During September-1992 (65 % surplus), SPI-1, *Z*-Score-1 and CZI-1, terminated the drought and defined the event under wet category. However EDI indicated this month as temporary escape from drought situation, which is quite obvious because the surplus rainfall during September-1992 was still not sufficient to meet the previously accumulated deficit and to be indicated as wet month.

During period October-1992 to June-1993, the rainfall is again either below normal or insignificant and EDI appropriately detected this rainfall deficit defining this period under drought. However, other indices detected this period as normal and the next onset of drought only during the month July-1993 (except SPI-1, which detected December-1992 also as drought month). Again, the rainfall during the period July-1993 to December-1993 is significantly below normal value and during the period January-1994 to May-1994 rainfall is more than average (non-significant in terms of its magnitude). During this period (July-1993 to May-1994) SPI-1, *Z*-Score-1, CZI-1 continue to define the combination of dry and wet events till the end of December-1994, while EDI continue to show the drought of moderate category till the end of May-1994 because of poor monsoon during the year 1993. The return to normal by EDI is shown at June-1994 with heavy rain. However in other indices that using 1-month time step, many normal conditions appeared during the drought period. Especially in May-92, where these DIs indicated the month as wet. In actual, EDI clearly showed the water deficit from July-1991 till May-1994, except temporal escape from drought situation (0.0>EDI>−1.0), in August-1991, May-1992, September-1992 and June-1993 months, and indicated it as long duration of drought, that is terminated only in June-94.

Above discussion indicates that application of SPI, CZI and *Z*-Score for 1-month time step may lead to erroneous assessment of drought situation. Further, it can be concluded from above discussion that the EDI has captured the real essence of drought situation of study area.

For comparison of various indices at 9-monthly time step the index values plotted for the same historical drought period are shown in Fig. 7(b). It can be seen from Fig. 7(b), that EDI time series curve is close to that of SPI-9, CZI-9 and *Z*-Score-9.

During, January-1991 to May-1991, the 9-monthly rainfall is more than the corresponding normal values, therefore, this period is well indicated as non-drought period by all DIs. Further, the 9-month rainfall from June-1991 to April-1993 is continuously below the 9-monthly average rainfall. This period is indicated as drought period except few temporal escapes by SPI-9, CZI-9 and *Z*-Score-9 during the month April-1992 and by EDI during May-1992 and September-1992. During April-1992, the magnitude of 9-month rainfall deficit is less as compared to previous deficit during September-1991 to March-1992. Therefore, SPI-9, *Z*-Score-9 and CZI-9 indicated temporary escape from drought in April-1992. However, the EDI, which depends on the 12 months' time dependent reduction function, indicated April-1992 as drought month because the effect of the surplus of rainfall at May-1991 has diminished due to the passing of time. At May-1992, EDI value shows slight increase because of surplus rainfall during this month and accumulated (but reduced on passing of time) effect of previous rainfall during the months from June-1991 to April-1992 (Fig. 7a). EDI considers the recent rainfall more important than the past rainfall because the major part of past rainfall might have diminished by evaporation, runoff and other losses. After May-1992, all indices indicated the similar pattern of increases and decreases of drought severity except for April-1993. In April-1993, small excess of rain made DIs above normal except EDI that show slight increase at June-1993.

The comparative analysis of various indices indicated that the period from June-1991 to May-1994 had been generally a drought period with one or two instances of above normal rainfall during this period. This period has been verified with the documented records of District Statistical Handbook. It is observed from the analysis that during the period of drought identified using EDI, the region suffered with huge crop loss. On the other hand, the wet periods identified using SPI, *Z*-Score and CZI did not help much in terminating the drought. It revealed that EDI seems to be more suitable in timely detection of long term as well as short term monthly rainfall deficiency as compared to other indices.

CONCLUSIONS

Following conclusions can be drawn from the study,

1. Comparison of EDI, SPI, *Z*-Score, CZI, RD and RDDI indicated that all these indices are highly correlated for same time steps and the correlation increases at higher time step with higher correlations at 9 and 12-month time steps. On the contrary, SPI, *Z*-Score, CZI, RD and RDDI are found to be poorly correlated with dissimilar time steps.
2. Historical drought analysis based on the EDI, SPI, *Z*-Score, CZI, RD and RDDI indicated that EDI showed the drought condition of the study basin more realistically than other DIs.
3. Selection of time step in identification of onset of any long term drought is very important. Drought indices compared at 1-month time step may lead to erroneous assessment of drought characteristics because sometimes a short term excess may terminate the long term drought and divide a prevailing drought event into two short events and this may not be appropriate. Therefore, while comparing drought indices, the drought severity should be computed using higher time step of the drought index. The time step should be chosen in such a way, so that at least one significant rainfall month is included in the time step.
4. EDI is found to be better correlated with other DIs for all time steps (except same time step of various DIs). The best correlation of EDI was found to be with 9-month time step of other indices. This indicates that 9-month time step of DIs is a better choice to detect the drought intensity for study basin.
5. Overall EDI is found to be good choice for assessment of drought characteristics and monitoring of drought condition, because of its capability of timely detection of drought onset and realistic quantification of severity of drought events in study area.

ACKNOWLEDGMENTS

A Ministry of Human Resources Development, GoI, fellowship through Indian Institute of Technology Roorkee at the Department of Hydrology, IIT Roorkee, India supported Dr. Vinit K. Jain. A Basic Science Research Program through the National Research Foundation of Korea (NRF) (Ministry of Education, NRF-2013R1A1A4A01007626) supported Dr. H. R. Byun.

REFERENCES

1. Akhtari, R., Morid, S., Mahdian, M.H., Smakhtin, V., 2009. Assessment of areal interpolation methods for spatial analysis of SPI and EDI drought indices. Int. J. Climatol. 29, 135–145.
2. Appa Rao, G., 1986. Drought Climatology. Jal Vigyan Samiksha. Publication of High Level Technical Committee on Hydrology, National Institute of Hydrology, Roorkee.
3. Barua, S., Ng, A.W.M., Perera, B.J.C., 2011. Comparative evaluation of drought indexes: case study on the Yarra

River catchment in Australia. J. Water Resour. Plan. Manag. 37, 215–226.

4. Bhalme, H.N., Mooley, D.A., 1980. Large-scale droughts/floods and monsoon circulation. Mon. Weather Rev. 108, 1197–1211.
5. Byun, H.R., Wilhite, D.A., 1999. Objective quantification of drought severity and duration. J. Clim. 12, 2747–2756.
6. Deo, R., Byun, H.R., 2014. Quantifying the severity, duration and intensity of drought in Australia using the effective drought index. J. Hydrol.
7. Dogan, S., Berktay, A., Singh, V.P., 2012. Comparison of multi-monthly rainfall-based drought severity indices, with application to semi-arid Konya closed basin, Turkey. J. Hydrol. 470–471, 255–268.
8. Edossa, D.C., Babel, M.S., Gupta, A.D., 2010 . Drought analysis in the Awash river basin, Ethiopia. Water Resour. Manag. 24, 1441–1460.
9. Edwards, D.C., Mckee, T.B., 1997. Characteristics of 20th century drought in theUnited States at multiple time scales. Atmos. Sci. Pap. 63, 1–30.
10. Gibbs, W.J., Maher, J.V.,1967. Rainfall deciles as drought indicators. Bur. Meteorol., Bulletin No. 48, Melbourne, Australia. Guttman, N.B., 1999. Accepting the standardized precipitation index: Acalculation algorithm. J. Am. Water Resour. Assoc. 35 (2), 311–322.
11. Kalamaras, N., Michalopoulou, H., Byun, H.R., 2010. Detection of drought events in Greece using daily precipitation. Hydrol. Res. 41 (2), 126–133.
12. Kendall, M.G., Stuart, A., 1977. The Advanced Theory of Statistics. Vol. 1, Distribution Theory. Charles Griffin Company, London, pp. 400–401.
13. Keyantash, J., Dracup, J.A., 2002. The quantification of drought: an evaluation of drought indices. Bull. Am. Meteorol. Soc. 83, 1167–1180.
14. Kim, D.W., Byun, H.R., 2009. Future pattern of Asian drought under global warming scenario. Theor. Appl. Climatol. 98, 137–150.
15. Kim, D.W., Byun, H.R., Choi, K.S., 2009. Evaluation, modification, and application of the effective drought index to 200-year drought climatology of Seoul, Korea. J. Hydrol. 378, 1–12.
16. Komuscu, A.U., 1999. Using the SPI to analyze spatial and temporal patterns of drought in Turkey. Drought Network News (1994-2001). Paper 49. pp. 7–13.
17. Lee, S.M., Byun, H.R., Tanaka, H.L., 2012. Spatiotemporal Characteristics of Drought Occurrences over Japan. J. Appl. Meteorol. Climatol. 51, 1087–1097.
18. Mckee, T.B., Doesken, N.J., Kleist, J., 1993. The relationship of drought frequency and duration to time scales. In: Proceedings of the 8th Conference on Applied Climatology,Anaheim, CA, USA, pp. 179–184.
19. Mishra, A.K., Singh, V.P., 2009. Analysis of drought severity-area-frequency curves using ageneral circulation model and scenario uncertainty. J. Geophys. Res. 114, D06120. http://dx.doi.org/10.1029/2008JD010986 2009.
20. Morid, S., Smakhtin, V., Bagherzadeh, K., 2007. Drought forecasting using artificial neural networks and time series of drought indices. Int. J. Climatol. 27, 2103–2111.
21. Morid, S., Smakhtin, V., Moghaddasi, M., 2006. Comparison of seven meteorological indices for drought monitoring in Iran. Int. J. Climatol. 26, 971–985.
22. Mpelasoka, F., Hennesy, K., Jones, R., Bates, B., 2008. Comparison of suitable drought indices for climate change impacts assessment over Australia towards resource management. Int. J. Climatol. 28, 1283–1292.
23. Narasimhan, B., Srinivasan, R., 2005. Development and evaluation of soil moisture deficit index (SMDI) and evapotranspiration deficit index (ETDI) for agricultural drought monitoring. Agric. For. Meteorol. 133, 69–88.
24. Ntale, H.K., Gan, T., 2003. Drought indices and their application to East Africa. Int. J. Climatol. 23, 1335–1357.
25. Palmer, W.C., 1965. Meteorological Drought. US Department of Commerce, Weather Bureau, Technical Paper 45. pp. 58.
26. Palmer, W.C., 1968. Keeping track of crop moisture conditions nationwide: the new crop moisture index. Weatherwise 21, 156–161.
27. Pandey, R.P., Dash, B.B., Mishra, S.K., Singh, R., 2008. Study of indices for drought characterization in KBK districts in Orissa (India). Hydrol. Process. 22, 1895–1907.
28. Patel, N.R., Chopra, P., Dadhwal, V.K., 2007. Analyzing spatial patterns of meteorological drought using standardized precipitation index. Meteorol. Appl. 14, 329–336.
29. Roudier, P., Mahe, G., 2010. Study of wa ter stress and droughts with indicators using daily data on the Bani River (Niger Basin, Mali). Int. J. Climatol. 30, 1689–1705.
30. Shafer, B.A., Dezman, L.E., 1982. Development of a surface water supply index (SWSI) to assess the severity of drought conditions in snowpack runoff areas. Western Snow Conference, Reno, NV, Colorado State University. pp. 164–175 (In preprints).
31. Smakhtin, V.U., Hughes, D.A., 2004. Review, Automated Estimation and Analyses of Drought Indices in South Asia. Working Paper 83, International Water Management Institute, Colombo, Sri Lanka. Smakhtin, V.U., Hughes, D.A., 2007. Automated estimation and analyses of meteorological drought characteristics from monthly rainfall data. Environ. Model. Softw. 22, 880–890.
32. Stricevic, R., Djurovic, N., Djurovic, Z., 2011. Drought classification in Northern Serbia based on SPI and statistical pattern recognition. Meteorol. Appl. 18, 60–69. http://dx.doi.org/10.1002/met.207.
33. Tsakiris, G., Pangalou, D., Vangelis, H., 2007. Regional drought assessment based on reconnaissance drought index (RDI). Water Resour. Manag. 21 (5), 821–833.
34. Tsakiris, G., Vangelis, H., 2004. Towards a drought watch system based on spatial PI. Water Resour. Manag. 18, 1–12.
35. Vicente-Serrano, S.M., Gonzalez-Hidalgo, J.C., de

Luis, M., Raventos, J., 2004. Drought patterns in the mediterranean area: the Valencia region (eastern Spain). Clim. Res. 26, 5–15.

36. Wilson, E.B., Hilferty, M.M., 1931. The Distribution of Chi-Square. Proc. Natl. Acad. Sci. USA 17, 684–688.

37. WMO (World Meteorological Organization), 1975. Drought and Agriculture. Technical Note No. 138, Report of the CAgM Working Group on Assessment of Drought, WMO, Geneva, Switzerland. pp. 127.

38. Wu, H., Hayes, M.J., Weiss, A., Hu, Q.I., 2001. An evaluation of the standardized precipitation index, the china-Zindex and the statistical Z-Score. Int. J. Climatol. 21, 745–758. http://dx.doi.org/10.1002/joc.658.

39. Zhai, J., Su, B., Krysanova, V., Vetter, T., Gao, C., Jiang, T., 2010. Spatial variation and trends in PDSI and SPI indices and their relation to streamflow in 10 large regions of China. J. Clim. 23 (3), 649–663.

CHAPTER 15

IMPACTS OF CLIMATE CHANGE ON SOIL EROSION

Jurgen D. Garbrecht, Mark A. Nearing, Jean L. Steiner, Xunchang J. Zhang, Mary H. Nichols

Keywords: Soil erosion, Conservation, Climate change, Climate projection, Soil productivity, Tillage

INTRODUCTION

Controlling accelerated soil erosion by water in agricultural environments has been a problem since man began plowing the ground. Climate change adds an additional dimension to already existing soil erosion problems. A vast body of knowledge and experience on soil erosion, conservation practices, and conservation tools has been built over the past century. Conservationists can rely on this knowledge and available tools to search for effective soil erosion controls under a changing climate. A brief historical background on soil erosion work and successes is given below to provide an overview of the existing foundation and framework within which this interdisciplinary investigation on soil erosion and climate change is conducted. In the following, the word "soil erosion" implies sheet and rill erosion by water. Soil erosion by wind is outside the scope of this study.

Conservation of soil resources in the United States made significant strides during the 20th century. Led by Hugh Hammond Bennett and conservationists of the time, the Soil Erosion Service was established within the US Department of Agriculture (USDA) in 1933 (Bennett, 1939), which later became the Soil Conservation Service and more recently the Natural Resources Conservation Service (NRCS). This agency promoted the use of soil conservation practices such as construction of terraces, grasses waterways, contour tillage, and gully control structures (Ayres, 1936 and Bennett, 1939) to control soil erosion, reduce sediment loss, and maintain soil productivity. The cultivation-revolution since the early 1980s brought about by the use of minimum tillage and no-till production systems, along with government programs, such as the Conservation Reserve Program that took highly erodible land out of crop production, led to additional significant improvements in soil conservation (US Department of Agriculture, 2013).

In 1982, the USDA conducted the first National Resource Inventory (NRI), which uses a statistical sampling technique to document land use, conservation practices, and erosion rates on non-federal lands across the United States. The survey has been repeated on an intermittent basis since that time. The NRI uses the Universal Soil Loss Equation (Wischmeier and Smith, 1978) to estimate soil loss rates at its sampling sites. The numbers reported indicate that soil erosion by water, on non-federal, cultivated croplands decreased by approximately 30% between 1982 and 2010 (based on data taken from US Department of Agriculture, 2013). These numbers only address sheet and rill erosion caused by water, and do not reflect soil erosion caused by water in ephemeral or permanent gullies, or by wind.

While conservation advances were made over the last century in the United States, the climate has also changed. Climate change impacts soil erosion rates by various pathways, of which primary ones include the drivers of rainfall, temperature, and atmospheric CO_2 concentrations, which impact biomass production and runoff, which in turn impact erosion rates (Fig. 1) (Walthall et al., 2012, Izaurralde et al., 2011 and Nearing et al., 2004). Temperature may have either a positive or negative impact on biomass production, depending on the plant and its response to the temperature changes (Walthall et al., 2012). Atmospheric CO_2 concentrations and rainfall amounts and intensities are generally expected

Citation: Jurgen D. Garbrecht, Mark A. Nearing, Jean L. Steiner, Xunchang J. Zhang, Mary H. Nichols, Impacts of climate change on soil erosion, http://dx.doi.org/10.1016/j.wace.2015.06.002.

to have a positive correlation to plant biomass production, though the impact of rainfall can be negative in cases when intense storm events during the early growing season remove seedlings or when rainfall causes excessive soil moisture conditions that may either influence the timing of planting or plant growth under waterlogged soil conditions (Bassu et al., 2014 and Walthall et al., 2012). Implicit in this generalized representation of the impacts of climate on erosion by water is that producers› response to climate change also will impact soil erosion rates (Delgado et al., 2013 and O'Neal et al., 2005). Under climate change farmers will alter planting and harvest dates, as well as cultivars or crops produced because of changes in temperatures, soil moistures, and rainfall patterns (Walthall et al., 2012, Southworth et al., 2002 and Pfeifer and Habeck, 2002).

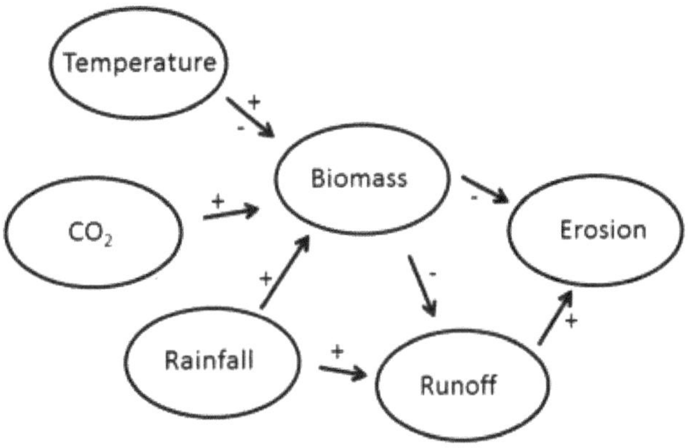

Figure 1: Primary pathways by which climate change may impact rainfall-driven soil erosion, with most common correlation trends specified as plus or minus.

Sustainable, high-yield crop production is critical to maintaining food security, especially during a time of climate change. Long-term sustainability of high crop yields requires soil management practices that promote soil function, soil quality, and soil health. Soil erosion adversely affects soil productivity by gradually depleting the soil of nutrients, fine soil particles, and water holding capacity. Degradation of soil aggregate stability also increases the risks of crusting and increased runoff. Soil conservation practices have proven effective in reducing soil erosion and maintaining soil productivity. However, climate change introduces a new dimension to the soil erosion problem. Soil erosion and conservation practices that were effective in the past may no longer be effective in the future. This question is examined for conditions in central Oklahoma.

Average temperature has increased over most of the contiguous United States in the last century and is expected to continue to do so this century and beyond (Melillo et al., 2014 and Karl et al., 2009). Annual precipitation is also changing. Trends in winter and spring precipitation are projected to rise for the northern United States and decline in the Southwest (Peterson et al., 2013; AMS, 2013, 2012). In the Southern Great Plains, rising air temperature and changes in timing and magnitude of rainfall events have already been observed (Peterson et al., 2013, Higgins and Kousky, 2013 and Groisman et al., 2012).

Rainfall amounts and daily rainfall intensities generally increased in the United States between years 1910 and 1996 (Karl and Knight, 1998). More than half of observed increases in total annual precipitation for the United States measured during that time were caused by increases in the frequency of heavy events, which were considered to be those in the upper 10 percentile of daily amount values. Also, the proportions of precipitation falling in heavy (>95th percentile), very heavy (>99th percentile), and extreme (>99.9th percentile) daily precipitation events increased during the years 1910 through 1999 by 1.7, 2.5, and 3.3% per decade, respectively, on average across the United States (Soil and Water Conservation Society, 2003). This is a pattern that appears to be occurring in many parts of the world (Groisman et al., 2005 and Meehl et al., 2007). Changes in the water cycle propagate through the watershed system and affect processes such as runoff, foliar and ground cover production, soil erosion, sediment transport, and land productivity. Changes in rainfall timing, amount, intensity, and frequency, and minimum and maximum air temperature will inevitably impact the agricultural landscape (O'Neal et al., 2005 and Karl and Knight, 1998). More intense and more frequent extreme rainfall events increase soil erosion, accelerate the degradation of soil quality, and diminish crop yields. Potentially large increases in soil erosion on cropland due to climate change are becoming a serious concern for farmers, land owners, and conservationist (Nearing et al., 2005, Zhang and Nearing, 2005 and Soil and Water Conservation Society, 2003). Existing conservation practices may prove inadequate to manage soil erosion in light of projected climate change impacts (Melillo et al., 2014).

Assessing the effectiveness of current conservation practices under changed climatic conditions that are assumed to prevail in the future is a complex task that involves many climate and agronomic drivers, physiographic variables, interdependencies, and feed-back mechanisms (Delgado et al., 2011 and Nearing et al., 20042005). Much research is being conducted to assess potential impacts of climate change scenarios on soil erosion and to establish the potential need for additional and/or more effective conservation practices to provide adequate protection. However, these efforts are hampered by sizable uncertainties in the projected future climate for which conservation practices are sought.

In this study, a typical system of climate, hydrologic, and soil erosion models is briefly reviewed and the source of uncertainties in model results is laid out. This is followed by an example of potential impacts of climate change and conservation practices on soil erosion by water from winter wheat crops in Central Oklahoma. Irrigation of winter wheat, such as encountered in the Texas and Oklahoma panhandles, was not considered because, first, soil erosion by irrigation runoff is minimal when irrigation is applied according to recommendations; and, second, irrigation of winter wheat is not a common practice in central Oklahoma where, during most years, there is sufficient rainfall to successfully grow a winter wheat crop. The impact of climate model uncertainties on simulated soil erosion is presented, and the effectiveness of several conservation options to offset the increase in projected soil erosion and associated uncertainty is investigated for a typical winter wheat field in Central Oklahoma. The goal of this study is to determine if and which traditional conservation practices would be necessary and sufficient to hold projected annual soil erosion by water, including uncertainties, at no more than today's soil erosion levels.

System of Climate Change and Conservation Assessment Models

Climate Change Models

Climate change models project future climate conditions. Several model selections must to be made before proceeding with the climate projections. Selections include a Green House Gas (GHG) emission scenario, a General Circulation Model (GCM), and a downscaling method (Fig. 2, upper right).

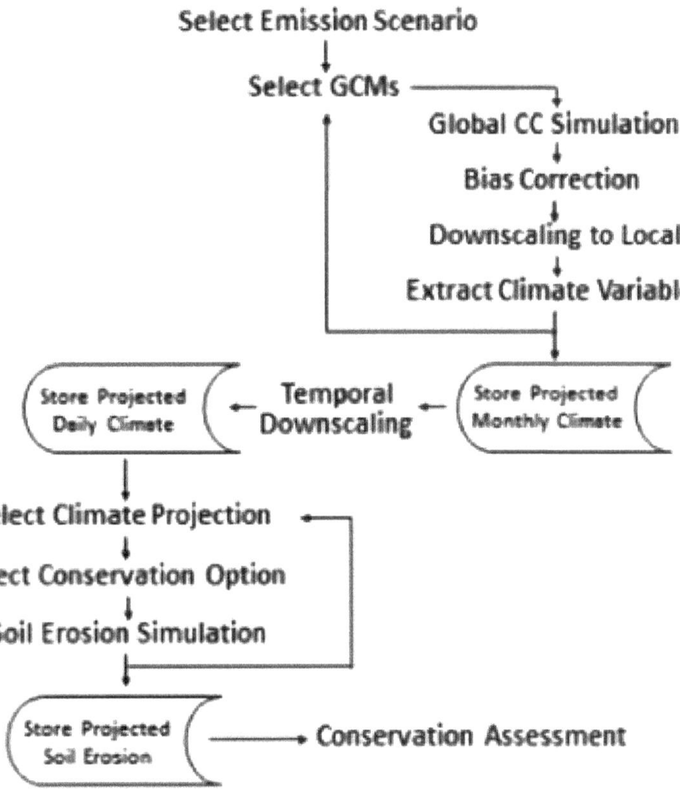

Figure 2: Schematic of the system of climate change and conservation assessment models.

The GHG emission scenario called Representative Concentration Pathway RCP8.5 was chosen for this study. It has a GHG emission path that produces the greatest GHG gas concentrations and leads to the greatest warming of the planet. By inference, this scenario is associated with the greatest change in climate and impacts on the hydrologic cycle and soil erosion. Thus, soil conservation practices that are shown to perform well under RCP8.5 are expected to perform well under scenarios of lower GHG emissions and lesser climate change. These conservation practices are said to be robust under climate uncertainty; that is, they are desirable under virtually any climate scenario (Wilby and Dessai, 2010).

Many GCMs are available to project the future climate for a given climate scenario. Different GCMs project different climates which, in turn, lead to different soil erosion estimates. A representative sample of ten GCMs was selected for this study to define the potential climate evolution over the next half century. The ten GCMs include a Canadian, three US, one European, one Australian, one English, two Japanese, and one German model (acronyms are not listed because of the limited information they convey). Primary climate change variables considered were monthly precipitation and min/max air temperature (Fig. 2, upper right).

The climate projections computed by the GCMs were bias corrected and spatially downscaled to a grid of approximately 12 by 12 km resolution as described by Maurer et al. (2007). For agricultural and soil conservation applications, a supplemental downscaling step was necessary to express the projected monthly climate in terms of daily weather data at a particular location within the downscaled GCM grid-box. Synthetic weather generator SYNTOR (Garbrecht et al., 2014) was used to perform this supplemental downscaling (Fig. 2, middle). SYNTOR is a stochastic weather generator patterned after WGEN (Richardson and Wright, 1984) that also accounts for user-provided seasonal climate forecasts or other monthly climate modulations. Precipitation is generated independently of the other variables by using a Markow chain-exponential model. The other three variables are generated by using a multivariate model with the means and standard deviations of the variables conditioned on the wet or dry day status of the day as determined by the precipitation model (Richardson, 1981). With appropriately adjusted weather generation variables, SYNTOR can generate long records of synthetic daily weather data representative of current and projected future climate conditions.

The bias corrected and spatially downscaled GCM monthly precipitation and air temperature projections (Fig. 2, upper right) are readily available for the contiguous United States from the archive of the World Climate Research Programme's (WCRP) Coupled Model Intercomparison Project Phase 5 (CMIP5) multi-model dataset (Maurer et al., 2007). This archive is referred to as the CMIP5 archive and the climate projections used in this study were downloaded from this archive (CMIP5 archive accessible at:

http://gdo-dcp.ucllnl.org/downscaled_cmip_projections/).

Watershed Description

Three unit experimental watersheds near El Reno, Oklahoma were used in the WEPP model calibration. Each watershed is 80-m wide and 200-m long with a drainage area of 1.6 ha. The longitudinal slope of the watershed is approximately 3–5%. Soils are primarily silt loam with an average of 23% sand and 56% silt in the tillage layer. An annual winter wheat–summer fallow cropping system with three contrasting tillage systems of no-till, conservation (disks) and conventional (moldboard) tillage systems was studied on each watershed between 1980 and 1996. Measured rainfall data, soil properties (particle size, bulk density, field capacity, and wilting point), surface runoff and soil loss were used to calibrate the hydrological and erosion components of the WEPP model (Zhang, 2004). The measured soil loss showed that the peak erosion occurred in summer months when the ground was bare and rainfall intensity was high. The unit watershed was used to simulate the erosion responses to climate changes under the selected tillage and cropping systems.

Tillage and Crop Systems

A regional cropping system of continuous annual winter wheat–summer fallow was simulated under climate change for seven tillage alternatives: conventional tillage with and without terraces, conservation tillage with and without terraces, double-crop system, no-till, and conversion of cropland to rangeland. Conventional tillage comprises one moldboard plow and 3 disks at a 30-day interval in summer. Conservation tillage comprises 3 disks in summer at a 45-day interval. No till only uses a drill planter. A no-till double- crop system (annual winter wheat- summer soybean) and a rangeland pasture are also simulated. The first tillage was made 2 weeks following wheat harvest. Wheat was planted on 20th October and harvested on 15th June under the baseline climate, while it was planted on 20th October and harvested on 20th May under climate change to accommodate enhanced growth and early maturation of the wheat due to increased temperature and increase CO_2 levels.

Soil Erosion Model

Application of the soil erosion model involved selection of conservation practices, and simulation of relevant watershed processes required for the estimation of soil erosion under future climate conditions (Fig. 2, lower half).

Soil erosion was simulated using the Water Erosion Prediction Project (WEPP) model Version 2004.7 (Laflen et al., 1997 and Flanagan and Nearing, 1995). Carbon dioxide (CO_2) component was added to capture the effects of increased CO_2 on plant growth. The WEPP model is driven by the SYNTOR generated synthetic daily weather of the current or projected climate. Major hillside rainfall-runoff processes simulated by WEPP include infiltration, evapotranspiration, vegetation growth, tillage and residue decomposition, effects of increased carbon dioxide, surface runoff, soil erosion, sediment transport and deposition, and sediment yield. The WEPP model was calibrated using measured hydrological and biological data at the Water Resources and Erosion Experimental (WRE) watersheds of the USDA, Agricultural Research Service, Grazinglands Research Laboratory, El Reno, Oklahoma (Zhang, 2004). The WEPP model was applied for each projected climate and each conservation practice. Baseline soil erosion was simulated under current climate conditions and conventional tillage operations. Soil erosion was simulated for all other combinations of projected climate and conservation practices (Fig. 2, lower half). Those combinations that resulted in soil erosion values at or below the baseline soil erosion were categorized as being effective at offsetting the expected increase in soil erosion under climate change conditions.

Uncertainty in Projected Climate and Soil Erosion

For this study, uncertainty is defined as imperfect knowledge, especially with regard to the probability of future climate outcomes (Runde, 1998). For example, the specific pathway that GHG emissions may take in the future represents an uncertain event. Likewise, climate models are not perfect and alternative models lead to differences in projected future climate characteristics, thereby introducing uncertainty as to which of the projected climates will be realized. These uncertainties in future climate propagate down the chain of models to ultimately result in uncertainties in simulated soil erosion. Sources of relevant climate model uncertainties are summarized below. Uncertainties associated with the WEPP model are outside the scope of this study. They are assumed to be independent of the climate and conservation scenarios.

Emission Scenario Uncertainty

Scenarios, in the context of this study, describe alternative possible, plausible, but not necessarily equally probable paths of future GHG concentrations. The likelihood that any one scenario will occur is generally not known (Tebaldi and Knutti, 2007). The problem of emission scenario uncertainty was bypassed in this study by selecting the scenario that produces more intense and more frequent storms that would result in the greatest increase in soil erosion (worst case scenario).

Climate Model Uncertainty

Different GCM models, subjected to the same emission scenario or external heating (i.e. an imbalance between incoming and outgoing solar radiation), simulate different responses in the climate, which is a form of model uncertainty. Model uncertainty is a result of different physical and numerical model formulations of global warming processes (Hawkins and Sutton, 2009). A multi-model approach is often used to provide an ensemble of responses. In this study, ten GCMs were selected resulting in ten different estimates of soil erosion distributions due to model uncertainty.

Climate-model Internal Variability

The variability of the climate system that arises in the absence of external heating is known as internal uncertainty (Deser at al., 2012). In this study, the lead time of the climate projection was 50 years and model internal variability played a subordinate role to scenario and model selection uncertainty (Hawkins and Sutton, 2009).

Downscaling uncertainty

The resolution of GCM grid-box climate projections is too large to assess climate impacts on agricultural systems (Wilby et al., 1998 and Tabor and Williams, 2010). In general, one cannot distinguish which downscaling model performs best, and the subjective selection among competing models leads to downscaling uncertainty.

For climate change impact assessment using agricultural models an additional downscaling to local and daily scales was necessary. Stochastic weather generator SYNTOR (Garbrecht et al., 2014) was used in this study to downscale the monthly GCM climate predictions to daily weather outcomes at the location of interest within the GCM grid-cell.

Storm intensification uncertainty

Frequency and intensity of heavy downpours have been increasing in recent decades across most of the continental United States (Melillo et al., 2014, Higgins and Kousky, 2013 and Groisman et al., 2004). Climate projections suggest that this trend towards heavier precipitation will continue (Kunkel et al., 2013a and Kunkel et al., 2013b). In this study, the percent increase in amount of precipitation falling in very heavy events (defined as the heaviest 1% of all daily events) in the Southern Great Plains from 1958 to 2012 (approximately 16% according to Melillo et al., 2014) were assumed to persist through the middle of this century and very heavy precipitation events of the projected climate were adjusted accordingly.

RESULTS

Baseline Weather and Climate

Baseline climate was defined as the weather observed at the WRE experimental watersheds of the USDA ARS Grazinglands Research Laboratory, El Reno, Oklahoma, and over the 1977-2012 time period. SYNTOR was calibrated with baseline weather data. The calibrated SYNTOR was used to generate 200 years of synthetic daily precipitation and min/max temperature for subsequent simulation of soil erosion under baseline climate conditions. The distribution of annual precipitation and annual average air temperature for the baseline climate are displayed by the thick line in Fig. 3 and Fig. 4, respectively.

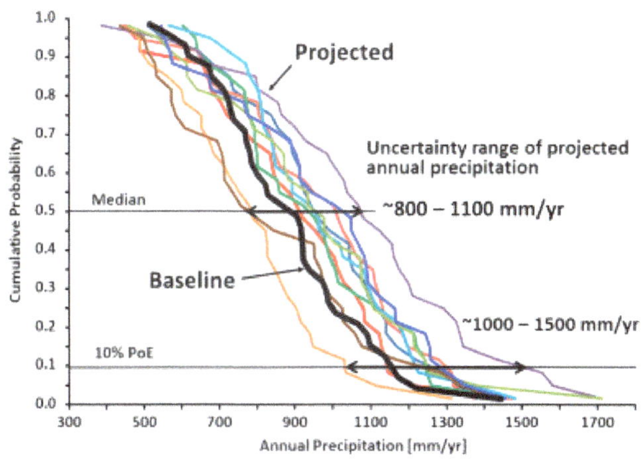

Figure 3: Distribution of baseline and projected annual precipitation. Projections are for high emission climate change scenario RCP8.5. The thick line represents the 1977–2012 baseline precipitation distribution, and the thin lines represent the 2041–2070 precipitation distributions for the 10 projected climates. The uncertainty range of annual precipitation at the median and at the 10% Probability-of-Exceedance (PoE) are given on the right side of the figure.

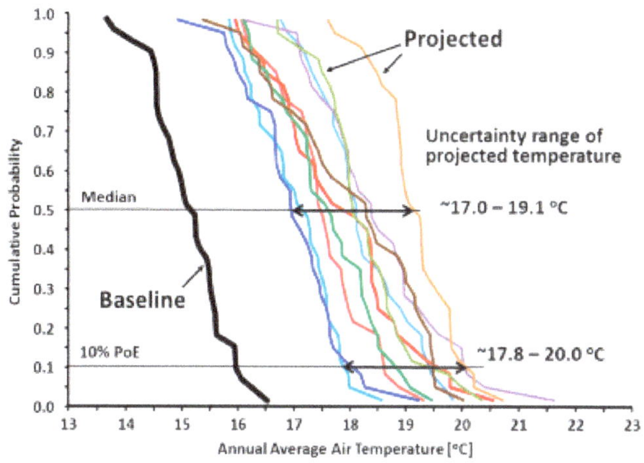

Figure 4: Distribution of baseline and projected annual average air temperature. Projections are for high emission climate change scenario RCP8.5. The thick line represents the distribution of the 1977–2012 baseline annual average air temperature, and the thin lines represent the 2041–2070 projected annual average air temperature distributions for the 10 projected climates. The uncertainty range of annual average air temperature at the median and at the 10% Probability-of-Exceedance (PoE) are given on the right side of the figure.

GCM Projected Climates

Ten different climate projections were generated by the ten selected GCMs. SYNTOR was first calibrated for each of the climate projections, and with the calibrated generation parameters 200 years of synthetic daily precipitation and min/max air temperature representative of the respective projected climate were generated. This produced ten different weather time series each representing potential future precipitation and air temperature outcomes for the 2041–2070 time period. The distribution of annual precipitation and annual average air temperature for the ten projected climates are displayed by the thin lines in Fig. 3 and Fig. 4, respectively.

Soil Erosion Projections

Baseline soil erosion from winter wheat cropland under conventional tillage operations was simulated by WEPP for the 200 years of generated baseline weather (Fig. 5, thick line). The average simulated annual soil erosion for baseline weather and conventional tillage operations was 7.1 t/ha/yr, and at the 10% Probability-of-Exceedance (PoE) level it was 21.1 t/ha/yr.

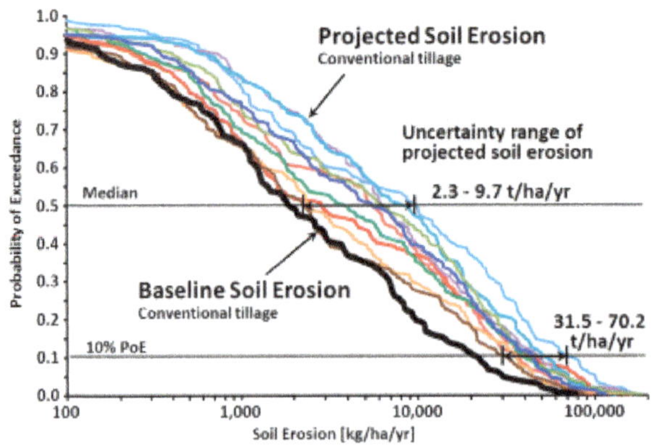

Figure 5: Distribution of baseline and projected annual soil erosion. Projections are for high emission climate change scenario RCP8.5. The thick line represents the distribution of the generated 1977–2012 baseline soil erosion, and the thin lines represent the distribution of the simulated 2041–2070 projected soil erosion for the 10 projected climates. The uncertainty range of annual soil erosion at the median and at the 10% Probability-of-Exceedance (PoE) are given on the right side of the figure.

Projected soil erosion was simulated for winter wheat under conventional tillage operations and for the 200 years of generated weather for each of the ten projected climates (Table 1 and Fig. 5, thin lines). The average of projected soil erosion corresponding to the ten projected climates was 15.3 t/ha/yr, and at the 10% PoE it was 44.3 t/ha/yr. This soil erosion simulation was repeated for each of the seven conservation practices. Relevant statistical characteristics of the simulated soil erosion for the ten projected climates and seven conservation practices are shown in Table 1 and plotted inFig. 6.

Table 1: Average, minimum, and maximum value of the mean and 10% Probability-of-Exceedance (PoE) values of the soil erosion distribution corresponding to each of the ten climate projections for 2041–2070 and for each of the seven conservation alternatives

Tillage alternative	Annual soil erosion in t/ha/yr for 2041–2070 projected climate					
	Means			10% PoE		
	Average	Min	Max	Average	Min	Max
Conventional	15.3	9.8	23.6	44.3	31.5	70.2
Conventional with terraces	8.5	5.7	12.4	22.3	15.5	32.6
Conservation	7.2	4.9	11.3	20.4	15.5	32.0
Conservation with terraces	4.8	3.4	6.8	11.9	8.6	17.5
Double cropping W. wheat-soybean	3.4	2.3	5.0	8.7	4.6	16.4
No till	0.6	0.4	0.9	1.4	1.0	1.7
Pasture/rangeland	~0.0	~0.0	~0.0	~0.0	~0.0	~0.0

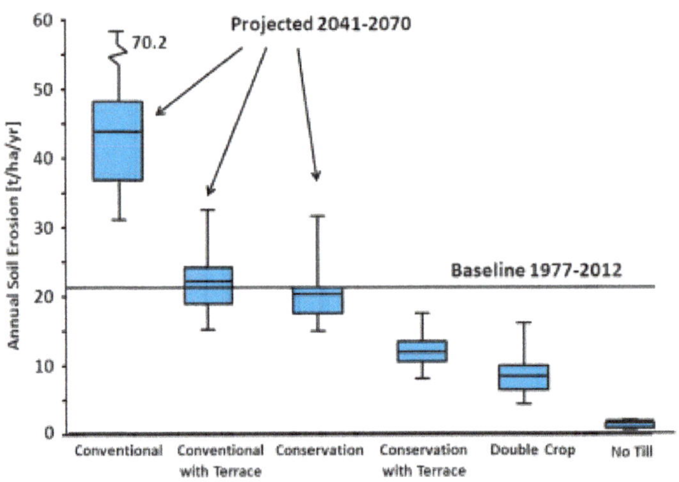

Figure 6: Basic statistical characteristics of the simulated soil erosion for the ten GCM projected climates and six conservation practices (pasture/rangeland is omitted because it had nearly zero soil erosion). Projections are for high emission climate change scenario RCP8.5. The whiskers of each conservation practice box represent the uncertainty range of projected soil erosion at the 10% Probability-of-Exceedance level. The horizontal line is the baseline soil erosion of 21.1 t/ha/yr at the 10% Probability-of-Excendance level.

Discussion of Results

The average simulated annual soil erosion for winter wheat under conventional tillage operations and under the 1977–2012 baseline climate conditions was 7.1 t/ha/yr. At the 10% Probability-of-Exceedance (PoE) level, it was 21.1 t/ha/yr. The high mean soil erosion in this baseline scenario emphasizes the importance of conservation practices to sustain soil resources under recent observed climate.

Future average annual soil erosion (2041–2070) simulated with the ten different projected climates ranged from 9.8 to 23.6 t/ha/yr with an average of 15.3 t/ha/yr (Table 1, first data record), which is about double that produced under baseline climate conditions, indicating an even greater need for conservation.

The projected soil loss is consistent with those reported by Zhang (2012) and Zhang et al. (2011), who simulated soil erosions for annual winter wheat in central Oklahoma using the WEPP model for the time period of 2010–2039 (next 30 years). These authors used three emission scenarios and four GCMs and a different spatiotemporal downscaling method. Their results showed that the projected mean precipitation would decrease during 2010–2039 by about 6%. However, the overall soil loss averaged over all tillage systems would increase by about 40% due to a sizable increase in heavy rainfall storms. The greater soil erosion obtained in this study is mostly due to simulating soil erosion under projected climates that are further in the future (2041–2070) and thereby exhibit a greater climatic change, as well as focusing on a high emission scenario.

Repeating the soil erosion simulations for conventional tillage with terraces, resulted in a future soil erosion of, on average, 8.5 t/ha/yr (Table 1, second data record). For conservation tillage the future soil erosion was, on average, 7.2 t/ha/yr (Table 1, third data record). When considering the average of the soil erosion produced by the ten different projected climates, both conventional tillage with terraces and conservation tillage without terraces would reduce future soil erosion close to the target baseline soil erosion of 7.1 t/ha/yr. The other four conservation practices (conservation with terraces, double cropping, no till, and conversion to pasture) produced future soil erosion values that were substantially lower than the target baseline soil erosion. As such, they were categorized as effective conservation alternatives that reduced the soil erosion below the baseline target soil erosion (Table 1). These results are in line with findings reported by Zhang (2012) and Zhang et al. (2011) for the time period of 2010–2039.

In the preceding, effectiveness of conservation practices was assessed in terms of average soil erosion. However, typically the largest 2% of storms accounted for 60–85% of total soil erosion in all the wheat production systems in the study region (Zhang and Garbrecht, 2002). Thus, focusing on large infrequent storm events provides a more appropriate basis for assessing the effectiveness of conservation practices because the bulk of the annual soil erosion is the result of these storms. In this case, if one were to use soil erosion values at the 10% PoE level in the conservation assessment (Table 1, right columns), one would reach similar conclusions as when using average soil erosion values. Thus, either conventional tillage with terraces (22.3 t/ha/yr) or conservation tillage (20.4 t/ha/yr) would be sufficient to reduce future soil erosion close to or below the target baseline soil erosion value of 21.1 t/ha/yr at the 10% PoE level.

The above considerations do not account for the uncertainty introduced by the spread of potential future climates resulting from the use of ten different GCMs. More effective conservation alternatives have to be considered to reach the target baseline soil erosion for the full range of projected future climate conditions. Referring to Table 1, the maximum average annual soil erosion for conventional tillage was 70.2 t/ha/yr at the 10% PoE level. This is about 3 times higher than the target baseline soil erosion of 21.1 t/ha/yr at the 10% PoE level. Thus, conservation tillage alone would not reduce projected soil erosion enough to overcome the soil erosion uncertainty range and reduce all projected future soil erosion to or below the target baseline soil erosion. Conservation tillage with terraces (17.5 t/ha/yr) or a summer cover crop (16.4 t/ha/yr) would be necessary to achieve soil erosion reductions to or below the target baseline value of 21.1 t/ha/yr.

No till and land conversion to pasture land produced the lowest soil erosion under all projected climate conditions. No till produced a maximum average soil erosion of less than 1 t/ha/yr, and a maximum soil erosion of 1.7 t/ha/yr at the 10% PoE level. These values are over an order of magnitude less than the corresponding target baseline soil erosion.

The sizeable reductions in soil erosion achieved by established and proven conservation practices attests to the wide range of available conservation tools and their effectiveness to hold soil erosion and soil erosion within tolerable levels. The progress in protecting our soil resource that was made recent decades is an important achievement for the long-term sustainability of the agricultural production system and for future food security. Improvements brought about by implementing conservation practices do not persist if the conservation practices are removed. In order to maintain the progress we have made in protecting the soil resource through conservation conditions, policies and incentives that fostered the positive changes over recent decades must be maintained and enhanced. With current trends in our changing climate, which point to the potential for those improvements to be reversed, this is not the time to become complacent about soil conservation.

Summary and Conclusions

Concerns have been expressed by the conservation community that climate change may increase soil erosion from cropland and that conservation practices that were effective in the past may no longer be adequate under future climate conditions. This question was examined for winter wheat cropland in central Oklahoma under the assumption of persistent high greenhouse gas emission scenario. The climate under this emission scenario was projected half a century into the future (2041–2070) by ten different Global Circulation Models. The resulting ten projected climates were used, one at a time, by a rainfall-runoff-erosion model to simulate soil erosion from winter wheat crops under various conservation practices. Seven alternative conservation practices were evaluated for their effectiveness to keep soil erosion at or below today's level. Findings and conclusions of this study are specific to winter wheat crops and physiographic conditions prevailing at the USDA ARS research watersheds at Fort Reno, Oklahoma, yet they are believed to be valid for the winter wheat production region in Central Oklahoma.

- Projected climate change will lead to an increase in soil erosion from winter wheat cropland. Under conventional tillage practices, the average projected soil erosion (15.3 t/ha/yr) was approximately double the rate of that under today's climate (7.1 t/ha/yr). This increase in projected soil erosion was largely attributed to the increase in intensity of heavy storm events.
- A switch from conventional to conservation tillage practices would be sufficient to offset the average increase in soil erosion under most projected climates.
- Conservation tillage with terraces or conservation tillage with summer cover crop (soybeans) would be required to control soil erosion under extreme conditions associated with the most severe climate projections.
- Implementation of no-till and land use conversion to grassland would produce soil erosion values under most projected climates that are a fraction of today's soil erosion.

The baseline scenario in this study, conventional tillage of monoculture winter wheat, indicates that existing conservation practices are absolutely essential to sustain the agricultural soil resource. The climate change scenarios indicate that even greater conservation will be required in the future. Fortunately, these findings show that agriculture in Central Oklahoma has a broad range of conservation tools with which it could offset projected future increases in soil erosion under assumed worst case climate change scenarios. Thus, the issue is not a lack of effective conservation tools, but one of adoption and implementation. The time-scale of climate change by far exceeds the time-scale of producers planning horizon and the uncertainty of the pathway climate will take in the future limits the urgency to change from the wide spread use of conventional tillage for winter-wheat crops in Central Oklahoma. Implementation of soil conservation practices are more likely to be driven by current climatic events. Specifically, the large year-to-year precipitation variability typical in the Southern Great Plains provide a great incentive to adopt soil conservation practices that ensure a sustainable utilization of agricultural soil resources and high crop yields. Increasing the implementation rate of today's conservation programs for today's problems would greatly contribute towards mitigation of projected increases in soil erosion and soil erosion due to climate change. And yes, conservation can trump impacts of climate change on long term mean soil erosion from winter wheat cropland in the central Oklahoma region.

ACKNOWLEDGEMENTS

We acknowledge the World Climate Research Programme's Working Group on Coupled Modelling which is responsible for CMIP5, and we thank the climate modeling groups for producing and making available their model output. For CMIP5, the U.S. Department of Energy's Program for Climate Model Diagnosis and Intercomparison provides coordinated support and led development of software infrastructure in partnership with the Global Organization for Earth System Science Portals.

REFERENCES

1. AMS (American Meteorological Society), 2012. AMS Information Statement on Climate Change. ⟨http://ametsoc.org/POLICY/2012climate change.html⟩.
2. AMS (American Meteorological Society), 2013. AMS Information Statement on Drought. ⟨http://ametsoc.org/POLICY/2013drought_amsstatement.html⟩.
3. Ayres, Q.C., 1936. Soil Erosion and its Control. McGraw-Hill Book Company, New York, p. 365.
4. Bassu, S., Brisson, N., Durand, J.-L., Boote, K., Lizaso, J., Jones, J.W., Rosenzweig, C., Ruane, A.C., Adam, M., Baron, C., Basso, B., Biernath, C., Boogaard, H., Conijn, S., Corbeels, M., Deryng, D., De Sanctis, G., Gayler, S., Grassini, P., Hatfield, J., Hoek,S., Izaurralde, C., Jongschaap, R., Kemanian, A.R., Kersebaum, K.C., Kim, S.-H.,Kumar, N.S., Makowski, D., Müller, C., Nendel, C., Priesack, E., Pravia, M.V., Sau,F., Shcherbak, I., Tao, F., Teixeira, E., Timlin, D., Waha, K., 2014. How do variousmaize crop models vary in their responses to climate change factors. Glob.Change Biol. 20, 2301–2320. http://dx.doi.org/10.1111/gcb.12520.
5. Bennett, H.H., 1939. Soil Conservation. McGraw-Hill Book Company, New York, p. 993.
6. Delgado, J.A., Nearing, M.A., Rice, C.W., 2013. Conservation practices for climatechange adaptation. Adv. Agron. 121, 47–115.
7. Delgado, J.A., Groffman, P., Nearing, M.A., Goddard, T., Reicosky, D., Lal, R., Kitchen,N., Rice, C., Towery, D., Salon, P., 2011. Conservation practices to mitigate and adapt to climate change. J. Soil Water Conserv. 66 (4), 118A–129A.
8. Deser, C., Phillips, A., Bourdette, V., Teng, H., 2012. Uncertainty in climate change projections: the role of internal variability. Clim. Dyn. 38, 527–546.
9. Flanagan, D.C., Nearing, M.A. (Eds.), 1995. USDA-water erosion prediction project: hillslope profile and watershed model documentation. NSERL Report, No. 10. USDA-ARS National Soil Erosion Research Laboratory, West Lafayette, Ind.
10. Garbrecht, J.D., Zhang, J.X., Steiner, J.L., 2014. Climate change and observed climate trends in the Fort Cobb experimental watershed. J. Environ. Qual. 43,1319–1327.
11. Groisman, P., Ya., R.W., Knight, T.R., Karl, D.R., Easterling, B., Sun, Lawrimore, J.M.,2004. Contemporary changes of the hydrological cycle over the contiguous United States: trends derived from in situ observations. J. Hydrometeorol. 5,64–85.
12. Groisman, P.Y., Knight, R.W., Easterling, D.R., Karl, T.R., Hegerl, G.C., Razuvaev, V.A. N., 2005. Trends in intense precipitation in the climate record. J. Clim. 18 (9), 1326–1350.
13. Groisman, P., Ya., R.W., Knight, Karl, T.R., 2012. Change in intense precipitation over the Central United States. J. Hydrometeorol. 13, 47–66.
14. Hawkins, E., Sutton, R., 2009. The potential to narrow uncertainty in regional climate predictions. Am. Meteorol. Soc. 90, 1095–1107.

15. Higgins, R.W., Kousky, V.E., 2013. Changes in observed daily precipitation over the United States between 1950–79 and 1980–2009. J. Hydrometeorol. 14, 105–121, doi:10.1175/jhm-d-12-062.1.
16. Izaurralde, R.C., Thomson, A.M., Morgan, J.A., Fay, P.A., Polley, H.W., Hatfield, J.L., 2011. Climate impacts on agriculture: Implications for forage and rangeland production. Agron. J. 103 (2), 371–381.
17. Karl, T.R., Melillo, J.T., Peterson, T.C., 2009. Karl, T.R., Melillo, J.T., T.C (Eds.), GlobalClimate Change Impacts in the United States. Cambridge University Press,p. 189.
18. Karl, T.R., Knight, R.W., 1998. Secular trends of precipitation amount, frequency, andintensity in the USA. Bull. Am. Meteorol. Soc. 79, 231–241.
19. Kunkel, K.E., Karl, T.R., Brooks, H., Kossin, J., Lawrimore, J.H., Arndt, D., Bosart, L., Changnon, D., Cutter, S.L., Doesen, N., Emanuel, K., Groisman, P. Ya, Katz, R.W.,
20. Knutson, T., O'Brien, J., Paciorek, C.J., Peterson, T.C., Redmond, K., Robinson, D., Rapp, J., Vose, R., Weaver, S., Wehner, M., Wolter, K., Wuebbles, D., 2013a. Monitoring and understanding trends in extreme storms: State of knowledge.Bull. Am. Meteorol. Soc. 94, 499–514. http://dx.doi.org/10.1175/BAMS-d-11-00262.1.
21. Kunkel, E.K., et al., 2013b. Monitoring and understanding trends in extreme storms:state of knowledge. Bull. Am. Meteorol. Soc. 94, 499–514.
22. Laflen, J., Elliot, W., Flanagan, D., Meyer, C., Nearing, M., 1997. WEPP-predicting water erosion using a process based model. J. Soil Water Conserv. 52–96, 102.
23. Maurer, E.P., Brekke, L., Pruitt, T., Duffy, P.B., 2007. Fine-resolution climate projections enhance regional climate change impact studies. EOS, Trans. Am. Geophys.Union 88 (47), 504.
24. Meehl, G.A., Stocker, T.F., Collins, W.D., Friedlingstein, P., Gaye, A.T., Gregory, J.M., Kitoh, A., Knutti, R., Murphy, J.M., Noda, A., Raper, S.C.B., Watterson, I.G., Weaver, M.A., Zhao, Z.-C., 2007. Global climate projections. In: Solomon, S., Qin, D., Manning, M., Chen, Z., Marquis, M., Averyt, K.B., Tignor, M., Miller, H.L. (Eds.),Climate Change 2007: The Physical Science Basis. Cambridge University Press, Cambridge, United Kingdom and New York, NY, USA, pp. 747–846.
25. Melillo, J.M., Richmond, T.C., Yohe, G.W., (Eds.), 2014. Climate Change Impacts in the United States: The Third National Climate Assessment. U.S. Global Change Research Program U.S. Government Printing Office, Washington, DC. 841 pp. doi:10.7930/J0Z31WJ2.
26. Nearing, M.A., Pruski, F.F., O'Neal, M.R., 2004. Expected climate change impacts on soil erosion rates: a review. J. Soil Water Conserv. 59 (1), 43–50.
27. Nearing, M.A., Jetten, V., Baffaut, C., Cerdan, O., Couturier, A., Hernandez, M., Le Bissonnals, Y., Nicols, M.H., Nunes, J.P., Renschler, C.S., Souchere, V., Van Oost, K.,2005. Modeling response of soil erosion and runoff to changes in precipitation and cover. Catena 61, 131–154.
28. O'Neal, M.R., Nearing, M.A., Vining, R.C., Southworth, J., Pfeifer, R.A., 2005. Climate change impacts on soil erosion in Midwest United States with changes in crop management. Catena 61, 165–184.
29. Peterson, T.C., Heim Jr., R.R., Hirsch, R., Kaiser, D.P., Brooks, H., Diffenbaugh, N.S., Dole, R.M., Giovannettone, J.P., Guirguis, K., Karl, T.R., Katz, R.W., Kunkel, K., Lettenmaier, D., McCabe, G.J., Paciorek, C.J., Ryberg, K.R., Schubert, S., Silva, V.B. S., Stewart, B.C., Vecchia, A.V., Villarini, G., Vose, R.S., Walsh, J., Wehner, M., Wolock, D., Wolter, K., Woodhouse, C.A., Wuebbles, D., 2013. Monitoring and understanding changes in heat waves, cold waves, floods and droughts in the United States: state of knowledge. Bull. Am. Meteorol. Soc. 94, 821–834.
30. Pfeifer, R.A., Habeck, M., 2002. Farm level economic impacts of climate change. In:Doering, O.C., Randolph, J.C., Southworth, J., Pfeifer, R.A. (Eds.), Effects of Climate Change and Variability on Agricultural Production Systems. Kluwer Academic Publishers, Boston, pp. 159–178, Chapter 8.
31. Richardson, C.W., 1981. Stochastic simulation of daily precipitation, temperature, and solar radiation. Water Resour. Res. 17 (1), 182–190.
32. Richardson, G.W., Wright, D.A., 1984. WGEN: A Model for Generating Daily Weather Variables. USDA Agricultural Research Service, Publication ARS-8, 8 pp.
33. (Available from U.S. Department of Agriculture, Agricultural Research Service, 808 E. Blackland Road Temple, Texas.).
34. Runde, J., 1998. Clarifying Frank Knight's discussion of the meaning of risk and uncertainty. Camb. J. Econ. 22 (5), 539–546. http://dx.doi.org/10.1093/cje/ 22.5.539.
35. Soil and Water Conservation Society, 2003. Conservation implications of climate change: soil erosion and runoff from cropland. A Report from the Soil and Water Conservation Society, 945 Southwest Ankeny Road, Ankeny, Iowa.
36. Southworth, J., Habeck, M., Pfeifer, R.A., Randolph, J.C., Doering, O., Johnston, J., Rao, G., 2002. Changes in soybean yields in the midwestern United States as a result of future changes in climate, climate variability, and CO2 fertilization. Clim. Change 53 (4), 447–475. http://dx.doi.org/10.1023/A:1015266425630.
37. Tabor, K., Williams, J.W., 2010. Globally downscaled climate projections for assessing the conservation impacts of climate change. Ecol. Appl. 29 (2), 554–565.
38. Tebaldi, C., Knutti, R., 2007. The use of the multi-model ensemble in probabilistic climate projections. Philos. Trans. R. Soc. A 365, 2053–2075.
39. U.S. Department of Agriculture, 2013. Summary Report: 2010 National Resources Inventory, Natural Resources Conservation Service, Washington, DC, and Center for Survey Statistics and Methodology, Iowa State University, Ames, Iowa. ⟨http://www.nrcs.usda.gov/Internet/FSE_DOCUMENTS/stelprdb1167354.pdf⟩.
40. Wilby, R.L., Dessai, S., 2010. Robust adaptation to climate change. Weather 65 (7), 180–185. http://dx.doi.org/10.1002/wea.543.

41. Walthall, C.L., Hatfield, J. Backlund, P., Lengnick, L., Marshall, E., Walsh, M., Adkins, S., Aillery, M., Ainsworth, E.A., Ammann, C., Anderson, C.J., Bartomeus, I., Baumgard, L.H., Booker, F., Bradley, B., Blumenthal, D.M., Bunce, J., Burkey, K.,Dabney, S.M., Delgado, J.A., Dukes, J., Funk, A., Garrett, K., Glenn, M., Grantz, D. A., Goodrich, D., Hu, S., Izaurralde, R.C., Jones, R.A.C., Kim, S.-H., Leaky, A.D.B., Lewers, K., Mader, T.L., McClung, A., Morgan, J., Muth, D.J., Nearing, M., Oosterhuis, D.M., Ort, D., Parmesan, C., Pettigrew, W.T., Polley, W., Rader, R., Rice, C., Rivington, M., Rosskopf, E., Salas, W.A., Sollenberger, L.E., Srygley, R., Stöckle, C.,Takle, E.S., Timlin, D., White, J.W., Winfree, R., Wright-Morton, L., Ziska, L.H., 2012. Climate Change and Agriculture in the United States: Effects and Adaptation. USDA Technical Bulletin 1935. Washington, DC. 186 pp.

42. Wilby, R.L., Wigley, T.M.L., Conway, D., Jones, P.D., Hewitson, B.C., Main, J., Wilks, D. S., 1998. Statistical downscaling of general circulation model output: a comparison of methods. Water Resour. Res. 34, 2995–3008.

43. Wischmeier, W.H., Smith, D.D., 1978. Predicting Rainfall Erosion Losses – A Guide to Conservation Planning. Government Printing Office, Washington, DC, p. 58, Agricultural Handbook No. 537.

44. Zhang, X.C., 2004. Calibration, refinement, and application of the WEPP model for simulating climatic impact on wheat production. Trans. ASAE 47, 1075–1085.

45. Zhang, X.C., 2012. Cropping and tillage systems effects on soil erosion under climate change in Oklahoma. Soil Sci. Soc. Am. J. 76, 1789–1797.

46. Zhang, X.C., Garbrecht, J.D., 2002. Precipitation retention and soil erosion undervarying climate, land use, and tillage and cropping systems. J. Am. Water Resour. Assoc. 38, 1241–1253.

47. Zhang, X.C., Liu, W.Z., Li, Z., Chen, J., 2011. Trend and uncertainty analysis of simulated climate change impacts with multiple GCMs and emission scenarios. Agric. For. Meteorol. 151, 1297–1304.

48. Zhang, X.C., Nearing, M.A., 2005. Impact of climate change on soil erosion, runoff, and wheat productivity in central Oklahoma. Catena 61 (2–3), 185–195.

CHAPTER 16

ASSESSMENT OF TEMPORAL AND SPATIAL CHANGES OF FUTURE CLIMATE

Rashid Mahmood, Mukand S Babel, Shaofeng JIA

Keywords: Climate change, Downscaling, Statistical downscaling, Jhelum River basin, Pakistan, India, Temperature, Precipitation

INTRODUCTION

The concentration of CO_2 and other greenhouse gases has been increasing dramatically since 1950, mostly because of industrialization (Gebremeskel et al., 2005). This increase has caused a global energy imbalance and has increased global warming. According to the Fifth Assessment Report of the Intergovernmental Panel on Climate Change (IPCC), global (land and ocean) mean surface temperature as calculated by a linear trend over the period 1880–2012, show a warming of 0.85 °C (0.65–1.06 °C). An alarming increase of 0.78 (0.72–0.85) °C has been observed for the period of 2003–2012 with respect to 1850–1900. Global mean surface temperatures is projected to increase by 0.3–1.7 °C, 1.1 to 2.6 °C, 1.4–3.1 °C, and 2.6–4.8 °C under RCP2.6, RCP4.5, RCP6.0 and RCP8.5, respectively, for 2081–2100 relative to 1986–2005 (IPCC, 2013).

This global warming can disturb the hydrological cycle of the world, and can pose problems for public health, industrial and municipal water demand, water energy exploitation, and the ecosystem (Chu et al., 2010 and Zhang et al., 2011).

In the last few decades, Global Climate Models (GCMs)—the most advanced and numerical-based coupled models representing the global climate system—have been used to examine future changes in climate variables such as temperature, precipitation, and evaporation (Fowler et al., 2007). However, their outputs are temporally and spatially very coarse (Gebremeskel et al., 2005), which makes them useful only at continental and global levels. Their application at local/regional levels, such as at the basin and sub-basin scales, to assess the impacts of climate change on the environment and hydrological cycle, is problematic due to a clear resolution mismatch (Hay et al., 2000 and Wilby et al., 2000). That is, GCMs cannot give a realistic presentation of local or regional scales due to parameterization limitations (Benestad et al., 2008). The local and regional scales are defined as 0–50 km and 50×50 km^2, respectively (Xu, 1999)

To overcome this problem, during the last two decades, many downscaling methods have been developed to make the large-scale outputs of GCMs useful at local/regional scales (Wetterhall et al., 2006). In the beginning, these methods were mostly implemented in Europe and in the USA but are now applied throughout the globe to examine changes in climate variables at the basin level (Mahmood and Babel, 2012).

Generally, downscaling techniques are divided into two main categories: Dynamical Downscaling (DD), and Statistical Downscaling (SD). In DD, a Regional Climate Model (RCM) of high resolution (5–50 km) (Chu et al., 2010), nested within a GCM, receives inputs from the GCM and then provides high resolution outputs on a local scale. Since the RCM is dependent on the boundary conditions of a GCM, there is a greater chance that systematic errors that belong to the driving fields of the GCM will be inherited by the RCM. In addition, simulations from RCMs are computationally intensive, and depend upon the domain size and resolution at which the RCMs are to run, which in turn limits the number of climate projections (Fowler et al., 2007).

In contrast, SD approaches, which establish a bridge among the large-scale variables (e.g., mean sea level pressure, temperature, zonal wind, and geopotential height) and local-

Citation: Rashid Mahmooda, Mukand S Babel, Shaofeng JIA, Assessment of temporal and spatial changes of future climate, http://dx.doi.org/10.1016/j.wace.2015.07.002.

scale variables (e.g., observed temperature and precipitation) by creating empirical/statistical relationships, are computationally inexpensive and much simpler than DD (Wetterhall et al., 2006). Moreover, SD approaches offer immediate solutions for downscaling climate variables and, accordingly, they have rapidly been adopted by a wider community of scientists (Wilby et al., 2000 and Fowler et al., 2007). The limitation of SD is that historical meteorological station data over a long period of time is required to establish a suitable statistical or empirical relationship with large-scale variables (Chu et al., 2010). This relationship is considered to be temporally stationary, which is the main assumption of this method (Hay and Clark, 2003). In addition, SD is mainly dependent on the level of uncertainties of the parent GCM(s). DD, therefore, is a good alternative for SD in basins where no historical data is available (Benestad et al., 2008).

To date, many SD models have been developed for downscaling, and among them Statistical Downscaling Model (SDSM) was selected for this study. SDSM—a combination of multiple linear regression and a stochastic weather generator—is a well-known statistical model, and is frequently used for downscaling important climate variables (e.g., temperature, precipitation, and evaporation). The downscaled variables are used to assess hydrological responses under changing climatic conditions (Diaz-Nieto and Wilby, 2005, Gagnon et al., 2005, Gebremeskel et al., 2005 and Wilby et al., 2006). SDSM has been widely used throughout the world in many impact assessment studies (Huang et al., 2011).

Several studies (Tripathi et al., 2006, Anandhi et al., 2007, Akhtar et al., 2008, Ghosh and Mujumdar, 2008, Mujumdar and Ghosh, S., 2008, Akhtar et al., 2009, Ashiq et al., 2010,Goyal and Ojha, 2010 and Opitz-Stapleton and Gangopadhyay, 2010) have used SD and DD techniques in South Asia. However, most of these studies have been conducted in Indian river basins. A couple of studies, such as by Akhtar et al., 2008 and Akhtar et al., 2009 and by Ashiq et al. (2010) using DD, and a study by Mahmood and Babel (2012)using SD, were conducted in Pakistan to investigate future changes in climate variables. In the study by Akhtar et al. (2008), a Regional Climate model, Providing Regional Climate for Impact Studies (PRECIS), and delta change were used to downscale mean temperature and precipitation in order to analyze hydrological responses to climate change in the Hindukush–Karakorum–Himalayas (HKH). However, future changes were only investigated under the A2 scenario and only for the period of 2071–2100. The study concluded that the simulations from PRECIS have several uncertainty sources. Ashiq et al. (2010) interpolated monthly precipitation outputs of PRECIS, run by Akhtar et al. (2008), from a coarse resolution (50×50 km^2) to a fine one (250×250 m^2) using some interpolation methods. Their study was carried out in the northwest region of the Himalayan Mountains and in the upper Indus plains of Pakistan. They also covered a small part of the Jhelum River basin, located inside Pakistan. The study's conclusion was that systematic errors associated with an RCM cannot be reduced by a simple interpolation methods.

Mahmood and Babel (2012) evaluated two sub-models of SDSM (annual and monthly sub-models) out of three sub-models (monthly, annual, and seasonal) in the Jhelum River basin. They concluded that the annual sub-model cannot be used to investigate the intra-annual (monthly or seasonal time series) variations in climate variables (like temperature and precipitation) without Bias Correction, although SDSM was able to predict mean annual values quite well. They also concluded that the monthly sub-model is capable of examining intra-annual variations without any Bias Correction.

No studies, to the best of our knowledge, have investigated the temporal and spatial future changes in maximum temperature, minimum temperature, and precipitation under A2 and B2 scenarios in the upper Jhelum River basin—one of the biggest tributaries of the Indus River basin and drains entirely into Mangla reservoir, the second biggest reservoir in Pakistan.

In the present study, the monthly sub-model of SDSM, as recommended by Mahmood and Babel (2012), was applied to analyze the temporal and spatial changes in maximum temperature, minimum temperature, and precipitation for the period of 2011–2099, under A2 and B2 scenarios of HadCM3. Since the study area is mountainous and has great altitudinal variations as shown in Fig. 1, an advanced spatial interpolation method, Kriging, was used to explore the spatial variability in the basin. For the sake of detailed analysis, the whole Jhelum basin was divided into two sub-basins, based on precipitation regimes. This study can be a guide for researchers who wish to apply statistical downscaling methods on mountainous regions in the Upper Indus basin, which is highly influenced by the monsoons.

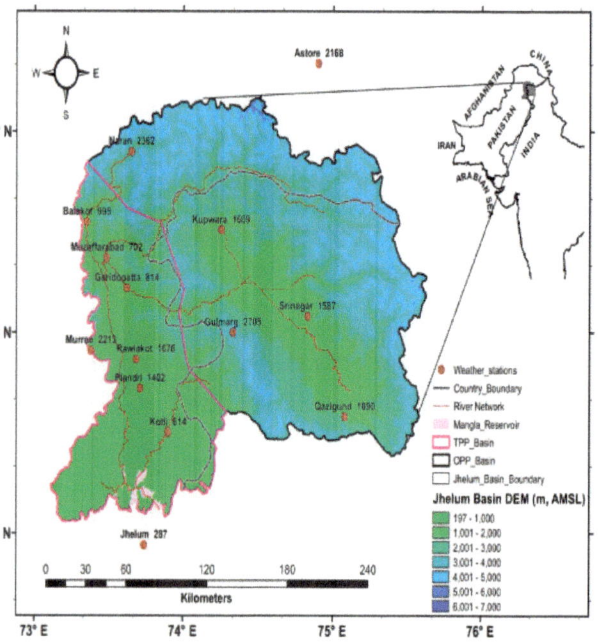

Figure 1: Location map of the study area with the climate stations.

STUDY AREA AND DATA DESCRIPTION

Study Area

The upper Jhelum River basin—located in Pakistan and India—stretches between 73–75.62 °E and 33–35 °N, as shown in Fig. 1. The Jhelum River, the second biggest tributary of the Indus River system (Ahmad and Chaudhry, 1988), has a drainage area of 33,342 km^2, with an elevation ranging from 235 to 6285 m. The whole basin drains into Mangla Reservoir, constructed in 1967, which is the second biggest reservoir in Pakistan. The main purpose of this dam is to provide irrigation water to 6 million hectares of land. Its secondary function is to produce hydropower; it has an installed capacity of 1000 MW, contributing 6% of the country's installed capacity (Archer and Fowler, 2008 and Qamer et al., 2008). For the purposes of the present study, the basin was divided into two parts, based on the precipitation regimes of the basin, as discussed inSection 2.3.

Data Description

Historical daily data of maximum temperature (Tmax), minimum temperature (Tmin), and precipitation (P) was collected from the Water and Power Development Authority of Pakistan (WAPDA), Pakistan Meteorological Department (PMD), and the Indian Meteorological Department (IMD). The dataset was obtained from 14 climate stations located in the basin (Fig. 1) for the period of 1961–2000. The historical data of Kupwara, Srinagar, Gulmarg, and Qazigund was obtained from the IMD; the data of Rawlakot, Plandri, Kotli, and Naran was obtained from the WAPDA; and, the data of Muzaffarabad, Jhelum, Garidopatta, Balakot, Murree, and Astore was obtained from PMD. Some part of the daily P data for Srinagar and Qazigund for the period of 1961–1970 was acquired from the National Climate Data Center (NCDC). Although Astore and Jhelum do not lie in the upper Jhelum River basin, these stations were included in the present study due to the lack of climate stations in the basin. In addition, both stations have good quality historical meteorological data. There were some missing values on some weather stations which was filled by multiple imputation method, using winMice software (Jacobusse, 2005)

The daily predictors of National Centers for Environmental Prediction (NCEP) for 1961–2001 and the daily predictors of HadCM3 (A2 and B2 scenarios) for 1961–2099 were collected from a Canadian website (http://www.cccsn.ec.gc.ca/?page=dst-sdi). A2 and B2 are the IPCC emission scenarios of HadCM3 and are mostly used for regional impact assessment studies. These predictors are specially prepared for SDSM. During the preparation, the NCEP predictors (2.5°×2.5°) were first interpolated to the grid resolution of HadCM3 (2.5°×3.75°) to eliminate any spatial mismatch. Then, the NCEP and HadCM3 predictors were normalized with mean and standard deviations of a long period (1961–1990) (CCCSN, 2012).

Climatic Condition in the Study Area

The mean monthly rainfall regimes of all the climate stations used in this study for the period of 1961–1990 are shown in Fig. 2(a). Naran, Srinagar, Kupwara, Gulmarg, Astore and Qazigund climate stations experience one big peak in March, and Kotli, Jhelum, Murree, Plandri, Rawlakot, Murree, Garidopatta, Muzaffarabad and Balakot experience two peaks, one small peak around March, and other big peak around July. According to the P regimes, the whole basin was divided into two sub-basins: (1) the One Peak Precipitation basin (OPPB), and (2) the Two Peak Precipitation basin (TPPB). The OPPB contains most of the northeast parts of the Jhelum basin, including Srinagar, Gulmarg, Qazigund and Astore weather stations, and some of the northwest parts of the basin, including Naran and Kupwara weather stations. The TPPB consists of the southwest parts of the basin, containing Jhelum, Kotli, Plandri, Rawlakot and Murree climate stations, and the northwest part of the basin that has climate stations like Garidopatta, Muzaffarabad, and Balakot. The locations of these stations are shown in Fig. 1. The one big peak (July) in the basin is the result of summer monsoon (June–July–August), which occurs due to the saturated south western winds from the Bay of Bengal and Arabian Sea. The second small peak (March) is the result of the Western disturbances (WDs) system in winter. Generally the whole Pakistan and particularly its northwestern pats receive precipitation due to the WDs. The WDs are natured by the depression over the Mediterranean regions that take precipitation to central southwest Asia in the months of December–March (Ahmad et al., 2015). More details about the Summer Monsoon and WDs are described in Ahmad et al. (2015).

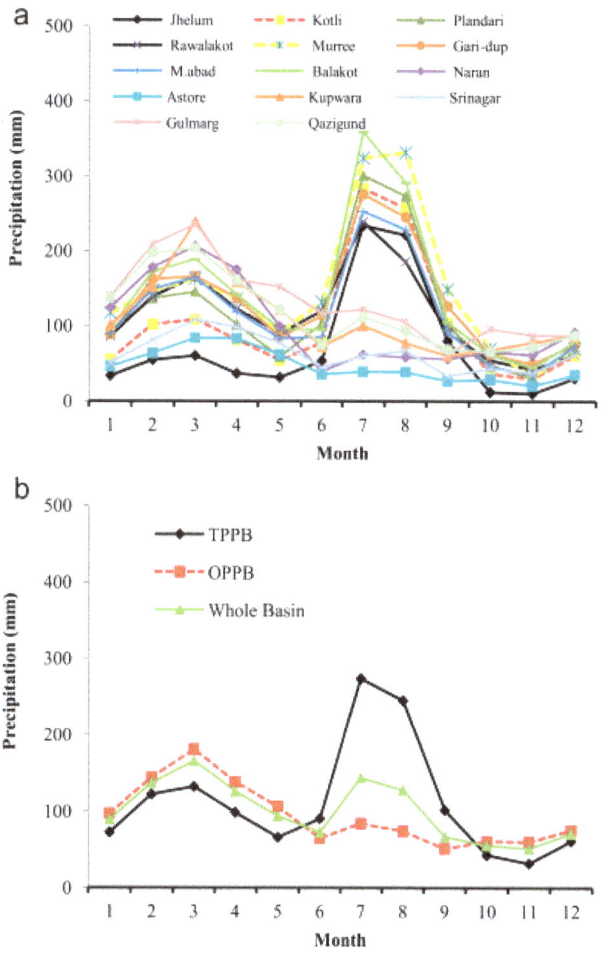

Figure 2: Distribution of mean monthly rainfall at (a) all the weather stations and (b) in the TPPB, OPPB, and the whole Jhelum basin for 1961–1990 (the baseline period).

Fig. 2 and Table 1 illustrate that the TPPB, with a mean annual P of 1341 mm, is relatively wetter than the OPPB, which has a mean annual P of 1139 mm. A mean annual P of 1202 occurs in the basin as a whole. July and March are the wettest months in the TPPB and OPPB respectively, and November and September are the driest months in the TPPB and OPPB respectively. As for climate stations, Murree (with a mean annual P of 1765 mm) and Kotli (with a mean annual P of 1249 mm) are the wettest and driest weather stations, respectively, inside the TPPB, and Balakot (1731 mm) and Srinagar (764 mm) are the wettest and driest climate stations, respectively, in the OPPB.

Table 1: Statistics of climate stations for the period of 1961-1990 in the Jhelum River basin.

	Station	Elevation	Annual precipitation (mm)	Mean T_{max}(°C)	Mean T_{min}(°C)	T_{mean}(°C)
	TPPB					
1	Jhelum	287	860	30.51	16.54	23.53
2	Kotli	614	1249	28.41	15.75	22.08
3	Plandri	1402	1459	22.73	11.30	17.01
4	Rawalakot	1676	1398	21.87	10.25	16.06
5	Murree	2213	1765	16.57	8.97	12.77
6	Garidopatta	814	1586	26.06	12.58	19.32
7	Muzaffarabad	702	1418	27.34	13.54	20.44
8	Balakot	995	1731	25.04	12.01	18.53
	Average	**1088**	**1341**	**26.21**	**13.50**	**19.85**
	OPPB					
9	Naran	2362	1217	11.14	1.15	6.14
10	Kupwara	1609	1314	19.64	6.30	12.93
11	Gulmarg	2705	1574	11.73	1.98	6.81
12	Srinagar	1587	764	19.76	7.37	13.52
13	Qazigund	1690	1395	19.11	6.57	12.80
14	Astore	2168	564	15.48	4.06	9.77
	Average	**2020**	**1139**	**16.88**	**5.04**	**10.93**
	Whole basin	**1487**	**1202**	**19.79**	**7.68**	**13.72**

T_{mean}=mean temperature, T_{max}=maximum temperature, T_{min}=minimum temperature

Fig. 3 and Table 1 show that the TPPB is hotter than the OPPB, with mean temperatures of 19.85 °C in the former and 10.93 °C in the latter. The whole basin has a mean temperature of 13.72 °C. The hottest month is June in the TPPB and July in the OPPB as well as in the whole basin. January is the coldest month in both sub-basins. As for climate stations, Kotli (with a mean temperature of 22 °C) and Naran (with a mean temperature of 6.14 °C) are the hottest and coldest stations, respectively, in the study area.

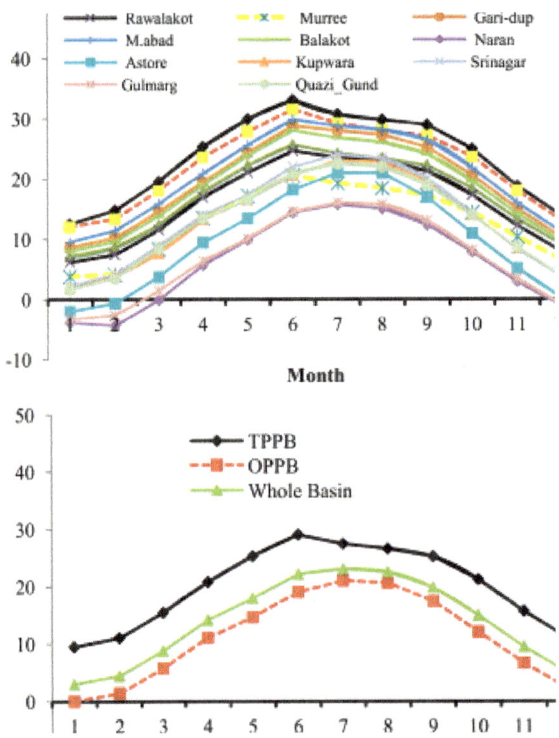

Figure 3: Distribution of mean monthly temperature at (a) all the weather stations (b) in the TPPB, OPPB, and the entire Jhelum basin for 1961–1990 (the baseline period).

METHODOLOGY

Description of SDSM

SDSM is a combination of the Stochastic Weather Generator (SWG) and Multiple Linear Regression (MLR). In MLR, statistical/empirical relationships between NCEP predictors and predictands are established, which leads to the production of some regression parameters. These regression parameters, along with NCEP or GCM predictors, are then used by SWG to generate a daily time series (Wilby et al., 2002 and Liu et al., 2009).

In SDSM, a combination of different indicators such as the correlation matrix, partial correlation, *P* value, histograms, and scatter plots can be used to select some suitable predictors through a multiple linear regression model. Multiple co-linearity, or the correlations between the predictors, must be considered during the selection of predictors. This can mislead during calibration of model. For example, if there is high co-linearity among the predictors during the calibration, there will be high value of coefficient of determination (R^2). This high value of R^2 can be due to co-linearity not due to correlation between predictand and predictors. Ordinary Least Squares (OLS) and Dual Simplex (DS) are two optimization methods. OLS was selected for this study, which is faster than DS and produces results comparable with DS (Huang et al., 2011). Three kinds of sub-models—monthly, seasonal, and annual—are available in SDSM to establish statistical/empirical relationships between predictands and predictors. In the monthly sub-model, 12 regression equations are developed—one for each month—during the calibration process. SDSM has two kinds of sub-models: conditional and unconditional. The conditional sub-model is used for dependent variables such as P and evaporation, and the unconditional sub-model is used for independent variables such as temperature during calibration (Wilby et al., 2002, Chu et al., 2010 and Mahmood and Babel, 2012).

Unlike temperature, P data is usually not normally distributed. So, P data is made normal by SDSM before it is used in regression equations (Khan et al., 2006). For example,Khan et al. (2006), Huang et al. (2011), and Mahmood and Babel (2012) have used the fourth root to turn P data into normal before using it in a regression equation. For the development of SDSM, two kinds of daily time series, observed and NCEP, are required. SDSM simulates daily time series as outputs, forcing NCEP or HadCM3 predictors (Huang et al., 2011). A full mathematical detail is presented in (Wilby et al., 1999).

Selection of Predictors

In statistical downscaling techniques, the first and most important process is the screening of large-scale variables (Wilby et al., 2002 and Huang et al., 2011). Four main indicators—explained variance, the correlation matrix, partial correlation, and *P* value— are used during the selection of predictors in SDSM. A combination of partial correlation and *P* value is generally used for the screening process, as it has been in studies like (Wilby et al., 2002, Chu et al., 2010, Hashmi et al., 2011, Huang et al., 2011 and Souvignet and Heinrich, 2011). The selection of predictors is more subjective and depends on the user's judgment. The main points, which must be considered during the selection of predictors, are multiple co-linearity among the predictors. In SDSM, the selection of the first and most prominent predictor is relatively easy and can be done with a simple correlation matrix. However, the selection of the second, third, fourth predictor and so on is more subjective. Therefore, in this study, a quantitative procedure used by Mahmood and Babel (2012) was applied for screening the predictors. In this procedure, a combination of the correlation matrix, partial correlation, *P* value, and percentage reduction in partial correlation was used. A set of 26 NCEP predictors, as described inTable 2, was regressed against each of predictands (T_{max}, T_{min}, and *P*) and some suitable predictors were screened out for calibration process. Multiple co-linearity was highly considered during screening process.

Table 2: NCEP predictors used in the screening process of SDSM

Predictors		Description	Predictors		Description
1	p_f	Surface airflow strength	14	r500	500 hPa relative humidity
2	p_u	Surface zonal velocity	15	p8_f	850 hPa airflow strength
3	p_v	Surface meridional velocity	16	p8_u	850 hPa zonal velocity
4	p_z	Surface vorticity	17	p8_v	850 hPa meridional velocity
5	p_th	Surface wind direction	18	p8_z	850 hPa vorticity
6	p_zh	Surface divergence	19	p8th	850 hPa wind direction
7	rhum	Surface relative humidity	20	p8zh	850 hPa divergence
8	p5_f	500 hPa airflow strength	21	r850	850 hPa relative humidity
9	p5_u	500 hPa zonal velocity	22	p500	500 hPa geopotential height
10	p5_v	500 hPa meridional velocity	23	p850	850 hPa geopotential height
11	p5_z	500 hPa vorticity	24	temp	Mean temperature at 2 m height
12	p5th	500 hPa wind direction	25	shum	Surface specific humidity
13	p5zh	500 hPa divergence	26	mslp	Mean sea level pressure

Calibration and validation

According to the available data, three data sets of periods 1961–1990, 1969–1990, and 1970–1990 were used for calibration, and a dataset from the period 1991–2000 for the validation of SDSM in this study. The monthly sub-model was used during calibration, utilizing the selected NCEP predictors for each of the predictands (T_{max}, T_{min}, and P), at each site. The unconditional sub-model was used for temperature without any transformation, and the conditional sub-model was set for P with fourth root transformation, before using it in regression equations. To check the performance of SDSM during calibration, two statistical indicators—the percentage of explained variance and standard error—were used in the present study, as they also have been in studies like Huang et al. (2011) and Wilby et al. (2002).

Temporal

For validation, Tmax, Tmin, and P datasets were simulated using NCEP, A2, and B2 predictors for 1991–2000. These simulated datasets were compared with observed datasets by calculating the correlation coefficient (R), the coefficient of determination (R^2), the root mean square error (RMSE), and the mean (μ). These indicators were first calculated for each weather station, and then, mean values were calculated from all the weather stations. The simulated data was also compared graphically with observed data to examine the variations in observed data captured by simulated data (Chu et al., 2010 and Huang et al., 2011).

Spatial

SDSM was also validated by comparing the spatial distribution of mean annual observed data with simulated data. For this purpose, spatial maps for each variable (T_{max}, T_{min}, and P) were built by converting the mean annual point data into raster data by the Kriging method, using ArcGis 9.3.

Description of Kriging Method

Kriging is an advanced, computationally intensive, geostatistical interpolation method (Buytaert et al., 2006) that takes care not only the distance but also the degree of variation among known data points when estimating values in unknown areas. The first step in kriging interpolation is to inspect the data in order to identify the spatial structure, using empirical semivariogram or covariance. A semivariogram is a diagram that represents the connection between the distance (the distance between all the pairs of available data points) and semivariance, and covariance, in statistics, shows the strength of the correlation between two or more random variables. A basic principle on which the semivariogram bases is the points (stations, things etc.) closer to each other are more alike than the points that are farther apart.

The next step is to compute the weights for points from the model variogram. These weights are based on the distance between the observed (measured) points and the prediction points, and also on the overall spatial relationships among the measured values surrounding the prediction location. In kriging, different statistical indicators such as Absolute Mean Error (AME), Root Mean Square Error (RMSE), Average Standard Error (ASE) and Root Mean Square Standardized Error (RMSSE) etc. and different graphs such as graph between predicted and measured values, error graph and QQ-plot are used to check the performance of model (Pokhrel et al., 2013). Mathematical equations of these indicators are described in (Dai et al., 2014). In the present study, AME and RMSSE were used to check the performance of model. RMSSE is the ratio of RMSE to ASE. RMSSE and AME should be close to 1 and 0 respectively for satisfactory results. After satisfactory model performance, finally the maps were created for the study area. Mathematical detail about kriging method is described in Pokhrel et al. (2013).

Future Changes

After satisfactory calibration and validation, Tmax, Tmin, and P data was simulated for three spells—the 2020s (2011–2040), the 2050s (2041–2070), and the 2080s (2071–2099). The datasets of each of these periods were compared to a baseline period to analyze future changes in the basin. In this study, the period from 1961 to 1990 was taken as the baseline period because this period has been used in the majority of climate change studies across the world (Huang et al., 2011). A 30-year period is also considered long enough to define local climate because it is likely to include dry, wet, cool, and warm periods. The IPCC also recommends such a length (of 30 years) for the baseline period (Gebremeskel et al., 2005).

For better explanation, the temporal and spatial changes are presented graphically in this paper, as they have also been in other studies (Chu et al., 2010 and Huang et al., 2011). Since Gulmarg and Qazigund weather stations lack data starting from 1961, the data for these stations was projected back by SDSM using NCEP predictors.

RESULTS AND DISCUSSION

Screening of Predictors

Table 3 shows the predictors selected after screening, along with their mean absolute partial correlation (P.r) for T_{max}, T_{min}, and P at a significance level of 0.05. It was determined that temperature at 2 m height (*temp*) was a super predictor for Tmax and Tmin in both the OPPB and TPPB sub-basins. A super predictor is the most prominent predictor and it displays maximum correlation with a predictand. For P, surface specific humidity (*shum*) and meridional wind velocity at 500 hPa (*p5_v*) were the most dominant predictors in the TPPB and OPPB respectively, for almost all weather stations. The selected predictors for P also expressed a logical sense because P depends mostly on water vapors (humidity) and wind direction. The selected predictors for the present study also match with the predictors used in several other studies such as Wilby et al. (2002),Chu et al. (2010), Huang et al. (2011), Hashmi et al. (2011), and Mahmood and Babel (2012).

Table 3: Selected predictors and their mean absolute partial correlation coefficients during the screening process of SDSM.

	Tmax	Abs P.r	Tmin	Abs P.r	Precipitation	Abs P.r
TPPB	**temp**	0.73	**temp**	0.82	**Shum**	0.21
	r500	0.22	rhum	0.18	p5_v	0.14
	P8_z	0.19	r500	0.12	r850	0.10
			P8_z	0.32	p8_v	0.15
					Mslp	0.12
OPPB	**temp**	0.76	**temp**	0.79	**p5_v**	0.23
	p_u	0.38	p_u	0.37	p8_v	0.14
	p_z	0.32	p_z	0.32	p5_z	0.11
	P8_z	0.25	p8_z	0.26	p8_z	0.08
	r500	0.17				

Words in bold are super predictors; *Abs P.r* is the absolute partial correlation coefficient

Calibration of SDSM

The explained variance (E), used as a performance indicator, ranged from 60% to 72% for T_{max}, 67% to 85% for T_{min}, and 8% to 32% for *P*. The standard error (SE) for T_{max},T_{min}, and *P* was 4.6 °C, 3.4 °C, and 0.45 mm/day respectively. These results are satisfactory and comparable with the results of some previous studies like Wilby et al. (2002), Nguyen (2005), Liu et al. (2009), Souvignet and Heinrich (2011), and Huang et al. (2011). In this study, the *E* values for *P* were much lower than the *E* values for T_{max} andT_{min}. Since *P* is a heterogeneous climate variable and difficult to simulate, the *E* of *P* is more likely to be lower than 40%, while the *E* of temperature is more likely to be greater than 70% (Wilby et al., 2002).

Validation of SDSM

Tabular

For the validation of SDSM, three datasets were simulated by forcing the NCEP, A2, and B2 predictors, along with their calibrated parameters, for the period of 1991–2000, and for each of the local scale variables (T_{max}, T_{min}, and *P*). The performance indicators (*R*,R^2, RMSE, and μ) were calculated using daily, monthly, and seasonal time series from observed and simulated T_{max}, T_{min}, and *P*. These are described in Table 4, Table 5 and Table 6 respectively.

Table 4: Validation (1991-2000) of SDSM for Tmax, using daily, monthly, and seasonal time series in the Jhelum River basin.

		R (%)		R^2 (%)		*RMSE* (°C)		μ (°C)	
		Range	Mean	Range	Mean	Range	Mean	Range	Mean
Daily	Obs							10.2–30.41	21.04
	NCEP	84–92	87.48	71–85	76.5	3.08–3.98	3.52	11.5–30.55	21.02
Monthly	Obs							10.12–30.01	21.00
	NCEP	95–98	96.48	89–97	93.1	1.55–2.41	1.88	11.47–30.52	20.98
	A2	94–98	95.96	88–96	92.1	1.70–3.33	2.12	11.29–30.49	21.09
	B2	93–98	00.96	89–96	92.0	1.73–3.25	2.17	11.32–30.52	21.16
Seasonal	Obs							10.12–30.01	21.00
	NCEP	96–99	97.67	92-98	95.4	1.06–2.04	1.40	11.47–30.52	20.99
	A2	95–99	97.65	92-98	95.3	1.11–2.75	1.47	11.29–30.49	21.09
	B2	95–99	00.97	91-98	94.6	1.22–2.74	1.59	11.32–30.52	21.16

R=Correlation coefficient, R^2=Coefficient of determination, *RMSE*=Root mean square error, and μ=mean

Table 5: Validation (1991–2000) of SDSM for T_{min}, using daily, monthly, and seasonal time series in the Jhelum River basin.

		R (%)		R^2 (%)		*RMSE* (°C)		μ (°C)	
		Range	Mean	Range	Mean	Range	Mean	Range	Mean
Daily	Obs							1.54–16.85	9.15
	NCP	88–95	92.27	77–89	85.2	2.32–3.26	2.642	1.18–16.51	8.98
Monthly	Obs							1.48–16.82	9.11
	NCEP	93–99	97.49	87–98	95.1	1.09–2.41	1.45	1.13–16.47	8.94
	A2	93–98	96.81	86–97	93.7	1.23–3.15	1.75	1.08–16.51	9.05
	B2	93–98	96.59	86–97	93.3	1.26–3.16	1.82	1.10–16.52	9.06
Seasonal	Obs							1.48–16.82	9.11
	NCEP	95–99	98.37	90–99	96.8	0.56–2.00	1.045	1.13–16.47	8.94
	A2	94–99	98.20	89–99	96.4	0.60–2.50	1.182	1.08–16.51	9.05
	B2	94–99	97.90	88–99	95.9	0.63–2.58	1.268	1.10–16.52	9.06

R=Correlation coefficient, R^2=Coefficient of determination, *RMSE*=Root mean square error, and μ=mean

Table 6: Validation (1991–2000) of SDSM for precipitation, using daily, monthly, and seasonal time series in the Jhelum River basin.

		R (%)		R^2 (%)		*RMSE* (mm)		μ (mm)	
		Range	Mean	Range	Mean	Range	Mean	Range	Mean
Daily	Obs							1.74–5.69	3.92
	NCEP	10–39	25.045	8–15	9.6	1.9–5.5	3.65	1.69–5.53	3.62
Monthly	Obs							53–174	119.77
	NCEP	43–85	62.70	21–73	42.0	24–82	50.67	50–166	110.41
	A2	29–72	51.40	18–51	28.1	43–12	86.64	50–154	109.87
	B2	25–68	43.91	13–46	22.3	47–13	89.89	48–149	105.13
Seasonal	Obs							53–174	119.30
	NCEP	50–93	70.90	25-87	61.7	17–55	35.40	50–165	110.21
	A2	40–81	66.00	30–67	55.2	27–85	53.71	50–154	109.87
	B2	26–77	55.27	22–60	50.8	33–95	57.88	48–149	105.13

R=Correlation coefficient, R^2=Coefficient of determination, *RMSE*=Root mean square error, and μ=mean

These tables show that the R^2 values for T_{max} and T_{min} were above 76%, 92% and 94% as calculated from daily, monthly, and seasonal time series respectively. The R^2 values for *P* were above 9%, 22%, and 50% as calculated from daily, monthly, and seasonal time series respectively. The results of Tmax and Tmin were better than the results of *P* with respect to all the indicators obtained from the daily, monthly, and seasonal time series, and Tmin results were slightly better than Tmax results. Although the R^2 values obtained by the daily time series (ranging 8–15%) were much lower than those obtained by monthly (21–73%) and seasonal (25–87%) time series when using NCEP predictors, these results are comparable with a previous study (Huang et al., 2011).

The main reason for lower values of R^2 for daily *P* is the occurrence/amount of *P*, which is a stochastic process. Therefore, simulation of *P* is always a difficult task (Huang et al., 2011). Several previous studies (Khan et al., 2006, Fealy and Sweeney, 2007 and Huang et al., 2011) have shown worse results obtained from daily time series, compared to results obtained from monthly and seasonal time series. SDSM shows better applicability using monthly and seasonal P time series, relative to the daily time series. It was observed that the results obtained from NCEP predictors were better than A2 and B2 predictors, and the simulated data from A2 gave slightly better results than B2. Since SDSM is calibrated with NCEP predictors, the calibrated parameters reflect some biases when the model is run with A2 and B2 predictors.

GRAPHICAL

The main purpose of graphical validation was to investigate the intra-annual variations in the TPPB, OPPB, and in the entire basin, as has been done in several studies (Wilby et al., 2002, Gebremeskel et al., 2005, Zhao and Xu, 2008, Chu et al., 2010 and Huang et al., 2011). For this, three datasets for Tmax, Tmin, and P were simulated by forcing the NCEP, A2, and B2 predictors into SDSM and the mean monthly, seasonal, and annual values were calculated and compared with the observed datasets. However, in the present study, only P comparisons are shown graphically in Fig. 4, which illustrates well the intra-annual variations in the TPPB, OPPB, and in the whole basin. The results ofT_{max} and T_{min} are not shown graphically in the present study because the results of $T_{max}$and T_{min} (Table 4 and Table 5) were quite satisfactory and much better than *P*. In addition, *P* is a heterogeneous variable and it is difficult to capture its variations. Furthermore, it is the most important climate variable among all climate variables

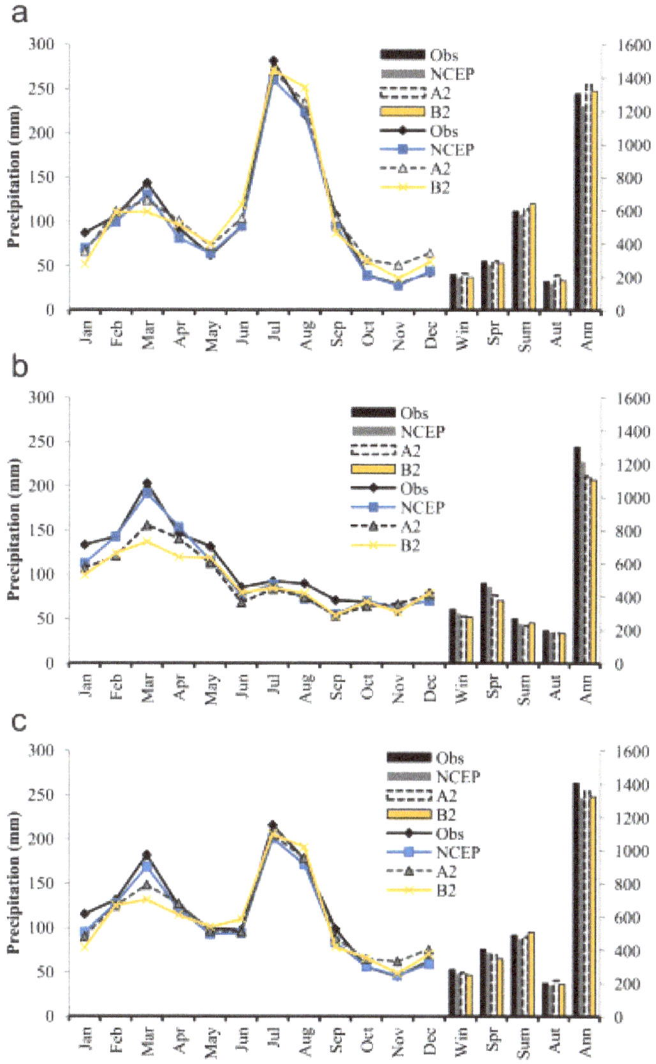

Figure 4: Observed versus simulated precipitation in the (a) TPPB, (b) OPPB, (c) whole Jhelum basin, for the validation (1991–2000) of SDSM. *Win, Spr, Sum, Aut and Ann* describe winter (December–February), spring (March–May), summer (June–August), autumn (September–November), and annual (January–December), respectively.

Fig. 4(a) shows that in the TPPB, the big peak occurs in July–August (the monsoon season), and this was reasonably well simulated by all three datasets (NCEP, A2, and B2). On the other hand, the small peak, occurring in March, was underestimated by all three datasets. On the whole, the patterns were well followed by all three data sets.

The mean seasonal precipitations simulated from NCEP and A2 predictors were underestimated and overestimated by 7% and 6% respectively, except in spring (March–May). B2 showed underestimations in winter (December–February) and spring and overestimations in summer (June to August) and autumn (September–November). Overall, B2 underestimated with an average *P* of 0.89%. The mean annual precipitation from NCEP was underestimated by 5.7% and from A2 and B2, it was overestimated by 4% and 1.2% respectively.

Fig. 4(b) indicates that mean monthly *P* was not well simulated by SDSM in the OPPB as it was in the TPPB. Nonetheless, the patterns were followed by all three data sets. The NCEP, A2 and B2 datasets underestimated P in all seasons, with an average seasonal *P*of 6.9%, 11.8%, and 12% respectively, and an average annual P of 6.6%, 13% and 15% respectively.

Fig. 4(c) shows that the big peak was well simulated by all three data sets. However, the small peak was well underestimated by the model. The average seasonal *P*, as simulated by NCEP, A2 and B2, was underestimated by about 6.5%, 2.3% and 6.2% respectively. NCEP, A2, and B2 datasets underestimated the mean annual *P* by 6%, 3.45% and 5.75% respectively.

On the whole, SDSM performed reasonably well in capturing the variations of observed mean monthly, seasonal, and annual *P*. The results from NCEP dataset were better than A2 and B2 because the model was calibrated with NCEP dataset. Nonetheless, results from A2 and B2 datasets were quite comparable with NCEP. These results are also comparable with other studies (Chu et al., 2010 and Huang et al., 2011)

Spatial Validation

To check the capability of SDSM in simulating spatial changes in the study area, three groups of maps (Fig. 5) for each variable (Tmax, Tmin, and P) were created by converting mean annual point data into raster data with the kriging interpolation method. As can be seen, each group has four maps; one was produced by using observed data and the other three by simulated data—one each from NCEP, A2 and B2 predictors. Two indicator, Absolute Mean Error (AME) and Root Mean Square Standardized Error (RMSSE) were utilized to check the performance of kriging method and results are described in Table 7. AME values for T_{min}, T_{max} and *P* ranged between 0.011 °C and 0.022 °C, 0.018 °C and 0.23 °C, and 7.2 mm and 11.6 mm respectively, in observed, NCEP, A2 and B2 maps, and RMSSE values for T_{min}, T_{max} and *P* were 0.99–1.05, 0.84–0.95, and 1.1–1.2 respectively, which were quite satisfactory.

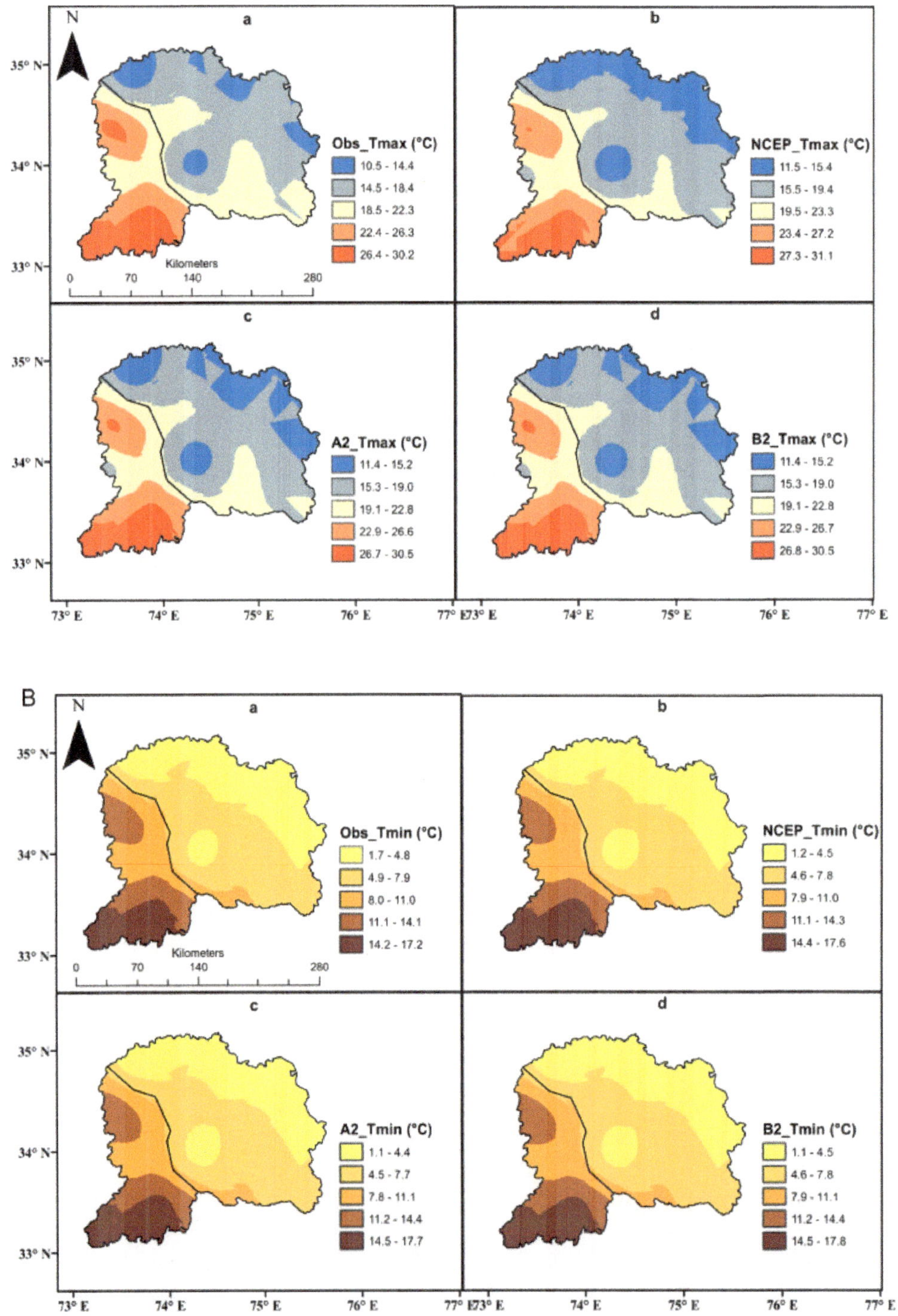

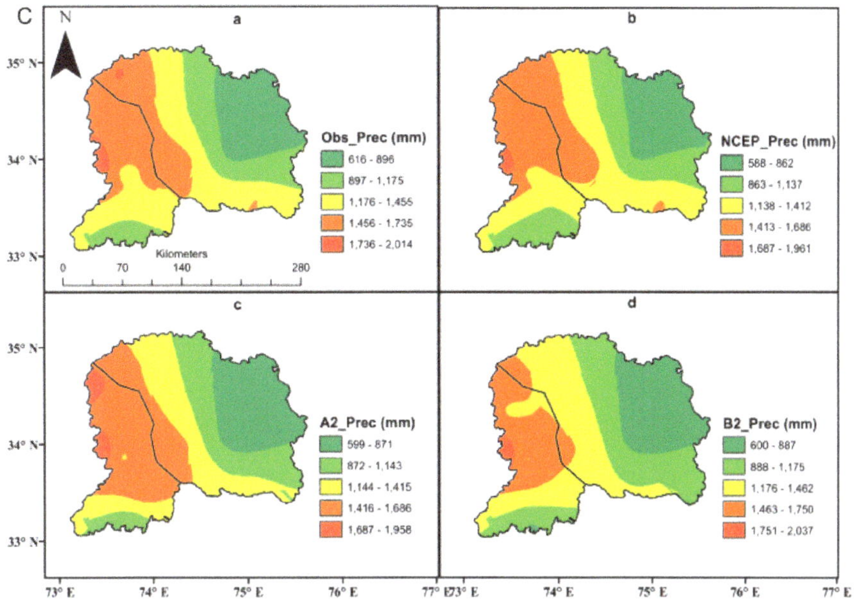

Figure 5: Spatial distributions of observed versus simulated (A) T_{max}, (B) T_{min}, and (C) Precipitation for the validation (1991–2000) of SDSM in the Jhelum River basin.

Table 7: AME and RMSSE between measured and predicted T_{max}, T_{min}, and Precipitation for the period of 1991–2000, during kriging interpolation, in the Jhelum River basin

	Obs	NCEP	A2	B2
$\boldsymbol{T_{min}}$				
AME (°C)	0.012	0.022	0.011	0.016
RMSSE	0.990	1.030	1.050	1.040
$\boldsymbol{T_{max}}$				
AME(°C)	0.022	0.023	0.018	0.019
RMSSE	0.840	0.920	0.940	0.950
Precipitation				
AME(°C)	7.300	11.600	7.200	7.900
RMSSE	1.150	1.100	1.200	1.160

AME=Absolute Mean Error, RMSSE=Root Mean Square Standardized Error

It was seen that the spatial variations in observed mean annual temperature (T_{max} andT_{min}) and *P* were well reflected by all three downscaled datasets, as is shown by Fig. 5. Spatial distributions of temperature and precipitation showed that temperature decreases from south to north in accordance with the increase in elevation, and *P*decreases from west to east. Spatial variability was reasonably better simulated by NCEP data than by A2 and B2 data, and it was better captured in the TPPB than in the OPPB. Nevertheless, the results from A2 and B2 were comparable with NCEP.

Thus, it was concluded that SDSM is more credible in simulating mean monthly, seasonal and annual future changes in Tmax, Tmin, and P rather than in simulating daily future changes in the Jhelum basin. The calibrated SDSM, with NCEP predictors, can be used to predict future changes in T_{max}, T_{min}, and *P* better with respect to the baseline period, than with A2 and B2 predictors.

TEMPORAL AND SPATIAL FUTURE CHANGES IN TEMPERATURE AND PRECIPITATION

After successful calibration and validation, daily time series of T_{max}, T_{min}, and *P* were simulated for the 2020s, 2050s and 2080s using A2 and B2 scenarios. Then, the calculated mean monthly, seasonal and annual T_{max}, T_{min}, and *P* data from the daily time series was compared with the baseline data to analyze future changes in the 2020s, 2050s and 2080s (Chu et al., 2010 and Huang et al., 2011).

Temporal Changes

Fig. 6 (A) shows mean annual and seasonal Tmax changes in the TPPB, OPPB, and in the whole Jhelum basin under A2 and B2 scenarios in the 2020s, 2050s and 2080s. Mean annual T_{max}, under both scenarios, is projected to increase by 0.3–1.2 °C in the TPPB, 0.14–0.74 °C in the OPPB, and 0.24–1.02 °C in the whole Jhelum basin in the 21st century. Kazmi et al. (2015) and Rajbhandari et al. (2015) also showed increasing trends over the whole Pakistan. However, Kazmi et al. (2015)

conducted study for the period of 2011–2013 and did not include the whole Jhelum basin because most of the Jhelum basin is situated in Jammu and Kashmir, and Rajbhandari et al. (2015) found out the projected changes over the Indus basin with Regional Climate Model. Similarly, all seasons show positive (increasing) changes under both scenarios in all three future periods (2020s, 2050s, and 2080s) except in winter in the OPPB. Under A2 scenario, in the TPPB and OPPB, the most affected seasons are spring and autumn, with a projected increase of 1.7 °C and 1.14 °C in the 2080s respectively. In the whole basin, spring, with a 1.4 °C increase in the 2080s, shows the most distinct change, relative to the other seasons. The same kinds of future changes are found under B2 but are relatively smaller in magnitude than the changes under A2.

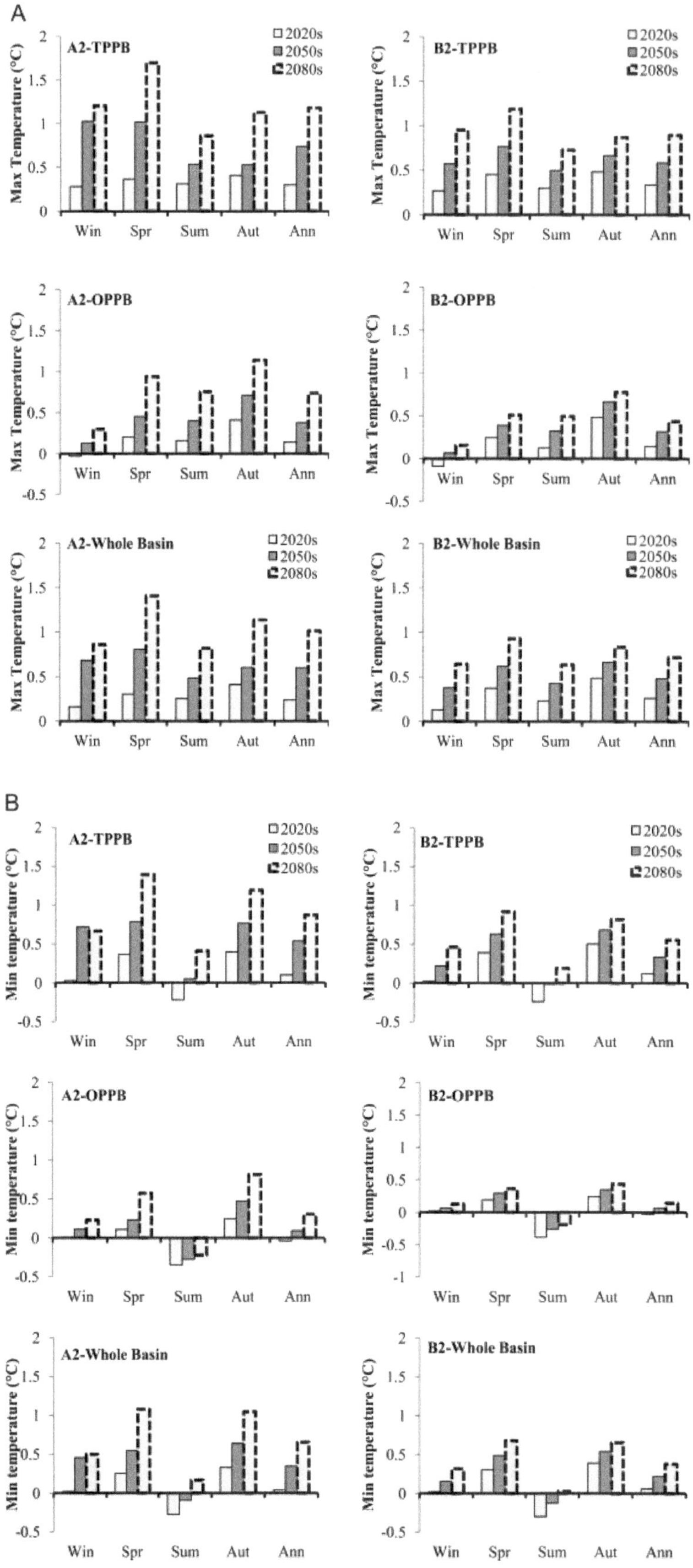

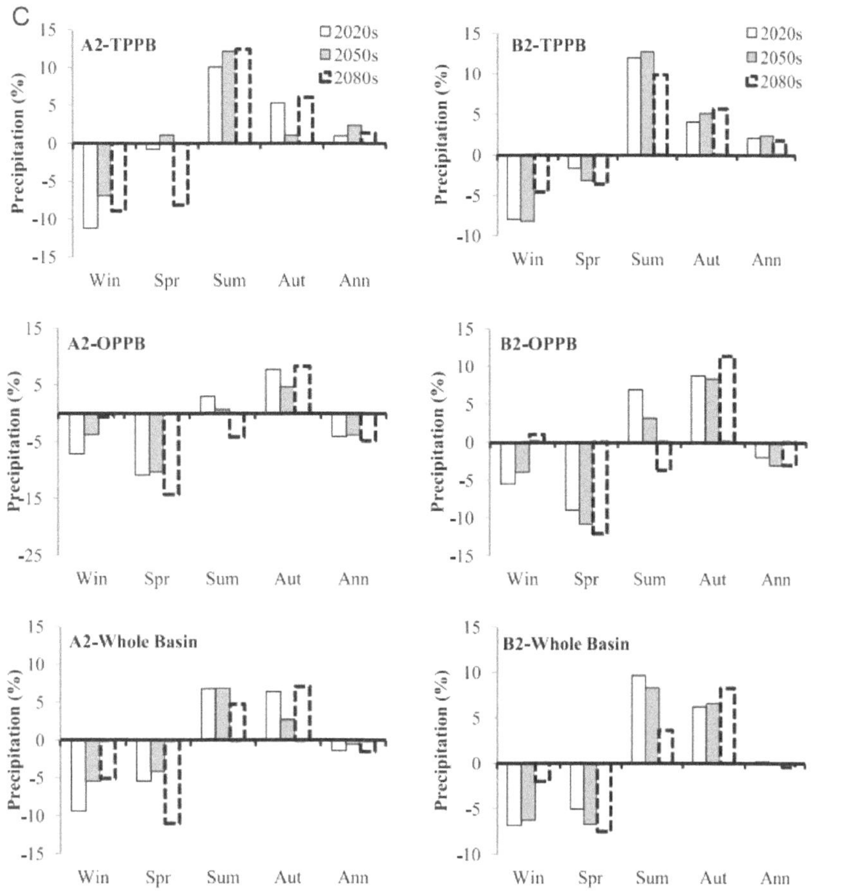

Figure 6: Future temporal changes in (A) T_{max}, (B) T_{min}, and (C) Precipitation, relative to the baseline period (1961–1990), under A2 and B2 in the TPPB, OPPB, and the whole Jhelum River basin.

Fig. 6 (B) illustrates the mean annual and seasonal changes in Tmin in the TPPB, OPPB, and in the whole basin under both scenarios in the three future periods. Under both scenarios, mean annual Tmin is predicted to rise by 0.12–0.87 °C in the TPPB, −0.03 to 0.31 °C in the OPPB, and 0.04–0.65 °C in the whole basin. These results are also supported by Kazmi et al. (2015) and Rajbhandari et al. (2015). In the case of seasonal changes, spring is affected more in the TPPB and autumn in the OPPB. The same is also the case of T_{max}. Future changes in T_{min} are predicted to be larger in magnitude under A2 than B2, reflecting the same pattern for Tmax projections.

Fig. 6 (C) presents the percentage changes in mean annual and seasonal P in the TPPB, OPPB, and in the whole basin under both scenarios. Mean annual P is estimated to increase by 1–3% in the TPPB and decrease by 2–5% in the OPPB, under both scenarios, with an overall decrease in the whole basin. As for seasonal changes in the TPPB basin, summer (the monsoon season in the basin) and autumn P show a definite increase of about 1–3% under A2 and 1–6% under B2 in the future, with the exception of the 2080s under B2. In contrast, in the TPPB, winter P is projected to decrease by 7–12% and 5–8% under A2 and B2, respectively. On the whole, in the TPPB, summer and autumn are more likely to receive increased P in the future with respect to the baseline values, and winter as well as spring are likely to receive less P in the future, relative to the baseline values. These results are also supported by Rajbhandari et al. (2015). They conducted study for winter and summer monsoon and explored that summer precipitation is projected to increase and winter decrease.

In the OPPB, only autumn presents a definite increase in *P* (by 5–12%) in all three periods under both scenarios. In summer, the changes are positive in the 2020s and 2050s but negative in the 2080s. *P* in spring (the peak season in the OPPB) is projected to decrease by 9–15% under both scenarios.

Therefore, it can be concluded that the peak in the OPPB is likely to go towards dryness in the future. In the whole basin, under both A2 and B2 scenarios, summer and autumn may be wetter than before, and winter and spring may be dryer, as compared to the baseline period. Since most of the meteorological stations are installed in the valleys and not on the high altitudes (on Peaks) and the number of stations available inside the basin is insufficient, as shown in Fig. 1, temperature and precipitation changes in the basin might be different if the high altitudes will be considered.

Spatial Changes

In this study, the mean annual changes in T_{max}, T_{min}, and *P* were also analyzed spatially in the Jhelum River basin under A2 and B2 scenarios, with the help of kriging method, as shown in Fig. 7. The AME and RMSSE between measured and predicted *Tmin*, T_{max}, and*P* are presented in Table 8. AME ranged from 0.0002 to 0.09 °C for Temperature ($T_{max}$and T_{min}) and 0.34 to 1.2 °C, respectively, and RMSSE ranged from 0.74 to 1.004 for Temperature (T_{max} and T_{min}) and 0.92 to 1.005,respectively, under A2 and B2 in 2020s, 2050s and 2080s in the Jhelum River basin.

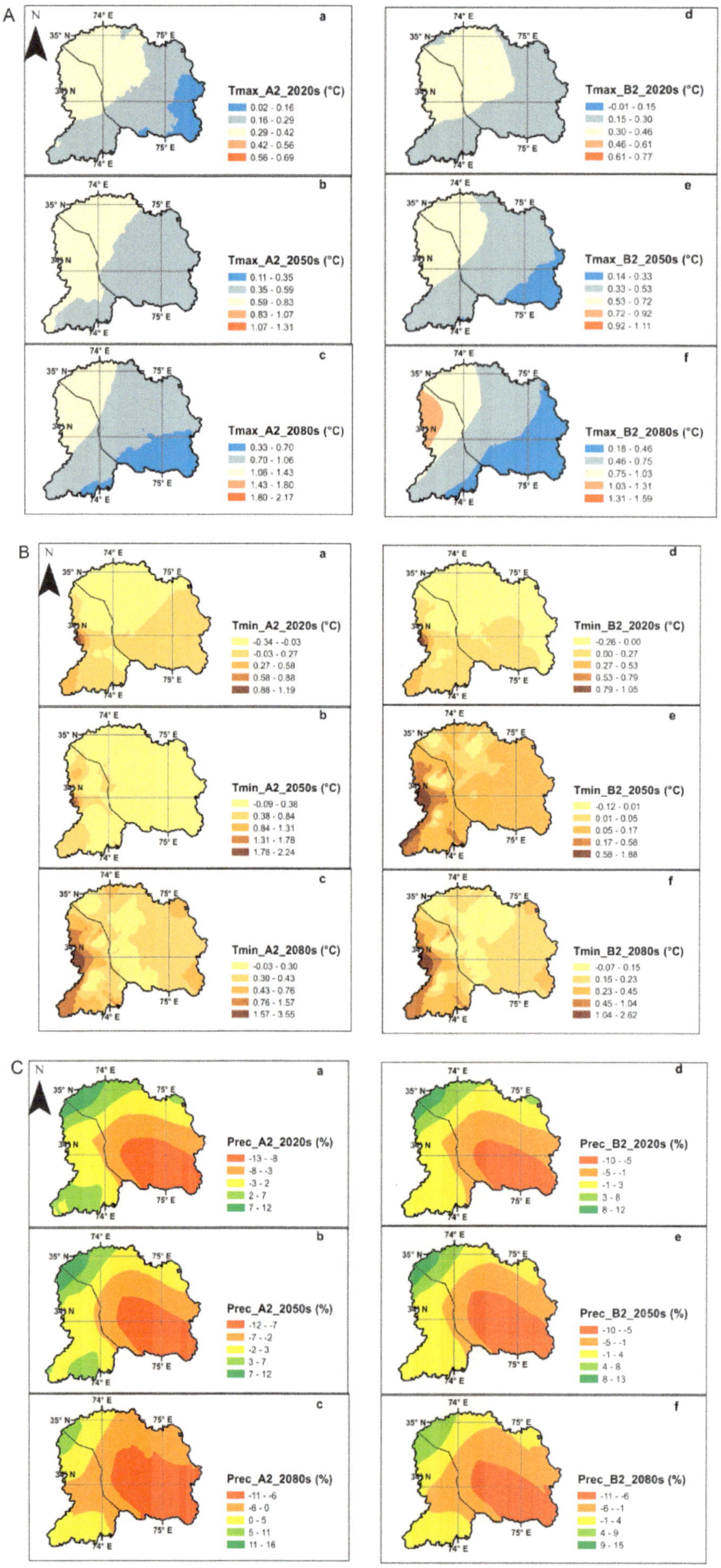

Figure 7: Future spatial changes in (A) T_{max} (°C), (B) T_{min} (°C), and (C) Precipitation (%) under scenarios A2 and B2, in the Jhelum River basin.

Table 8: AME and RMSSE between predicted Tmax, Tmin, and Precipitation for the period of 2020s, 2050s and 2080s, during kriging interpolation, in the Jhelum River basin

	2020s		2050s		2080s	
T_{min}	A2	B2	A2	B2	A2	B2
AME (°C)	0.0300	0.0100	0.0900	0.0500	0.0120	0.0800
RMSSE	0.9000	0.8611	0.9155	0.9259	0.8750	0.7595
T_{max}						
AME(°C)	0.0200	0.0200	0.0002	0.0200	0.0056	0.0240
RMSSE	1.0043	1.0909	0.9211	0.9333	0.9574	0.9189
Precipitation						
AME(°C)	0.6100	0.4300	0.4500	0.3400	1.2000	0.7800
RMSSE	0.9281	0.9455	0.9400	0.9180	1.0047	0.9644

AME=Absolute Mean Error, *RMSSE*=Root Mean Square Standardized Error

Fig. 7 (A) shows the spatial distribution of mean annual changes in T_{max} in the 2020s, 2050s and 2080s under A2 and B2 scenarios, relative to the baseline. The northwest part of the basin is projected to be the most affected in all three future periods, with a definite increase in T_{max}. The southeast part of the basin is predicted to be the least affected. T_{max} only shows a small increase in this region. The results also indicate that the changes are mostly increasing moving from southeast to northwest side under both scenarios.

Fig. 7 (B) presents the spatial distribution of mean annual changes in T_{min} in the three future periods under A2 and B2 scenarios, relative to the baseline. Most of the basin shows decrease in T_{min} under A2 in 2020s and 2050s and most of the basin presents increase in 2080s. Under B2, almost half of the basin gives indication about decrease in T_{min} in 2020s. However, most of the basin shows increase T_{min} in 2050s and 2080s.

Fig. 7 (C) provides the spatial distribution of percentage changes in mean annual *P*, as compared to the baseline under A2 and B2 scenarios in all three future periods. As can be seen, on the whole, the changes are spread between −13% and 16% over the whole basin under both scenarios, in all three future periods. The simulated decrease in P occurs in eastern part and increases in the west of the basin. It can be seen that almost half of the basin shows decreasing *P* in the 2020s. However, in the 2080s, it is projected that more than half of the basin will have decreased *P* under both scenarios. Under both scenarios, similar kinds of spatial distribution patterns of mean annual precipitation changes can be seen, for all three future periods; however, the changes (positive or negative) are higher under A2 than under B2.

CONCLUSIONS

In this study, a well-known decision support tool, the Statistical Downscaling Model (SDSM), was applied to analyze the temporal and spatial future changes in maximum temperature, minimum temperature and precipitation in the sub-basins, TPPB and OPPB, and in the whole Upper Jhelum River basin under A2 and B2 scenarios. The downscaling of these variables is very important in order to study the impacts of climate change on the hydrological cycle of the basin.

SDSM showed good capability in simulating Tmax and Tmin in daily, monthly, and seasonal time series, during the calibration and validation. However, it presented good results only for monthly and seasonal precipitation. However, only seasonal and annual future changes, with respect to baseline, were presented in this study. Seasonal data was simulated better than monthly and daily data.

Mean annual Tmax and Tmin were projected to increase in both parts of the Jhelum basin, the TPPB and the OPPB, under both scenarios, and in all three future periods. This simulated increase in temperature was higher under A2 than B2 in both the sub-basins and it was higher in the TPPB than OPPB. For seasonal changes in mean annual temperature (Tmax and Tmin), the maximum increase was projected for spring under both scenarios (A2 and B2) in the whole basin. Spring in the TPPB and autumn in the OPPB were projected to be the most affected seasons as far as rise in temperature was concerned.

Mean annual precipitation was predicted to increase by 1–3% in the TPPB and decrease by 2–5% in the OPPB under both scenarios, with an overall decrease in the whole basin. Summer and winter were the most affected seasons in the TPPB under both scenarios and in all three future periods. In the OPPB, autumn showed increase in precipitation, and spring showed decrease under both scenarios. As for the whole basin, summer and autumn were projected to receive more precipitation, and winter as well as spring to receive lesser amounts of precipitation in the future, as compared to baseline period.

The spatial distribution of mean annual *Tmax* showed a rise in almost all parts of the basin in the future periods, relative to the baseline period. However, northwestern parts of the basin were projected to face a higher increase than southeastern parts under both scenarios. T_{min} was projected to decrease in some parts of the basin but a majority of the basin will experience a rise in T_{min} under both scenarios especially in 2050s and 2080s, as compared to the baseline values. It was seen that almost half of the basin showed decreasing precipitation in the 2020s, but in the 2080s, decreasing precipitation was likely to be experienced in most parts of the basin under both scenarios. Both scenarios presented similar kinds of spatial distribution patterns of mean annual Tmax, Tmin, and precipitation changes in all three future periods, but with different magnitudes. The changes under A2 were observed higher than under B2.

ACKNOWLEDGMENTS

This study is a part of the first author's doctoral research work, conducted at the Asian Institute of Technology Thailand. The authors wish to acknowledge and offer gratitude to Pakistan Meteorological Department (PMD), the Water and Power Development Authority(WAPDA) of Pakistan, the India Meteorological Department (IMD), and the National Climate Data Center (NCDC), for providing important and valuable data for the research. Heartfelt gratitude is also extended to the Higher Education Commission (HEC) of Pakistan and AIT for providing financial support to the first author for his doctoral studies at AIT.

REFERENCES

1. Ahmad, I., Ambreen, R., Sun, Z., Deng, W., 2015. Winter-Spring Precipitation Variability in Pakistan. Am. J. Clim. Change 4 (1), 115–139. http://dx.doi.org/10.4236/ajcc.2015.41010.
2. Ahmad, N., Chaudhry, G.R., 1988. Irrigated agriculture of Pakistan S. Nazir, Lahore.Akhtar, M., Ahmad, N., Booij, M.J., 2008. The impact of climate change on the water resources of Hindukush–Karakorum–Himalaya region under different glaciercoverage scenarios. J. Hydrol. 355 (1-4), 148–163. http://dx.doi.org/10.1016/j.jhydrol.2008.03.015.
3. Akhtar, M., Ahmad, N., Booij, M.J., 2009. Use ofregional climate model simulationsas input for hydrological models for the Hindukush-Karakorum-Himalaya region. Hydrol. Earth Syst. Sci. 13 (7), 1075–1089. http://dx.doi.org/10.5194/hess-13-1075-2009.
4. Anandhi, A., Srinivas, V.V., Nanjundiah, R.S., Nagesh Kumar, D., 2007. Downscalingprecipitation to river basin in India for IPCC SRES scenarios using support vector machine. Int. J. Climatol. 28, 401–420. http://dx.doi.org/10.1002/joc.1529.
5. Archer, D.R., Fowler, H.J., 2008. Using meteorological data to forecast seasonalrunoff on the River Jhelum, Pakistan. J. Hydrol. 361 (1-2), 10–23. http://dx.doi.org/10.1016/j.jhydrol.2008.07.017.
6. Ashiq, M., Zhao, C., Ni, J., Akhtar, M., 2010. GIS-based high-resolution spatial interpolationof precipitation in mountain–plain areas of Upper Pakistan for regional climate change impact studies. Theor. Appl. Climatol. 99 (3), 239–253.http://dx.doi.org/10.1007/s00704-009-0140-y. . http://dx.doi.org/10.1007/s00704-009-0140-y.
7. Benestad, R.E., Hansen-Bauer, I., Chen, D., 2008. Empirical-statistical Downscaling.World Scientific, New York, p. 215.
8. Buytaert, W., Celleri, R., Willems, P., Bièvre, B.D., Wyseure, G., 2006. Spatial andtemporal rainfall variability in mountainous areas: a case study from the South Ecuadorian Andes. J. Hydrol. 329 (3–4), 413–421. http://dx.doi.org/10.1016/j.jhydrol.2006.02.031. CCCSN, 2012. Statistical Downscaling Input: HadCM3 Predictors: A2 and B2Experiments.
9. Chu, J., Xia, J., Xu, C.Y., Singh, V., 2010. Statistical downscaling of daily mean temperature,pan evaporation and precipitation for climate change scenarios inHaihe River, China. Theor. Appl. Climatol. 99 (1), 149–161. http://dx.doi.org/10.1007/s00704-009-0129-6.
10. Dai, F., Zhou, Q., Lv, Z., Wang, X., Liu, G., 2014. Spatial prediction of soil organicmatter content integrating artificial neural network and ordinary kriging inTibetan Plateau. Ecol. Indic. 45 (0), 184–194. http://dx.doi.org/10.1016/j.ecolind.2014.04.003.
11. Diaz-Nieto, J., Wilby, R.L., 2005. A comparison of statistical downscaling and climate change factor methods: impacts on low flows in the River Thames, United Kingdom. Clim. Change 69 (2), 245–268. http://dx.doi.org/10.1007/s10584-005-1157-6.
12. Fealy, R., Sweeney, J., 2007. Statistical downscaling of precipitation for a selection of sites in Ireland employing a generalised linear modelling approach. Int. J. Climatol. 27 (15), 2083–2094. http://dx.doi.org/10.1002/joc.1506.
13. Fowler, H.J., Blenkinsop, S., Tebaldi, C., 2007. Linking climate change modelling to impacts studies: recent advances in downscaling techniques for hydrological modelling. Int. J. Climatol. 27 (12), 1547–1578. http://dx.doi.org/10.1002/joc.1556.
14. Gagnon, S., Singh, B., Rousselle, J., Roy, L., 2005. An application of the statistical downscaling model (SDSM) to simulate climatic data for streamflow modelling in Québec, Canada. Water Resour. J. 30 (4), 297–314. ttp://dx.doi.org/10.4296/ cwrj3004297.
15. Gebremeskel, S., Liu, Y.B., de Smedt, F., Hoffmann, L., Pfister, L., 2005. Analysing the effect of climate changes on streamflow using statistically downscaled GCM scenarios. Int. J. River Basin Manag. 2 (4), 271–280. http://dx.doi.org/10.1080/
16. Ghosh, S., Mujumdar, P., 2008. Statistical downscaling of GCM simulations to streamflow using relevance vector machine. Water Resour. 31, 132–146.
17. Goyal, M.K., Ojha, C.S.P., 2010. Robust weighted regression as a downscaling tool in temperature projections. Int. J. Glob. Warm. 2 (3), 234–251. http://dx.doi.org/10.1504/IJGW.2010.036135.
18. Hashmi, M.Z., Shamseldin, A.Y., Melville, B.W., 2011. Comparison of SDSM andLARS-WG for simulation and downscaling of extreme precipitation events in awatershed. Stoch. Env. Res. Risk A 25 (4), 475–484. http://dx.doi.org/10.1007/s00477-010-0416-x.
19. Hay, L.E., Clark, M.P., 2003. Use of statistically and dynamically downscaled atmospheric model output for hydrologic simulations in three mountainous basins in the western United States. J. Hydrol. 282 (1–4), 56–75. http://dx.doi.org/ 10.1016/s0022-1694(03)00252-x.
20. Hay, L.E., Wilby, R.L., Leavesley, G.H., 2000. A comparison of delta change and downscaled GCM scenarios for three mountainous basins in the United States. J. American Water Resour. Ass. 36 (2), 387–397. http://dx.doi.org/10.1111/ j.1752-1688.2000.tb04276.x.
21. Huang, J., et al., 2011. Esti mation of future precipitation change in the Yangtze River basin by using statistical downscaling method. Stoch. Env. Res. Risk A. 25 (6), 781–792. http://dx.doi.org/10.1007/s00477-010-0441-9.
22. IPCC, 2013. Climate Change 2013: The Physical Scien ce Basis. Contribution of Working Group I to the Fifth

Assessment Report of the Intergovern–Mental Panel on Climate Change Cambridge University Press, Cambridge, United Kingdom and New York, NY, USA.

23. Jacobusse, G., 2005. WinMICE User's manual. Kazmi, D., et al., 2015. Statistical downscaling and future scenario generation of temperatures for Pakistan Region. Theor. Appl. Climatol. 120 (1–2), 341–350. http://dx.doi.org/10.1007/s00704-014-1176-1.

24. Khan, M.S., Coulibaly, P., Dibike, Y., 2006. Uncertainty analysis of statistical downscaling methods. J. Hydrol. 319 (1-4), 357–382. http://dx.doi.org/10.1016/j.jhydrol.2005.06.035.

25. Liu, J., Williams, J.R., Wang, X., Yang, H., 2009. Using MODAWEC to generate daily weather data for the EPIC model. Environ. Model. Softw. 24 (5), 655–664. http: // dx.doi.org/10.1016/j.envsoft.2008.10.008.

26. Mahmood, R., Babel, M., 2012. Evaluation of SDSM d eveloped by annual and monthly sub-models for downscaling temperature and precipitation in the Jhelum basin, Pakistan and India. Theor. Appl. Climatol., 1–18. http://dx.doi.org/10.1007/s00704-012-0765-0.

27. Mujumdar, P., Ghosh, S., P., 2008. Modeling GCM and scenario uncertainty using apossibilistic approach: application to the Mahanadi River, India. Water Resour. Res. 44, W06407-1.

28. Nguyen, V., 2005. Downscaling methods for evaluating the impacts of climate change and variability on hydrological regime at basin scale. In: Proceedings of the International Symposium on Role of Water Sciences in Transboundary River Basin Management, Thailand, pp. 1–8.

29. Opitz-Stapleton, S., Gangopadhyay, S., 2010. A non-parametric, statistical downscaling algorithm applied to the Rohini River Basin, Nepal. Theor. Appl. Climatol. 103 (3–4), 375–386. http://dx.doi.org/10.1007/s00704-010-0301-z.

30. Pokhrel, R.M., Kuwano, J., Tachibana, S., 2013. A kriging method of interpolation used to map liquefaction potential over alluvial ground. Eng. Geology. 152 (1), 26–37. http://dx.doi.org/10.1016/j.enggeo.2012.10.003.

31. Qamer, F.M., Saleem, R., Hussain, N., Raza, S.M., 2008. Multi-scale watershed database of Pakistan. In: Proceedings of the 10th International Symposium on High Mountain Remote Sensing Cartography. Rajbhandari, R., Shrestha, A.B., Kulkarni, A., Patwardhan, S.K., Bajracharya, S.R., 2015. Projected changes in climate over the Indus river basin using a high resolutionregional climate model (PRECIS). Clim Dyn. 44 (1-2), 339–357. http://dx.doi.org/10.1007/s00382-014-2183-8.

32. Souvignet, M., Heinrich, J., 2011. Statistical downscaling in the arid central Andes:uncertainty analysis of multi-model simulated temperature and precipitation. Theor. Appl. Climatol. 106 (1-2), 229–244. http://dx.doi.org/10.1007/s00704-011-0430-z.

33. Tripathi, S., Srinivas, V.V., Nanjundiah, R.S., 2006. Downscaling of precipitation for climate change scenarios: a support vector machine approach. J. Hydrol. 330 (3–4), 621–640. http://dx.doi.org/10.1016/j.jhydrol.2006.04.030.

34. Wetterhall, F., Bárdossy, A., Chen, D., S.H., C.-Y , X., 2006. Daily precipitation downscaling techniques in three Chinese regions. Water Resour. Res. 42, W11423. http://dx.doi.org/10.1029/2005WR004573.

35. Wilby, R.L., Dawson, C.W., Barrow, E.M ., 2002. SDSM — a decision support tool for the assessment of regional climate change impacts. Environ. Model. Softw. 17 (2), 145–157. http://dx.doi.org/10.1016/s1364-8152(01)00060-3.

36. Wilby, R.L., et al., 2000. Hydrological responses to dynamically and statistically downscaled climate model output. Geophys. Res. Lett. 27 (8), 1199. http://dx. doi.org/10.1029/1999GL006078.

37. Wilby, R.L., Hay, L.E., Leavesley, G.H., 1999. A comparison of downscaled and rawGCM output: implications for climate change scenarios in the San Juan Riverbasin, Colorado. J. Hydrol. 225 (1–2), 67–91. http://dx.doi.org/10.1016/s0022-1694(99)00136-5.

38. Wilby, R.L., et al., 2006. Integrated modelling of climate change impacts on waterresources and quality in a lowland catchment: River Kennet, UK. J. Hydrol. 330 (1–2), 204–220. http://dx.doi.org/10.1016/j.jhydrol.2006.04.033.

39. Xu, C.Y., 1999. Climate change and hydrologic models: a review of existing gaps andrecent research developments. Water Resour. Manag. 13 (5), 369–382. http: //dx.doi.org/10.1023/a:1008190900459.

40. Zhang, X.C., Liu, W.Z., Li, Z., Chen, J., 2011. Trend and uncertainty analysis of simulated climate change impacts with multiple GCMs and emission scenarios. Agric. For. Meteorol. 151 (10), 1297–1304. http://dx.doi.org/10.1016/j.agrformet.2011.05.010.

41. Zhao, F.F., Xu, Z.X., 2008. Statistical downscaling of future temperature change in source of the Yellow River Basin. Plateau Meteorol. 27 (1), 153–161 (in Chinese with English abstract).

CHAPTER 17

MANAGEMENT OF FRESHWATER ECOSYSTEMS IN THE CONTEXT OF CLIMATE CHANGE

Junguo Liu, Giri Kattel, Hans Peter H. Arp, Hong Yang

Keywords: Ecological threshold, Ecosystem management, Ecosystem services, Water management, Climate change

INTRODUCTION

Freshwater ecosystems, defined as aquatic systems with average salinities less than 0.5 parts per thousand (Moss, 2009), provide a diverse range of essential services such as food products, clear water, waste recycling, nutrient cycling, carbon storage, as well as cultural and recreational amenities. However, freshwater ecosystems are changing rapidly due to factors like urbanization and climate change. Over the past 50 years, economic and population growth have resulted in more rapid changes in the structure and function of freshwater ecosystems than in any other comparable time period of human history (MEA, 2005). Populations of freshwater species in North America, Europe, Australia, and New Zealand are estimated to have declined on an average by 50% between 1970 and 2000 (MEA, 2005). Freshwater ecosystems in urban areas are among the most affected (Kozlowski and Bondallaz, 2013), in that urbanization has led to dramatic changes in freshwater ecosystems throughout the globe (Alberti et al., 2007 and Kozlowski and Bondallaz, 2013).

Adding to the ongoing burdens of intensive agriculture, industrialization and urbanization, climate change is an additional serious threat to freshwater ecosystems and biodiversity worldwide. Climate change can alter freshwater ecosystems via various direct and indirect mechanisms (Chu et al., 2005 and Vörösmarty et al., 2010). Rising temperature and changing precipitation directly influence shifts in habitats and seasons, and also the physiological adaptation and phenology of freshwater species, thereby altering food web structure and ecosystem dynamics (Doak and Morris, 2010 and Walther et al., 2002). Climate change can also indirectly affect freshwater ecosystems via geomorphological alterations of lake and river systems, changes in nutrient and ionic loads (leading towards alteration of photosynthetic rates, eutrophication, acidification, salinization) as well as enhancing the impacts of prevalent diseases, chemical pollutants, biological invasions, and changes in predation and competition among species (IPCC, 2007). There have been a large number of discussions on the challenges and solutions facing human interventions to freshwater ecosystems (e.g., Chen et al., 2014), considering experimental and theoretical ecological thresholds (e.g., Horan et al., 2011), and how to best implement risk management of aquatic ecosystems (e.g., Chen et al., 2011 and Chen et al., 2013). However, it remains rare to explicitly discuss the freshwater ecological thresholds associated with climate change. There is an urgent need for integrating the scientific understanding of the diverse, complex and interrelated impacts of climate change on the thresholds of freshwater ecosystems. Here we review freshwater ecosystem thresholds in the context of climate change, and suggest the need for collaborative efforts across scientists, decision makers and stakeholders at all levels.

WHAT ARE ECOLOGICAL THRESHOLDS?

There are several definitions for the term "ecological threshold". Most of these definitions commonly emphasize the non-linearity of ecological or biological responses to pressures caused by human interventions or natural processes. As defined by Groffman et al. (2006), an ecological threshold is the point, or "tipping point", at which there is an abrupt change in an ecosystem quality, property or phenomenon, or where small changes in environmental drivers can lead to dramatic changes

Citation: Junguo Liu, Giri Kattela, , Hans Peter H. Arp, Hong Yang, Management of freshwater ecosystems in the context of climate change, http://dx.doi.org/10.1016/j.ecolmodel.2014.09.010.

to an ecosystem. Thresholds and their associated stability towards different environmental drivers can be conceptualized within a coupled socioeconomic–ecological system (Horan et al., 2011). Any restoration of losses in ecosystem services after a threshold is crossed could be difficult or costly (Groffman et al., 2006).

Identifying ecological thresholds related to climate change is complex, as various climate-change control variables (e.g., atmospheric and surface water temperature) can be related differently to changes in the ecosystem service responses (e.g., fish supply, volume of potable water) (Rockström et al., 2009).

Fig. 1 presents a general schematic diagram on how to view the interactions between climate change control variables and ecosystem service response variables using a threshold-based approach. Ecosystem services can be linearly dependent, non-linearly dependent, or independent of the climate control variables. When a shift in a climate control variable causes a substantial depletion in a dependent ecosystem service, a "threshold", point F_1, may be crossed, after which the drop in this service can become depleted rapidly with slight increases in the climate control variable. Often it is difficult to predict when a threshold occurs, or even ascertain the relationship between the climate change variable and the ecosystem service. Thus, a range can be assigned as a "zone of uncertainty". The "ecosystem boundary", point F_0 in Fig. 1 refers to the lowest value of a climate change control variable (compared to the present value) in which the probability of passing a "threshold" becomes higher than an assigned margin of safety (Rockström et al., 2009). Sometimes, ecosystems exhibit resilience (Kattel et al., 2013). After the thresholds are crossed, ecosystem services can recover either due to the climatic and other environmental variable reversing to an earlier level or some delayed feedback or buffering mechanism. This restoration is usually not instantaneous. It often requires the control-variable to decrease to a level substantially below the threshold, which is a phenomenon, referred to as ecosystem hysteresis (Scheffer et al., 2001). However, if ecosystem thresholds are crossed and restored periodically (e.g., from forest fires), this behavior would be "cyclical". However, if the change in a climatically sensitive and controlled variable becomes extreme, a "point of no return" can be reached, indicated as point F_2 in Fig. 1, i.e., a 'catastrophic' transition where the ecosystem can no longer return to its original state (e.g., Kattel et al., 2013). The condition between F_1 and F_2 can have profound implications for 'early warning signals' for ecosystem management (Wang et al., 2012). However, this is often unnoticed (Scheffer et al., 2001 and Folke et al., 2004).

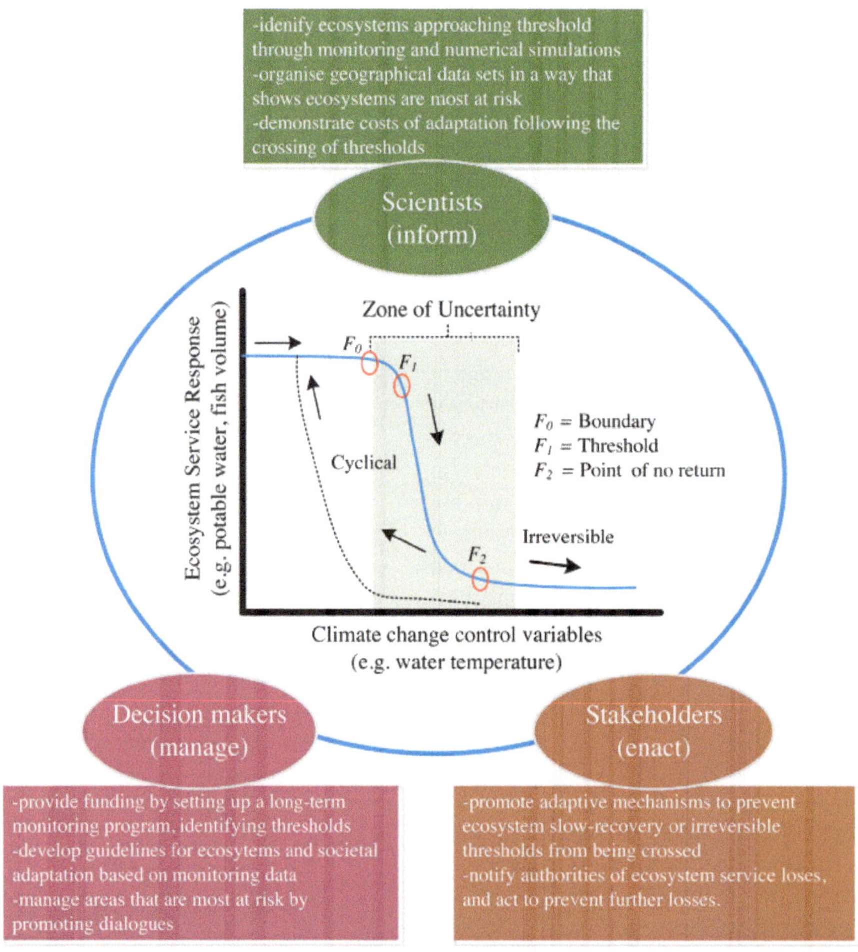

Figure 1: Conceptualization of threshold-based ecosystem change. Climate change control variables are on the *x*-axis and the response variables on ecosystem services are on the *y*-axis. The climate change variables could be one or any combination of parameters that are directly or indirectly affected by climate change (such as temperature, number of invasive species, nutrient concentration). The ecosystem service could refer to any required by the local population (such as potable water, fish population). Conceptualizing ecosystem change in this way could serve as a basis for communication in ecosystem management, where scientists inform of susceptibility to threshold changes, decision makers manage appropriately, and stakeholders get involved through enacting mitigation measures to prevent loss of services.

A key goal of ecosystem management is to avoid thresholds with slow recovery or points of no return. Though difficult for scientists to anticipate, boundaries and thresholds are highly useful for mobilizing policy makers or stakeholders. For instance, the use of 350 ppm CO_2 being a safe level to avoid a tipping point for global ecosystem change has been successful for mobilizing stakeholders and policy efforts (Hansen et al., 2008), as has been the prediction that a warming of 1–2.5 °C above pre-industrial levels could be a threshold value where substantial loss of Arctic summer ice or the Greenland ice sheet could occur (IPCC, 2007).

EFFECTS OF CLIMATE CHANGE AND THRESHOLD TENDENCIES IN FRESHWATER ECOSYSTEMS

The most practical thresholds for use by policy makers and stakeholders are those based on measurable control variables that have a straightforward link to an ecosystem response (Table 1). The easier the climate control variable is to quantify, as well as its effect on ecosystem services, the more practical it is to implement policy and effective monitoring measures. However, in practice direct links between climate change control variables and ecosystem responses are difficult to establish. Anthropogenic climate change is quantified by several parameters that could be used as control variables for defining the thresholds of freshwater ecosystems, e.g., atmospheric and surface water temperature, the length and onset of seasons, and precipitation (Table 1). However, in most cases, multiple control variables acting together must be considered. For example, riverine ecosystems are strongly and simultaneously influenced by variables like streamflow, erosion rates, temperature, and concentrations of micro-pollutants; though how to account for all the variables in a threshold type model is not always straightforward (Groffman et al., 2006). Other control variables that are indirectly influenced by climate change as well as other anthropogenic processes include water volume, salinity, dissolved oxygen (DO) and pH. All of these parameters, in isolation or in combination, could influence ecosystem services. In Section 4, we review the state-of-the-art understanding of how shifts towards ecological thresholds for freshwater ecosystems can be influenced directly or indirectly by climate change. A schematic overview of this presentation is provided in Fig. 2.

Table 1: Example freshwater ecosystem monitoring approaches with possible scientific, policy and stakeholder involvement to assess or prevent possible thresholds and loss of socio-economic services

Environmental parameter monitoring	Potential affected threshold response monitoring[a]	Potentially affected socio-economic services monitoring[a]
Climate: surface to bottom water temperature, evapotranspiration rates, precipitation, humidity season onset and length	Population of algae and plants, migration of fish and other species, habitat range	Changes in local economy due to change in products (prices of food, water, plants, real estate), as well as quality of life (health, opinion pools, recreational land area)
Hydrology: water level, water volume, water resident time	Population of algae and plants, areas of excess flooding, drowning, loss of water habitat	
Light: UV radiation, visible water depth	Population of algae and plants (planktonic and benthic)	
Oxygen: dissolved oxygen (DO) levels, lake stratification.	Phytoplankton blooms, aerobic benthic species, vertically migrating fish species, oxygen sensitive species,	
Geomorphology: erosion rates, sedimentation, siltation	Riparian fish habitat, formation of anoxic zones	
Water quality: pH, micropollutants, dissolved organic carbon (DOC), nutrients including total nitrogen (TN), total phosphorus (TP) and chlorophyll a (chl a)	Loss of sensitive species, change in species dynamics,population of algae and plants	
Species diversity: presence of invasive species, changes in foodweb structure, biodiversity losses	Loss of locally important species	

a. The threshold like responses and socio-economic service may be affected by a combination of climate and local anthropogenic impact parameters (e.g., pollution levels, water over usage, urbanization).

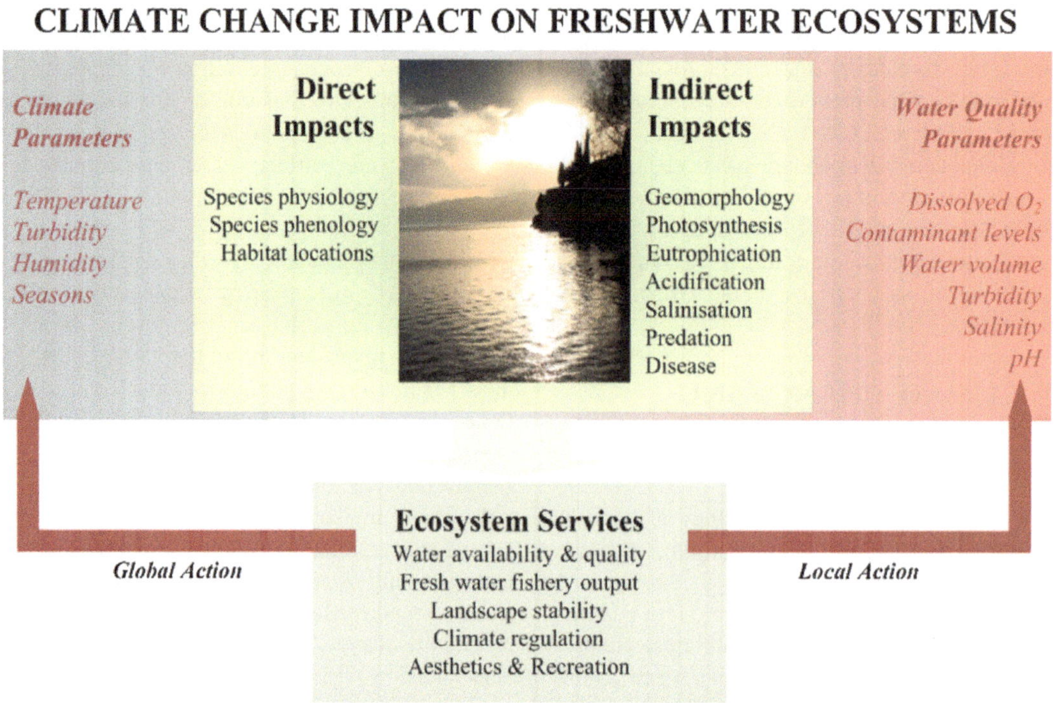

Figure 2: Overview of ecosystem services that are directly and indirectly impacted by climate change and local anthropogenic impacts. The figure is an illustration of the complex, cyclical nature of how the use of ecosystem services can, through direct and indirect mechanisms, affect those same ecosystem services. Appropriate global action to reduce harmful climate change impacts and local action to reduce harmful disturbances in water quality can help mitigate losses of ecosystem services.

THRESHOLDS ASSOCIATED WITH DIRECT EFFECTS OF CLIMATE CHANGE

Over the past decades, freshwater ecosystems have exhibited direct responses to changes in temperature, precipitation, and atmospheric humidity (Hader et al., 2007), particularly in regard to the phenology and physiology of freshwater species, as well as their habitats. These responses become more pronounced with an increasing global surface temperature (Schippers et al., 2004).

Phenological Thresholds

Climate change has directly influenced the phenology of aquatic biota. Changes in air temperature influence water temperature and the timing and duration of the stratified period, while changes in precipitation can influence water residence time. These changes in turn influence the seasonal dynamics of the biota present (Peeters et al., 2007). Phytoplankton phenology in lentic systems is directly affected by an increased water temperature, changes in the onset and the duration of thermal stratification, and earlier ice break up (Meis et al., 2009). For instance, phytoplankton blooms in the US lakes and Taihu Lake, the third largest lake in China, have begun to occur earlier due to the warmer spring (Qin et al., 2010 and Winder and Schindler, 2004). In addition, high air temperatures and low wind speeds have directly led to an early onset of the spring phytoplankton bloom in Upper Lake Constance of the western European Alps (Peeters et al., 2007).

Changes in temperature and seasons influence multiple levels of biological organization, including foraging behavior, phenology and metabolic rates, potentially leading to profound shifts in the stoichiometry of elemental fluxes between consumers and resources at the base of the food web (Woodward et al., 2010). A rise in surface-water temperature and a regional decrease in wind velocity in central Africa's Lake Tanganyika, which provides 25–40% of the animal protein for surrounding populations, has lead to a reduction in primary productivity by ~20%, implying a ~30% decrease in fish yields (O'Reilly et al., 2003). Arctic lake and river systems, however, experiencing the opposite effect, as longer growing seasons have led to increased primary productivity (Smol et al., 2005). Such differences in phenology and life cycle cues can induce strong variance in community composition (Burgmer et al., 2007), trophic levels and fish production, ultimately changing ecosystem services (Ficke et al., 2007).

Physiological Thresholds

Some species that are exposed to warmer temperatures for longer periods will show biochemical variations at the cellular level. Heatwaves, which are likely to increase in severity due to the changing climate (Schar et al., 2004), may influence freshwater gastropods and mussels by affecting the release of the stress-associated enzymes (e.g., protein kinases), altering metabolic activity and causing hypoxia in these animals (Anestis et al., 2010). Physiological processes in aquatic organisms, including rates of oxygen uptake, movement, feeding, developmental rate, and immune function are strongly driven by temperature (Helmuth, 2009). Warmer surface temperatures of freshwater lakes can lead directly (through water currents) and indirectly (through affecting plant and algae growth as discussed above) to the alteration of thermal stratification in temperate lakes, which can lead to an increased dissolved organic carbon (DOC) concentrations in the hypolimnion, limiting the DO level for vertically migrating fish species, potentially causing hypoxia (Hader et al., 2007). A mismatch between the demand for oxygen and the capacity of oxygen supply to the tissues can deviate the physiological adaptation of aquatic species and restrict tolerance of these animals to a range of climatic extremes (Pörtner and Knust, 2007). Decreased oxygen in lakes also decreases the processing of organic matter, recycling of nutrients, and the microbial breakdown of pollutants (Carpenter et al., 2011).

To place this in the context of a freshwater ecosystem threshold, a previous modeling study suggests that a doubling of CO_2 concentrations would cause earlier onset of lake stratification and a 1–7 °C increase in summer epilimnetic temperature, consequently limiting the vertical migration of zooplankton in addition to physiological and respiration failures in cold water fish species (De Stasio et al., 1996). This example and other examples of freshwater ecosystem thresholds related to climate change are presented inTable 2.

Table 2: Thresholds for freshwater ecosystem regime shift

Regime-shift	Parameter	Threshold	Location	References
Physiological related				
Early lake stratification, limiting vertical migration and causing physiological failure in cold fish	Atmospheric CO2 concentrations	A doubling of CO2 concentrationfrom the 1980s level	North temperate lakes	De Stasio et al. (1996)
Habitat related				
Reduction in fish habitats by 50%	Summer mean temperature	2–6°C	US lakes	Smol et al. (2005)
Shift of small bass and yellow perch habitat 500km north	Mean air temperature	4°C	US lakes and rivers	Shuter and Post (1990)
Substantial loss of ice sheet	Temperature	1–2.5°C increase above pre-industrial levels	Arctic, Antarctic, Greenland and other mountain areas	IPCC (2007)
Water quality related				
Shift in diatom community	Nitrogen deposition loads	1.4kgNha −1y−1	Lakes around the eastern Sierra Nevada and the Greater Yellowstone National Park, US	Saros et al. (2011)
Shift in biotic assemblages, water quality and carbon budget	Dissolved organic carbon	91% increase	22 upland lakes and streams, UK	Evans et al. (2005)
From clear to turbid state 1)	Orthophosphate phosphorus concentration in the water	10–12μg/L	26 temperate lakes, Northern Italy	Chiaudani and Vighi (1974)
From clear to turbid state	Total phosphorus in water	150μg/L	Lake Velume, Netherland	

From turbid to clear state	Total phosphorus in water	100μg/L	Lake Velume, Netherland	
From clear to intermediate state 2)	Total phosphorus in water	61μg/L	Gehu Lake, China	Tao et al. (2012)
From intermediate to turbid state 3)	Total phosphorus in water	115μg/L	Gehu Lake, China	Tao et al. (2012)
From clear- to turbid-state	Total phosphorus in water	70–100mgm−3	46 small and medium lakes along the mid-lower Yangtze River	Wang (2007)
From turbid- to clear-state	Total phosphorus in water	20–30mgm−3	46 small and medium lakes along the mid-lower Yangtze River	Wang (2007)
From turbid to clear state 4)	Phosphorus loading	0.3gm−2y−1	Lake Mendota, Wisconsin, USA	Carpenter and Lathrop (2008)
From clear to turbid state	Dynamic linear model standard deviation (DLM SD) of phosphorus in soil, water and sediment	Around 1gm−2	Lake Mendota, Wisconsin, USA	Carpenter and Brock (2006)

Note: Although different words are used for the states of freshwater ecosystems, we used the terms clear and turbid states (widely accepted now); the original words for the states were listed: (1) from meso- to eutrophic; 2) from grass-stable to grass-algae intermediate state; (3) from grass-algae intermediate to algae-stable state; (4) from persistent eutrophy to oligotrophy.

Habitat-associated thresholds

Climate change also directly influences fish populations, as changes in climate lead to changes in the hydrological regimes and habitat niches that endemic fish have evolved to cope with (Ficke et al., 2007). Global warming has been attributed with decreasing the thermal niche and habitat availability for cold water fish, such as galaxid and salmonid fishes (Carpenter et al., 2011). A 3 °C change in mean annual temperature corresponds to a shift in approximately 300–400 km latitudinal and 500 m altitudinal isotherms (Smol et al., 2005). One study estimates that global warming could shift natural habitat ranges by approximately 6.1 km per decade towards the poles (Parmesan and Yohe, 2003). This change, though subtle, is enough to impact fish communities and ecosystem adaptation in the higher latitudes. To place habitat changes in the context of climate change related thresholds, a coupled CO_2–climate model suggests that an increase in 2–6 °C summer mean temperature will reduce 50% of the suitable habitats for cold water fish in the US (Eaton and Scheller, 1996). A further 4 °C increase in mean air temperature will extend the habitat of small bass mouth and yellow perch as far as 500 km towards the north (Shuter and Post, 1990).

THRESHOLDS ASSOCIATED WITH INDIRECT EFFECTS OF CLIMATE CHANGE

Climate change can affect freshwater ecosystems indirectly via a range of positive and negative feedbacks. These indirect effects may be gradual and subtle in the short term but could be catastrophic in the long term, leading to slowly recovering or irreversible threshold type losses. Changes in geomorphology, water depth, flow regimes, siltation and sediment flux in lake and river systems can occur as a result of indirect effects of climate change (Lake et al., 2000). Biotic and abiotic interactions with climate can further lead to unprecedented ecological conditions, including the alteration of photosynthetic rates, eutrophication, acidification and salinization (Schindler, 2001 and Woodward et al., 2010). Widespread invasion from exotic flora and fauna, and changes in predation and competition among species are also exacerbated by indirect interactions between climates and human activities (Baron et al., 2002).

Below, we take the trophic state threshold as an example of indirect effects of climate change. Other thresholds related to indirect effects are described in Supplementary information, and they include morphological, photosynthetic, acidification, salinity, pollutant thresholds, disease, biological invasion, as well as predation and competition-associated thresholds.

Decreased precipitation and lower water flows imply smaller dilution volumes and thus, higher concentrations of nutrients downstream from point pollution source, particularly untreated sewage in developing countries. On the other hand, the increased occurrence of storm events can cause an overflow and discharge of nutrient into receiving rivers or lakes (Whitehead et al., 2009). Severe storms can entrain large amount of deeper and anoxic sediments, particularly when the water level is low. As an example, a storm in 1996 that occurred at Lake Võrtsjärv, Estonia, resulted in a greater release of sediments on one stormy day compared to the total nutrient inflows over 1 y (Nõges and Kisand, 1999). The effects of wind-mixing on the vertical transfer of nutrients from sediment are more important in small lakes than the large lakes. The relationship between North Atlantic Oscillation Winter Index (NAOw) and phosphate has been observed in some European lakes (e.g., George et al., 2004 and Weyhenmeyer, 2004). In temperate Europe, positive values of NAOw imply milder winter, higher water levels and lower spring concentrations of phosphate, while negative values indicate drier winters, low water levels and larger amounts of phosphate released from sediment. In cold areas, the extension of the ice-free period can cause marked increases in phosphorus concentration, as has been observed in the hypolimnion and epilimnion in Lake Erken, Sweden (Pettersson et al., 2010). This additional phosphorus is released annually from the sediment zone to the water column as a result of increased temperature and low oxygen concentration in the bottom water (Malmaeus and Rydin, 2006). Increased temperature and precipitation intensity and changes in winter precipitation are expected to enhance phosphorus loading in the temperate freshwater lakes and reduce the loading in Mediterranean lakes (Jeppesen et al., 2009).

Nutrient loads are predicted to increase under climate change in many areas such as Denmark and UK (Andersen et al., 2006 and Whitehead et al., 2006). Higher temperature increases the release of nitrogen from soil, and lower stream flows decrease the dilution capacity of rivers (Whitehead et al., 2006). However, the impact on eutrophication is not straightforward due to the complex interaction between nutrient, light, temperature, residence time and flow conditions (Jeppesen et al., 2005 and Yang and Flower, 2012). Increased surface water temperature can indirectly lead to stronger stratification and regeneration of water column nutrients, which can intensify eutrophication (Rabalais et al., 2009). Climate change has exacerbated eutrophication through nutrient dynamics of the north temperate lakes due to a longer growing season, depending on epilimnion and mineral mixing (Smol et al., 2005). The analyses of 103 Chinese lakes indicated that the mean precipitation was one of the main predictors of eutrophication (Liu et al., 2010). A 41-year data series (1968–2008) from Blelham Tarn UK, where eutrophication has reduced hypolimnetic DO, indicates negative effects that are likely to be exacerbated by changes in climate and the thermal structure of lakes (Foley et al., 2012). The interactive effects of future eutrophication and climate change on harmful cyanobacterial blooms are difficult to understand, though much of the current knowledge suggests that climate change will likely enhance the magnitude and frequency of these events (O'Neil et al., 2012). Recently eutrophication in lacustrine environments has been adopted as a way to define ecological and societal thresholds in space and time (Carpenter et al., 1999). Experimental management programs have been replicated among several lakes based on these thresholds (Carpenter et al., 1999).

The phosphorus thresholds for the regime shift between clear and turbid states have been suggested for different lakes (Table 2). Compared to the primary drivers of eutrophication such as nutrients, the role of climate change to modulate eutrophication thresholds is not well understood. However, increasing temperature, changing precipitation and atmospheric circulation can indirectly influence the nutrient dynamics, and increase the likelihood of exceeding thresholds of nutrients in freshwater ecosystems (Battarbee, 2000).

IDENTIFYING THRESHOLD RESPONSES FOR MANAGING FRESHWATER ECOSYSTEMS

Identifying and predicting threshold responses in ecological systems is a challenging, though useful task. Thresholds related to lake eutrophication causing regime shift is a well studied area (Scheffer et al., 2001) and one that can be used to understand thresholds from other climate-change control variables. Excessive nutrient levels, particularly phosphorus, have been widely accepted as the leading factor of eutrophication (Schindler, 1974 and Carpenter et al., 1999). The orthophosphate phosphorus concentration 10–12 $\mu g\ L^{-1}$ in water was suggested as the threshold from clear to turbid states in the lakes of Northern Italy (Chiaudani and Vighi, 1974) (Table 2). Higher concentrations were also found to be thresholds for the regime shift in Dutch and Chinese lakes (Tao et al., 2012). Although the phosphorus concentration in water can be controlled to some extent by lowering nutrients emissions, such as by controlling fertilizer usage; internal sediment release of phosphorus in addition to temporal changes in external catchment input can cause thresholds to be exceeded e.g., the level of 0.3 $g\ m^{-2}\ y^{-1}$ external phosphorus input in Lake Mendota, Wisconsin for threshold crossing benchmark (Carpenter and Lathrop, 2008). With many dynamic influential factors, changing non-linearly with time, the use of Bayesian forecasting through Dynamic Linear Models (DLM), which allows influential parameters to have a prior distribution (West and Harrison, 1989), is regarded as a more reasonable approach for modeling changes in release of phosphorus from the lake sediment (Cottingham et al., 2000), than steady linear models. DLM standard deviation (DLM SD) of phosphorus load in soil, water and sediment was also suggested as a way to anticipate a threshold of lake regime shift (Carpenter and Brock, 2006).

Thresholds from other climate-control variables, similar to those caused by phosphorus induced eutrophication are not only related to external driving forces, but are also mediated by internal changes in the ecological regimes (Walker and Meyers, 2004). The direct and indirect impacts listed above do not occur in isolation but simultaneously, and are often indicated by a change in several quantifiable parameters (Fig. 2). Often these changes are inconsistent and may behave non-linearly across and within ecosystems. Even a minor disturbance may move a system to a new regime, for instance, a gradual increase in nutrients can transform an oligotrophic lake into an eutrophic one (Limburg et al., 2002). Approaches to identifying the thresholds of fresh water ecosystems should involve the use of continuous monitoring, paleolimnological data and simulations, and thereby account for changes in the present, past and future.

Implementing Effective Monitoring Programs

Environmental policies designed to maintain freshwater quality are often dependent on identifying "threshold" dose–response relationships (Groffman et al., 2006). Effective monitoring is useful for tracking ongoing ecological system shifts that are driven by external environmental forces or internally-mediated drivers, and thus for predicting threshold type responses, and devising priorities and practical strategies for biological conservation (Doak and Morris, 2010 and Lindenmayer and Likens, 2009). For example, a time series analysis of DOC in 22 upland lakes and streams of the UK obtained during a 15 y monitoring program indicated a 91% increase in DOC concentration has triggered a shift in biotic assemblages, water quality and the carbon budget in the region (Evans et al., 2005). A 20-year monitoring study of water conductivity of shallow lakes across the Arctic region (in combination with paleolimnological data) has provided a description of thresholds related to wetland permanency as a result of climate warming, in which evaporation/precipitation ratios are a suitable climate-control variable (Smol and Douglas, 2007). For all monitoring programs, it is essential to establish an archive of environmental data, such as the Vannmiljø system used in Norway (http://vannmiljo.miljodirektoratet.no/) for analyzing trends and predicting threshold responses.

Key variables, such as precipitation, temperature, total nitrogen (TN), total phosphorus (TP) and chlorophyll *a* (chl*a*) should be continuously monitored based on field sampling and laboratory measurement. In addition, advancement of technologies including remote sensing will improve techniques of monitoring environmental forces over a regional scale. A few examples of further monitoring approaches to obtain relevant data are presented in Table 1.

Reconstructing Historical Data

Palaeolimnological techniques can help identify thresholds of ecosystem change. For example, critical loads of nitrogen deposition were determined for alpine lake ecosystems in the Western US using fossil diatom assemblages in lake sediment cores (Saros et al., 2011). A transfer function technique, where calibration training sets are constructed to show relationships between individual species and one or more quantifiable (climate change related) parameters, followed by a reconstruction of changes in ecosystem functions and services based on a biological data, is useful for identifying ecological regime shifts caused by environmental perturbations (Battarbee, 2000). As an example, sub-fossil assemblages of diatoms in lakes around the eastern Sierra Nevada and the Greater Yellowstone National Park suggests that 1.4 kg Nitrogen ha $^{-1}$ y^{-1} can cause a regime shift in the diatom community (Saros et al., 2011). The advancement of dating techniques and the use of multiple species approaches for lake sediments can further help improve climate reconstruction and define climate thresholds of freshwater ecosystems (Battarbee, 2000), but it is still necessary to disentangle the relative role of all influential factors using a combination of both limnological and paleolimnological data (Battarbee et al., 2012). It also needs to be pointed out that the historically caused perturbations result primarily from internal shifts and climate change, whereas current human induced perturbation (e.g., land use, changed nutrient budgets) is now superimposed on internal shifts and climate-related changes, making it difficult to use past changes as direct analogs of current and future thresholds.

Developing Appropriate Numerical tools and Models

Precision in ecosystem projections will increase when quantitative estimates of thresholds are backed by reliably measured data (Woodward et al., 2010). Scenario-based models can provide decision-makers a range of possibilities, as well as integrate natural and social science based strategies for adaptation (IPCC, 2007). The Intergovernmental Panel on Climate Change (IPCC) has implemented a variety of scenario-based climate-change models, including models associated with historical freshwater ecosystems (IPCC, 2007). However, models to describe thresholds that influence individual freshwater ecosystems have not yet been developed, due to the uncertainty in future climate projections as well as scale issues, as small lakes cannot be adequately represented using coarse resolution climate models (Groffman et al., 2006). Some attempts have been made for improved modeling of the ecological thresholds such as the trophic cascade model for game fish (Carpenter and Brock, 2004) and the water quality model for DO concentrations in North American lakes (Stefan et al., 1993). Advancement of computer software and numerical modeling will continue to improve the knowledge of threshold based ecological management strategies.

Several types of models, e.g., statistical, rule-based, the previously mentioned DLM models, and process-based models, have been used to guide ecosystem management under global change. Process-based models are based on a scientific understanding of relevant ecological processes, and provide a valuable framework to include responses to the changed environmental conditions (Cuddington et al., 2013). The current understanding of the role of thresholds is often based upon conceptual models of how ecosystems work (Dennison et al., 2007). Process-based threshold models can be specifically designed to anticipate ecological consequences of human activities on freshwater ecosystems, and should play a key role in setting conservation targets by water resources managers.

Testing Climatological Hypotheses

Laboratory or field-based experimental designs have the potential to characterize ecological thresholds (Adler et al., 2009). For instance, a laboratory experiment on burrowing mayfly, *Hexagenialimbata*, from the Lower Mobile River, Alabama, suggests that the nymphs of this animal can survive at elevated salinity levels (6.3%) only for temperature below 18 °C, however, when the salinity level is reduced to 2.4% they can survive at higher temperatures up to 28 °C, which was further supported by field observations (Chadwick and Feminella, 2001). Similarly, an experiment which reduced the pH of a small lake from 6.1 to 4.7 was found to cause a substantial decline in species richness of many freshwater taxa, reducing the ecosystem's adaptive capacity (Hogsden et al., 2009). These studies assist with understanding which species within ecosystems are the most vulnerable to climate change, and how the disappearance of these species will influence the broader ecosystem services.

Relating threshold-Associated costs to Ecological and Societal Adaptation

Ecosystem service changes also influence the societal structure of the population dependent on those services. Loss of ecosystem services, including water quality, water quantity, fishery resources, and recreational amenities, can lead to the crossing additional thresholds associated with social systems and unacceptable societal costs. Integrating key components of societal development can promote ecological resilience, including information management, cultural integrity, technological development, institutional responsibility, and education (Falkenmark, 2003). Consideration of a probabilistic framework with a range of components can offer policy and management responses to emerging crisis from freshwater ecosystem changes worldwide (Vörösmarty et al., 2010). Specifying the thresholds related to time, location, species, disturbance and scale, and understanding their economic and societal implications can help set up regulations, improve adaptive capacity of freshwater ecosystems, and facilitate societal adaptation (Naiman and Turner, 2000). The socio-economic costs of crossing ecological thresholds in freshwaters need to be placed clearly into the context of threshold-based management program (Table 1). This involves a careful cost-benefit analysis of the services affected before and after crossing thresholds.

Certain threats to freshwater ecosystems may be addressed through investments and advances in infrastructure, science and technology (Vörösmarty et al., 2010). Though this may be mainly a question of political will in richer countries, poorer countries often lack the resources to invest in freshwater infrastructure (UNFCCC, 2007). It is estimated that the United States currently would have to put in $60 billion dollars for the restoration of its ecosystems. In regions like the Middle East and North Africa, water pollution and over exploitation of water is expected to incur environmental damage costs equivalent to 2.1–7.4% of their gross domestic product(GDP). Similarly, the cost of water related crises in China was 2.3% of China's GDP in 2003.

Assigning an economic value to the crossing of ecological thresholds is an effective way of communicating the societal and economical costs to decision makers and stakeholders (Balmford et al., 2002). For example, the use of artificial drainage in South Ontario marshes for the purpose of increasing agricultural productivity was found to be less beneficial economically than not draining (van Vuuren and Roy, 1993). The cost-benefit analysis suggested that the naturally intact marsh ecosystems would cost less (3700 USD ha^{-1}) than the artificially drained marshes (8800 USD ha^{-1}). Thus the use of scenario-based numerical models of freshwater systems coupled with economical models is strongly encouraged, as it can help implement threshold-based management approaches of freshwater ecosystems effectively, and thereby underscore the services that intact ecosystems can provide to society (de Groot et al., 2002). Risk analysis should be conveyed together with threshold analysis, in which models are used to predict the risk and benefits (environmental, economical, societal) from thresholds being crossed under different climate change scenarios, as well as the risk and benefits from potential mitigation efforts.

THRESHOLD-BASED MANAGEMENT IN FRESHWATER ECOSYSTEMS

The goal of ecosystem management is to restore or preserve some main attributes of an ecosystem that are desirable for humans (Mayer and Rietkerk, 2004). Due to the inherent complexity of ecosystems, this can only be achieved through a framework (e.g.,Fig. 1) involving long-term collaborations between scientists, policy makers and stakeholders. The main goal of such collaboration would be to set up a warning and prevention system from approaching thresholds (near F_0 or F_1), as well as mitigation strategies to prevent such thresholds being reached. In this effort, scientists with the assistance of local stakeholders would monitor relevant environmental parameters and potential threshold responses (examples are presented in Table 1). Decision makers would set up a program to encourage such monitoring efforts, as well as be prepared in advance to enforce appropriate mitigation strategies as needed (e.g., lowering fish quotas, capping contaminated areas, preventing excess nutrients from entering the local area). Mitigation strategies would ideally have to be planned far well in advance based on previously conducted simulations and risk analysis. Clear communication with stakeholders regarding risks from crossing thresholds is critical, particularly in cases where long term and short term interests collide (e.g., over fishing, urbanization). Further, stakeholders should be encouraged to notify authorities in case a rapid loss in ecosystem services occurs (potentially indicating a threshold F_1 being met)(Fig. 1), to get involved in monitoring programs (e.g., species counting, environmental parameter logging), as well as be presented with opportunities to act in a way to mitigate risks from boundaries F_0being reached.

Disagreement may often occur amongst actors on practices, remediation and logistic issues. Scientists being too cautious about the position of F_0, may risk losing trust by decision makers and stakeholders, particularly if no change in a response variable is evident after F_0 is crossed (Fig. 1). On the other hand, if scientists are under cautious, predicting an F_0 that is close to F_2, then trust in scientists would also be lost alongside the ecosystem service.

Ecological systems are not simple to understand, neither are threshold responses. After a threshold is crossed, the situation may become restored by various feedback loops (Fig 1). The prudent option is, however, to assume that this does not happen (Briske et al., 2010). Avoiding disagreements amongst actors is essential when ecosystem is at a risk to cross thresholds. Disagreements are resolved through regular interactions by organizing public involvement (e.g., through fairs, festivals, workshops, panel discussions, news media), as well as research projects and societal awareness programs implemented amongst ecologists, modelers, economists, sociologists, resource managers, stakeholders and decision makers, activists and reporters (Carpenter et al., 2009 and Lindenmayer and Likens, 2009). Bringing these actors in a common, formerly established platform can be a successful approach for freshwater ecosystems globally (Carpenter et al., 2009 and MEA, 2005). This reflects an adaptive management approach, where solutions to the problems are proposed and implemented, and the management strategies are constantly reviewed over the course of ecological response to climate change (Williams, 2011). Due to multiple spatial and temporal scales of ecosystems, Côté and Darling (2010) argued that the ecosystem management to control local anthropogenic disturbance, for example nutrient input, and to reverse the ecosystem degradation will inadvertently lower resilience to climate disturbance. Ongoing research on freshwater ecosystem management is needed, especially in regards to integrating rare or extreme events in freshwater ecology (e.g., see Fuentes et al., 2006 and Denny et al., 2009), and evaluating evidence and uncertainty in threshold concepts and models (Katharine and Richard, 2009). Establishing programs that seek to identify or prevent thresholds from being crossed, and which have the involvement of the scientific, policy and stakeholder communities, are a way to effectively manage threats of local impacts and global climate change on freshwater ecosystem services.

ACKNOWLEDGEMENTS

This study was supported by the International S&T Cooperation Program from the Ministry of Science and Technology of China (No. 2012DFA91530), Special Fund for Forestry Scientific Research in the Public Interest (No. 201204204), Projects of International Cooperation and Exchanges NSFC (41161140353, 91325302), the 1st Youth Excellent Talents Program of the Organization Department of the Central Committee of the CPC, the Fundamental Research Funds for the Central Universities (TD-JC-2013-2), and the NGI GBV Fund (H.P.H. Arp). We also thank Potsdam Institutes of Climate Change Impacts (PIK) in Germany for supporting J. Liu's visits, and the University of Leeds for providing J. Liu a Cheney Senior Fellowship. We thank Steve Carpenter from the University of Wisconsin and Elena Bennett from McGill University for sharing discussions through the preparation of the manuscript.

REFERENCES

1. Adler, P.B., Leiker, J., Levine, J.M., 2009. Direct and indirect effects of climate changeon a prairie plant community. PLos One 4, e6887.
2. Alberti, M., Booth, D., Hill, K., Coburn, B., Avolio, C., Coe, S., Spirandelli, D., 2007. The impact of urban patterns on aquatic ecosystems: an empirical analysis in Puget lowland sub-basins. Landsc. Urban Plann. 80 (4), 345–361.
3. Andersen, H.E., Kronvang, B., Larsen, S.E., Hoffmann, C.C., Jensen, T.S., Rasmussen, E. K., 2006. Climate-change impacts on hydrology and nutrients in a Danish lowland river basin. Sci. Total Environ. 365, 223–237.
4. Anestis, A., Portner, H.O., Michaelidis, B., 2010. Anaerobic metabolic patterns relatedto stress responses in hypoxia exposed mussels Mytilus galloprovincialis. J. Exp. Mar. Biol. Ecol. 394, 123–133.
5. Balmford, A., Bruner, A., Cooper, P., Costanza, R., Farber, S., Green, R.E., Jenkins, M., Jefferiss, P., Jessamy, V., Madden, J., Munro, K., Myers, N., Naeem, S., Paavola, J.,Rayment, M., Rosendo, S., Roughgarden, J., Trumper, K., Turner, R.K., 2002. Economic reasons for conserving wild nature. Science 297, 950–953.
6. Baron, J.S., Poff, N.L., Angermeier, P.L., Dahm, C.N., Gleick, P.H., Hairston, N.G., Jackson, R.B., Johnston, C.A., Richter, B.D., Steinman, A.D., 2002. Meeting ecological and societal needs for freshwater. Ecol. Appl. 12, 1247–1260.
7. Battarbee, R.W., Anderson, N.J., Bennion, H., Simpson, G.L., 2012. Combining limnological and palaeolimnological data to disentangle the effects of nutrient pollution and climate change on lake ecosystems: problems and potential.Freshwater Biol. 57, 2091–2106.
8. Battarbee, R.W., 2000. Palaeolimnological approaches to climate change, with special regard to the biological record. Quat. Sci. Rev. 19, 107–124.
9. Briske, D.D., Washington-Allen, R.A., Johnson, C.R., Lockwood, J.A., Lockwood, D.R., Shugart, T.K.S., 2010. Catastrophic thresholds: a synthesis of concepts, perspectives and applications. Ecol. Soc. 15, 37.
10. Burgmer, J., Hillebrand, H., Pfenninger, M., 2007. Effects of climate-driven temperature changes on the diversity of freshwater macroinvertebrates. Oecologia 151, 93–103.
11. Côté, I.M., Darling, E.S., 2010. Rethinking ecosystem resilience in the face of climatechange. PLoS Biol. 8.
12. Carpenter, S.R., Brock, W.A., 2004. Spatial complexity, resilience, and policy diversity: fishing on lake-rich landscapes. Ecol. Soc. 9.
13. Carpenter, S.R., Brock, W.A., 2006. Rising variance: a leading indicator of ecological transition. Ecol. Lett. 9, 308–315.
14. Carpenter, S.R., Lathrop, R.C., 2008. Probabilistic estimate of a threshold foreutrophication. Ecosystems 11, 601–613.
15. Carpenter, S., Ludwig, D., Brock, W., 1999. Management of eutrophication for lakes subject to potentially irreversible change. Ecol. Appl. 9, 751–771.
16. Carpenter, S.R., Mooney, H.A., Agard, J., Capistrano, D., DeFries, R.S., Diaz, S., Dietz, T.,Duraiappah, A.K., Oteng-Yeboah, A., Pereira, H.M., Perrings, C., Reid, W.V., Sarukhan, J., Scholes, R.J., Whyte, A., 2009. Science for managing ecosystem services: beyond the millennium ecosystem assessment. Proc. Natl. Acad. Sci. U.S. A. 106, 1305–1312.
17. Carpenter, S.R., Stanley, E.H., Vander Zanden, M.J., 2011. State of the world'sfreshwater ecosystems: physical, chemical, and biological changes. Annu. Rev. Environ. Resour. 36, 75–99.
18. Chadwick, M.A., Feminella, J.W., 2001. Influe nce of salinity and temperature on the growth and production of a freshwater Mayfly in the lower mobile river, Alabama. Limnol. Oceanogr. 46, 532–542.
19. Chen, S.Q., Fath, B.D., Chen, B., 2011. Information-based network environ analysis: a system perspective for ecological risk assessment. Ecol. Indic. 11 (6), 1664–1672.
20. Chen, S.Q., Chen, B., Fath, B.D., 2013. Ecological risk assessment on the system scale: a review of state-of-the-art models and future perspectives. Ecol. Modell. 250, 25–33.
21. Chen, S.Q., Chen, B., Fath, B.D., 2014. Urban ecosystem modeling and global change: potential for rational urban management and emissions mitigation. Environ. Pollut. 190, 139–149.
22. Chiaudani, G., Vighi, M., 1974. The N:P ratio and tests with Selenastrum to predict eutrophication in lakes. Water Res. 8, 1063–1069.

23. Chu, C., Mandrak, N.E., Minns, C.K., 2005. Potential impacts of climate change on the distributions of several common and rare freshwater fishes in Canada. Divers. Distrib. 11, 299–310.
24. Cottingham, K.L., Rusak, J.A., Leavitt, P.R., 2000. Increased ecosystem variability and reduced predictability following fertilisation: evidence from palaeolimnology. Ecol. Lett. 3, 340–348.
25. Cuddington, K., Fortin, M.-J., Gerber, L.R., Hastings, A., Liebhold, A., O'Connor, M., Ray, C., 2013. Process-based models are required to manage ecological systems in a changing world. Ecosphere 4 (2), art20.
26. De Stasio, J.B.T., Hill, D.K., Kleinhans, J.M., Nibb elink, N.P., Magnuson, J.J., 1996.Potential effects of global climate change on small north-temperate lakes: physics, fish, and plankton. Limnol. Oceanogr. 41, 1136–1149.
27. Dennison, W.C., Lookingbill, T.R., Carruthers, T.J.B., Hawkey, J.M., Carter, S.L., 2007. An eye-opening approach to developing and communicating integrated environmental assessments. Front. Ecol. Environ. 5 (6), 1540–9295.
28. Denny, M.W., Hunt, L.J.H., Miller, L.J., Harley, C.D.G., 2009. On the prediction of extreme ecological events. Ecol. Monogr. 79, 397–421.
29. Doak, D.F., Morris, W.F., 2010. Demographic compensation and tipping points in climate-induced range shifts. Nature 467, 959–962.
30. Eaton, J.G., Scheller, R.M., 1996. Effects of climate warming on fish thermal habitat in streams of the United States. Limnol. Oceanogr. 41, 1109–1115.
31. Evans, C.D., Monteith, D.T., Cooper, D.M., 2005. Long-term increases in surface water dissolved organic carbon: observations, possible causes and environmental impacts. Environ. Pollut. 137, 55–71.
32. Falkenmark, M., 2003. Freshwater as shared between society and ecosystems: fromdivided approaches to integrated challenges. Philos. Trans. R. Soc. London Ser. B Biol. Sci. 358, 2037–2049.
33. Ficke, A.D., Myrick, C.A., Hansen, L.J., 2007. Potential impacts of global climate change on freshwater fisheries. Rev. Fish Biol. Fish. 17, 581–613.
34. Foley, B., Jones, I.D., Maberly, S.C., Rippey, B., 2012. Long-term changes in oxygen depletion in a small temperate lake: effects of climate change and eutrophication. Freshwater Biol. 57, 278–289.
35. Folke, C., Carpenter, S., Walker, B., Scheffer, M., Elmqvist, T., Gunderson, L., Holling, C.S., 2004. Regime shifts, resilience, and biodiversity in ecosystem management. Ann. Rev. Ecol. Evol. Syst. 35, 557–581.
36. Fuentes, M., Kittle, T.G.G., Nychka, D., 2006. Sensitivity of ecological models to their climate drivers: statistical ensembles for forcing. Ecol. Appl. 16, 99–116.
37. George, D., Maberly, S., Hewitt, D., 2004. The influence of the North Atlantic sscillation on the physical: chemical and biological characteristics of four lakes in the English Lake district. Freshwater Biol. 49, 760–774.
38. Groffman, P., Baron, J., Blett, T., Gold, A., Goodman, I., Gunderson, L., Levinson, B.,Palmer, M., Paerl, H., Peterson, G., Poff, N., Rejeski, D., Reynolds, J., Turner, M., Weathers, K., Wiens, J., 2006. Ecological thresholds: the key to successful environmental management or an important concept with no practicalapplication? Ecosystems 9, 1–13.
39. Hader, D.P., Kumar, H.D., Smith, R.C., Worrest, R.C., 2007. Effects of solar UV radiation on aquatic ecosystems and interactions with climate change. Photochem. Photobiol. Sci. 6, 267–285.
40. Hansen, J., Sato, M., Kharecha, P., Beerling, D., Berner, R., Masson-Delmotte, V., Pagani, M., Raymo, M., Royer, D.L., James, C., 2008. Target atmospheric CO2: where should humanity aim? Open Atmos. Sci. 2, 217–231.
41. Helmuth, B., 2009. From cells to coastlines: how can we use physiology to forecast the impacts of climate change? J. Exp. Biol. 212, 753–760.
42. Hogsden, K.L., Xenopoulos, M.A., Rusak, J.A., 2009. Asymmetrical food web responses in trophic-level richness, biomass, and function following lake acidification. Aquat. Ecol. 43, 591–606.
43. Horan, R.D., Fenichel, E.P., Drury, K.L., Lodge, D.M., 2011. Managing ecological thresholds in coupled environmental-human systems. Proc. Natl. Acad. Sci. 108 (18), 7333–7338.
44. IPCC, 2007. Climate Change 2007: The Physical Science Basis. Cambridge University, Cambridge pp. 1–18.
45. Jeppesen, E., Søndergaard, M., Jensen, J.P., Havens, K.E., Anneville, O., Carvalho, L., Coveney, M.F., Deneke, R., Dokulil, M.T., Foy, B., Gerdeaux, D., Hampton, S.E., Hilt,S., Kangur, K., Köhler, J., Lammens, E.H.H.R., Lauridsen, T.L., Manca, M., Miracle, M.R., Moss, B., Nõges, P., Persson, G., Phillips, G., Portielje, R., Romo, S., Schelske,C.L., Straile, D., Tatrai, I., Willén, E., Winder, M., 2005. Lake responses to reduced nutrient loading – an analysis of contemporary long-term data from 35 case studies. Freshwater Biol. 50, 1747–1771.
46. Jeppesen, E., Kronvang, B., Meerhoff, M., Søndergaard, M., Hansen, K.M., Andersen, H.E., Lauridsen, T.L., Liboriussen, L., Beklioglu, M., Ozen, A., Olesen, J.E., 2009. Climate change effects on runoff, catchment phosphorus loading and lakeecological state, and potential adaptations. J. Environ. Qual. 38, 1930–1941.
47. Katharine, N.S., Richard, J.H., 2009. Threshold models in restoration andconservation: a developing framework. Trends Ecol. Evol. 24 (5), 271–279.
48. Kattel, G.R., Elkadi, H., Meikle, H., 2013. Developing a complementary framework for urban ecology. Urban For. Urban Greening 12, 498–508.
49. Kozlowski, G., Bondallaz, L., 2013. Urban aquatic ecosystems: habitat loss and depletion of native macrophyte diversity during the 20th century in four Swiss cities. Urban Ecosyst. 16 (3), 543–551.
50. Lake, P.S., Palmer, M.A., Biro, P., Cole, J., Covich, A.P., Dahm, C., Gibert, J., Goedkoop, W., Martens, K., Verhoeven, J.O.S., 2000. Global change and the biodiversity of freshwater ecosystems: impacts on linkages between above-sediment and sediment biota. BioScience 50, 1099–1107.

51. Limburg, K.E., O'Neill, R.V., Costanza, R., Farber, S., 2002. Complex systems and valuation. Ecol. Econ. 41, 409–420.

52. Lindenmayer, D.B., Likens, G.E., 2009. Adaptive monitoring: a new paradigm for long-term research and monitoring. Trends Ecol. Evol. 24, 482–486.

53. Liu, W., Zhang, Q., Liu, G., 2010. Lake eutrophication associated with geographic location, lake morphology and climate in China. Hydrobiologia 644, 289–299.

54. MEA, 2005. Ecosystems and Human Well-beings: Wetlands and Water Synthesis. World Resources Institute, Washington, DC.

55. Malmaeus, J., Rydin, E., 2006. A time-dynamic phosphorus model for the profundal sediments of Lake Erken, Sweden. Aquat. Sci. Res. Across Boundaries 68, 16–27.

56. Mayer, A.L., Rietkerk, M., 2004. The dynamic regime concept for ecosystem management and restoration. BioScience 54, 1013–1020.

57. Meis, S., Thackeray, S.J., Jones, I.D., 2009. Effects of recent climate change on phytoplankton phenology in a temperate lake. Freshwater Biol. 54, 1888–1898.

58. Moss, B.R., 2009. Ecology of fresh waters: man and medium, past to future. Blackwell, Oxford.

59. Nõges, P., Kisand, A., 1999. Forms and mobility of sediment phosphorus in shalloweutrophic Lake Võrtsjärv (Estonia). Int. Rev. Hydrobiol. 84, 255–270.

60. Naiman, R.J., Turner, M.G., 2000. A future perspective on North America's freshwater ecosystems. Ecol. Appl. 10, 958–970.

61. O'Neil, J.M., Davis, T.W., Burford, M.A., Gobler, C.J., 2012. The rise of harmful cyanobacteria blooms: the potential roles of eutrophication and climate change. Harmful Algae 14, 313–334.

62. O'Reilly, C.M., Alin, S.R., Plisnier, P.D., Cohen, A.S., McKee, B.A., 2003. Climate change decreases aquatic ecosystem productivity of Lake Tanganyika, Africa. Nature 424, 766–768.

63. Pörtner, H.O., Knust, R., 2007. Climate change affects marine fishes through the oxygen limitation of thermal tolerance. Science 315, 95–97.

64. Parmesan, C., Yohe, G., 2003. A globally coherent fingerprint of climate change impacts across natural systems. Nature 421, 37–42.

65. Peeters, F., Straile, D., Lorke, A., Livingstone, D.M., 2007. Earlier onset of the spring phytoplankton bloom in lakes of the temperate zone in a warmer climate. Global Change Biol. 13, 1898–1909.

66. Pettersson, K., George, G., Nõges, P., Nõges, T., Blenckner, T., 2010. The impact of thechanging climate on the supply and re-cycling of phosphorus. In: Glen, G. (Ed.), The Impact of Climate Change on European Lakes. Springer, Dordrecht, Heidelberg, London, New York.

67. Qin, B., Zhu, G., Gao, G., Zhang, Y., Li, W., Paerl, H.W., Carmichael, W.W., 2010. A drinking water crisis in Lake Taihu, China: linkage to climatic variability andlake management. Environ. Manage. 45, 105–112.

68. Rabalais, N.N., Turner, R.E., Diaz, R.J., Justic, D., 2009. Global change and eutrophication of coastal waters. Ices J. Mar. Sci. 66, 1528–1537.

69. Rockström, J., Steffen, W., Noone, K., Persson, Å., Chapin, F.S., Lambin, E.F., Lenton, T. M., Scheffer, M., Folke, C., Schellnhuber, H.J., Nykvist, B., De Wit, C.A., Hughes, T., Van Der Leeuw, S., Rodhe, H., Sörlin, S., Snyder, P.K., Costanza, R., Svedin, U., Falkenmark, M., Karlberg, L., Corell, R.W., Fabry, V.J., Hansen, J., Walker, B.,Liverman, D., Richardson, K., Crutzen, P., Foley, J.A., 2009. A safe operating spacefor humanity. Nature 461, 472–475.

70. Saros, J.E., Clow, D.W., Blett, T., Wolfe, A.P., 2011. Critical nitrogen deposition loads inhigh-elevation lakes of the western US inferred from paleolimnological records.Water Air Soil Pollut. 216, 193–202.

71. Schar, C., Vidale, P.L., Luthi, D., Frei, C., Haberli, C., Liniger, M.A., Appenzeller, C., 2004.The role of increasing temperature variability in European summer heatwaves. Nature 427, 332–336.

72. Scheffer, M., Carpenter, S., Foley, J.A., Folke, C., Walker, B., 2001. Catastrophic shifts in ecosystems. Nature 413, 591–596.

73. Schindler, D.W., 1974. Eutrophication and recovery in experimental lakes: implications for lake management. Science 184, 897–899.

74. Schindler, D.W., 2001. The cumulative effects of climate warming and other human stresses on Canadian freshwaters in the new millennium. Can. J. Fish. Aquat. Sci. 58, 18–29.

75. Schippers, P., Lurling, M., Scheffer, M., 2004. In crease of atmospheric CO2 promotes phytoplankton productivity. Ecol. Lett. 7, 446–451.

76. Shuter, B.J., Post, J.R., 1990. Climate, population viability, and the zoogeography of temperate fishes. Trans. Am. Fish. Soc. 119, 314–336.

77. Smol, J.P., Douglas, M.S.V., 2007. Crossing the final ecologicalthreshold in high Arctic ponds. Proc. Natl. Acad. Sci. U. S. A. 104, 12395–12397. Smol, J.P., Wolfe, A.P., Birks, H.J.B., Douglas, M.S.V., Jones, V.J., Korhola, A., Pienitz, R., Ruhland, K., Sorvari, S., Antoniades, D., Brooks, S.J., Fallu, M.A., Hughes, M., Keatley, B.E., Laing, T.E., Michelutti, N., Nazarova, L., Nyman, M., Paterson, A.M., Perren, B., Quinlan, R., Rautio, M., Saulnier-Talbot, E., Siitonen, S., Solovieva, N., Weckstrom, J., 2005. Climate-driven regime shifts in the biological communities of Arctic lakes. Proc. Natl. Acad. Sci. U. S. A. 102, 4397–4402.

78. Stefan, H.G., Hondzo, M., Fang, X., 1993. Lake water quality modeling for projected future climate scenarios. J. Environ. Qual. 22, 417–431.

79. Tao, H., Pan, J.Z., Shen, Y.L., Li, W.C., Huang, F., Zhao, H.G., 2012. Phosphorus threshold for the shift between grass-and algae-stable states in Dahong Harbor of Gehu Lake. Chin. J. Appl. Ecol. 23, 264–270.

80. UNFCCC, 2007. Climate Change: Impacts, Vulnerabiliti es and Adaptation in Developing Countries. United Nations Framework Convention on Climate Change, Bonn, Germany 64 pp..

81. Vörösmarty, C.J., McIntyre, P.B., Gessner, M.O., Dudgeon, D., Prusevich, A., Green, P., Glidden, S., Bunn, S.E., Sullivan, C.A., Liermann, C.R., Davies, P.M., 2010. Global threats to human water security and river biodiversity. Nature 467, 555–561.

82. Walker, B., Meyers, J.A., 2004. Thresholds in ecological and social-ecologicalsystems: a developing database. Ecol. Soc. 9, 3.

83. Walther, G.R., Post, E., Convey, P., Menzel, A., Parmesan, C., Beebee, T.J., Fromentin, J.M., Hoegh-Guldberg, O., Bairlein, F., 2002. Ecological responses to recent climate change. Nature 416, 389–395.

84. Wang, R., Dearing, J.A., Langdon, P.G., Zhang, E., Yang, X., Dakos, V., Scheffer, M., 2012. Flickering givers early warning signals of a critical transition to a eutrophic lake state. Nature 492, 419–422.

85. Wang, H.J., 2007. Predictive Limnological Researches on Small- to Medium-sized Lakes Along the Mid-lower Yangtze River. Institute of Hydrobiology, ChineseAcademy of Sciences, Wuhan, China. West, M., Harrison, P., 1989. Bayesian Forecasting and Dynamic Models. SpringerVerlag,New York.

86. Weyhenmeyer, G.A., 2004. Synchrony in relationships between the North Atlantic oscillation and water chemistry among Sweden's largest lakes. Limnol. Oceanogr. 49, 1191–1201.

87. Whitehead, P.G., Wilby, R.L., Butterfield, D., Wade, A.J., 2006. Impacts of climate change on in-stream nitrogen in a lowland chalk stream: an appraisal ofadaptation strategies. Sci. Total Environ. 365, 260–273.

88. Whitehead, P., Wilby, R., Battarbee, R., Kernan, M., Wade, A.J., 2009. A review of the potential impacts of climate change on surface water quality. Hydrol. Sci. J. 54, 101–123.

89. Williams, B.K., 2011. Adaptive management of natural resources—framework and issues. J. Environ. Manage. 92, 1346–1353.

90. Winder, M., Schindler, D.E., 2004. Climatic effects on the phenology of lake processes. Global Change Biol. 10, 1844–1856.

91. Woodward, G., Perkins, D.M., Brown, L.E., 2010. Climate change and freshwater ecosystems: impacts across multiple levels of organization. Philos. Trans. R. Soc. B Biol. Sci. 365, 2093–2106.

92. Yang, H., Flower, R.J., 2012. Effects of light and substrate on the benthic diatoms in an oligotrophic lake: a comparison between natural and artificial substrates. J. Phycol. 48, 1166–1177.

93. de Groot, R.S., Wilson, M.A., Boumans, R.M.J., 2002. A typology for the classification, description and valuation of ecosystem functions, goods and services. Ecol.Econ. 41, 393.

94. van Vuuren, W., Roy, P., 1993. Private and social returns from wetland preservation versus those from wetland conversion to agriculture. Ecol. Econ. 8, 289–305.

CHAPTER 18

THE DROUGHT RISK ATLAS: ENHANCING DECISION SUPPORT FOR DROUGHT RISK MANAGEMENT IN THE UNITED STATES

Mark D. Svoboda, Brian A. Fuchs, Chris C. Poulsen, Jeff R. Nothwehr

Keywords: Drought indices, Drought climatology, Drought planning, Drought risk management, Drought frequency, Drought time series

INTRODUCTION

Decision makers have continuously asked for better tools and resources to help them assess their risks related to climate variability and extremes (Jacobs, 2002). Drought is one of the risks they face, and the need for better drought risk tools and resources has also been expressed (GSA, 2007). Drought is a natural phenomenon that impacts every location and climate regime around the world. Understanding how droughts develop (and have developed), evolve, propagate and affect us is vital to knowing how to better prepare and plan for them and mitigate their impacts. The Multihazard Mitigation Council (2005) estimated that every dollar spent by the Federal Emergency Management Agency on hazard mitigation provides the nation with approximately $4 in future benefits. By its very nature, drought is typically characterized by slow onset and slow recovery. Since drought is usually not associated with widespread structural damage or loss of life in the United States, it typically does not receive the same attention as other disasters. According to the National Climatic Data Center (NCDC) (2014), eighteen droughts between 1980 and 2013 resulted in just over $253 billion in damages. More recently, the ongoing droughts of 2011–2014 in the central/southern Great Plains and California have already surpassed $50 billion (NCDC, 2014) in damages, and this number will continue to grow as more losses are assessed and calculated.

Given the importance of and need for a current online drought atlas resource, the National Drought Mitigation Center (NDMC) began a process to create an enhanced web-based Drought Risk Atlas (DRA), building on the efforts of the first National Drought Atlas (NDA) and recent work at the state level via the *Hydrologic Drought Atlas for Texas*(Rajsekhar et al., 2014). With the DRA's emphasis on providing usable information, this work also compliments other delivery systems such as the U.S. Drought Monitor (USDM) (http://droughtmonitor.unl.edu) (Svoboda et al., 2002), National Integrated Drought Information System (NIDIS) Portal (http://drought.gov), NDMC's portal (http://drought.unl.edu), Global Integrated Drought Monitoring and Prediction System (GIDMaPS) (Hao et al., 2014 and Momtaz et al., 2014) and the U.S. Surface Water Monitor (Wood and Lettenmaier, 2006).

The NDA was developed in 1994 (Willeke et al.) by the U.S. Army Corps of Engineers (ACE), the National Oceanic and Atmospheric Administration (NOAA) and IBM. The NDA analyzed 1,036 stations from the United States Historical Climatology Network (USHCN) (Karl et al., 1990), which contained data from the Cooperative Observer Program (COOP) of the National Weather Service (NWS). The NDA focused primarily on hydrology and the Palmer Drought Severity Index (PDSI), which was calculated from monthly precipitation and temperature totals.

Developing a strategy to mitigate drought risk and manage water resources for any location is dependent upon understanding the climate regimes and drought climatology for the specific area of interest. The challenge moving forward is to have the best available data and assessment tools for decision makers—especially producers, water resource managers and planning practitioners—to adequately plan and prepare for drought events. With a changing climate, the ability to compare trends over the last several decades to historical values will give decision makers the ability to contrast current drought events with historical drought events, with a goal of

Citation: Mark D. Svoboda, Brian A. Fuchs , Chris C. Poulsen , Jeff R. Nothwehr, The drought risk atlas: Enhancing decision support for drought risk management in the United States, http://dx.doi.org/10.1016/j.jhydrol.2015.01.006.

making better informed management decisions to account for, and reduce, drought risk in the future.

Climate data and various drought indicators and indices are key components used to monitor drought. From a historical archive perspective, the PDSI suite and Standardized Precipitation Index (SPI) drought indices have been computed each month on a climate division (multi-county) scale for the entire United States by the NCDC with data that extends back to 1895. For a generalized perspective on drought, this approach is adequate. Problems associated with using information on a coarse climate division scale can be eliminated with other techniques that allow for county and sub-county level analyses. With each new drought event, the question comes up as to how the current drought compares to others historically with regard to frequency, severity, duration and spatial extent for a specific location, county or basin. From the NDMC's DRA database, queries can be made to help users find answers to these questions and more for a given location or region. The results and data are then freely available for downloading or viewing through a number of visualization tools made available through the DRA's web interface.

The NDMC approached the idea of an enhanced drought risk atlas by expanding upon the work in the NDA by including more stations, using stations with a longer period of record than what was available to the NDA in the early 1990s (along with an additional two decades worth of data), using a weekly time step instead of a monthly period, and calculating more drought indices and providing access to the tools and outputs in a user-friendly interface by implementing GIS techniques and delivering data and information digitally via the web. The DRA consists of more than one billion index records and will house more than 500,000 grids once fully complete, and all of this data will be cached and made readily available, with the data running through 2012. We assigned each station a unique start date based upon the characteristics of that station.

OBJECTIVES

The NDMC's DRA was built with the intention of providing a wide range of decision makers and users with locally and historically tailored drought information via a web-based tool and database in order to visualize and assess their risk to drought. At this time, few tools are available for decision makers and scientists to use in evaluating drought characteristics, or climatology, on a localized scale for the United States. The DRA fills a gap by providing a much-needed mechanism for research, decision making and planning perspectives on both past and future drought episodes by providing historical climate data and drought indices at a more localized level (using a station-based approach) and on a more frequent time step (weekly instead of monthly for several indices). One of the primary goals of this tool is to increase users' capacity to understand their drought history and to identify past and present trends along with past, present and future vulnerabilities to drought in order to make more informed decisions aimed at reducing risk to future droughts.

The DRA is focused on increasing the capacity of users and decision makers (including policy makers and planners, as well as producers and water and natural resource managers) to analyze their potential risk to drought in any particular location for any time of the year. The regionalization techniques developed for the first drought atlas allowed for the estimation of drought frequencies from several locations instead of just one, by calculating frequencies using L-moments statistical techniques. The NDMC has developed methods to do all of this work in an ArcGIS environment, allowing for the rapid assessment of results and graphical output via a web interface available at http://droughtatlas.unl.edu. The data, maps, graphs and other derivative output products are freely available to be shared with the scientific research community and in support of the NIDIS and NWS-Climate Services Division (CSD) efforts, along with the general public.

Tailoring the DRA to the needs of planners, producers, university extension agents, natural resource decision makers and NOAA–NWS field personnel through a series of workshops and webinars during development allowed for the integration of their feedback throughout the process. These efforts led to the improvement of analyzing drought as an extreme event by providing data and visualization tools to help them better understand the frequency, historical context, magnitude, spatial extent and trends of drought at the local (station), basin, tribal, regional or state level. The DRA is already populated for the entire continental United States and can serve as an immediate application tool within any of the NIDIS regional drought early warning system (RDEWS) pilot regions. The tool's built-in spatial flexibility will provide producers, natural/water resource managers and planners with a valuable resource to help them better inform their constituents on how to cope with climate variability and change at all scales and under various levels of risk or uncertainty.

DATA AND METHODS

Station Screening

More than 12,000 stations were initially screened and run through a quality control process by the NDMC and High Plains Regional Climate Center (HPRCC) using the Applied Climate Information System (ACIS) database. Of these initial stations, 3059 stations (nearly triple the number analyzed in the first drought atlas) collecting precipitation from NOAA's–NWS COOP network housed within ACIS were identified by the NDMC as meeting our specific selection criteria, making them the best long-term climatic stations found within the USHCN. All of the stations in the selected subset have at least 40 years of daily data (with an average period of record of nearly 70 years) with no more than two consecutive months of missing data occurring at any point in the period of record. Of these 3059 stations, 2569 have both temperature and precipitation data, which are needed for calculating several of our chosen indices such as the PDSI suite and the Standardized Precipitation Evaporation Index (SPEI).

After final extraction of the COOP station data from ACIS (accomplished with our partners at the HPRCC), the following criteria were established for a station's inclusion in the drought risk atlas: (1) the station had to currently be open and taking regular observations; (2) the station had to have a minimum of 40 years of data available, going backward from 2012; and (3) the station could not have more than 2 consecutive months of missing data at any time within the period of record.

Based upon the above criteria, a unique period of record (POR) was established for each station. Using the 12,000+

stations found within the COOP database, two screenings of the data were done. The first screening looked for stations that had at least 30 years of data in their record and had 80% or more of the daily data available. A report for each station was made showing the months that had missing data for each year. By looking at the most current data and going back in time, we determined for each station how far back the data met our criteria. Gaps in the records were investigated to determine if the data existed somewhere else and was not present in the digital data archive within ACIS. The National Climatic Data Center (NCDC) provided access to scanned copies of the COOP forms in their archive via the Web Search Store Retrieve Display (WSSRD) system, which has transitioned over to the Environmental Document Access and Display System (EDADS). By comparing the data that was available on paper with the gaps in ACIS, we identified 288 stations for which the data gaps could be rectified and entered into the digital ACIS archive. HPRCC staff keyed in the data for the gaps we identified using Datzilla, and these stations were then incorporated into our DRA database and eventually into the NCDC archives as well. Although a 30-year historical record is a good recommended starting point, periods of 40–60 years of monthly data would be better (Guttman, 1994) in order to give a reasonable sample size used in the approximation of the computed means, so we arbitrarily adjusted the initial station query to include only those stations having 40 or more years of data.

Next, a secondary screening (gap) analysis was conducted on each station to identify where the gaps in the POR fell within each station that passed the initial screening as a way of helping us determine the unique starting date for the POR associated with that station. All stations have an ending period through December 2012. Many stations have records that go back to the late 1800s, but the earlier data is often incomplete and contains many gaps. By taking these years of missing data out of the record and establishing a unique starting point that differs from the date the station first began taking observations, a unique and complete POR was established. By providing each station with a unique starting date, the longest periods of record were made available by allowing the record to stop when the established criteria were no longer met, but an adequate climatic time period was established. To aggregate the daily data into a weekly time step, weeks had to be established to provide a summation of precipitation for each week. We determined that Week 1 starts on March 1 and includes seven days. The last week of the year ends in February and, depending on the year, may include additional days in the summation to account for leap years.

After this daily summation exercise, we presented the information to the American Association of State Climatologists (AASC) at their annual meeting and through interactions with the AASC list server. We asked for their assistance in determining whether our chosen stations were of high quality and also if they knew of additional stations missing from our screenings that would potentially meet our criteria. Twenty-two state climatologists and one regional climate center offered feedback, including notification that there were other candidate stations we should consider. It was determined that another screening of the full COOP dataset was needed. Using the established criteria, we looked at the data going backward and stopped our search the first time a two-month gap was found. Using the 80% of available data criteria threshold excluded a number of stations where the data period had large gaps early in the record, but were otherwise fairly complete. The 80% criteria also excluded stations that are primarily "event reporters" in that they only record precipitation when it takes place and do not record zeros very often. By screening the data a second time and comparing the stations to those we already had, we were able to add 915 stations into our catalog, bringing the total number of stations in the DRA up to a total of 3059, which is where it currently stands as of this writing (see Table 1). Fig. 1 shows the spatial distribution of those COOP stations recording precipitation and having the minimum amount of 40 years across the lower 48 states. The most notable gaps in station density (Fig. 2) were, as expected, found in the West across parts of the Great Basin, Four Corners region and within Wyoming. East of the Rocky Mountains, station density and distribution is quite good with most stations falling within 20–30 miles of one another. For the country as a whole, virtually all stations were no farther than 75 miles apart, expect for parts of the West as noted above.

Table 1: Final number of stations meeting at least the minimum DRA selection criteria

Number of stations	Years of data	Percent of stations (%)
3059	40	100.00
2462	50	81.04
1733	60	57.04
1170	70	38.51
827	80	27.22
537	90	17.68
349	100	11.50

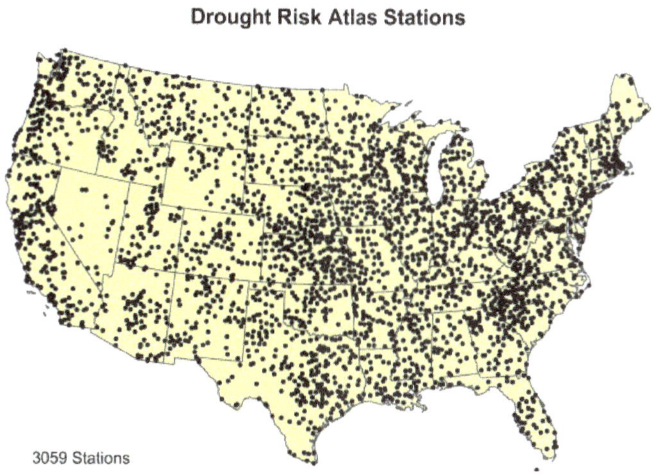

Figure 1: All COOP precipitation collection stations meeting the DRA selection criteria and having a minimum of 40 years of record.

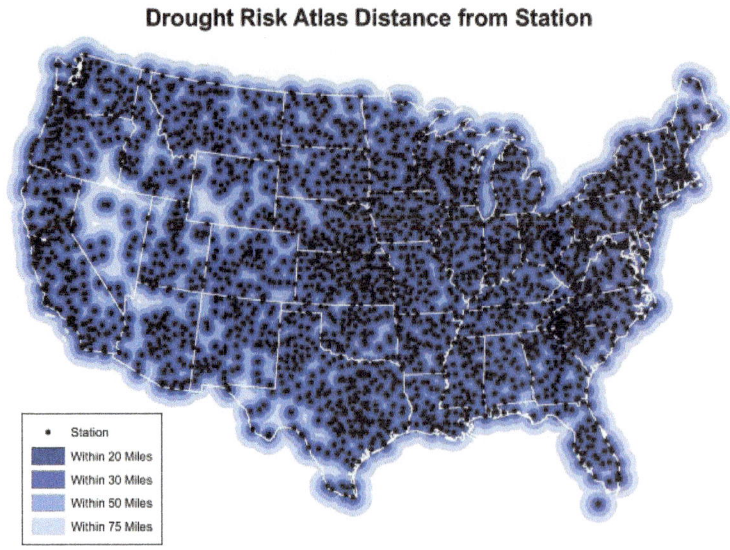

Figure 2: GIS analysis shows the density of the DRA network of precipitation stations. Note that one should be able to find a similarly behaving station within at least 75 miles from any given point of interest for all but parts of the West. Density and distribution is quite good east of the Rocky Mountains where most stations are within 20–30 miles of another station.

Cluster Analysis

For the DRA, a regional frequency analysis technique was applied for the 3059 stations based upon an L-moments approach (Hosking and Wallis, 1997) in which a cluster analysis was performed to develop unique climate regions in which stations within the same cluster showed similar climatic and statistical behaviors with regards to precipitation. The concept, given the lack of ideal station density and spatial distribution around the country, is aimed at providing a homogeneous region of similar climatic stations whereby users can choose from multiple stations having similar drought behaviors to use in the decision-making process instead of a single point. This is important given the fact that in many cases users will have to choose a station that is not their hometown, or preferred, station. Using the cluster analysis, return frequencies were generated for the cluster and for each station within the unique climate region using the L-moments techniques.

The cluster analysis process consisted of grouping stations together based upon each station's latitude, longitude, elevation and precipitation characteristics. A GIS widget was developed by the NDMC that aided in the process of determining where to place stations that could fit in more than one cluster. For each cluster, precipitation characteristics were developed for each season (winter, spring, summer and autumn), giving us the ability and flexibility to cluster the stations based upon data from any of the seasons. We decided the summer season, because of the widespread convective nature and consequent variability of precipitation, would be our best choice to determine our clusters. Utilizing the final cluster analysis based upon the station attributes and the summer precipitation characteristics for each station, 139 unique clusters (Fig. 3) were developed. Each cluster was tested for homogeneity and discordancy, with less than 3% of the points being discordant for any particular season and only nine of the clusters determined to be inhomogeneous for the summer months. From these established clusters, individual drought/climate regimes were established in which frequency and return periods could be calculated.

Figure 3: Total number of clusters in the DRA as determined from L-moments and homogeneity testing.

The average cluster size contained 22 stations and the range of cluster populations varied between 5 and 49 stations. Fig. 4 represents a fairly typical cluster in terms of size, number of stations and distribution. The stations falling within each cluster share common drought and statistical behavior. After the 139 clusters were decided upon, discordancy testing was done on each of the 139 clusters and for each season. Only 37 points (1.21%) within our clusters were considered discordant for the summer precipitation season. Our reasoning for this is explained below in the homogeneity testing discussion. Following the Hosking and Wallis L-moments methodology (1997), discordancy was tested using the clusters developed for the summer season, but with the precipitation characteristics of the other seasons used instead. For autumn, there were 80 discordant points (2.62%); for spring, 84 discordant points (2.75%); and for winter, 75 discordant points (2.45%). These results were considered acceptable as we discovered that several points were discordant regardless of the cluster they were put into, so they were subjectively assigned into a cluster that made the most sense geographically. We could have also chosen to dismiss these discordant stations from the DRA, but given the fact that the stations were of such high quality, we decided against that course of action.

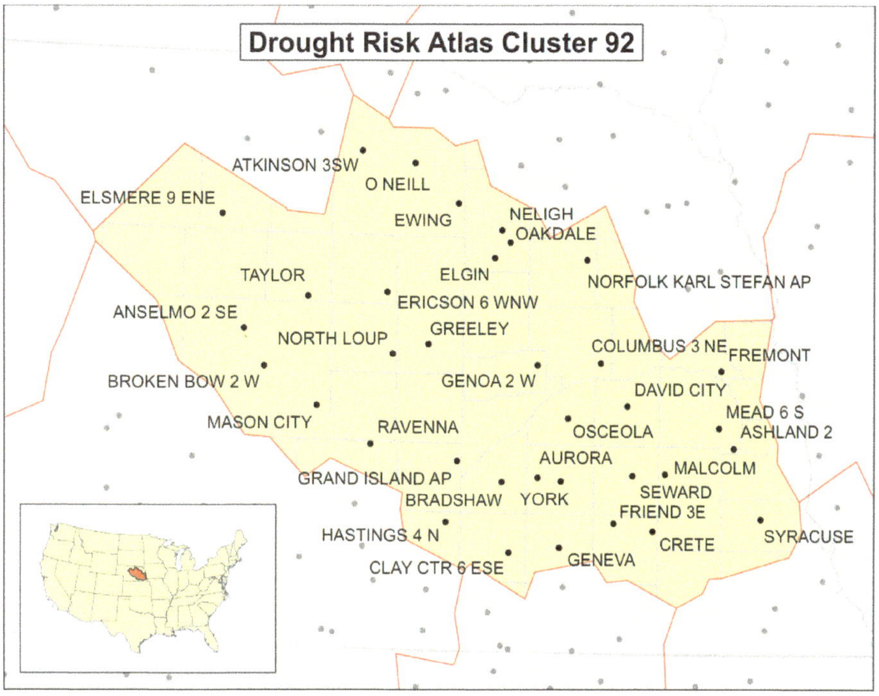

Figure 4: Example of a DRA cluster and station distribution. Stations in clusters exhibit similar statistical and precipitation behavior allowing a user multiple options for assessing drought history in their point of interest.

Homogeneity Tests

Homogeneity tests (H1) (Hosking and Wallis, 1997) were conducted upon each cluster using the clusters derived from the summer precipitation characteristics. A Monte Carlo simulation was used, with 2000 simulations being conducted for each season. For the summer season, 9 of the 139 clusters failed the homogeneity test (6.45%). Of the clusters that failed, many were located in the western United States where the summer season is dry and not the most ideal period of the year to look for similarity in station records, as a precipitation event is an anomaly. For autumn, 35 clusters did not pass the homogeneity test (25.2%), spring had 20 failures (14.4%) and winter produced 16 failures (11.5%). It also makes sense climatologically that the transitional seasons of spring and autumn would have the most variability for most of the country and have the most in-homogeneities compared to the summer season. Only 3 of the 139 clusters failed the homogeneity tests for all four seasons. This was not surprising as some clusters were developed in terrains and landscapes that have tremendous variability, and regardless of how the stations were clustered they would not become homogeneous. In addition, these three particular clusters contained fewer reporting stations and covered large arid areas in the Desert Southwest, which also contains contrasting elevations over short distances. During development of the cluster regions, the clusters were made available for review and critiquing by the state climatologists and Regional Climate Centers. Feedback from these groups resulted in adjustments to the clusters, with some clusters being divided into two or more clusters.

The L-moments were calculated for each season and for various probabilities. The quantiles calculated were 0.01, 0.02, 0.05, 0.10, 0.20, 0.50, 0.80, 0.90, 0.95, 0.98, and 0.99. In calculating the L-moments for each cluster, several distributions were available to use, including: Exponential, Gamma, Generalized extreme-value, Generalized logistic, Generalized normal, Generalized Pareto, Bumbel, Kappa, Normal, Pearson Type-III and Wakeby.

From our analysis, we found the Pearson Type III to perform and best fit our overall needs given the fact that no one distribution will typically fit all sites. The other distributions were tested and our results were not changed, or improved, in any significant manner such that it warranted changing from the use of the Pearson Type III.

Data and Indices

Millions of daily temperature and precipitation values were collected (see Table 2) and aggregated in order to be able to calculate each of the indices across various time steps.

Table 2: The total number of climate records used in computing the indices

Data types	Total records	Start date
Raw data (daily data)	81,861,427	10/2/1849
Serially complete (daily data)	80,464,757	1/1/1908
Aggregated data (weekly, monthly, etc.)	40,574,010	n/a

For the estimation and data filling component, work was conducted at the HPRCC to generate the serially complete datasets for both precipitation and temperature for those stations having missing daily data. The HPRCC has developed and refined several methodologies (You et al., 2008) over the years aimed at providing data estimates as part of their data quality control and assurance programs. Using a nearest-neighbor approach along with inverse distance weighting and spatial regression testing techniques, the HPRCC generated temperature and precipitation estimates for the missing DRA data, thereby giving us a serially complete daily dataset to complement the raw daily set. Given the quality screening and resultant completeness of those stations meeting the screening criteria, the amount of daily data that ended up being estimated from a total of nearly 82 million records was minimal at just 1.02%. In the interface, the user has the option to analyze, display and/or download the data using either the raw or serially complete dataset.

For each station meeting the screening criteria, drought indices were computed using a weekly or monthly time period and utilizing both the raw and serially complete (estimated) datasets. From these daily raw and serially complete datasets, the following indices, indicators and climatic information were calculated and are made available through the DRA web interface: the Standardized Precipitation Index (SPI), Standardized Precipitation Evapotranspiration Index (SPEI), Palmer Drought Severity Index (PDSI), Self-Calibrated Palmer Drought Severity Index (scPDSI), Deciles, analog climate and indices data and historical rankings, frequency and return periods by location, and the U.S. Drought Monitor (USDM).

Using the selected stations, batch programming was done in order to calculate the SPI, SPEI, and Deciles for the 1–12-, 15-, 18-, 24-, 36-, 48-, 60-, 72-, 84- and 96-month periods using a weekly and/or monthly moving window time step. Gridded maps (0.25°) for each week from 1887 to 2012 are still being generated for the indices where there are enough data to support them.

As shown in Table 3, in order to compile the indices for the atlas, millions of records were calculated for each index resulting in more than one billion records generated and archived in our database. Efforts will be made to periodically update the database over time in order to keep the statistics and information current and relevant.

Table 3: The total number of records generated by each drought index

Index	Total records
SPI (weekly)	554,211,720
SPEI (monthly with weekly coming)	46,156,540
PDSI (monthly)	2,125,528
Self-calibrated PDSI (monthly)	2,125,520
Deciles (weekly)	538,915,840

The SPI (McKee et al., 1993) is available for all 3059 stations in the DRA and was calculated using both weekly and monthly aggregates of precipitation data using either raw or serially complete data, although fewer stations (2569) are available for any serially complete analyses.

Calculation of the SPEI (Vicente-Serrano et al., 2010), PDSI (Palmer, 1965) and scPDSI (Wells et al., 2004) requires precipitation and temperature inputs along with serially complete data, thus it is available for computing on just those 2569 stations containing both precipitation and temperature data within the DRA. The SPEI is currently calculated using monthly aggregates of data although work is underway to allow for weekly moving window calculations within the SPEI, as is currently the case with the SPI. For this project we used the original version of the SPEI as coded by Vicente-Serrano, which uses a Thornthwaite calculation for the PET values.

The Deciles (Gibbs and Maher, 1967) rankings were also calculated for all 3059 stations in the DRA. Like the SPI and SPEI, Deciles are relatively straightforward to compute and interpret, with a value that falls within the lowest 10% of the record said to be in the first Decile, and so on. For the purposes of determining when a region is in drought, the first or second Decile is typically used as the breaking point. Any values that fall within the first decile are considered to be in drought. Like the SPI, Deciles can be calculated using multiple time-steps. This is done by determining the length of the time-step and summing the precipitation values over the entire time-step. A given value for one time-step is then compared to the historical record of values for that same time-step. These rankings were completed using both weekly and monthly aggregates of data, and the 1st through 10th Deciles periods were computed.

Although the USDM (Svoboda et al., 2002) was not calculated specifically for the DRA, it is included in our database and made available through the tool so that the DRA indice archives can be compared to the state-of-the-art system used to monitor drought today. The key limitation with the USDM is that the archive only goes back to January 2000.

RESULTS AND DERIVATIVE PRODUCTS

A combination of tabular and visualization tools can be used to assess a location's drought history through a web-based interface at http://droughtatlas.unl.edu. The DRA tool was designed to deliver data and products in a way that allows both novice and expert users to benefit from the information housed within. Using database management techniques, the output includes: user defined and interactive time series, heat maps, index comparison charts, analog rankings, drought periods, raw and serially complete tabular data, drought frequency, intensity and magnitude, and more than 500,000 gridded index maps.

The DRA decision support tool allows for data sorting (and extraction) and the viewing of multiple indices over a user-defined period of time. The time series are interactive and the user can highlight and choose, within the time series chart, subset periods of interest by clicking and dragging on the *x*-axis of the time series graphic. This flexibility is critical in that users and decision makers can assess and compare various drought indices to see both their previous and potential behavior as a trigger for impacts and/or decision making, which can then ultimately be tied to a drought mitigation plan or framework within the context of a Drought Early Warning System (DEWS).

The DRA's Map Viewer displays gridded drought indices and other basic information such as the locations of the stations used in the Atlas. The climate layers mainly consist of the different drought indices and also the basic atlas stations and clusters. Each of the drought indices layers provides the option to select a specific time period and time step. Multiple geospatial base layers can be displayed on the map. These include counties, climate divisions, NWS county warning areas, RMA regions, river basins, congressional districts, tribal lands and federal lands. These layers will be displayed underneath the climate data. The data viewing tool has three main functions: The ability to select a station, the ability to view datasets, and the ability to download datasets. There are three ways to select a station: by state, by keyword or ID, and by radius. Data queried and chosen for downloading are made available in either CSV or TXT file format. Geospatial information and products generated and provided by the DRA will be made discoverable and available in Open Geospatial Consortium (OGC)-compliant web services formats.

Users frequently want to know how the severity of a current drought/time period compares to other historical droughts or if the current period is the driest (or wettest) on record. By providing the analog data, using a weekly and/or monthly time step, comparisons can be made and users can answer such questions using the tool. The ability to rank order the various climatic indicators and drought indices allows decision makers to have a better idea of how their climate, or drought, profile (i.e. climatology) has changed over time. Fig. 5 illustrates the flexibility that the SPI offers through time series analyses at various time scales. The 3-month SPI allows for tracking drought (or wet) periods at the seasonal level (Fig. 5a), which is more indicative of a growing season, or agricultural, drought. Hydrological droughts typically evolve more slowly and lag behind agricultural drought before presenting impacts to water resources making annual (Fig. 5b) 12-month SPI or multi-year (Fig. 5c) 24-month SPI more applicable.

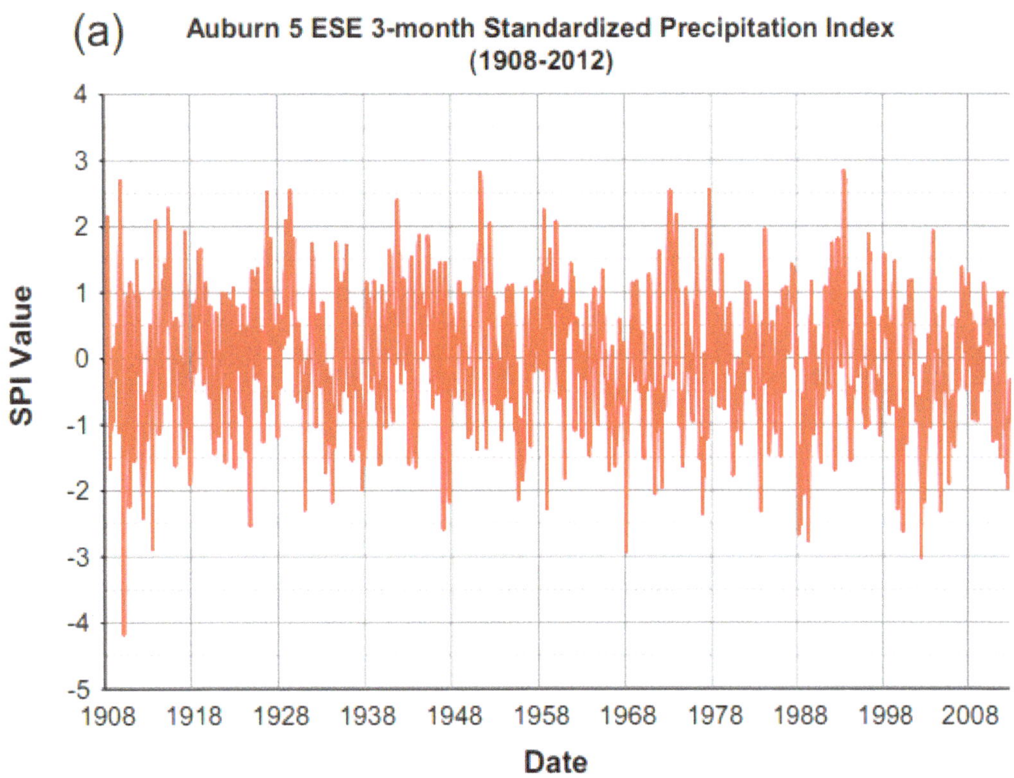

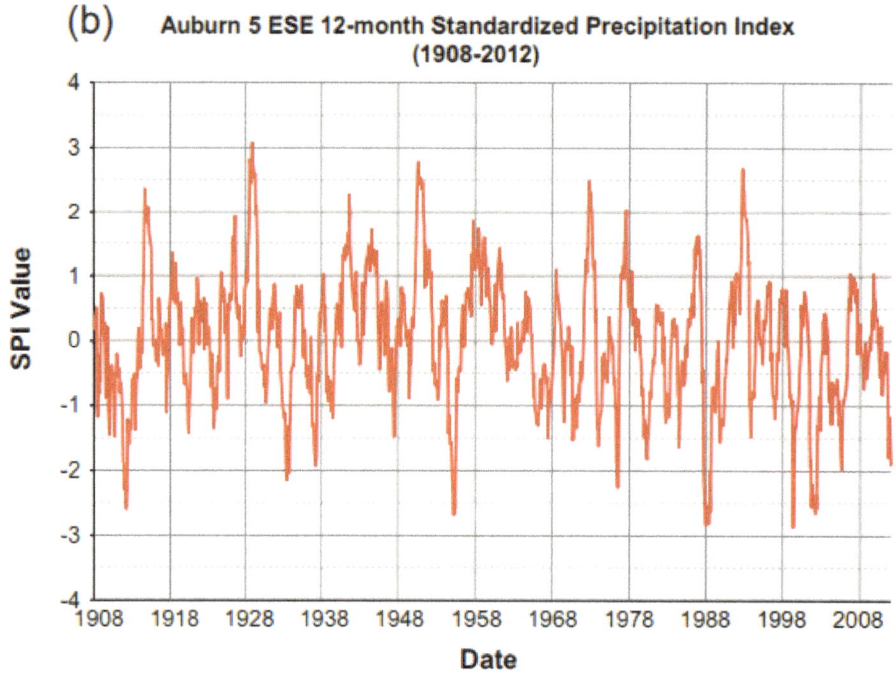

Fig. 5 *(continued)*

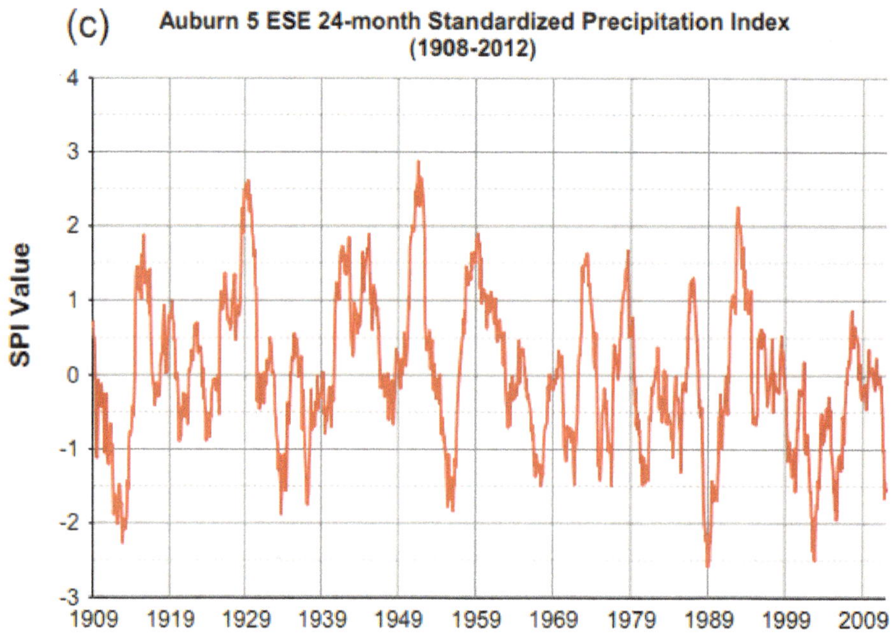

Figure 5: (a). Example of a short-term, seasonal 3-month SPI time series for Auburn 5 ESE, Nebraska. (b). Example of an annual 12-month SPI time series for Auburn 5 ESE, Nebraska. (c). Example of a long-term 24-month SPI time series for Auburn 5 ESE, Nebraska.

Other visualization tools within the DRA offer up ways of illustrating drought magnitude by various thresholds (Fig. 6). Severity and duration are both key to better understanding and comparing droughts and drought impacts over a station's period of record. A long lasting, lower intensity drought may end up causing more impacts than a very intense, very short-lived drought depending on timing and other factors that vary by region and season. A quick look at drought periods results in tables showing, depending on the user query, either the driest or wettest periods (Fig. 7) as determined by each index in the DRA database. This allows for a quick look at which years stand out from the rest. Drought frequency is yet another parameter of great interest by users and the DRA delivers this in a couple of ways. The user can see the number of occurrences for each index threshold (either positive/wet or negative/dry) value to see the distribution and how the tails (extremes) behave on both ends of the spectrum, both wet and dry (Fig. 8). In addition, return periods (based on years) show the user (Fig. 9) how often you can expect to see a particular drought index value occur based on the station history of that particular location. For example, the user can see how often a SPI (or any other index within the DRA) value of -2 (or any other threshold) occurs, expressed as once in every *n*-amount of years.

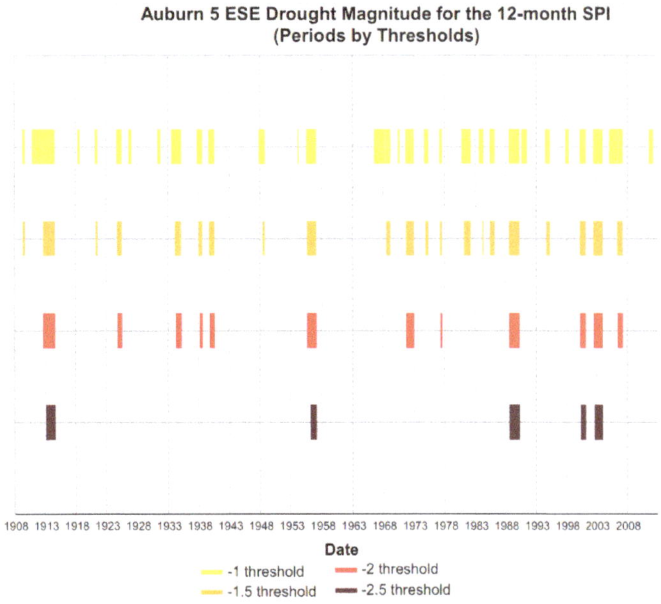

Figure 6: Drought periods (duration) and magnitude by various 12-month SPI thresholds for Auburn 5 ESE, NE.

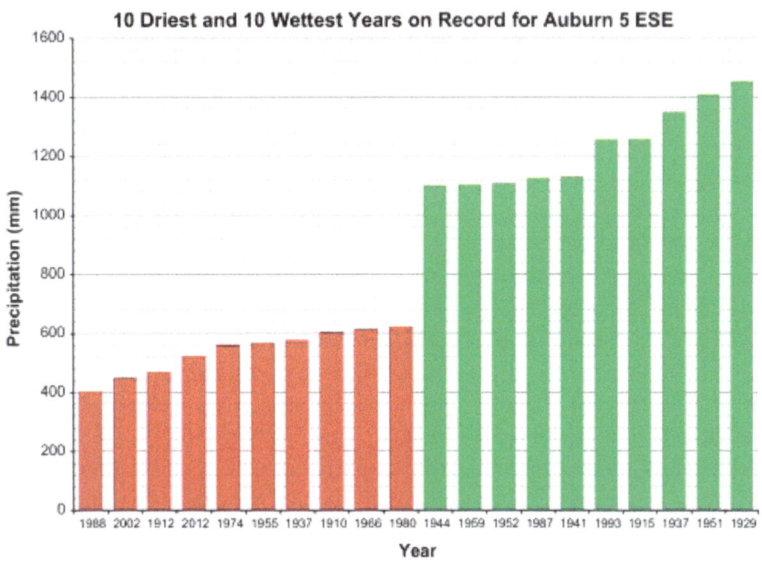

Figure 7: Sample DRA query for the driest and wettest years on record for Auburn 5 ESE, NE.

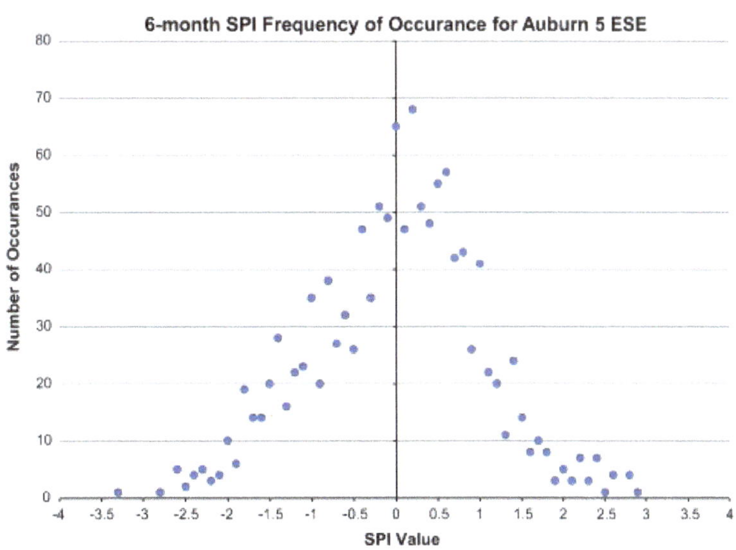

Figure 8: Sample frequency plot of all 6-month SPI threshold values for Auburn 5 ESE, NE.

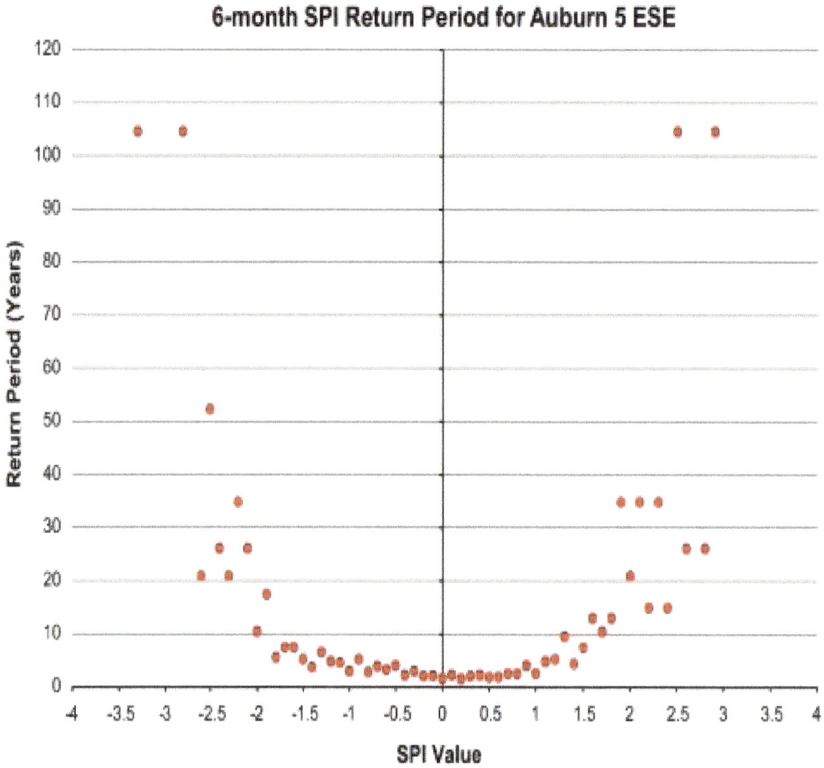

Figure 9: Return periods (in years) for various 6-month SPI values for Auburn 5 ESE, NE.

More than 500,000 gridded raster maps are being generated using ArcGIS for all indices and for all weeks/ months within the DRA's database period of record where enough data and derived indice values are available and can support it. A user map interface has been created for this application as well. Currently, not very many historical maps are available for any drought index using a station-based approach, especially at the weekly time scale. The DRA will provide, for each index (and all time frames computed for each index), a gridded map for each time step using the available stations for each particular week/month for all years back to the late 1800s where there is enough data to support it. With the current number of stations available, the optimal grid will have a radius of 25 km from the data point. The older maps do not have as many points available for the gridded surface as compared to more recent maps where more stations are readily available, which can result in some white (no data) areas on the respective grid. With each station having at least 40 years of data, every map from 1970 to present will include all data points for each week of each year. In addition to standard mapping tools, the mapping interface will allow users to overlay various geographic and political boundaries and have an interactive platform for customizing the map. These visual depictions can be used to determine the spatial extent of drought events and how they impacted a region. While transferring these data to users, specific geographic overlays will be made available, such as the NWS county warning areas for each Weather Forecast Office (WFO); 2-, 4-, 6-, and 8-digit HUCS; climate divisions; congressional districts; and tribal and federal lands. Fig. 10a and b are examples of the .25° gridded maps that will be generated for all weeks, or months depending on the index, back to the early 1900s. Various time steps can be utilized and both drought and wet patterns can be easily seen when mapped in this format. This also allows for a variety of ways to quickly visualize and package data to a particular time or area of interest. Fig. 10a represents more of an annual snapshot showing the hydrological drought in the West during 2002 whereas the 6-month SPI for September 2011 shows more of a growing season drought that afflicted the southern Great Plains and parts of the Four Corner region, eastern New Mexico and southeastern Colorado. A user can use the map feature to identify hot spots, which can then be followed up by zooming into regional or state templates (still to come within the DRA) for a closer look and then down to the cluster, or station, level for even more tailored and specific drought information queries and data retrieval. Fig. 11a shows that drought is not just a western phenomenon and that the tool can also be used in identifying and assessing extreme wetness as is clearly depicted in Fig. 11b given the heavy rains (and resultant high/positive SPI values) that led to a lot of flooding on the Missouri River during summer 1993.

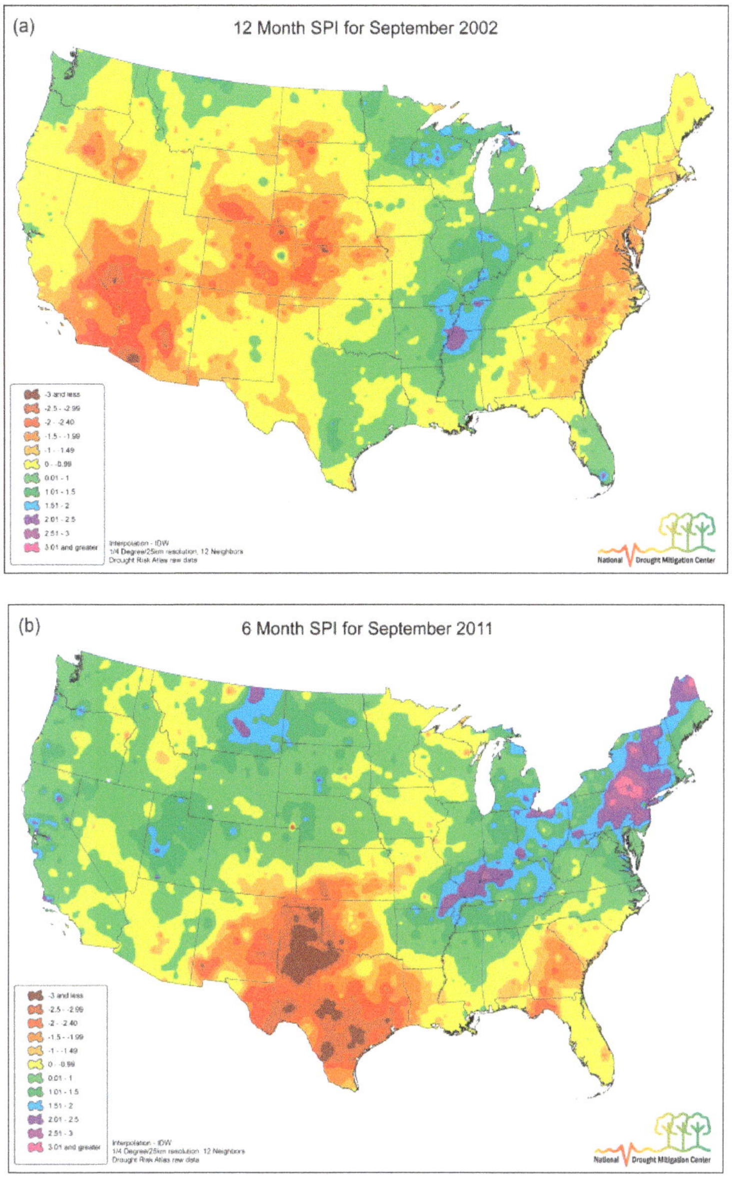

Figure 10: (a). Gridded 12-month SPI map for September 2002. The red (warm) colors represent dryness and the green and blues (cool) represent wet areas. (b). Gridded 6-month SPI map for September 2011. The red (warm) colors represent dryness and the greens and blues (cool) represent wet areas.

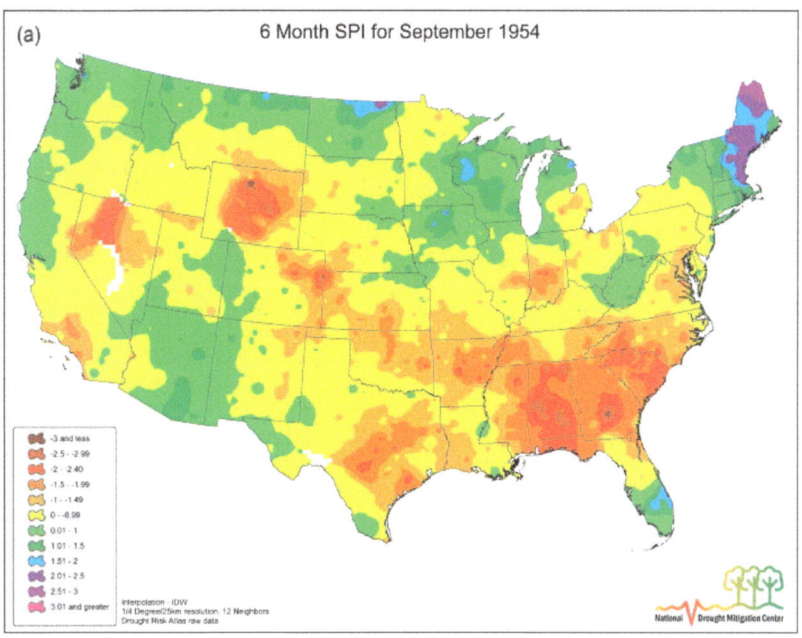

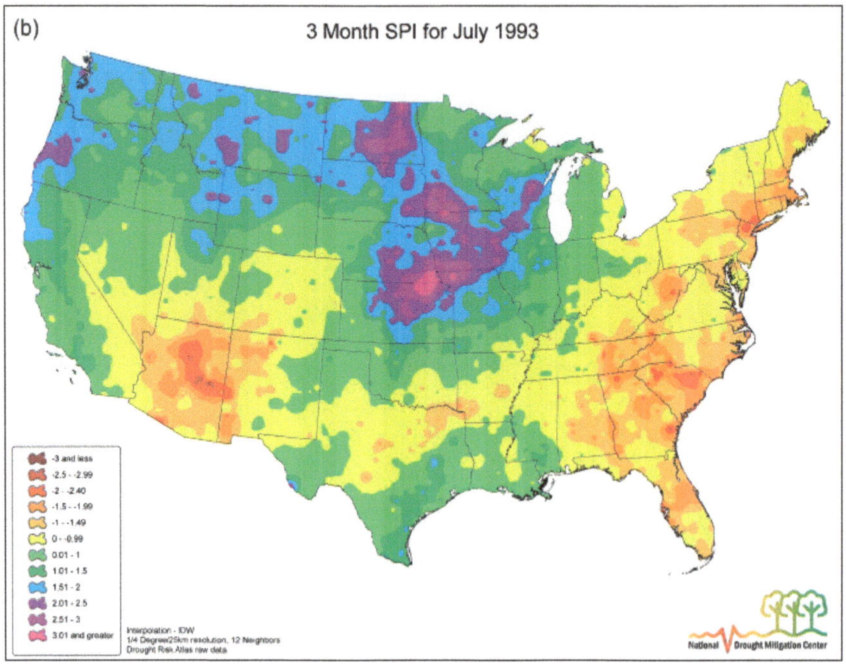

Figure 11: (a). Gridded 6-month SPI map for September 1954. The red (warm) colors represent dryness and the greens and blues (cool) represent wet areas. (b). Gridded 3-month SPI map for July 1993. The red (warm) colors represent dryness and the greens and blues (cool) represent wet areas.

DISCUSSION

As we continue to face an unknown and changing climate, drought will continue to be a prominent natural disaster within any resultant climate regime. The more that is known of past drought events and their impacts, the better prepared we can be as a society to face and plan for drought in future (Woodhouse and Overpeck, 2000). It has been shown that not all drought indicators work well for all locations, and providing a drought risk atlas for decision makers will allow for better and more informed choices by producers, decision makers and the drought planning and services community at the local level. This is especially true for water resource managers, who will need to adapt to supplying water to users in a climatic environment that has changed over the last several decades. As the climate regimes are changing across the United States, the DRA will give water resource managers the ability to "drill down" to their basin, watershed or local area of interest to find detailed information on the climate and historical drought trends for that area.

The ability for users to query and tailor the DRA to their needs should allow for an increased capacity in analyzing drought by providing data and visualization tools to help them better understand the frequency, historical context, magnitude and trends of drought at the local, basin, tribal, regional or state levels. The tool's built-in spatial flexibility provides natural and water resource planners with a valuable resource for better informing their constituents about how to better cope with climate variability and change at all scales and under various levels of risk or uncertainty.

In addition, data and information from the DRA will be made available within the NWS's developing Local Climate Analysis Tool (LCAT). We also intend to work closely with NOAA's CSD to provide tailored training and education on drought (and various drought indices) by specifically making them aware of the potential uses and applications of the DRA tool and its extensive database and derivative products. The DRA tool has already been integrated into the NIDIS Portal at drought.gov.

Continued outreach and collection of feedback will be done by the NDMC team through the provision of webinars and/or hands-on training on how to use and get the most out of the DRA tool. With each webinar, for example, the technology is available to archive these sessions while developing a "Frequently Asked Questions" informational page that would be contained in the drought risk atlas web delivery system. Feedback from webinars, surveys, workshops and professional meetings will allow for enhancement of the DRA. Ideally, the users will continually help us identify and determine what can be added to make the tool better.

SUMMARY AND FUTURE WORK

Dissemination of the DRA tool will be broad, given the NDMC's linkages with our NOAA and USDA partners (including the new Regional Climate Hubs) and the vast number of other partners such as the Regional Climate Centers, state climate offices and university research community (including extension, which the NDMC is engaged with locally, regionally, nationally and even internationally). Our strategy includes free distribution via the NDMC web site (drought.unl.edu) and the NIDIS Portal mentioned above, which allows for direct access by all existing NIDIS RDEWS pilot region communities and any future pilot regions/basins that may come on-line in the future. The NIDIS Engaging Preparedness Community will also be tapped into, as this is a large body of practitioners dealing with drought planning and preparedness around the country. The U.S. Drought Monitor network of authors and experts is yet another collection of drought experts (now numbering more than 350 people) that serves as a direct conduit to users around the country, along with our networks involving the American Planning Association (APA) and the state drought contact network maintained at the NDMC.

The DRA tool has the potential to serve as a vehicle for assessing risk to extreme drought events, which should help planners to determine their vulnerability and deal with drought proactively. One such potential application is through the integration of DRA deliverables into a framework like the Drought Ready Community (DRC) program (DRC, 2011), which is aimed at helping users develop and understand their drought climatology, risk and potential trends when undertaking a drought planning exercise. In addition, recent work between the NDMC and the APA led to the development of a Planning Advisory Service (PAS) on drought planning (APA, 2013). The PAS is geared toward exposing planning practitioners to decision support tools and best practices in the drought planning arena that they can utilize in their planning work. Finally, the NDMC continually educates users through workshops, webinars and other interactions about the merits and potential information available with tools like the DRA. Providing and maintaining a tool such as the DRA will allow us to offer a stable and timely climate service that is well equipped to address mitigation and adaptation measures for producers, decision makers and planners alike.

It is a given that the COOP data records constantly change, and therefore the criteria need to be investigated again over the entire period of record for all 12,000+ stations. Working with our partners in the HPRCC and utilizing their ACIS database, we will look for the stations in the COOP network that have a long enough period of record (meeting stringent quality-control data requirements) to use in the DRA. With continual updates to the COOP data as well as corrections, quality control and data recovery projects, the database changes often enough that routine investigation is necessary to determine if more long-term COOP stations are reaching the criteria for inclusion in the DRA.

There is also a need to assess the current criteria and station locations to determine if more long-term stations could be utilized if the current criteria were modified with less restrictive screening steps. The reason for modifying the current criteria is to try to fill in spatial gaps across the United States while maintaining an adequate climate record for long-term analyses. If it is determined that areas not represented by an adequate number of stations could be better represented, the strict criteria could be amended for those regions to find the best climatic stations available in the COOP network. Additionally, with each year that passes, some stations may stop recording and others may reach our minimum requirement of 40 years, so there is a continual need to update the database and statistics along with the various derivative index products, including the grids.

Two other significant DRA enhancements are already under way as work has begun on identifying and bringing screened stations for Alaska and Hawaii into the DRA, along with all of the derivative products. Work also continues on lengthening the hydrology locations period of record in order to bring the hydrology data back into the final DRA. Streamflow locations void of human management are harder to come by, and budget cuts and stations being taken off-line are also concerns. All of these factors hurt the history and availability of high quality, long-term gauge stations that meet the DRA's screening criteria.

With the database in place, we are just beginning to research and assess changes and/or trends in drought frequency (and return periods) along with the intensity and various characteristics of drought behavior across the United States. The DRA allows for an in-depth analysis of the entire period of record, by decade(s), or by other user-defined periods, allowing for the detection of potential trends in both frequency and intensity of drought along with comparisons between current and past droughts for every COOP station and cluster contained within the DRA.

The NDMC DRA project team also realizes that not every drought index is ideal for every location or season. By providing several different indices at multiple time steps, the DRA gives users a vast menu of options to study and investigate drought for their region aimed at helping them find which indicators and time steps are most suitable for their location. Using these techniques, the ability for decision makers to concentrate on time periods that are specifically relevant to them and their application is possible.

ACKNOWLEDGEMENTS

This work was funded under a grant from the *Sectoral Applications Research Program (SARP) of the NOAA-Climate Program Office*. Additional funding was provided by the*NIDIS Program Office* (Coping with Drought Program) and the *USDA-Risk Management Agency* (*RMA*). The authors would also like to thank the anonymous reviewers of this article as well as Deborah Wood for her editing prowess and all of the students who have helped us out along the way. Additionally, the author team would like to sincerely thank Ned Guttman (retired, previously with the NOAA-National Climatic Data Center and who was involved in helping develop the first National Drought Atlas) for his invaluable guidance during the early stages of development. Finally, the authors would like to thank the editor, special journal editors and anonymous reviewers who took the time to review and provide guidance on this paper.

REFERENCES

1. American Planning Association (APA), 2013. Planning and Drought. Planning Advisory Service Report Number 574. APA Planning Advisory Service, p. 144.<https://www.planning.org/store/product/?ProductCode=BOOK_P574>.
2. Drought Ready Communities (DRC), 2011. Drought Ready Communities: A Guide toCommunity Drought Preparedness. National Drought Mitigation Center, p. 53. <http://drought.unl.edu/Planning/PlanningProcesses/DroughtReadyCommunities.aspx>.
3. Gibbs, W.J., Maher, J.V., 1967. Rainfall deciles as drought indicators. Bureau of Meteorology Bulletin No. 48, Commonwealth of Australia, Melbourne.
4. Geological Society of America (GSA), 2007. Managing Drought: A Roadmap for Change in the United States. A conference report from Managing Drought and Water Scarcity in Vulnerable Environments, 18–20 September 2006, Longmont, CO. http://www.geosociety.org/meetings/06drought/roadmapHi.pdf.

5. Guttman, N.B., 1994. On the sensitivity of the sample L moments to sample size. J. Climate 7 (6), 1026–1029.
6. Hao, Z., AghaKouchak, A., Nakhjiri, N., Farahmand, A., 2014. Global integrated drought monitoring and prediction system. Sci. Data 1, 140001. http://dx.doi.org/10.1038/sdata.2014.1.
7. Hosking, J.R.M., Wallis, J.R., 1997. Regional Frequency Analysis, An Approach Based on L-Moments. Cambridge University Press, Cambridge, UK.
8. Jacobs, K.L., 2002. Connecting Science, Policy and Decision-Making: A Handbook for Researchers and Science Agencies. National Oceanic and AtmosphericAdministration, Office of Global Programs, Silver Spring, MD, p. 2, <http:// www.ogp.noaa.gov/mpe/sci/doc/hdbk.pdf>.
9. Karl, T.R., Williams, Jr., C.N., Quinlan, F.T., Boden, T.A., 1990. United States Historical Climatology Network (HCN) Serial Temperature and Precipitation Data, Environmental Science Division, Publication No. 3404, Carbon DioxideInformation and Analysis Center, Oak Ridge National Laboratory, Oak Ridge, TN, 389pp.
10. McKee, T.B., Doesken, N.J., Kleist, J., 1993. The Relationship of Drought Frequency and Duration to Time Scales. In: 8th Conference on Applied Climatology, pp. 179–184. January 17–22, Anaheim, California.
11. Momtaz, F. et al., 2014. Toward a drought cyberinfrastructure system. Eos, Trans.Am. Geophys. Union 95 (22), 182–183. http://dx.doi.org/10.1002/2014EO220002.
12. Multihazard Mitigation Council, 2005. Natural Hazard Mitigation Saves: An Independent Study to Assess the Future Savings from Mitigation Activities. National Institute of Building Sciences, Washington, DC.
13. National Climatic Data Center (NC DC), 2014. Billion Dollar U.S. Weather Disasters, 1980–2012. <http:// www.ncdc.noaa.gov/oa/reports/billionz.html>.
14. National Drought Mitigation Center (NDMC), 2014: <http://drought.unl.edu>. NIDIS Drought Portal, 2014. <http://drought.gov>.Palmer, W.C., 1965. Meteorological Drought. Research Paper No. 45, U.S. Dept. of Commerce Weather Bureau, Washington, DC. Rajsekhar, D., Singh, V., Mishra, A., 2014. Hydrologic drought atlas for Texas. J. Hydrol. Eng., 05014023 http://dx.doi.org/10.1061/(ASCE)HE.1943- 5584.0001074.
15. Svoboda, M., LeComte, D., Hayes, M., Heim, R., Gleason, K., Angel, J., Rippey, R., Tinker, R., Palecki, M., Stooksbury, D., Miskus, D., Stephens, S., 2002. The droughtmonitor. Bull. Am. Meteorol. Soc. 83 (8), 1181–1190.
16. Vicente-Serrano, S.M., Beguería, S., López-Moreno, J., 2010. A multiscalar drought index sensitive to global warming: the standardized precipitationevapotranspiration index. J. Climate 23 (7), 1696–1718.
17. Wells, N., Goddard, S., Hayes, M.J., 2004. A self-calibrating Palmer drought severityindex. J. Climate 17 (12), 2335–2351.
18. Willeke, G., Hosking, J.R.M., Wallis, J.R., Guttman, N.B., 1994. The National Drought Atlas. Institute for Water Resources Report 94-NDS-4, U.S. Army Corps of Engineers.
19. Wood, A.W., Lettenmaier, D.P., 2006. A test bed for new seasonal hydrologic forecasting approaches in the western United States. Bull. Am. Meteorol. Soc.87, 1699–1712.
20. Woodhouse, C.A., Overpeck, J.T., 2000. 2000 years of drought variability in the central United States. Bull. Am. Meteorol. Soc. 79 (12), 2693–2714.
21. You, J., Hubbard, K.G., Goddard, S., 2008. Comparison of methods for spatially estimating station temperatures in a quality control system. Int. J. Climatol. 28, 777–787. http://dx.doi.org/10.1002/joc.1571.

CHAPTER 19

DROUGHT-RELATED VARIABLES OVER THE BÂRLAD BASIN (EASTERN ROMANIA) UNDER CLIMATE CHANGE SCENARIOS

Sorin Ionuţ Dascălu, Mădălina Gothard, Roxana Bojariu, Marius-Victor Birsan,Roxana Cică, Ruxandra Vintilă, Mary-Jeanne Adler, Viorel Chende, Rodica-Paula Mic

Keywords: Climate change, Eastern Romania, Palmer Drought Severity Index, EURO-CORDEX, Available water capacity of soils, Potential evapotranspiration

INTRODUCTION

The Bârlad catchment covers 16% of the Siret river basin — an important tributary of the Danube (Fig. 1). The main characteristic is the high homogeneity of the landforms, over 92% of the area illustrating the plateau morpho-structural unit. As a consequence, the mean elevation is of only 211 m, with a low standard deviation (88 m). A relative large area (45%) is characterized by low slopes. Usually, the snow melting and the rain provide the main supply for the surface waters in the Bârlad river catchment (Ujvari, 1972 and Butelcă, 2012). In this context, the assessments of influences expected from changes in temperature and precipitation on the local water resources are important.

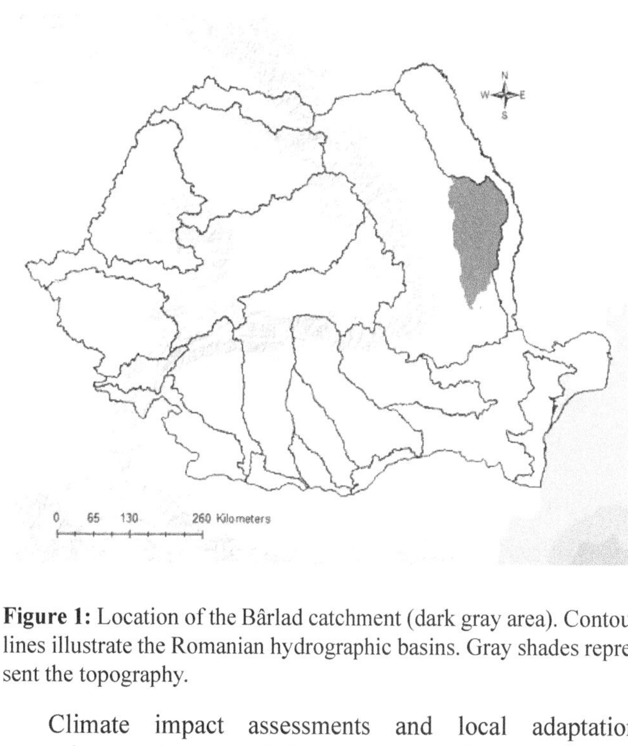

Figure 1: Location of the Bârlad catchment (dark gray area). Contour lines illustrate the Romanian hydrographic basins. Gray shades represent the topography.

Climate impact assessments and local adaptation strategies require analysis based on numerical experiments

Citation: Sorin Ionuţ Dascălu , Mădălina Gothard , Roxana Bojariu, Marius-Victor Birsan, Roxana Cică, Ruxandra Vintilă, Mary-Jeanne Adler, Viorel Chendeş, Rodica-Paula Mic, Drought-related variables over the Bârlad basin (Eastern Romania) under climate change scenarios, http://dx.doi.org/10.1016/j.catena.2016.02.018.

using climate models with very high spatial resolution under scenarios of global climate change, and robust evaluation of the results within the limits of reasonable uncertainty. For now, no detailed analysis of climate change projections in the Bârlad basin has been done, taking into account the new scenarios assessed within the Fifth Report of the Intergovernmental Panel on Climate Change (IPCC, 2013). The main purpose of this study is the assessment of the current and future climate conditions that will affect the droughts in the Bârlad basin by using recent results made available thanks to the European component of the COordinated Regional Climate Downscaling Experiment (EURO-CORDEX) program (Jacob et al., 2014). These results are based on regional climate experiments under the Representative Concentration Pathway (RCPs) scenarios (van Vuuren et al., 2011).

We use here the Palmer Drought Severity Index (PDSI) to analyze the droughts (Palmer, 1965). The observations of soil moisture (directly related to droughts) are limited in space and time. Therefore, the use of available data to build soil moisture-related indices is a pragmatic approach. On the other hand, in the climate models, errors in initializing and representing soil processes tend to cause errors in soil moisture prediction and projections. In this context, the PDSI provides a simple tool having the advantage to keep the drought metric close to the real world through the combination of climate variables easily available from the observation system, which are also relatively well simulated by the climate models (such as air temperature and precipitation).

DATA AND METHODS

In this study we have used monthly temperature and precipitation from observations and model results covering the Bârlad basin (which has the outlet at Tecuci gauging station). The results of the climate model used here have been derived under the RCP scenarios. The RCP scenarios describe temporal evolutions in the global concentrations of the greenhouse gases (GHGs) illustrating the radiative forcing within the period 2006–2100. For instance, the radiative forcing caused by increased levels of the GHGs reaches in the year 2100 a value around 4.5 (8.5) W/m^2 above the pre-industrial level in the RCP 4.5 (RCP 8.5) scenario (van Vuuren et al., 2011).

The recent CMIP5 results have a global coverage with horizontal resolutions of about 100 km, which are suitable for global and continental scale analysis. However, these resolutions are not appropriate for detailed characterization of the local processes like those taking place in river catchments such as the Bârlad basin. Therefore we use regional climate models from the COordinated Regional Climate Downscaling Experiment (CORDEX) driven by global models from the CMIP5. As part of CORDEX framework, EURO-CORDEX initiative provides regional climate projections for Europe at horizontal resolutions of around 50 km (EUR-44) and 12.5 km (EUR-11). In this study, we have extracted the available EURO-CORDEX results having very high resolution (EUR-11). Table 1 presents the regional and global climate models analyzed here.

Table 1: Regional and global climatic models used for assessing the influence of climate change on the future evolution of temperature and precipitation in the Bârlad basin

No.	Regional climatic modeling center	Regional model	Global model
1	CLMcom(CLMcom Consortium)	CLM 4-8-17	MPI-ESM-LR
2	DMI(Danish Meteorological Institute)	HIRHAM5	ICHEC-EC-EARTH
3	KNMI(Royal Netherlands Meteorological Institute)	RACMO22E	ICHEC-EC-EARTH
4	MPI-CSC(Max-Planck Institute — Climate Service Center, Hamburg, Germany)	REMO2009	MPI-ESM-LR
5	SMHI(Swedish Meteorological and Hydrological Institute)	RCA4	ICHEC-EC-EARTH

A Romanian gridded dataset (ROCADA) provides observations of air temperature and precipitation amounts with a horizontal resolution of 10 km (Birsan and Dumitrescu, 2014a and Dumitrescu and Birsan, 2015). In our area of interest there are 5 meteorological stations but the interpolation method of kriging in space and time uses data from a larger number of locations, some of them are located nearby the edges of the Bârlad basin (about 12 stations).

We have spatially averaged the observed and simulated data across the basin of the Bârlad River and then compared them to validate the results from the models for the present time›s climate (1971–2000). The analysis of the present time›s climate and future projections is based on the multiannual monthly means of the periods 1971–2000, 2021–2050 and 2071–2100.

The impact of the air temperature and precipitation exerted on the droughts affecting the Bârlad basin under climate change is investigated by computing the Palmer Drought Severity Index (PDSI). Palmer (1965) calibrated his index using data from 9 stations located mainly in the North-American Mid-West region raising the issue of the PDSI behavior at various locations. Wells et al. (2004) developed a self-calibrated version of PDSI that automatically calibrates the behavior of the index at any location by replacing empirical constants in the index computation with dynamically calculated values.

The National Oceanic and Atmospheric Administration (NOAA, USA) provides on their web site (http://www.esrl.noaa.gov/psd/) monthly values of PDSI with global coverage. Details about deriving these datasets are presented in Dai et al. (2004). Jacobi et al. (2013) developed a MATLAB tool to compute the PDSI and Ficklin et al. (2015)presented an addition to this tool that includes an option to use the Penman–Monteith evapotranspiration as an input to the PDSI calculations. We used here the self-calibrated PDSI and code of Wells et al. (2004) modified to include either Thornthwaite or Penman–Monteith potential evapotranspiration.

Another issue discussed in the literature is the choice of the baseline period to define and calibrate the PDSI moisture categories (e.g. Trenberth et al., 2014). We selected as a baseline the reference period 1971–2000 for the present time›s climate which seems to be a natural choice as the common period of the analyzed CMIP5 and EURO-CORDEX experiments, under current climate forcing, starts on the 1st of January 1971 and lasts until the 31st December 2005. On the other hand, the 30-year period is also appropriate from the standpoint of the availability of streamflow observations.

The 5-member ensemble means of monthly temperature and precipitation amounts from the EURO-CORDEX models provide the input data for the PDSI calculation (Table 1). We consider the use of ensemble means in PDSI computation a reasonable choice due to the low-frequency variability involved in the drought-related processes and our interest in the climate change signal.

In addition to air temperature and precipitation, in the PDSI computation we have also used the available water capacities (AWCs) of the soils from the analyzed river basin. In order to assess the sensitivity of the basin-averaged PDSI to the AWC values we have derived two AWC datasets. One consists of estimated topsoil and subsoil AWC values extracted from the European Soil Database (ESDB) (Panagos et al., 2012, Hiederer, 2013a and Hiederer, 2013b) for the studied area. In this case, we have spatially averaged the sum of topsoil and subsoil AWCs from the ESDB on the 12.5 km by 12.5 km square cells centered in the model grids to provide the soil constants used as input data for the PDSI calculation in each EURO-CORDEX grid point of the Bârlad basin.

The other AWC dataset is computed using several pedotransfer rules, which have been developed at the Joint Research Centre (JRC) of the European Commission based on the work of Thomasson (1995). The pedotransfer rules have been applied in the 12.5 km EURO-CORDEX grids to the data that characterize the soils covering the area of the Bârlad basin. These input soil data have been made freely available by JRC, from the Soil Geographical Database of Eurasia at the scale 1:1,000,000 (SGDBE) version 4.0 (http://eusoils.jrc.ec.europa.eu/esbn/SGDBE.html).

The PDSI calculation is based on the Palmer representation of hydrological balance. The validity of this representation has been checked at the basin scale by correlating the basin-averaged components of the PDSI derived from observations – such as the moisture index, and potential runoff – with the observed streamflow at the Tecuci station (the basin outlet).

Finally, we have analyzed the linear trends of the basin-averaged PDSI to assess the impact of the climate change on droughts under medium and high RCP scenarios within the period 1971–2100. The slope estimate of the linear trend is obtained with the nonparametric Kendall–Theil method (also known as the Theil–Sen slope estimate), which is less affected by non-normal data and outliers (Helsel and Hirsch, 1992). Sensitivity of these results to the use of either Thornthwaite formula (1948) or Penman–Monteith parametrization (Penman, 1948 and Monteith, 1981) has been investigated using the one model out of the entire ensemble with available potential evapotranspiration (RCA4 driven by ICHEC-EC-EARTH). The computation of RCA4 potential evapotranspiration has followed the definition proposed by FAO, which was derived from the Penman–Monteith method (Allen et al., 1998 and Samuelsson et al., 2015). We have extracted these values from the EURO-CORDEX archive and use them instead of the Thornthwaite›s ones by modifying accordingly the C ++ code of Wells et al. (2004) for the self-calibrated computation of PDSI.

RESULTS AND DISCUSSION

Future Regional Projections of Temperature and Precipitation under RCP 4.5 and RCP 8.5 Scenarios

In general, the multiannual seasonal cycle of the observed temperature averaged over the Bârlad basin is well simulated by the CMIP5 and EURO-CORDEX models used here (Table 1), within the reference interval 1971–2000.

However, there are clear underestimations in May and June and an overestimation in August (Fig. 2). The underestimation of temperature in May and June is a feature shared by 4 of the 5 regional models and seems to be related to the poor representation of some local processes in both regional and global climate models presented in Table 1. The multimodel monthly averages are closer to the observed values compared with the averages provided by any single model run from the regional model ensemble. On the other hand, Fig. 2 shows improvements in the simulation of the winter temperatures revealed by the ensemble means of the regional numerical experiments as compared with the ensemble means of the driving global runs. Furthermore, the general deficiency of underestimating the temperatures in May and June is slightly reduced in the case of the ensemble means of the regional numerical experiments (Fig. 2).

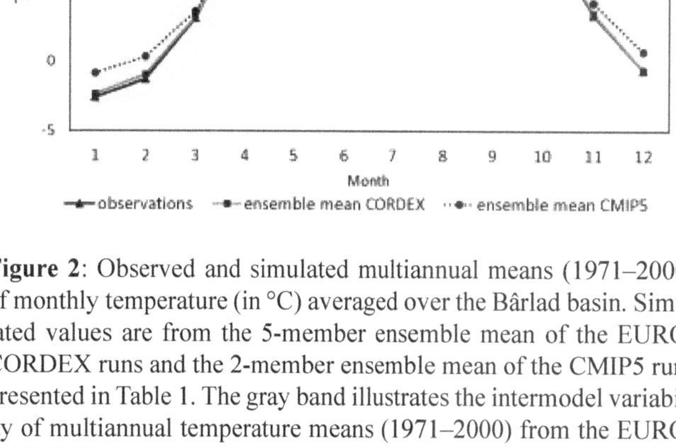

Figure 2: Observed and simulated multiannual means (1971–2000) of monthly temperature (in °C) averaged over the Bârlad basin. Simulated values are from the 5-member ensemble mean of the EURO-CORDEX runs and the 2-member ensemble mean of the CMIP5 runs presented in Table 1. The gray band illustrates the intermodel variability of multiannual temperature means (1971–2000) from the EURO-CORDEX numerical experiments used here.

In the case of precipitation, the results of numerical experiments generally show overestimated values in the winter, autumn, and spring seasons, and underestimated values during the summer as compared with observations within the reference interval 1971–2000 (Fig. 3). Fig. 3 shows that, as in the case of temperature, any single model cannot solely simulate the observed annual cycle of the precipitation and the use of the ensemble means brings the simulated values closer to the observed ones. On the other hand, one could see some improvements when using the ensemble means of regional runs. They simulate the annual precipitation cycle closer to observations compared with the global ensemble means. Exceptions in this regard are the months of July and August (Fig. 3).

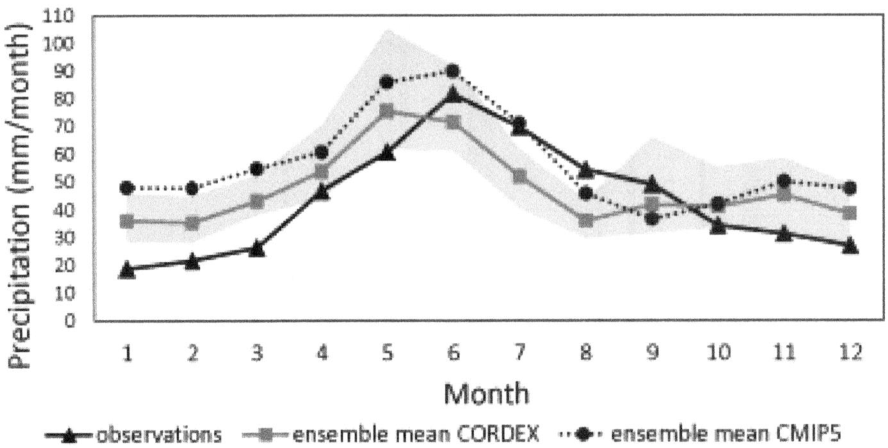

Figure 3: Observed and simulated multiannual means (1971–2000) of monthly precipitation amounts(in mm) averaged over the Bârlad basin. Simulated values are from the 5-member ensemble means of the EURO-CORDEX runs and the 2-member ensemble means of the CMIP5 runs presented in Table 1. The gray band illustrates the intermodel variability of multiannual precipitation means (1971–2000) from the EURO-CORDEX numerical experiments used here.

One may conclude that the multiannual temperature and precipitation cycles provided by the RCMs resemble to some extent the observed ones providing a certain level of confidence for the analysis of precipitation, temperature and related indices in the area of interests, under climate change scenarios. The temperature differences of multimodel monthly means between the future periods (2021–2050 and 2071–2100) and the reference one (1971–2000) reveal a stronger increase in temperatures during the summer and winter seasons under both RCP scenarios analyzed here. The increase in temperature is strongest within the time interval 2071–2100 under the RCP 8.5 scenario. In winter, the maximum temperature increase in the regional ensemble mean reaches 4.9 °C and is detected in the month of January. As for the summer, the maximum temperature increase in the regional ensemble mean (4.3 °C) is found in the month of August (Fig. 4).

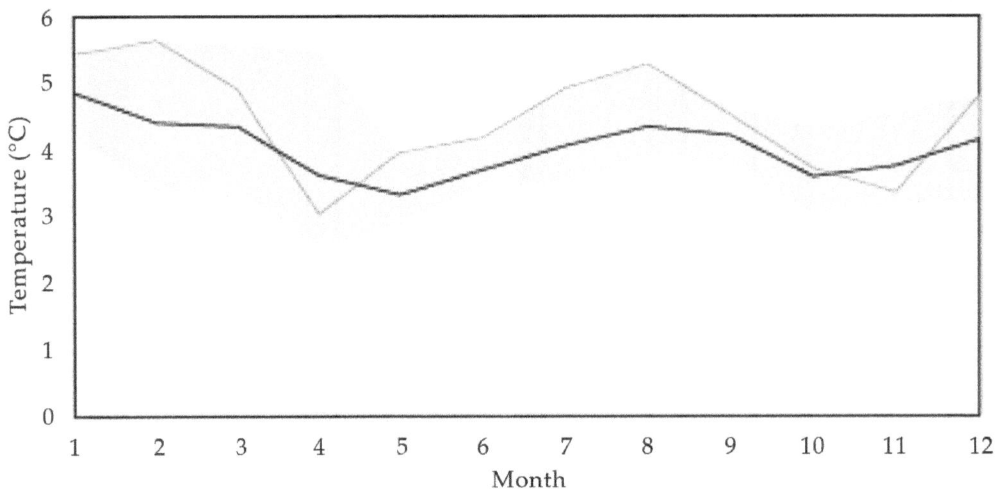

Figure 4: Mean monthly temperature difference (in °C) between the period 2071–2100 and 1971–2000, averaged over Bârlad river basin under RCP 8.5 scenario, from five different RCMs (light gray-colored band), multimodel mean of RCM ensemble (black line), and ensemble member RCA4 (gray line) (Table 1).

In summer, under the RCP 4.5 scenario, the difference between the evolutions of the first and second future period is of an order of 1 °C, while under the RCP 8.5 scenario the same difference is around 2 °C to 3 °C. From the second half of the 21st century, the analyzed scenarios increasingly diverge in terms of global concentrations of greenhouse gases, resulting in significant differences in their associated radiative forcing which lead to larger differences in the temperature increase.

The evolutions of the multiannual means of monthly precipitation generally suggest a decrease from May to September under climate change. This signal is stronger in the late 21st century under the conditions of RCP 8.5 scenario (Fig. 5). In this case, the largest monthly decrease (in percent) of the ensemble mean of precipitation amount occurs in September and is up to 13% for the interval 2071–2100. The largest decrease in the ensemble mean expressed in mm/month is in July and reaches around 6 mm for the interval 2071–2100 under the RCP 8.5 scenario (Fig. 5).\

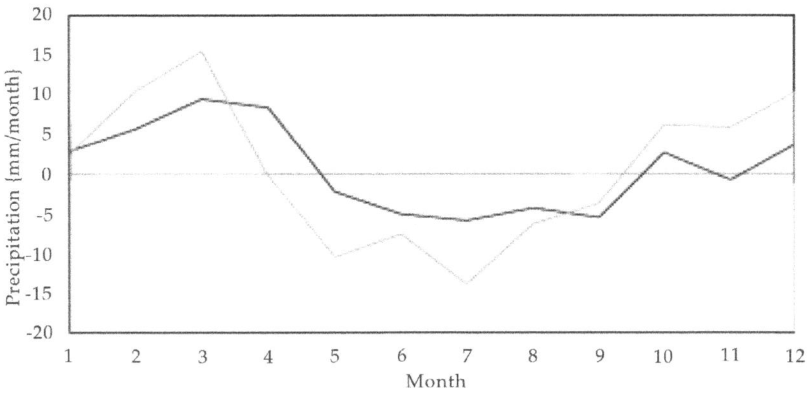

Figure 5: Mean monthly rainfall amount difference (mm) between the period 2071–2100 and 1971–2000, averaged over Bârlad river basin under RCP 8.5 scenario, from five different RCMs (light gray-colored band), multimodel mean of RCM ensemble (black line), and ensemble member RCA4 (gray line) (Table 1).

Climate Change Impact on Drought over the Bârlad basin

In this study we have used as a drought indicator the Palmer Drought Severity Index (PDSI), which measures the cumulative effect of monthly precipitation deficit/surplus with respect to a value which is Climatologically Appropriate For Existing Conditions (CAFEC) in a given region (Palmer, 1965). The computation of the PDSI requires precipitation, air temperature, soil characteristics (i.e. available water capacity) and the latitude of the location to estimate the length of day over which the solar radiation is received (for deriving potential evapotranspiration). In order to calculate the PDSI for a certain month (i), one has first to determine the moisture anomaly index $ZIND_i$ for that month (i):

$$ZIND_i = k\,(P-\alpha PET-\beta PR-\gamma PRO+\delta PL) \qquad (1)$$

$$PDSI_i = PDSI_{i-1} + ZIND_i/3 - 0.103\,PDSI_{i-1} \qquad (2)$$

where: k is an empirical weighting factor, specific for each region; α, β, γ, and δ are coefficients for evapotranspiration, soil water recharge, runoff and water loss from the soil, computed to link the potential quantities and real ones; P, PET, PR, PRO, and PL represent the observed precipitation, Thornthwaite potential evapotranspiration (Thornthwaite, 1948), potential recharge, potential runoff and potential water loss from the soil. Palmer (1965) built the index based on the simple representation of the components shaping the hydrological balance in a given area. Potential Evapotranspiration (PET) is the maximum evapotranspiration in the given environmental conditions, when soil moisture is not a limiting factor. Potential Recharge (PR) is the amount of moisture required to bring the soil to its AWC from the available moisture at the beginning of the month. Potential Run-Off (PRO) is defined as the difference between the potential precipitation and the potential recharge. Runoff is assumed to occur if the Palmer soil model reaches its available moisture capacity, AWC. Potential Loss (PL) is the amount of moisture that could be lost from the soil provided that the monthly precipitation is zero (Palmer, 1965).

One can view the PDSI as the output of a two-level "bucket" model of the soil. The top layer of soil is assumed to hold 25.4 mm of moisture. The amount of moisture that can be held by the two-layered soil is a soil-dependent value – available water capacity (AWC) – which must be provided as an input parameter, too. The soil can hold more water if the soil particles are finer (as in soils high in clay content), but a plant has to work harder to draw moisture out of the soil. The opposite conditions occur for coarser particles (as in sandy soils): the soil holds less moisture, but more water is available for plant. In a comprehensive review, Thomasson (1995) noticed that AWC has been defined for the water held between two suctions in 1 m depth of soil (or other fixed depth), without concern about how this water can actually be available through the root system. Thomasson introduced a characteristic that he called "soil water available to plants (SWAP)" and defined it as the total amount of the moisture "that can be considered to be extractable by the roots". Despite this, AWC continues to be used in environmental modeling and assessment mainly due to measurement constraints (Wells et al., 2004 and Lal and Shukla, 2005).

In the studied area, we have used two datasets of AWCs to compute the PDSI: (1) the sum of topsoil and subsoil AWCs (at 1 km × 1 km resolution) extracted from the European Soil Database (ESDB) (Hiederer, 2013a,b) and averaged over the 12.5 km × 12.5 km grid cells centered in the model grids and (2) AWCs computed in EURO-CORDEX grid points using pedotransfer rules (Thomasson, 1995 and Baruth et al., 2006) and soil characteristics from the Soil Geographical Database of Eurasia at the scale 1:1,000,000 (SGDBE), which also includes local information on the presence of shallow impermeable layers (between the depths of 40 cm and 80 cm, in the area of interest) (Fig. 6). The pedotransfer rules used here take into account a standardized soil depth (1 m or to an obstruction like an impermeable layer) and the water holds between − 5 kPa suction (field capacity) and − 1500 kPa suction (wilting point). The second AWC dataset has generally lower values compared with the first one. This is reflected in the averages of the two AWC datasets over the basin (152 and 171).

Figure 6: Available water capacities (AWCs) (in mm water column/1 m soil) in the Bârlad basin computed in the 12.5 km EURO-CORDEX grids (black circles) using pedotransfer rules and soil characteristcs, provided by the European Soil Data Centre (managed by the JRC) from the Pedotransfer Rules Database (PTRDB) and the Soil Geographical Database of Eurasia at the scale 1:1,000,000 (SGDBE) (Baruth et al., 2006) (black numbers). Gray numbers illustrate AWCs obstained as the sum of topsoil and subsoil AWCs (at 1 km × 1 km resolution) extracted from the European Soil Database (ESDB) (Hiederer, 2013a,b) and averaged over the 12.5 km × 12.5 km grid cells centered in the model grids. Light gray lines are the tributaries of the Bârlad River.

We have investigated the spatial variability of the PDSI using the Empirical Orthogonal Function (EOF) decomposition of its gridded values over the time. The first EOF mode has a spatial pattern with the same sign and almost the same magnitude for all grids within the basin (not shown). This mode represents from 83% to 90% of data variability (depending on the climate scenario and AWC dataset) so we consider that the temporal analysis of the spatial-averaged PDSI over the entire river catchment is a valid approach.

The averaged PDSIs over the Bârlad basin corresponding to the two AWC datasets have been validated by correlating PDSI components derived from observations such as the moisture index (ZIND) and potential runoff (PRO) with the observed monthly means of the streamflow (QMED) at Tecuci (the basin outlet). The correlation coefficients between the Palmer indices and QMED are almost the same in the two cases. We illustrate here only the correlation coefficients estimated with the pedotransfer functions applied to local soil characteristics (the second AWC dataset). Furthermore, we have correlated the difference between precipitation and potential evapotranspiration with the QMED.

Fig. 7 shows that the PDSI represents reasonably well the local process taking place in the Bârlad catchment under current climate conditions (1971–2000). The correlation coefficients between the basin-averaged ZIND derived from monthly observations and the monthly mean of streamflow (QMED) at the basin outlet (Tecuci gauging station) are statistically significant at 95% confidence level except for the months of January and February. Strong correlation coefficients (> 0.7) between the ZIND and the QMED are present in the summer months (Fig. 7 and Fig. 8). Correlation coefficients between the monthly PRO and QMED at Tecuci are also statistically significant (except for August) with higher values from January to February (e.g. Fig. 7 and Fig. 9) and from October to November.

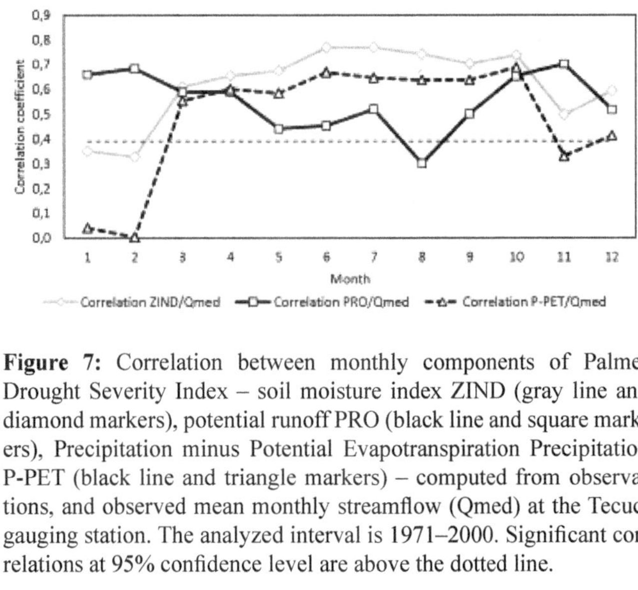

Figure 7: Correlation between monthly components of Palmer Drought Severity Index – soil moisture index ZIND (gray line and diamond markers), potential runoff PRO (black line and square markers), Precipitation minus Potential Evapotranspiration Precipitation P-PET (black line and triangle markers) – computed from observations, and observed mean monthly streamflow (Qmed) at the Tecuci gauging station. The analyzed interval is 1971–2000. Significant correlations at 95% confidence level are above the dotted line.

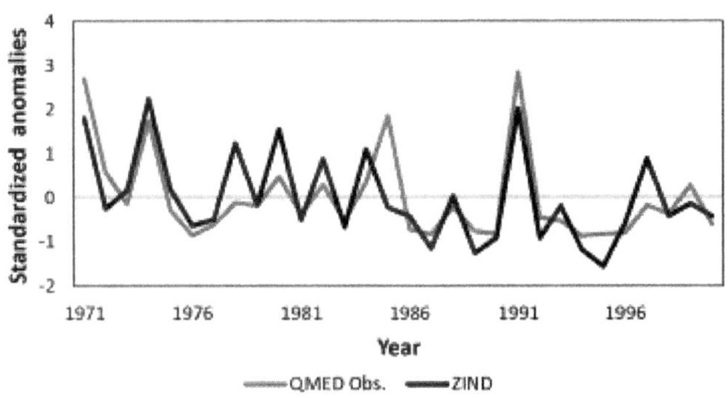

Figure 8: Time evolutions of observation-derived ZIND and QMED averaged over the Bârlad basin, in the months of July within the period 1971–2000. The correlation coefficient is 0.77.

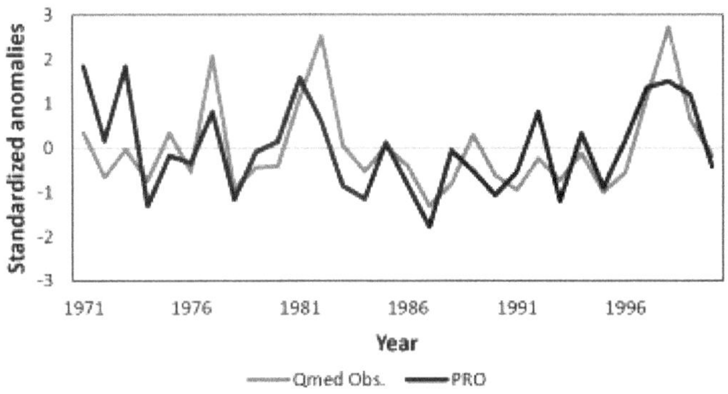

Figure 9: Time evolutions of observation-derived PRO and QMED averaged over the Bârlad basin, in the months of January within the period 1971–2000. The correlation coefficient is 0.66.

In winter months, the observed soil moisture and QMED are more or less decoupled from the precipitation despite what the Palmer model indicates due to the processes related to snow falls, snow cover and frozen soils, which are not included in the Palmer hydrological balance (van der Schrier et al., 2007). In this case, the correlation coefficients between the Palmer moisture anomaly (depending on monthly precipitation) and the QMED will be lower. Winter episodes with ice along the river are also lowering the correlation which links the QMED and ZIND.

In this context it is also interesting to note that, referring to the months of January and February of the period 1971–2000, when correlation coefficients between the PRO (ZIND) and QMED are high (low), we recorded a multiannual mean number of days with snow cover varying from 12 to 17 and a multiannual mean snow depth from 5.5 cm to 7.4 cm per day with snow cover at the 4 stations within the Bârlad basin. In winter, time intervals without snow cover could be related to some extent to warmer and dryer conditions brought by westerly atmospheric circulations (Bojariu and Giorgi, 2005,Bojariu and Dinu, 2007, Birsan and Dumitrescu, 2014a, Birsan and Dumitrescu, 2014b and Dumitrescu and Birsan, 2015). Thus the wintertime is often an alternation of relatively warm and dry spells and cold and snowy episodes that makes possible a correlation between the Palmer-defined runoff and the runoff from the (partially) melted snow, the latter being physically coupled with the river flow variability at the monthly time scales.

On the other hand, the low value of correlation coefficients between the PRO and QMED in August could be due to the poor infiltration of water from convective rainfall into the soil which leads to underestimation of the runoff in the Palmer model. Reverse conditions occur in the spring and autumn months, when the infiltration into the soil of the water from steady rains is more effective.

The correlation between the precipitation minus potential evapotranspiration and QMED is generally lower than that between the ZIND and QMED showing the added value of the Palmer water balance model on monthly scale (Fig. 7). We conclude that, in general, the PDSI represents reasonably well drought-related processes within the Bârlad basin. In this context, we could use the PDSI to assess the impact of projections of temperature and precipitation on drought-related conditions over the Bârlad catchment under climate change scenarios.

The input data for computing the PDSI future projections are the 5-member ensemble means of monthly temperature and precipitation from the EURO-CORDEX models driven by the CMIP5 models presented in Table 1.

The use of the ensemble means in the PDSI computation is a reasonable choice due to long-range time scales and low frequency variability which shape the drought trend. The linear trend analysis of the basin-averaged PDSI shows statistically significant tendences for drought conditions within the period 1971–2100, over the Bârlad basin, for both concentration scenarios analyzed here. As we have expected, the slopes of the linear trends are larger for the higher concentration scenario (RCP 8.5). The basin-averaged PDSI is reduced in summer with about 5.1 (4.4) over the interval 1971–2100 under the RCP 8.5 (RCP 4.5) when the second AWC dataset is used. The confidence limits at 99% level are from − 5.8 to − 3.2. A 4-unit decrease in PDSI is consistent with the transition from normal conditions to extreme drought (Palmer, 1965). Downward PDSI trends are almost the same for the two AWC datasets, with slope differences in tenth of a PDSI unit. These differences are very small and not statistically significant. The PDSI method applied to the Bârlad basin seems to have low sensitivity to the soil characteristics when drought trends are investigated.

An issue which adds more uncertainties related to the assessment of future drought tendences is the way in which potential evapotranspiration is parametrized in the Palmer model. The water balance proposed by Palmer uses the Thornthwaite parametrization, which is solely based on the air temperature and the solar radiation contribution is empirically derived under the current climate conditions. Under the climate change, the Thornthwaite empirical approach is questionable (e.g. Dai, 2011 and Trenberth et al., 2014). The Penman–Monteith method uses a more physically-oriented parametrization to estimate the potential evaporation explicitly based on temperature, net radiation, air pressure, air humidity and wind data (Allen et al., 1998). We have investigated how the drought trends (expressed by the PDSI tendences) change when the potential evapotranspiration is computed either by the Thornthwaite or by the Penman–Monteith approach using the results from the one model out of the entire ensemble, which has available data in this regard (RCM RCA4 driven by the GCM ICHEC-EC-EARTH).

Our results reveal that the PDSI which incorporates the Penman–Monteith potential evapotranspiration show also a downward trend in the summer months that is more pronounced in the last decades of the 21st century and under the high concentration scenario. However, the trend magnitude is significantly reduced compared with the case in which the Thornthwaite approach is used (Table 2) as in the study of Dai (2011). In our case, the correlation coefficients between the Thornthwaite and Penman–Monteith potential evapotranspiration are around 0.9 under both climate scenarios, consistent with the results of van der Schrier et al. (2011). However, the Penman–Monteith values increase slower than the Thornthwaite ones from 1971 towards the end of the 21st century, explaining the smaller magnitude of aridity tendences revealed by the corresponding version of PDSI (Table 2). Almost all trends of the basin-averaged PDSI and potential evapotranspiration illustrated in Table 2 are statistically significant, according to the Mann–Kendall test (Mann, 1945 and Kendall, 1975). The exception is in the case of the PSDI version based on the Penman–Monteith potential evapotranspiration under the moderate concentration scenario.

Table 2: Changes in summer averaged PDSI (dimensionless units) and potential evapotranspiration (PET, mm/month) computed from linear trends using the Thornthwaite and Penman–Monteith approaches applied to the results of RCA4 driven by ICHEC-EC-EARTH for the interval 1971–2100.

Approach	Climate scenario	PDSI change in 130 years	PDSI change confidence limits at 99%	PET change in 130 years	PET change confidence limits at 99%
FAO Penman–Monteith	RCP 4.5	−0.4	−2.1–1.2	11.7	−2.0–26.0
	RCP 8.5	−1.9	−3.7 to −0.1	37.1	21.8–51.5
Thornthwaite	RCP 4.5	−1.4	−2.8–0.0	18.4	9.9–27.6
	RCP 8.5	−4.0	−5.6 to −2.5	44.9	34.4–55.3

CONCLUSIONS

The regional modeling approach provides added value on the global simulations, especially for the multiannual cycle of basin-averaged temperature and precipitation computed as the ensemble means. These results provide a certain level of confidence when analyzing the future evolution of the precipitation, temperature and related indices in the Bârlad basin under climate change scenarios.

The analyzed concentration scenarios (RCP 4.5 and RCP 8.5) indicate an increase in the multimodel average of temperature in all months, and for both analyzed periods (2021–2050 and 2071–2100) as compared with the reference interval (1971–2000). The increase is stronger in the summer and winter seasons, under the scenario with the highest value of the GHG concentrations towards the end of the 21st century. The increase in ensemble mean of temperature is close to 4.9 °C for the January multiannual value of the period 2071–2100 under the RCP 8.5 scenario. The highest increase in summer is 4.3 °C and is recorded for the August multiannual value of the period 2071–2100, under the RCP 8.5 scenario. The precipitation change in the Bârlad basin occurs mostly in the late 21st century (2071–2100), under both analyzed scenarios (RCP 4.5 and RCP 8.5). As compared with the reference interval (1971–2000), the largest monthly decrease in the ensemble mean of precipitation occurs in September for the interval 2071–2100 and gets up to 13% under the RCP 8.5 scenario.

The correlations between observed streamflow at the Bârlad basin outlet and PDSI-related indices show that the PDSI represents reasonably well the drought-related processes taking place in the Bârlad catchment. This allows us to use the PDSI based on the basin-averaged ensemble means of monthly temperature and precipitation for the assessment of climate change impact on droughts in the area of interest.

Generally, the linear trend analysis reveals that the basin-averaged PDSI goes lower in summer showing an aridity tendence over the Bârlad catchment under the climate change. In the Palmer classification, depending on the specific climate scenario and parametrization of the potential evapotranspiration, droughts that were deemed as incipient, mild or severe towards the end of the 20th century, will become a normal summer feature towards the end of the 21st century.

The analysis of how the potential evapotranspiration parametrization affects the drought projections indicates that, under the climate change conditions, the Thornthwaite approach will lead to significant overestimation of the aridity tendence compared with the Penman–Monteith method.

On the other hand, downward PDSI trends are almost the same for the two AWC datasets indicating that the PDSI method applied to the Bârlad basin seems to show low sensitivity to soil characteristics when the drought trends are investigated. However, this is only a glimpse in the role of the soil when the drought variability and change are locally assessed. In the context of a better soil moisture assimilation and improved parametrization of the soil processes in the climate models, approaches integrating depth data of rooting zone – for all vegetation classes covering the land, including crops along the seasons – would also improve the representation of the local hydrological balance and consequently the assessment of the climate change impact on the regional droughts.

Our results show that, under the climate change, the tendence towards more severe summer droughts is a significant feature of the Bârlad basin in the long run despite the uncertainties related to global and regional models, parametrization of potential evapotranspiration in the Palmer model and the soil data. This tendence is a result of a basin-wide decrease in the precipitation and increase in the potential evaporation, which have been also identified as causes for the summer time drying in mid-latitudes in the CMIP 3 model results (Meehl et al., 2007 and Sheffield and Wood, 2008).

We need more studies that couple the local climatic information with their hydrologic impact in soil and atmosphere to limit the uncertainties in the drought analysis and to provide the tools for the water resource assessment, under the climate change conditions, in terms of adaptation planning and sustainable development.

ACKNOWLEDGMENTS

This study was funded by the Executive Agency for Higher Education, Scientific Research, Development and Innovation Funding (UEFISCDI, Romania) through the research project CLIMHYDEX 'Changes in climate extremes and associated impact in hydrological events in Romania' (PNII-ID-2011-2-0073). The authors are grateful to the anonymous reviewer who helped us to substantially improve this study. Gelu G. Bojariu helped us to modify the C ++ code of Wells et al. (2004) to compute the self-calibrated PDSI with Penman–Monteith potential evapotranspiration. We acknowledge the World Climate Research Programme's Working Group on Regional Climate, and the Working Group on Coupled Modelling, former coordinating body of CORDEX and responsible panel for CMIP5. We thank the climate modeling groups mentioned in the Table 1 for producing and making available their model output. Also, we acknowledge the Earth System Grid Federation infrastructure, an international effort led by the U.S. Department of Energy's Program for Climate Model Diagnosis and Intercomparison, the European Network for Earth System Modelling and the partners in the Global Organisation for Earth System Science Portals (GO-ESSP).

REFERENCES

1. Allen, R.G., Pereira, L.S., Raes, D., Smith, M., 1998. Crop Evapotranspiration: Guidelines for Computing Crop Water Requirements. FAO Irrigation and Drainage Paper 56. FAO 92-5-104219-5.
2. Baruth, B., Genovese, G., Montanarella, L. (Eds.), 2006. Redefinition of New Soil PhysicalGroupsNew Soil Information for the MARS Crop Yield Forecasting System. Office for Official Publications of the European Communities, Luxembourg. ISBN: 92-79-03376-X, pp. 49–71 (22499 EN).
3. Birsan, M.V., Dumitrescu, A., 2014a. ROCADA: Romanian Daily Gridded Climatic Dataset (1961–2013) V1.0. Administratia Nationala de Meteorologie, Bucuresti, Romania http://dx.doi.org/10.1594/PANGAEA.833627.
4. Birsan, M.V., Dumitrescu, A., 2014b. Snow variability in Romania in connection to largescale atmospheric circulation. Int. J. Climatol. 34, 134–144. http://dx.doi.org/10.1002/joc.3671.
5. Bojariu, R., Dinu, M., 2007. Snow variability and change in Romania. In: Strasser, U., Vogel,M. (Eds.), Proceedings of the Alpine Snow Workshop, Munich, 5–6 October 2006. Berchtesgaden National Park Report 52, pp. 64–68.
6. Bojariu, R., Giorgi, F., 2005. The North Atlantic Oscillation signal in a regional climate simulationfor the European region. Tellus 57A, 641–653. http://dx.doi.org/10.1111/j.1600-0870.2005.00122.x.
7. Butelcă, D., 2012. Geomorphological Processes and Their Influence on Soils in the Upper Bârlad Catchment Upstream Băceşti (in Romanian) (PhD Thesis) Alexandru Ioan Cuza University, Iaşi.
8. Dai, A., 2011. Drought under global warming: a review. Wiley Interdiscip. Rev. Clim.Chang. 2, 45–65. http://dx.doi.org/10.1002/wcc.81.
9. Dai, A., Trenberth, K.E., Qian, T., 2004. A global data set of Palmer Drought Severity Index for 1870–2002: relationship with soil moisture and effects of surface warming. J. Hydrometeorol. 5, 1117–1130.
10. Dumitrescu, A., Birsan, M.V., 2015. ROCADA: a gridded daily climatic dataset over Romania (1961–2013) for nine meteorological variables. Nat. Hazards 78 (2), 1045–1063. http://dx.doi.org/10.1007/s11069-015-1757-z.
11. Ficklin, D.L., Letsinger, S.L., Gholizadeh, H., Maxwell, J.T., 2015. Incorporation of the Penman–Monteith potential evapotranspiration method into a Palmer Drought Severity Index tool. Comput. Geosci. 85 (A), 136–141. http://dx.doi.org/10.1016/j. cageo.2015.09.013 (December 2015).
12. Helsel, D.R., Hirsch, R.M., 1992. Statistical Methods in Water Resources. Elsevier, Amsterdam.
13. Hiederer, R., 2013a. Mapping soil properties for Europe — spatial representation of soil database attributes. EUR26082EN Scientific and Technical Research Series. Publications Office of the European Union, Luxembourg http://dx.doi.org/10.2788/ 94128 (127 pp.).
14. Hiederer, R., 2013b. Mapping soil typologies — spatial decision support applied to European Soil Database. EUR25932EN Scientific and Technical Research Series. Publications Office of the European Union, Luxembourg http://dx.doi.org/10.2788/ 87286.
15. IPCC, 2013. In: Stocker, T.F., Qin, D., Plattner, G.-K., Tignor, M., Allen, S.K., Boschung, J.,Nauels, A., Xia, Y., Bex, V., Midgley, P.M. (Eds.), Climate Change 2013, The Physical Science Basis. Contribution of Working Group I to the Fifth Assessment Report of the Intergovernmental Panel on Climate Change. Cambridge University Press, Cambridge,United Kingdom and New York, NY, USA (1535 pp.).
16. Jacob, D., et al., 2014. EURO-CORDEX: new high-resolution climate change projections for European impact research. Reg. Environ. Chang. 14 (2), 563–578. http://dx.doi.org/ 10.1007/s10113-013-0499-2.
17. Jacobi, J., Perrone, D., Duncan, L.L., Hornberger, G., 2013. A tool for calculating the Palmer drought indices. Water Resour. Res. 49, 6086–6089.
18. Kendall, M.G., 1975. Rank Correlation Methods. Charles Griffin, London. Lal, R., Shukla, M.K., 2005. Principles of Soil Physics. Taylor & Francis e-Library 0-203- 02123-1, pp. 268–278 (pp. 126).
19. Mann, H.B., 1945. Nonparametric tests against trend. Econometrica 13, 245–259.
20. Meehl, G.A., et al., 2007. Global Climate Projections. In: Solomon, S., et al. (Eds.), Climate
21. Change 2007: The Physical Science Basis. Contribution of Working Group I to the Fourth Assessment Report of the Intergovernmental Panel on Climate Change. Cambridge Univ. Press, Cambridge, U. K., pp. 747–846.
22. Monteith, J.L., 1981. Evaporation and surface temperature. Q. J. R. Meteorol. Soc. 107, 1–27.
23. Palmer, W.C., 1965. Meteorological drought. Research Paper No. 45. U.S. Weather Bureau. NOAA Library and Information Services Division, Washington, D.C., p. 20852.
24. Panagos, P., Van Liedekerke, M., Jones, A., Montanarella, L., 2012. European Soil Data Centre: response to European policy support and public data requirements. Land Use Policy 29 (2), 329–338.
25. Penman, H.L., 1948. Natural evaporation from open water, bare soil and grass. Proc. Roy. Soc. London 193A, 120–145. http://dx.doi.org/10.1098/rspa.1948.0037.

26. Samuelsson, P., Gollvik, S., Jansson, C., Kupiainen, M., Kourzeneva, E., van de Berg, W.J., 2015. The surface processes of the Rossby Centre regional atmospheric climate model (RCA4). Meteorologi 157 (SMHI, 42 pp. ISSN 0283-7722).

27. van der Schrier, G., Efthymiadis, D., Briffa, K.R., Jones, P.D., 2007. European alpine moisture variability for 1800–2003. Int. J. Climatol. 27, 415–427.

28. van der Schrier, G., Jones, P.D., Briffa, K.R., 2011. The sensitivity of the PDSI to the Thornthwaite and Penman–Monteith parameterizations for potential evapotranspiration. J. Geophys. Res. 116, D03106. http://dx.doi.org/10.1029/2010JD015001.

29. Sheffield, J., Wood, E.F., 2008. Projected changes in drought occurrence under future global warming from multi-model, multi-scenario, IPCC AR4 simulations. Clim. Dyn. 31, 79–105. http://dx.doi.org/10.1007/s00382-007-0340-z.

30. Thomasson, A.J., 1995. Assessment of soil water reserves available for plants (SWAP): a review. In: D., King, R.J.A., Jones, A.J., Thomasson (Eds.), European Land Information Systems for Agro-environmental Monitoring 16232. Office for Official Publications of the European Community, EUR, Luxembourg, pp. 115–129.

31. Thornthwaite, C.W., 1948. An approach toward a rational classification of climate. Geogr. Rev. 38, 55–94.

32. Trenberth, K.E., Dai, A., van der Schrier, G., Jones, P.D., Barichivich, J., Briffa, K.R., Sheffield, J., 2014. Global warming and changes in drought. Nat. Clim. Chang. 4 (1), 17–22. http://dx.doi.org/10.1038/nclimate2067.

33. Ujvari, I., 1972. Geography of Romanian Waters (in Romanian). Editura Ştiinţifică, Bucharest. van Vuuren, D.P., et al., 2011. The representative concentration pathways: an overview. Clim. Chang. 109, 5–31.

34. Wells, N., Goddard, S., Hayes, M., 2004. A self-calibrating Palmer Drought Severity Index. J. Clim. 17, 2335–2351. http://dx.doi.org/10.1175/1520-0442(2004)017b2335:ASPDSIN2.0.CO;2.

CHAPTER 20

AN EXPANDABLE WEB-BASED PLATFORM FOR VISUALLY ANALYZING BASIN-SCALE HYDRO-CLIMATE TIME SERIES DATA

Joeseph P. Smith, Timothy S. Hunter, Anne H. Clites, Craig A. Stow, Tad Slawecki, Glenn C. Muhr, Andrew D. Gronewold

Keywords: Data visualization, Great lakes, Data access, Analytics, Software, Object oriented

INTRODUCTION

The North American Laurentian Great Lakes comprise the largest system of freshwater lakes in the world. In response to changes in the Great Lakes ecosystem, natural resource availability, and hydrology (Morris and Gill, 1994, Ricciardi and Rasmussen, 1998, Wilcox et al., 2002 and Gronewold and Stow, 2014), and amidst growing regional and international awareness of global climate change (Stocker et al., 2013) and urban development (Martine et al., 2008), research and monitoring on the Great Lakes system has intensified over the past couple of decades. Consequently, large amounts of data have been and continue to be produced for this region (Bulatewicz et al., 2014). For instance, the National Oceanic and Atmospheric Administration (NOAA), the Canadian Hydrographic Service (CHS), and the Detroit District of the United States Army Corps of Engineers (USACE) are three, among many other, agencies that develop or maintain large amounts of Great Lakes regional data. Developed or maintained data include water level and water budget model simulations, observations, and forecasts, as described inGronewold et al. (2011) and Deacu et al. (2012). Given the recent acceleration of regional research on the Great Lakes system (e.g. projects funded by the Great Lakes Restoration Initiative), we have observed a coincident rise in demand for these agencies to make their data and model output readily available to the public.

In and outside of the Great Lakes research community, these needs have been documented and attempts have been made to fulfill them – providing access to, and visual analysis of, multi-dimensional data (Jeong et al., 2006). Two notable efforts include the Consortium of Universities for the Advancement of Hydrologic Science, Inc. Hydrologic Information System (CUAHSI-HIS, Horsburgh et al. (2009)) and the Great Lakes Observing System (GLOS, Read et al. (2010)). Both projects focus on providing access to and documentation for point observation data, as opposed to aggregate (e.g. averages – over temporal, spatial domains, or both) data. Access is realized through web services, with CUAHSI HIS implementing a propietary Observations Data Model (ODM,Horsburgh et al. (2008)) via a Relational Database Management System (RDBMS), and GLOS utilizing tools compliant with Open Geospatial Consortium (OGC, see Web References section) standards.

Beyond the need for basic hydrological data management and dissemination, there is a strong demand for simple, inexpensive or free, and accessible tools facilitating exploratory analysis of regional environmental and aggregate data. Such applications could be used to assess initial quality, identify candidate supporting information, and provide decision support (Jakeman et al., 2008). Demand for these tools is enhanced by the existence of rapidly advancing and

Citation: Joeseph P. Smith, Timothy S. Hunter, Anne H. Clites, Craig A. Stow, Tad Slawecki, Glenn C. Muhr, Andrew D. Gronewold, An expandable web-based platform for visually analyzing basin-scale hydro-climate time series data, http://dx.doi.org/10.1016/j.envsoft.2015.12.005.

increasingly sophisticated software packages, such as those employed by the user accessible WeatherSpark.com, as well as Aquatic Informatics› AQUARIUS and KISTERS' WISKI (see Web References section). These web and commercial software applications, respectively, streamline the data discovery, fetching, visualization, and analysis process, making a wide variety of tools and data easily available to users. While CUAHSI HIS, GLOS, and other existing distribution platforms (for example, see Hendler et al., 2012) serve data, they often do not demonstrate that data are readily available and easy to access and analyze. For instance, while web services may provide data in standard formats to users, data insights are not realized without expertise in web scraping, programming, or other technical skill.

In this paper, we introduce the Great Lakes Dashboard (GLD, Fig. 1), a free web application that provides user-friendly tools to explore aggregated time series data, model output, and forecasts for multiple variables potentially affecting the Great Lakes along with background information. Produced by NOAA›s Great Lakes Environmental Research Laboratory (NOAA-GLERL) and the Cooperative Institute for Limnology and Ecosystems Research (CILER), the GLD is one of very few, if any, available tools that aggregate multiple time series data sets on the Great Lakes in a basin-wide spatial domain and present them in an accessible interface.

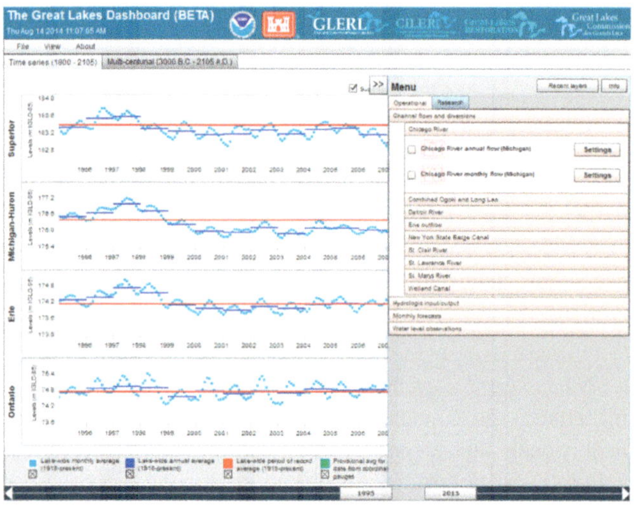

Figure 1: Screenshot of the Great Lakes Dashboard on initial load.

The GLD is driven by a robust, expandable Time-Series Core Object Model (GLD-TSCOM). As a generic framework and programming model similar to MapReduce (Dean and Ghemawat, 2008), programmers can develop applications similar to the dashboard using the GLD-TSCOM. In that sense, it can also be adapted to new requests from users, applied to regions other than the Great Lakes or other topical areas of interest, and be framed around advances in available technology. As an example, instead of focusing on sub-basins within the Great Lakes (see top of Fig. 2), the GLD-TSCOM could work with hydrological inputs, outputs, and storage within subregions of California, defined by the United States Geological Survey (USGS), currently experiencing extreme drought conditions (Swain et al., 2014).

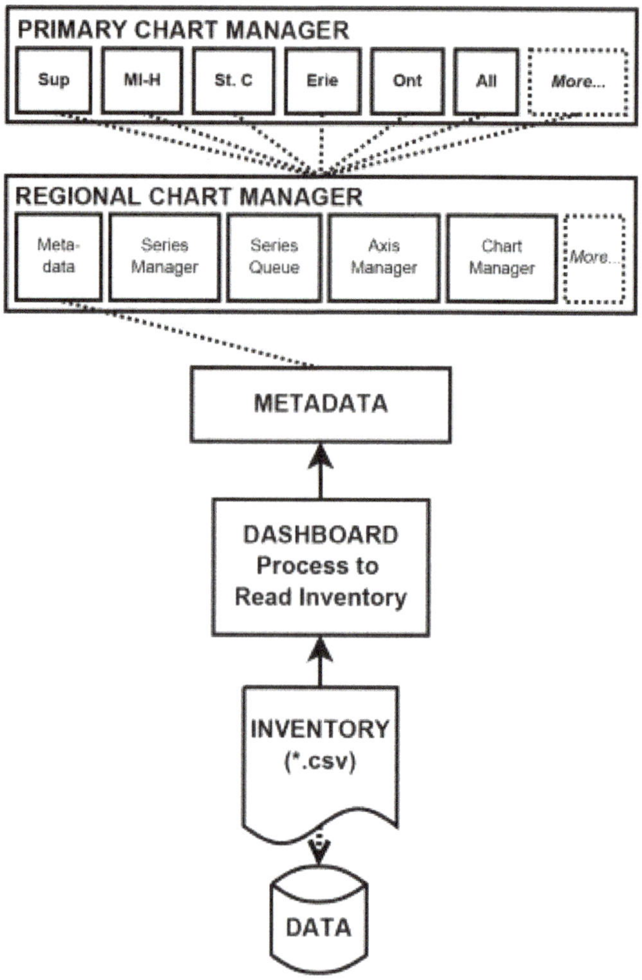

Figure 2: The GLD-TSCOM and its data management model. The dashboards read comma-delimited inventories, splitting them by their inventory items and transforming those into metadata objects. The metadata objects are then grouped by region and sent to a regional chart manager object, each of which has a set of objects for configuring the charts and the series displayed on them. A primary chart manager has domain over all regional chart managers created, allowing for administration of user requests and facilitating communication between the user and the dashboard.

METHODS

Predecessors

The GLD and the GLD-TSCOM were preceded by the Great Lakes Water Level Dashboard, or GLWLD (Gronewold et al., 2013) and the Great Lakes Hydro-Climate Dashboard, or GLHCD (Clites et al., 2014). Prior to their release, many previously available data services advertised as Great Lakes data or containing Great Lakes data were exclusive to the United States, leaving out Canadian data, which are invaluable for Great Lakes basin-wide data and model output. Other previously available applications focused strictly on point measurements, most on land and not over water. Lastly, a subset of project stakeholders desired an application to quickly generate plots for presentations through various media. The GLWLD and GLHCD were designed to begin filling in those gaps in available software applications and web data services.

The GLWLD and GLHCD were developed with a focus on ease of use and ability of the general public to understand data being presented without being overwhelmed (Kelleher and Wagener, 2011 and Spiegelhalter et al., 2011). WeatherSpark. com›s graphical user interface design was, therefore, a good model for development. The GLWLD and GLHCD were used to test certain features as well as communicate to the public general characteristics of Great Lakes hydrology. Programming methodology for these was originally functional and linear, with the side effect of the data model being tightly interwoven into the code. As a result, the addition of data sets required production of several lines of programming, and as data and improvement requests increased, this development paradigm quickly became unsustainable. The code written to drive these initial dashboards is traditionally called "spaghetti code" (Mikkonen and Taivalsaari, 2007). "Spaghetti code" is programming that is not well organized, contains processes with many unnecessary routes to completion, and is difficult to break up into independent, functional parts for reuse and modification. While such code yields timely results and is useful for prototyping and trial releases, creating a sustainable and expandable application requires careful programming analysis and redesign.

Due to the above issues and additional requests to include better communication of displayed data, an Object Oriented Programming (OOP) approach was adopted, and the GLD and GLD-TSCOM were developed amidst initial success with and widespread adoption of the GLWLD and GLHCD. In addition, the GLD-TSCOM was implemented in new versions of the GLWLD and GLHCD to increase their sustainability and demonstrate the model›s applicability to more than one web application.

GLD-TSCOM

The GLD and its predecessors were built using the free, open-source Apache Flex (or Flex) framework (Apache Software Foundation – Forest Hill, Maryland, USA) under the Adobe Flash Builder (Adobe Systems – San Jose, California, USA) and JetBrains' Intellij IDEA (Czech Republic). Flex is a high level, event-driven (Etzion and Niblett, 2011) framework providing graphical user interface (GUI) objects or components and other behind-the-scenes services (i.e. data, web connectivity) which can be implemented at a high level of programming abstraction. Event-driven frameworks, analogous to a series of dominoes set up to fall once one domino is tipped towards others, can be very efficient and fast depending on the quality of engineering. The front-end user interface can be expressed in MXML - Macromedia Extensible Markup Language, a modification to XML – the Extensible Markup Language (Bray et al., 2006), with similar behavior to HTML for web pages. Programmatically-generated user interface components along with efficient responses to user actions are expressed in the object-oriented ActionScript (Gamma et al., 2002 and Smith, 2011). Together, packages of MXML and ActionScript in a Flex project are compiled, producing a Small Web Format (SWF) file for display on the web through the client-side Adobe Flash Player, which is available for all popular desktop computer operating systems - a major target platform for the dashboards.

We will describe at a high level the GLD-TSCOM›s mechanics. This model may be used as a template for other data series formats (e.g. categorical series). We note that the design of the GLD-TSCOM is heavily influenced by the underlying Flex framework. Manager objects described below and depicted in Fig. 2 manipulate framework-provided chart, series, and axis objects. The properties and behavior of those and other framework-provided objects influenced the design of the managers and ultimately the GLD-TSCOM. Objects in this context are groups of variables, functions, and data structures that are utilized for a single general task.

When a dashboard loads, at least one time series data inventory is read in. Each inventory item is stored as an individual metadata object. Once an inventory is read in completely, the dashboard is notified of that event, and the collection of metadata is broken up by basins, currently defined as Superior, Michigan-Huron, St. Clair, Erie, Ontario, and the entire Great Lakes basin. A regional chart manager object is created for each region, and a single primary chart manager object is given domain over all the regional managers, facilitating communication between the user and the dashboard platform. This allows for administration of user requests, from visible time scale changes to data set selection. User events can be processed via a single call to the primary chart manager, which calls the regional managers to process the request.

The regional chart manager objects contain the metadata for its assigned region, the chart component to display the region›s data, a series manager object for handling data sets, a queue for processing data series into the series manager as series objects, and an axis manager object for managing the vertical axes that describe data sets of various units. The series queue keeps an archive of previously processed data in a user session so that costly data processing does not repeatedly occur. Only data initially set to 'on' are processed when the dashboards load. Additionally, as the number of series available can be expanded, the axis manager can be configured to handle many vertical axis types. As of this publication, axes are configured for water level data, percentage data such as ice cover over a lake›s surface, hydrologic input and output data, flow rate data for channel flows and diversions to and from the lakes, temperature data, velocity data such as wind speeds, and global climatological pattern or oscillation indices such as the Multivariate El Niño Southern Oscillation (ENSO) Index.

Data Management

Creating a virtual wall of separation to isolate the growing amount of available data sets and the dashboards is critical to ensuring sustainability for the software. We have, therefore, developed a comprehensive data management model that isolates the growth and complexity of Great Lakes data from the web application interfaces. In this section, we begin by describing the formats of data parsed and displayed by the dashboards, followed by a description of their inventories, ending with discussion of a taxonomic organization scheme employed by the GLD's inventories. An overview of data organization schemes in addition to those described here can be found in Garshol (2004).

Data Formats

Time-series data are copied from their source, processed, and stored in the 'data' directory of the dashboard as delimited text files in a variety of formats, with the common delimiter being a comma (a Comma Separated Values file, or CSV). As a majority of the data appropriate for the dashboard can be expressed in tabular format, delimited text files are sufficient and simple, compared to formats more appropriate for non-tabular structured data such as XML or JSON – JavaScript Object Notation (Bray, 2014). Delimited text data can easily be used in a variety of ways by users who download the data as there is a wide selection of text editors, spreadsheet applications, and data analysis tools available to work with the data. String parsing methods in a dashboard data factory are written to handle the delimited formats, and more methods can be written to handle other formats, including other delimiter formats (tabbed, for example). Currently the following formats are supported:

- Singleton: a number, usually representing a standard or statistic, followed by the starting year for which that number applies. We assume that the single number applies up to the present day
- Per month records and averages: 12 lines, with a number, and if they are not averages, the year the extreme record was set (maximum or minimum)
- Monthly data grid: line for every year of data available, starting with the year, followed by 12 data points for data from January through December
- Time series: multiple lines which have either a year or date in MM/DD/YYYY format, followed by a single data point
- Time series of ranges: multiple lines which have either a year or date in MM/DD/YYYY format, followed by two data points, used often for forecast ranges
- Time series deviation from average: multiple lines which have either a year or date in MM/DD/YYYY format, followed by three data points - the observation for that date, the period of record average, and the deviation of the observation from that average
- Time series high, low, open, close (HLOC): multiple lines which likely have a year, followed by four data points: the starting data for that year, ending data, the maximum data point for that year, and the minimum for that year. These data are often based off of a set of monthly averages
- Data format 6, time series deviation from average, allows viewing of monthly scale hydrological input/output total data with the same data at an annual scale. Because the ranges significantly differ where the monthly total data would be essentially hidden under the full scale of the plain annual total data, it works conveniently to use the deviation from average statistic for annual total data.

Inventories

• Time series data inventories consist of line items that represent individual data sets and even data points (i.e. the period of record average water level for a given lake). Each line item contains the following elements, most of which follow case-sensitive controlled vocabularies for machine readability and processing into essential dashboard components:

- Lake, basin, or region covered by the data
- General descriptor which can act as an identifier of the data in memory
- Label for the data tip a user will see when they hover over a data point with their mouse cursor
- Web directory location (URL) of the delimited text data set
- Format of the delimited text data set
- Number of lines to prime the data when reading (to ignore header info)
- Number of decimal places to display to the user (i.e. for data tips)
- Metric units of the data
- Target vertical axis for the data
- Default display state when the dashboard initially loads
- How the data set is to be plotted (dots, step-wise, etc.)
- Plot default fill and/or stroke color and opacity
- Plot point radius and/or stroke weight

We note that the web directory location can either be specified as a URL relative to the dashboard directory or an absolute URL. This implies that not only can the inventories point to data processed into the dashboard 'data' directory, but they can also point to data external to the dashboard directory. Pointing to external data would require either 1) the data be in an already established format, or 2) the development of a new method to parse a common format that the data are in.

Inventories act as high level tools to control how data are communicated to the public. The dashboards' programming and inventories are intrinsically flexible enough that we can expand these line items to include other information if additional communication is necessary. For example, the GLD inventories' line items have an element for a watermark to display behind a data series item in the menu, indicating the origin of the data (i.e. NOAA, USACE, etc.). Lastly, delimited text files are currently sufficient for inventory data. Spreadsheet applications and text editors are widely available to maintain the inventory, and using additional formats would create an unnecessary conceptual barrier to future maintenance. However, a switch to more structured formats such as XML or JSON could easily be done.

Data Taxonomy

To support a wide range of Great Lakes data, the GLD features a three-tier, alphabetically-organized menu generated from inventory items. By clarifying whether a data set is operational or produced in research, and what the data describe, the GLD's menu aids in improving communication and organization as well. It is thus useful for the GLD to expand the general descriptors in inventory line items to include information about data's operational status, category, and potential subcategory. This organization scheme can be visualized as a taxonomy for the data (Fig. 3), whose four levels are used in varying combinations to establish data categories and sub-categories:

- Status: operational or research
- Lake basin or region
- Variable of interest
- Temporal scale

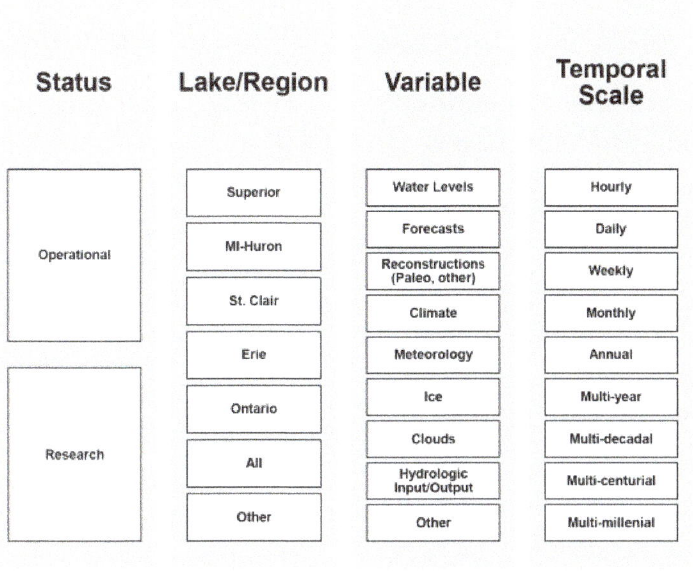

Figure 3: Taxonomic structure used to help define primary and secondary categories in the GLD data inventory.

For example, in the Operational (level 1) tab of the GLD data menu, under the main category of 'Water level observations' (level 3), there are subcategories for annual and monthly average data (level 4), along with a subcategory for statistics and standards over the period of record (level 3). Level 2 is used in placing the data on a specific lake's chart, and not often attached to a single set of data's label. We note that basins or regions can be defined in a custom fashion (thus the 'Other' subcategory in Fig. 3) as done, for example, by the Great Lakes – St. Lawrence River Basin Water Resources Compact and Great Lakes St. Lawrence River Basin Sustainable Water Resources Agreement (see Web References section). They combine the sub-basins of St. Clair and Erie, modify the Ontario basin, and have the St. Lawrence River basin as an additional focus region.

If certain data are not given a sub-category in the inventory, they fall under a generic sub-category labeled 'General' in the GLD menu.

RESULTS

The expandable GLD-TSCOM makes many options and features available on the GLD interface (Fig. 4). The GLD's capability to interchange axes of different types facilitates time series visualizations for a wide variety of variables of interest on the Great Lakes, unlike its predecessors. If multiple axes are required, they will appear on both the left and right sides of the dashboard, with the right-hand axes visible when users fold in the data menu (Fig. 4C) using the '>>' button on its top left corner.

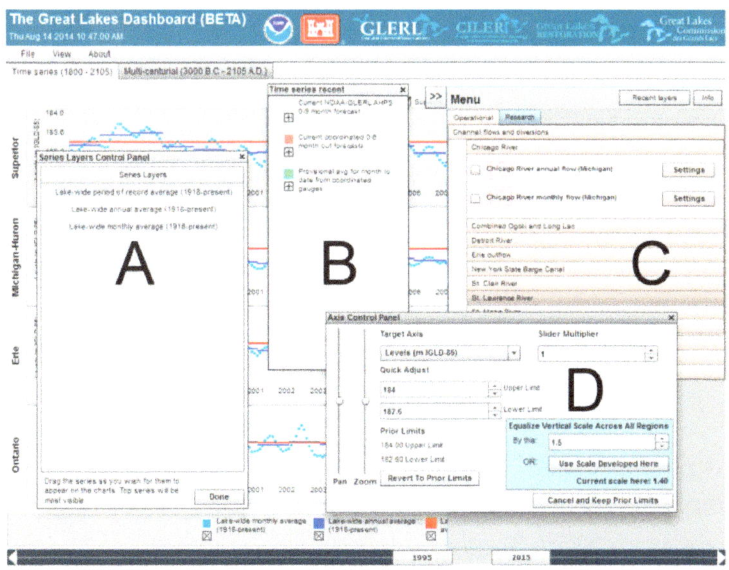

Figure 4: Visual synopsis of features for the new GLD. A.) The series layers control panel enables users to drag and drop series by their label to configure the order in which they are overlayed. This is aided by the Series Manager in the GLD-TSCOM. B.) Users may open a window to reactivate series which were previously visible in the session. C.) Three-tier, alphabetically organized data menu generated from dashboard metadata fed in via inventories D.) By clicking on a basin's chart, users can adjust the vertical axis scale for active axes via the GLD-TSCOM's Axis Manager. The chart of focus is shaded in gray, which in this case is the chart that represents data from Lake Superior.

Here, we demonstrate the GLD's ability to allow users to explore data sets from two distinct studies.

Visualizing the Impacts of Water Temperature and over-lake Evaporation on Water Levels

The water levels of the Great Lakes rise and fall in regular seasonal patterns corresponding with rainfall, snow melt, and evaporation. Unlike a typical lake, the surface area of each of these massive inland seas comprises about a third of its drainage area, causing total evaporation to be comparable in magnitude to total precipitation on the lake surface. When the upper Great Lakes (Superior, Michigan, and Huron) experienced a precipitous drop in water levels beginning in 1997, hydrologists sought to explain the sudden decline (Gronewold and Stow, 2014). These lakes stayed below their monthly averages, at times significantly below, for a period of 15 years, including an all-time record low set in January 2013 on Lakes Michigan and Huron.

The GLD can be used (Fig. 5) to compare annual average water levels from 1950 to 2013 for these upper lakes. Lakes Michigan and Huron are shown on one panel because they are considered one lake in terms of water level, joined at the north by the Straits of Mackinac. Annual average modeled surface water temperature and annual over-lake simulated evaporation's deviation from average, from 1950 to 2013, are also displayed. One interesting parallel is the correlation between water temperature and over-lake evaporation, which is especially apparent for Lake Superior. The late 1990's and early 2000's were periods with high water temperatures, dramatically high over-lake evaporation, and consistently low water levels on the northern Great Lakes – particularly for Lakes Michigan and Huron. The GLD makes this time series analysis straight forward.

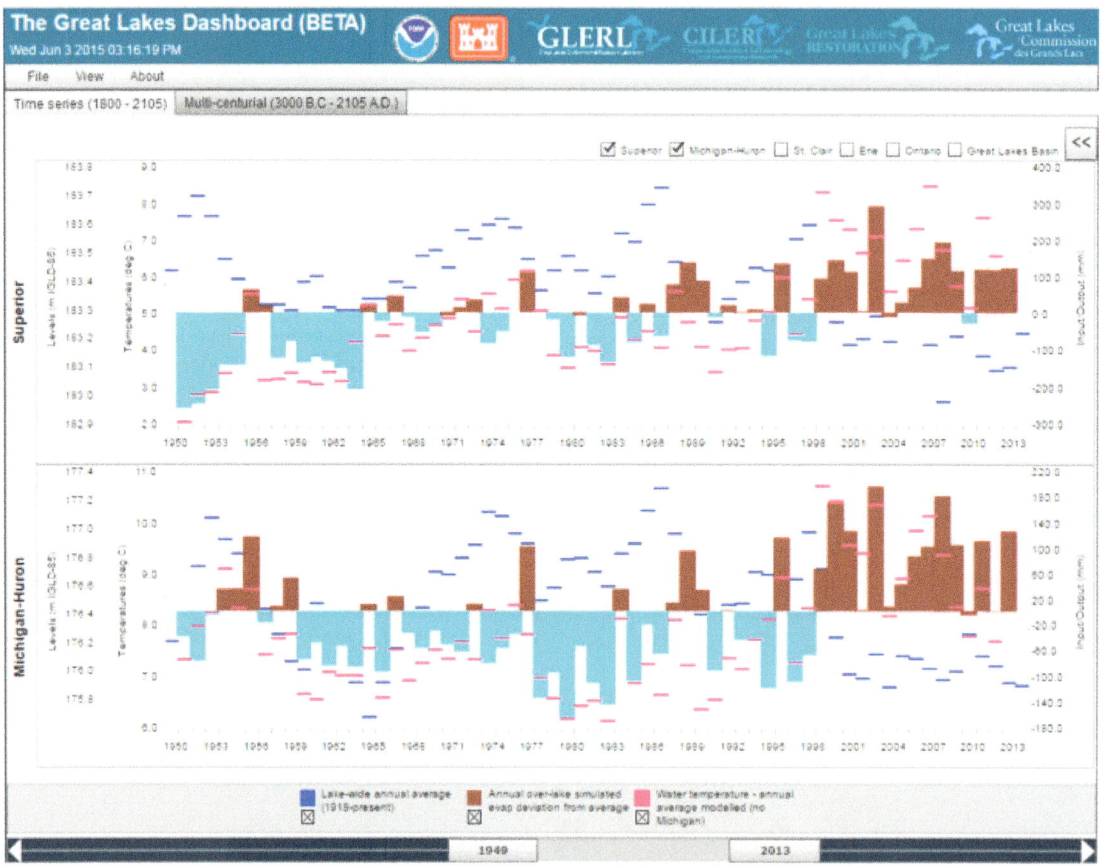

Figure 5: The GLD displaying datasets – including lake-wide water levels, surface water temperature and evaporation – found in Gronewold and Stow (2014).

Relating a Major Atmospheric Oscillation to Regional Climate Response using the GLD

Climatologists have suggested that the decline in water levels in 1997 and 1998 were caused by a particularly strong El Niño coincident with a warm winter in those years (Van Cleave et al., 2014). The GLD makes it easy to overlay any major climate oscillation with, for example (Fig. 6), Lake Superior water temperature and over-lake evaporation estimates. The timing of both the large El Niño and significant Pacific Decadal Oscillation (PDO, plotted), corresponds with the transition from low to high water temperatures, low to high evaporation, and average to low water levels.

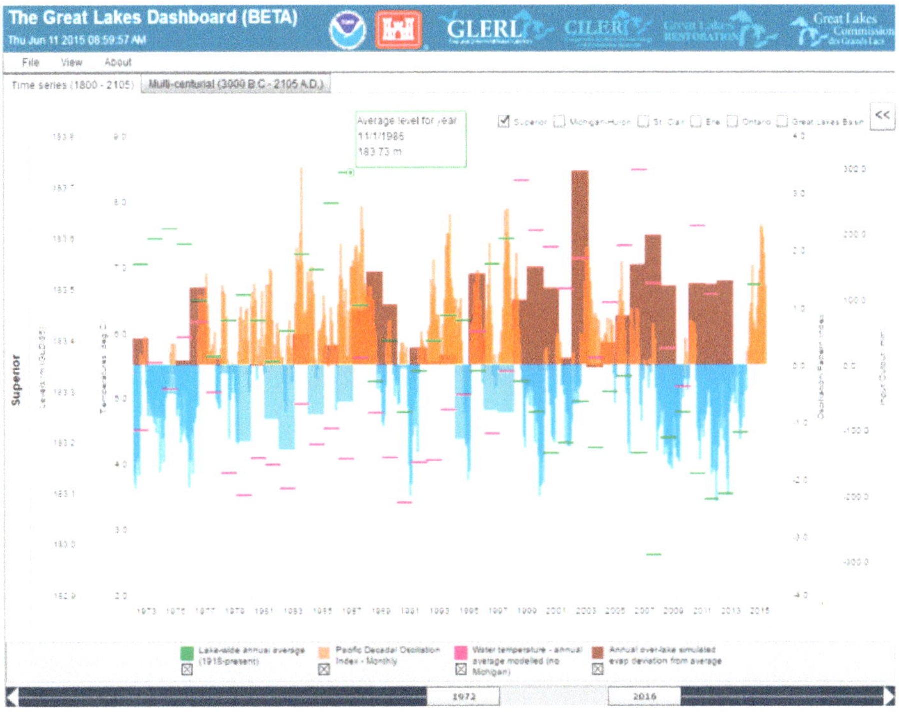

Figure 6: The GLD displaying datasets also used in Van Cleave et al. (2014), including lake-wide water levels, surface temperature, evaporation, and Pacific Decadal Oscillation (PDO). The "data tip", or small box at the top of the plot, is shown to clarify the green lines are for annual average water level.

CONCLUSIONS

Amidst growth in collected data, establishment of large, central data discovery and distribution services, and demand for simple access and visualization of aggregate data, we have developed applications based on a single model that fulfill needs within the Laurentian Great Lakes data user communities. The GLD and its predecessors are flexible, powerful, expandable applications for easy visual analysis of aggregated time series data. We have considered and incorporated the demands of many stakeholders, researchers, and managers in the Great Lakes community. These applications represent the leading edge of big data efforts (Vitolo et al., 2015) – aggregating and visualizing measurements and model outputs delivered from data catalogs and services. Instead of accessing data through scripts or data services of variable accessibility, users of the GLD can visualize and download data through a few clicks and drags of a computer mouse.

Adopting an Agile development approach (Ahalt et al., 2014), the GLD-TSCOM can grow in its scope of data and in its capabilities to serve users. This web framework's expandability and its resulting graphical front-end functionalities are useful in light of recent implementations of adaptive management practices (Linkov et al., 2006), such as NOAA's Next Generation Strategic Plan. The flexibility built into the platform leaves open the possibility of utilizing large data services such as CUAHSI-HIS and GLOS – displaying aggregate versions of their data on the dashboards. Additionally, these tools are helpful in the participatory modeling process between stakeholders and modellers as discussed by Voinov and Bousquet (2010), especially when it comes to model analysis, discussion, and planning improvements to models. This platform provides a means by which users can gain perspective and insights that could not be achieved without the juxtaposition of potentially related data.

During the time that the dashboard was developed, available server capabilities and resources were limited both in terms of installed software and support staff. Consequentially, text-based data and inventories were the best option, as updating and maintaining them does not require a specific skillset beyond spreadsheet manipulation. Recent developments, including the instantiation of a PostgreSQL service, may present opportunities to migrate the inventories and data to a more structured, more scalable, and robust system. Such developments would make options, such as the ability to download only the actively displayed data, simple to implement. However, adopting server-entrenched technologies may render the dashboard less transferable. Thus, as of this publication, we have a more lightweight, portable, and easily deployable data system than data infrastructures such as the CUAHSI-HIS.

While the dashboards are built under the Flash platform due to 1.) the availability of powerful, easy to use Integrated Developement Environments for the platform and 2.) initial developer experience with application programming, there is a trend towards doing web development under the relatively new HTML5 platform. This is driven by growth in mobile platform adoption, Flash's general incompatibility with mobile devices, and recent exposure of critical and severe security issues in the platform (see the Symantec link in the Web References section). Early in the research and development process, we considered the JavaScript package Flot. While we previously drafted version of the GLWLD using jQuery and Dan Vanderkam's Dygraphs, the resulting performance was insufficient and unable to match what is achieved with the Flex framework. However, using the GLD-TSCOM as a guide, we have developed a draft HTML5 version of the GLD using some of the pseudo object-oriented features of the HTML5 platform (see the Software Availability section). The draft dashboard has a limited feature set since the platform composed of JavaScript and HTML5 is not as innately rich as Flash. Additionally, as there are limitations to the Dygraphs package, we employ a different visualization paradigm - users can view multiple variables' data for a single Great Lakesbasin with one chart per variable, or view a single variable type for all the Great Lakes basins with one chart per basin. The variable types available are the same as the axis types employed in the Flash GLD. Despite the limitations, the HTML5 dashboard is capable of performing the similar analyses that its Flash counterparts can, and do them on mobile platforms such as Apple Inc.'s iOS (Cupertino, California, USA).

ACKNOWLEDGMENTS

Funding for this research and development was provided through CILER by NOAA, the Great Lakes Restoration Initiative (No. DW1392377301-0, administered by USEPA), the International Joint Commission International Upper Great Lakes Study (No. P14100186), and the USACE (No. W56MES32262531). The authors would like to thank NOAA's Climate.gov dashboard team for initial advice, including Viviane Silva, David Herring, and Mark Phillips. For technical and communication support, we would like to thank NOAA-GLERL's Margaret Lansing and Cathy Darnell. For general guidance and partnership, we thank the United States Army Corps of Engineers - Detroit District, specifically Jim Lewis, John Allis, and Keith Kompoltowicz, along with Heather Stirratt and Brandon Krumwiede at NOAA's Office for Coastal Management. Special thanks to all who took time to review this paper, including Brent Lofgren of NOAA-GLERL and John Bratton of LimnoTech, and all anonymous reviewers.

The use of product names, commercial and otherwise, in this paper does not imply endorsement by NOAA, NOAA-GLERL, CILER, or any other contributing agency or organization.

REFERENCES

1. Ahalt, S., Band, L., Christopherson, L., Idaszak, R., Lenhardt, C., Minsker, B., Palmer, M., Shelley, M., Tiemann, M., Zimmerman, A., 2014. Water Science Software Institute: Agile and open source scientific software development.Comput. Sci. Eng. 16 (3), 18e26.
2. Bray, T., 2014. The JavaScript Object Notation (JSON) Data Interchange Format. Available at: http://tools.ietf.org/html/rfc7159 (accessed 13.06.14.).
3. Bray, T., Paoli, J., Sperberg-McQueen, C.M., Maler, E., Yergeau, F., Cowan, J., 2006. Extensible Markup Language (XML) 1.1, second ed. Available at: http://www.w3.org/TR/2006/REC-xml11-20060816 (accessed 13.06.14.).
4. Bulatewicz, T., Adresen, D., Auvenshine, S., Peterson, J., Steward, D., 2014. A distributed data component for the open modeling interface. Environ. Model. Softw. 57, 138e151.
5. Clites, A.H., Smith, J.P., Hunter, T.S., Gronewold, A.D., 2014. Visualizing relationships between hydrology, climate, and water level fluctuations on Earth's largest system of lakes. J. Gt. Lakes. Res. 40 (3).
6. Deacu, D., Fortin, V., Klyszejko, E., Spence, C., Blanken, P.D., 2012. Predicting the net basin supply to the Great Lakes with a hydrometeorological model. J. Hydrometeorol. 13 (6), 1739e1759.
7. Dean, J., Ghemawat, S., Jan. 2008. MapReduce: Simplified data processing on large clusters. Commun. ACM 51 (1), 107e113.
8. Etzion, O., Niblett, P., 2011. Event Processing In Action. Manning, Stamford, CT. Gamma, E., Helm, R., Johnson, R., Vlissides, J., 2002. Design Patterns: Elements of Reusable Object-Oriented Software. Addison-Wesley, Boston, MA.
9. Garshol, L.M., 2004. Metadata? Thesauri? Taxonomies? Topic maps! Making senseof it all. J. Inf. Sci. 30 (4), 378e391. Gronewold, A.D., Clites, A.H., Hunter, T.S., Stow, C.A., 2011. An appraisal of the Great Lakes advanced hydrologic prediction system. J. Gt. Lakes. Res. 37 (3), 577e583.
10. Gronewold, A.D., Clites, A.H., Smith, J.P., Hunter, T.S., 2013. A dynamic graphical interface for visualizing projected, measured, and reconstructed surface water elevations on the Earth's largest lakes. Environ. Model. Softw. 49, 34e39.
11. Gronewold, A.D., Stow, C.A., 2014. Water loss from the Great Lakes. Science 343 (6175), 1084e1085.
12. Hendler, J., Holm, J., Musialek, C., Thomas, G., 2012. US government linked open data: semantic.data.gov. IEEE Intell. Syst. 27 (3), 25e31.
13. Horsburgh, J.S., Tarboton, D.G., Maidment, D.R., Zaslavsky, I., 2008. A relational model for environmental and water resources data. Water Resour. Res. 44 (5),w05406.
14. Horsburgh, J.S., Tarboton, D.G., Piasecki, M., Maidment, D.R., Zaslavsky, I.,Valentine, D., Whitenack, T., 2009. An integrated system for publishing environmental observations data. Environ. Model. Softw. 24 (8), 879e888.
15. Jakeman, A.A., Voinov, A.A., Rizzoli, A.E., Chen, S.H., 2008. Environmental Modelling,Software, and Decision Support e State of the Art and New Perspectives. Elsevier, Amsterdam, Netherlands.
16. Jeong, S., Liang, Y., Liang, X., 2006. Design of an integrated data retrieval, analysis, and visualization system: application in the hydrology domain. Environ. Model. Softw. 21 (12), 1722e1740.
17. Kelleher, C., Wagener, T., 2011. Ten guidelines for effective data visualization inscientific publications. Environ. Model. Softw. 26 (6), 822e827.
18. Linkov, I., Satterstrom, F., Kiker, G., Batchelor, C., Bridges, T., Ferguson, E., 2006. Fromcomparative risk assessment to multi-criteria decision analysis and adaptivemanagement: recent developments and applications. Environ. Int. 32 (8),1072e1093.
19. Martine, G., McGranahan, G., Montgomery, M., Fernandez-Castilla, R., 2008. TheNew Global Frontier: Urbanization, Poverty, and Environment in the 21st Century. Routledge Earthscan, New York, NY.

20. Mikkonen, T., Taivalsaari, A., 2007. Web Applications e Spaghetti Code for the 21st Century. Tech. Rep.. Sun Microsystems.

21. Morris, C.S., Gill, S.K., 1994. Variation of Great Lakes water levels derived from GEOSAT altimetry. Water Resour. Res. 30 (4), 1009e1017.

22. Read, J., Klump, V., Johengen, T., Schwab, D., Paige, K., Eddy, S., Anderson, E., Manninen, C., 2010. Working in freshwater: the Great Lakes observing system contributions to regional and national observations, data infrastructure, and decision support. Mar. Technol. Soc. J. 44 (6), 84e98.

23. Ricciardi, A., Rasmussen, J.B., 1998. Predicting the identity and impact of future biological invaders: a priority for aquatic resource management. Can. J. Fish Aquatic Sci. 55 (7), 1759e1765.

24. Smith, B., 2011. AdvancED ActionScript 3.0: Design Patterns. Apress, New York, NY. Spiegelhalter, D., Pearson, M., Short, I., 2011. Visualizing uncertainty about the future. Science 333 (1393), 1393e1400.

25. Stocker, T.F., Qin, D., Plattner, G.-K., Tignor, M.M., Allen, S.K., Boschung, J., Nauels, A., Xia, Y., Bex, V., Midgley, P.M., 2013. Climate Change 2013: the Physical Science Basis. Contribution of Working Group I to the Fifth Assessment Report of the Intergovernmental Panel on Climate Change. Cambridge University Press, New York, NY.

26. Swain, D.L., Tsiang, M., Haugen, M., Singh, D., Charland, A., Rajaratnam, B.,Diffenbaugh, N.S., 2014. The extraordinary California drought of 2013/2014: character, context, and the role of climate change. Bull. Am. Meteorol. Soc. 95 (9), S3.

27. Van Cleave, K., Lenters, J.D., Wang, J., Verhamme, E.M., 2014. A regime shift in lake superior ice cover, evaporation, and water temperature following the warm El Nino winter of 1997e1998. Limnol. Oceanogr. 59 (6), 1889e1898.

28. Vitolo, C., Elkhatib, Y., Reusser, D., Macleod, C.J., Buytaert, W., 2015. Web technologiesfor environmental big data. Environ. Model. Softw. 63 (0), 185e198.

29. Voinov, A., Bousquet, F., 2010. Modelling with stakeholders. Environ. Model. Softw.25 (11), 1268e1281.

30. Wilcox, D.A., Meeker, J.E., Hudson, P.L., Armitage, B.J., Black, M.G., Uzarski, D.G., 2002. Hydrologic variability and the application of index of biotic integrity metrics to wetlands: a Great Lakes evaluation. Wetlands 22 (3), 588e615.

INDEX